T0319260

HOMEOSTASIS AND TOXICOLOGY OF ESSENTIAL METALS

This is Volume 31A in the

FISH PHYSIOLOGY series

Edited by Chris M. Wood, Anthony P. Farrell and Colin J. Brauner

Honorary Editors: William S. Hoar and David J. Randall

A complete list of books in this series appears at the end of the volume

HOMEOSTASIS AND TOXICOLOGY OF ESSENTIAL METALS

Edited by

CHRIS M. WOOD
Department of Biology
McMaster University
Hamilton, Ontario
Canada

ANTHONY P. FARRELL
Department of Zoology and Faculty of Land and Food Systems
The University of British Columbia
Vancouver, British Columbia
Canada

COLIN J. BRAUNER
Department of Zoology
The University of British Columbia
Vancouver, British Columbia
Canada

ELSEVIER

AMSTERDAM • BOSTON • HEIDELBERG • LONDON • OXFORD
NEW YORK • PARIS • SAN DIEGO • SAN FRANCISCO
SINGAPORE • SYDNEY • TOKYO
Academic Press is an imprint of Elsevier

Academic Press is an imprint of Elsevier
32 Jamestown Road, London NW1 7BY, UK
225 Wyman Street, Waltham, MA 02451, USA
525 B Street, Suite 1800, San Diego, CA 92101-4495, USA

First edition 2012

British Library Cataloguing-in-Publication Data
A catalogue record for this book is available from the British Library

Library of Congress Cataloging-in-Publication Data
A catalog record for this book is available from the Library of Congress

ISBN: 978-0-12-378636-4
ISSN: 1546-5098

For information on all Academic Press publications
visit our website at www.elsevierdirect.com

Typeset by MPS Limited, a Macmillan Company, Chennai, India
www.macmillansolutions.com

Printed and bound in United States of America
Transferred to Digital Printing, 2011

CONTENTS

3.   Zinc
Christer Hogstrand

4.   Iron
Nicolas R. Bury, David Boyle and Christopher A. Cooper

8. Molybdenum and Chromium
    Scott D. Reid

9. Field Studies on Metal Accumulation and Effects in Fish
    Patrice Couture and Greg Pyle

CONTENTS OF
HOMEOSTASIS AND TOXICOLOGY OF
NON-ESSENTIAL METALS, VOLUME 31B

3.   Cadmium
      James C. McGeer, Som Niyogi and D. Scott Smith

4.   Lead
      Edward M. Mager

CONTRIBUTORS

The numbers in parentheses indicate the pages on which the authors' contributions begin.

RONNY BLUST *(291), University of Antwerp, Antwerp, Belgium*

DAVID BOYLE *(201), University of Plymouth, Plymouth, UK*

NICOLAS R. BURY *(201), King's College London, London, UK*

CHRISTOPHER. A. COOPER *(201), University of Guelph, Guelph, Ontario, Canada*

PATRICE COUTURE *(253, 417), University of Quebec, Quebec, Canada*

MARTIN GROSELL *(53), University of Miami, Miami, Florida, USA*

CHRISTER HOGSTRAND *(135), King's College London, London, UK*

DAVID M. JANZ *(329), University of Saskatchewan, Saskatoon, Canada*

GREG PYLE *(253, 417), Lakehead University, Thunder Bay, Ontario, Canada*

SCOTT D. REID *(375), University of British Columbia, Okanagan Campus, Kelowna, British Columbia, Canada*

CHRIS M. WOOD *(1), McMaster University, Hamilton, Ontario, Canada, and University of Miami, Miami, FL, USA*

PREFACE

We are pleased to present this two-volume book on the homeostasis and toxicology of metals to the *Fish Physiology* series, the brainchild of Bill Hoar and Dave Randall, which has become the bible of our field since its inception more than 40 years ago. Physiology and toxicology are particularly closely linked in the aquatic sciences, and all three editors are practitioners of both fields. Indeed, we prefer to work at the interface of the two fields where physiological understanding of mechanisms explains toxic response, and toxicological phenomena illuminate physiological theory. We believe the book captures this interface. We trust it will appeal to the regular readers of the *Fish Physiology* series, as well as to a much broader audience including nutritional physiologists, toxicologists, and environmental regulators.

The motivation for this two-volume book has two origins:

Firstly, there has been an explosion of new information on the molecular, cellular, and organismal handling of metals in fish in the past 15 years. While most of the research to date has focused on waterborne metals, there is a growing realization of the importance of diet-borne metals. These elements are no longer viewed by fish physiologists as evil "heavy metals" (an outdated and chemically meaningless term) that kill fish by suffocation. Rather, they are now viewed as interesting moieties that enter and leave fish by specific pathways, and which are subject to physiological regulation. These regulatory pathways may be ones dedicated for essential metal uptake (e.g., copper-specific, iron-specific, zinc-specific transporters) or ones at which metals masquerade as nutrient ions ("ionic mimicry" e.g., copper and silver mimic sodium; cobalt, zinc, lead, strontium, and cadmium mimic calcium; nickel mimics magnesium). Internally, homeostatic mechanisms include regulated storage and detoxification (e.g., metallothioneins, glutathione, granule formation) and protein vehicles for transporting metals around the body in the circulation (e.g., ceruloplasmin, transferrin).

Molecular and genomic techniques have allowed precise characterization of these pathways, and how they respond to environmental challenges such as metal loading and deficiency. Bioaccumulation of metals is now widely studied in both the laboratory and the field, but interpretation of the data remains controversial. New techniques such as subcellular fractionation and modeling of metal-sensitive and metal-insensitive pools are providing clarification and new pathways for further research.

Secondly, this same period has seen a progressively increasing concern about the potential toxicity of metals in the aquatic environment. At present, the European Union, the United States, Canada, Australia/New Zealand, China, several Latin American countries, and diverse other jurisdictions around the world are all in the process of revising their ambient water quality criteria for metals. Coupled to this has been a sharp growth in toxicological research on metal effects on fish. Much of this research has focused on the physiological mechanisms of uptake, storage, and toxicity, and from this various modeling approaches have evolved which have proven very useful in the regulatory arena. For example, tissue residue models, to relate internal metal burdens to toxic effects, and biotic ligand models (BLMs), to relate gill metal burdens in different water qualities to toxic effects, are two physiological models that are now being considered by regulatory authorities in setting environmental criteria for metals (e.g., residue models for selenium and mercury regulations; BLMs for copper, zinc, silver, cadmium, and nickel criteria).

This work was conceived as a single book to cover all the metals for which a sizeable database exists. Its division into two published volumes (Vol. 31A dealing with essential metals, Vol. 31B dealing with non-essential metals) was solely for practical reasons of size, stemming from each metal being dealt with in a uniform and comprehensive manner. Regardless, the two volumes are fully integrated by cross-referencing between the various chapters, and they share a common index.

Three chapters in particular tie the package together with real-world scenarios and applications: Chapter 1 of Vol. 31A on *Basic Principles* serves as an Introduction to the whole book, while Chapter 9 of Vol. 31A on *Field Studies on Metal Accumulation and Effects in Fish* and Chapter 9 of Vol. 31B on *Modeling the Physiology and Toxicology of Metals* serve as integrative summaries dealing with both essential and non-essential metals.

The other 15 chapters each deal with specific metals, and authors were strongly urged to adopt a unified format which is explained in Chapter 1 of Vol. 31A. This format includes consideration of the following topics:

1. Chemical Speciation in Freshwater and Seawater
2. Sources of Metals and Economic Importance

As a result, the book should serve as a one-stop source for a synthesis of current knowledge on both the physiology and toxicology of a specific metal, and selective readers should be able to quickly find the specific information they require. Furthermore, the chapters should help guide future research by pointing out significant data gaps for particular metals.

This book would not have been possible without a vast contribution of time and effort from many people. First and foremost, our gratitude to the authors of the chapters, who represent some of the leading experts in the world in metals physiology and toxicology. Not only did these researchers sacrifice nights, weekends, and holidays to craft their chapters, they also constructively reviewed many of the other chapters. In addition, more than 20 anonymous external peer-reviewers contributed greatly to the quality of the chapters. Pat Gonzalez, Kristi Gomez, Caroline Jones, and Charlotte Pover at Elsevier provided invaluable guidance and kept the project on track. Finally, special thanks are due to Sunita Nadella at McMaster University, who proofread and corrected every chapter before submission to Elsevier.

This book is dedicated to the memory of Rick Playle, a good friend and a superb scientist who pioneered physiological understanding and modeling of the effects of metals on fish.

Chris M. Wood
Anthony P. Farrell
Colin J. Brauner
May 2011

1

AN INTRODUCTION TO METALS IN FISH PHYSIOLOGY AND TOXICOLOGY: BASIC PRINCIPLES

CHRIS M. WOOD

1. Background
2. Structure of the Book
3. Chemical Speciation in Freshwater and Seawater
4. Sources of Metals and Economic Importance
5. Environmental Situations of Concern
6. Acute and Chronic Ambient Water Quality Criteria
7. Mechanisms of Toxicity
   7.1. Acute Toxicity
   7.2. Chronic Toxicity
8. Essentiality or Non-Essentiality of Metals
9. Potential for Bioconcentration and/or Biomagnification of Metals
10. Characterization of Uptake Routes
    10.1. Gills
    10.2. Gut
    10.3. Other Routes
11. Characterization of Internal Handling
    11.1. Biotransformation
    11.2. Transport through the Bloodstream
    11.3. Accumulation in Specific Organs
    11.4. Subcellular Partitioning of Metals
    11.5. Detoxification and Storage Mechanisms
    11.6. Homeostatic Controls
12. Characterization of Excretion Routes
    12.1. Gills
    12.2. Kidney
    12.3. Liver/Bile
    12.4. Gut
13. Behavioral Effects of Metals
14. Molecular Characterization of Metal Transporters, Storage Proteins, and Chaperones
15. Genomic and Proteomic Studies
16. Interactions with Other Metals

Homeostasis and Toxicology of Essential Metals: Volume 31A
FISH PHYSIOLOGY

A brief history of metals, their early investigation in fish by physiologists and toxicologists, and current terminology are presented. The conceptual basis for the topics explored in each of the metal-specific chapters of these two volumes is then described. These include sources of metals, their economic importance, environmental situations of concern, essentiality or non-essentiality, bioconcentration or lack thereof, and the overarching importance of chemical speciation in understanding their effects on fish. The techniques used to derive ambient water quality criteria for metals are explained. Key mechanisms of acute and chronic toxicity are reviewed, as well as recent findings on the mechanisms and sites of uptake, internal handling, biotransformation, subcellular partitioning, detoxification, storage, and excretion. Important new research fronts focus on behavioral effects, molecular and omic analyses of cellular responses, and the effects of interacting metals in fish. Similarities and differences among the metals dealt with in these volumes are highlighted.

1. BACKGROUND

Of the 94 naturally occurring elements, 70 are metals, broadly defined as elements which are good conductors of electricity and heat, which form cations by loss of electrons, and which yield basic oxides and hydroxides. A few others, including Se and As, are honorary metals ("metalloids"), sharing some but not all properties of true metals. The terms "heavy metal" (universally viewed in a negative light by the general public) and "light metal" (often positively viewed) are outdated and chemically meaningless (Duffus, 2002; Hodson, 2004). Various other classifications have been proposed, of which the most scientifically defensible appears to be that based on their Lewis acid behavior (Lewis, 1923), as articulated by Nieboer and Richardson (1980). In this scheme, "hard" class A metals (e.g. Na, Mg, K, Ca, Rb, Li, U, Al) tend to bond ionically with oxygen donors, while "soft" class B metals (e.g. Ag, Hg, Pb, Cu) tend to bond covalently with sulfur donors. Unfortunately, this classification has proven unpopular with aquatic toxicologists, perhaps because so many important metals (e.g. Co, Cd, Ni, Cr, Fe, Zn) fall between the cracks as borderline or intermediate class metals, and their classification is controversial.

Metals have been long prized by humans for their generally attractive appearance (lustrous and shiny), malleability when heated, and hardness when cold, especially when blended in alloys, which gives them great practical utility for the making of tools, machines, weapons, and structures. The computer on which this manuscript was typed contains 30–40 different metals. The exploitation of metals by humans goes back to at least 6000 BC (the end of the Neolithic period or perhaps even earlier), but amazingly, up until the end of the Dark Ages (around AD 1400), only seven metals had been firmly identified and were in common use: Au, Cu, Ag, Pb, Sn, Hg, and Fe. With respect to the latter, Pliny the Elder wrote, "the ores of iron provide a metal which is at once the best servant of mankind–but the blame for death must be credited to man and not to nature". Many of the metals that we take for granted today (e.g. Co, Mn, Mo, Zn, Cd, Ni, Cr, Al) were only discovered in the eighteenth and nineteenth centuries. Indeed, only in the twentieth century was it realized that many of these same metals (Cu, Fe, Mn, Mo, Zn, Cr, and Co, plus probably Ni, Ge, Rb, and V in some organisms) are "essential", i.e. absolutely required in trace amounts for biological life owing to their participation in metabolic reactions as cofactors or integral parts of enzymes (Jeffery, 2001). There are no known biological functions for "non-essential" metals, which means that physio-logical mechanisms for specifically taking up such metals into an organism theoretically should not have evolved.

As elements, metals can be neither created nor destroyed, so once they are extracted from ores, they are ultimately dispersed into the environment. The vast majority of this extraction and dispersion ("production") has occurred since 1900, with production rates increasing in a quasi-exponential fashion throughout the last century. On a global basis, anthropogenically driven metal fluxes through the environment account for approximately half of all metal fluxes, and most metals are being "produced" at a rate that is orders of magnitude higher than the natural rate of renewal in the Earth's crust (i.e. by molten core upwelling and meteorite deposition) (Rauch and Graedel, 2007; Rauch and Pacyna, 2009). For a range of commonly used metals, *cumulative* world production by the year 2000 had reached levels many times higher than estimated levels in the year 1900: for example, Cr (643 ×), Ni (110 ×), Cu (25 ×), Zn (22 ×), Cd (18 ×), and Hg (7.3 ×) (Han et al., 2002). One notable exception is Pb (only 2–3 ×), a metal that was heavily exploited in "preindustrial" times, and whose production in the latter part of the twentieth century was greatly curtailed owing to health concerns and efficient recycling. If dispersed homogeneously throughout the world's soils and sediments, this cumulative anthropogenic production would have increased the levels of most metals to two- to ten-fold above natural background, as illustrated in Fig. 1.1(A). Note that the largest increases are for Hg, mainly owing to prolific burning of coal in which it is a trace contaminant. The smallest increases are for Cr and Ni, for which exploitation only started in about 1950. Of course, dispersion is a slow, non-homogeneous process, so while there are regions where natural background concentrations still persist, yet other areas have metal levels that are orders

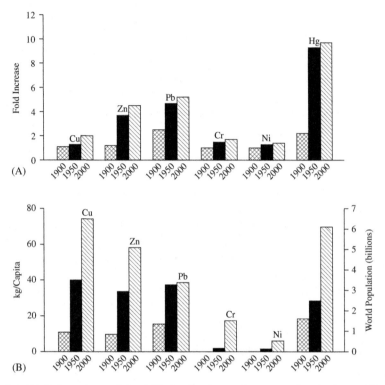

Fig. 1.1. (A) Estimated cumulative fold elevation in metal concentrations in the world's soils and sediments, above historical levels, by the years 1900, 1950, and 2000, due to anthropogenic production, assuming homogeneous dispersal. (B) Estimated cumulative world metal burden per capita due to anthropogenic production. Data calculated from Han et al. (2002) and Rauch and Pacyna (2009).

of magnitude higher, owing to local anthropogenic contamination. Dividing estimated cumulative production by population is another way of putting these data into perspective; note the large increases in the *cumulative* metal burden per capita, despite the greater than four-fold increase in world population from 1900 to 2000 (Fig. 1.1B).

While there were some important early studies in the aquatic toxicology of metals (e.g. Jones, 1939; Holm-Jensen, 1948), prior to about 1950, there was general belief in "better living through chemistry", and relatively little public and scientific concern about the dispersion of metals in the environment or their toxicological effects. The growth of such concern paralleled the growth of the environmental movement, catalyzed by the publication of *Silent Spring* by Rachel Carson (1962). This landmark book focused mainly on organic pollutants, especially pesticides and herbicides, rather than on metals. Nevertheless, it fundamentally shifted the landscape towards environmental awareness for all potential pollutants. The establishment of national environmental protection agencies [e.g. US Environmental Protection Agency (EPA) in 1970, Environment Canada in 1971] in many jurisdictions ensued in the following two decades, together with efforts to establish national water quality guidelines for various contaminants, including metals. Simultaneously, there was a massive surge in aquatic toxicological research, which has provided the data critical for developing water guidelines and criteria for metals, many of which remain in use today.

Some remarkable studies in fish toxicology from this era blurred the traditional boundaries with both physiology and geochemistry, by addressing mechanisms of toxicity and showing that the impacts of metals depended on what else was present in the water (Lloyd and Herbert, 1962; Brown, 1968; Zitko et al., 1973; Brown et al., 1974; Pagenkopf et al., 1974; Zitko and Carson, 1976; Chakoumakos et al., 1979). The interests of physiologists and geochemists were thereby piqued. The international journals *Aquatic Toxicology* and *Environmental Toxicology and Chemistry* were founded just 30 years ago, in 1981 and 1982, respectively. Another important driving force was the acid rain crisis of the 1970s and 1980s, when focused research revealed that many of the effects originally attributed to the acidity of the water alone were in fact due to metals which became dissolved and/or more toxic at low pH (see Couture and Pyle, Chapter 9). This was particularly true of Al (see Wilson, Chapter 2, Vol. 31B). There followed a surge of mechanistic research which continues to this day, and which forms the basis for these two volumes. Much of this research has been sponsored by government agencies and various metal-producing industries, often in cooperation, because of common interests in regulatory issues. In this regard, the European Union (EU), the USA, Canada, Australia/New Zealand, China, and several Latin American countries have recently revised, or are in the process of revising, their ambient water quality criteria (AWQC) for metals, making the present volumes timely.

Since about 1990, there has been an explosion of new information on the molecular, cellular, and organismal handling of metals in fish. Much of this research has focused on the physiological mechanisms of metal uptake, toxicity, and excretion. Internally, homeostatic mechanisms have been characterized which include regulated storage and detoxification (e.g. metallothioneins, ferritin, glutathione) and vehicles for transporting metals around the body in the circulation (e.g. ceruloplasmin, transferrin). New molecular, genomic, and proteomic techniques are now facilitating precise characterization of these pathways, and how they respond to environmental challenges. All these topics are major themes in the various metal-specific chapters. In turn, this information has proven useful in interpreting the responses of fish populations in the wild to chronic metal contamination (see Couture and Pyle, Chapter 9). Very importantly, new physiological and geochemical information has now been captured in a number of widely used modeling approaches (see Paquin et al., Chapter 9, Vol. 31B).

These elements are no longer viewed by fish physiologists as evil "heavy metals" (an outdated and chemically meaningless term) (Duffus, 2002; Hodson, 2004) that kill fish by suffocation (except at concentrations that are only relevant in an industrial "end-of-pipe" context). Instead, they are studied as interesting moieties that enter and leave fish by specific pathways, and which are subject to physiological regulation. These regulatory pathways may be ones dedicated for essential metal uptake, e.g. Cu-specific, Zn-specific, and Fe-specific transporters as detailed by Grosell (Chapter 2), Hogstrand (Chapter 3), and Bury et al. (Chapter 4), respectively. Alternately or additionally, they may be ones where metals masquerade as nutrient ions ("ionic mimicry") (Clarkson, 1993; Busselburg, 1995; Bury et al., 2003). In general, most of the research to date has focused on waterborne metals, but there is a growing realization of the importance of diet-borne metals (e.g. Dallinger and Kautzky, 1985; Dallinger et al., 1987; Clearwater et al., 2002; Meyer et al., 2005; Mathews and Fisher, 2009; Couture and Pyle, Chapter 9), a topic that is addressed in each metal-specific chapter. Indeed, for some metals such as Se (Janz, Chapter 7), Hg (Kidd and Batchelar, Chapter 5, Vol. 31B), and As (McIntyre and Linton, Chapter 6, Vol. 31B), trophic transfer (i.e. through the food chain) appears to be the major route of uptake.

2. STRUCTURE OF THE BOOK

This two-volume book consists of 15 metal-specific chapters and three integrative chapters. The integrative chapters are designed to provide background and general principles (Wood, this chapter), to take laboratory-derived information back to the field so as to interpret impacts on wild fish populations (Couture and Pyle, Chapter 9), and to illustrate the advances that have been made in using the laboratory and field information for predictive modeling (Paquin et al., Chapter 9, Vol. 31B). In the metal-specific chapters, the metals featured are those about which there has been most public and scientific concern, and therefore they are those most widely studied by fish researchers. Cu, Zn, Fe, Ni, Co, Se, Mo, and Cr are either proven to be or strongly suspected to be essential in trace amounts, yet are toxic in higher doses, and are the focus of specific chapters in Volume 31A. In contrast, Ag, Al, Cd, Pb, Hg, As, Sr, and U, which have no known nutritive function in fish at present, but which are toxic at fairly low levels, are considered in specific chapters in Volume 31B. Thus, this two-volume book is simply divided according to our present understanding of essentially of metals in fish, but with three chapters transcending both volumes.

Macronutrient metals (e.g. Na, Ca, K, Mg) have been excluded as they are commonly reviewed as physiological and nutritive parameters. Thallium, tin, manganese, lithium, cesium, lanthanum, bismuth, antimony, platinum, palladium, and rhodium have also been excluded. These metals are of increasing concern in ecotoxicology, but as yet are too data poor to justify review. In addition, metals in nanoparticles have not been considered; their principles of uptake and toxicity appear to be fundamentally different from those of dissolved metals. The current status of the nanoparticle field with respect to effects on fish has been captured in several excellent reviews (Handy et al., 2008a,b; Klaine et al., 2008; Shaw and Handy, 2011). Research in this area is increasing exponentially, and should be ripe for a future volume in the *Fish Physiology* series in a few years' time.

In each of the metal-specific chapters, authors were requested to address each of the subsequent topic headings, with recognition that some deviation from this basic plan might be necessary because of the particular properties of an individual metal. The remainder of this chapter provides the context for each of these topics.

3. CHEMICAL SPECIATION IN FRESHWATER AND SEAWATER

For metal physiology and toxicology, the importance of chemical speciation cannot be overstated. Perhaps the simplest feature of speciation is whether the metal is in the dissolved or particulate form. Originally, environmental regulations were based on total metals present in the water as assayed by hot acid digestion of the samples. However, there has been a gradual change in many jurisdictions to regulations based on the dissolved component only (see Section 6). This reflects the general recognition that particulate metals exhibit negligible toxicity and bioavailability to aquatic organisms relative to dissolved metals. The definition of "dissolved" is an operational one, with most jurisdictions and practicing toxicologists accepting the definition that the dissolved component is not retained by a 0.45 µm filter, although occasionally a 0.22 µm filter is used. While metals in nanoparticles are not considered in the current volumes, it is worth noting that the increasing environmental dispersion of nanoparticles will soon require a reassessment of these criteria, as most will pass through 0.45 µm and 0.22 µm filters.

At present, the best practice in both field monitoring and laboratory experimentation is to measure both the total metal present and the dissolved component after 0.45 μm filtration, together with as many features of water chemistry as possible, i.e. pH, alkalinity (by titration to pH 4.0), dissolved organic matter (DOM), and major ions, particularly the hardness cations (Ca^{2+} and Mg^{2+}). Indeed, in journals specializing in aquatic toxicology and environmental science, it is virtually impossible today to publish experimental work without these measurements. "Nominal" concentrations (i.e. concentrations calculated from the amount added) have become meaningless because they almost invariably overestimate both the total and dissolved metal concentrations present in a test system. This is because metals are notoriously sticky, quickly adsorbing to walls of containers (even those used to introduce the metal into the test system), plumbing, surfaces of test organisms, particles of food, and mucus and feces given off by the animals.

Even within the dissolved component, there can be massive differences due to speciation. A simple example will suffice. Toxicity is generally quantified as an LC50 value: the concentration of the toxicant that will kill 50% of the test organisms in a given period. In 7 day toxicity tests with juvenile trout, Ag was 15,000-fold more toxic (i.e. the 7 day LC50 was 15,000-fold lower) when tested as silver nitrate ($AgNO_3$) than as silver thiosulfate [$Ag(S_2O_3)_n^-$] (Hogstrand et al., 1996) and the concentration-specific uptake rate of Ag into internal organs of the fish was about 1000-fold greater for $AgNO_3$ (Hogstrand and Wood, 1998). This remarkable situation occurred despite the fact that both salts were fully dissolved. The explanation is that Ag remains tightly bound to thiosulfate in solution, whereas $AgNO_3$ dissociates freely in solution, yielding large amounts of the "free metal ion". The latter is usually portrayed as Ag^+ (a short-hand notation employed throughout these volumes), but in reality, it is the hydrated metal ion or aquo complex, i.e. $Ag(H_2O)_x^+$, because there are no bare metal ions in aqueous solutions. There are hundreds of similar examples in the literature for other metals. The general principle is that free metal ions are by far the most toxic and most bioavailable species, because they are most bioreactive with sites on the gills (such as proteinaceous enzymes, transporters, and channels).

This principle can be traced back at least as far as the study of Zitko et al. (1973) on Cu toxicity to juvenile Atlantic salmon, and was cemented by the classic conceptual paper of Pagenkopf (1983) formulating the Gill Surface Interaction Model (GSIM). The GSIM proposed that Cu, Cd, Pb, and Zn toxicity to fish resulted from free metal cations binding to a fixed number of anionic "interaction sites" on the gill surface, and that the availability of free Cu^{2+}, Cd^{2+}, Pb^{2+}, and Zn^{2+} in fresh water was dictated strongly by pH and alkalinity. The GSIM also recognized that other free cations such as Ca^{2+}, Mg^{2+}, and H^+ could offer protection by competing with the free metal cations for these interaction sites. Despite the fact that Zitko et al. (1973) had made Cu^{2+} ion activity measurements in various humic solutions, the original GSIM curiously overlooked organic complexation, but noted that inorganic anions present in solution would complex cationic metals, decreasing their bioavailability. Shortly thereafter, Morel (1983) formulated the Free Ion Activity Model (FIAM), which focused on the binding of free metal cations to algae. While similar to the GSIM, it additionally recognized that DOM could protect by complexing metal cations, and that other metal species might also bind to interaction sites on the algae, albeit less strongly. Together, the GSIM and FIAM provided the theoretical framework for the modern biotic ligand models (BLMs) (Paquin et al., 2000, 2002; McGeer et al., 2000; Di Toro et al., 2001; Santore et al., 2001; Niyogi and Wood, 2004a; Paquin et al., Chapter 9, Vol. 31B).

A common feature of these three models (GSIM, FIAM, BLM) is that they use geochemical principles to characterize the reactions that occur in the exposure water. At equilibrium (which is always assumed), these reactions are described by conditional equilibrium stability constants. These are the negative logarithms of the dissociation constants for each reaction, and are commonly termed log K or log K_D values. The higher the log K value, the stronger the binding. The advent of computer-based geochemical modeling programs, such as MINTEQA2 (Allison et al., 1991) and MINEQL+ (Schecher and McAvoy, 1992), in which most common log K values are available, has greatly facilitated this approach. Most of these constants are taken from the US National Institute of Standards and Technology (NIST) database. The critically important evolutionary step between the pioneering GSIM and FIAM, and the modern BLMs was the practical work of Playle and colleagues (1993a,b). These workers employed geochemical speciation programs together with inorganic competition and organic complexation experiments and measurements of short-term gill metal burdens to quantify the strength (log K_D) and molar density (i.e. concentration or B_{max}) of metal binding sites on fish gills (Fig. 1.2A) (Playle, 1998). Therefore, these values could also be entered into the modeling programs. When coupled with toxicity data linking the amount of short-term gill metal binding to the amount of longer term toxicity (e.g. the 4 day or 7 day LC50) (MacRae et al., 1999; Morgan and Wood, 2004) (Fig. 1.2B), a prediction could be made as to how toxic a metal would likely be in water of differing compositions.

In fresh water, the speciation chemistry of different metals varies greatly, but in general lower pH increases the free ion concentration, thereby increasing toxicity, whereas alkalinity (i.e. HCO_3^- and CO_3^{2-}) and inorganic anions tend to complex metal ions, thereby decreasing toxicity. The hardness cations (Ca^{2+} and Mg^{2+}) as well as Na^+ and K^+ (and sometimes H^+) may also

10   CHRIS M. WOOD

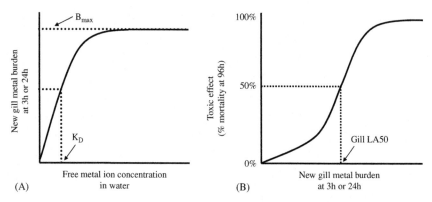

Fig. 1.2. Key principles of the biotic ligand model (BLM). (A) Relationship between the free metal ion concentration in the water versus the amount of new metal (i.e. above background) bound to the gill in a short period (usually 3 h or 24 h) before appreciable damage occurs. The concentration on the gill when all high-affinity binding sites are saturated is the site density (B_{max}). The free metal ion concentration that yields 50% B_{max} is the affinity (K_D). (B) Relationship between short-term binding of new metal to the gills (as determined in panel A) versus eventual percentage mortality, usually determined at 96 h of exposure when constructing acute BLMs for fish. The gill new metal burden (accumulation predictive of 50% lethality) is the gill LA50, which appears to vary considerably among species. In practice, BLMs are often now derived simply from relationships between free metal ion concentrations (estimated from geochemical modeling programs) in different water chemistries versus observed mortality data. In modeling, the LA50 may be arbitrarily adjusted to fit available toxicity data.

decrease toxicity by competing for metal binding sites on the gills. However, in many natural waters, the most effective agent of protection against most metals (Fe is a notable exception; Bury et al., Chapter 4), and the one that remains most poorly characterized, is DOM. In general, modeling approaches for geochemical speciation of metals in the presence of DOM rely on the Windemere Humic Aqueous Model (WHAM) (Tipping, 1994).

Complications arise from the fact that natural DOM molecules are extremely heterogeneous, both internally and among different sources (Thurman, 1985). Different parts of a single DOM molecule may have many different binding sites with different apparent log K values for both metals and protective cations. Allochthonous (also called terrigenous) DOM comes from the land and is produced by the degradation of lignins and other plant-based molecules. It tends to have greater molecular size, to be more darkly colored owing to more aromatic groups, and to be generally more protective against metal toxicity. Autochthonous DOM is produced in open lakes and oceans by algal photosynthesis, and by the eventual breakdown of terrigenous DOM by microbial activity and photodegradation. It tends to have smaller molecular size, to be more lightly colored, and to be less protective against metal toxicity than allochthonous DOM. DOM is usually measured by combustion or other oxidation methods as dissolved organic carbon (DOC), with the carbon atom constituting about 50% by mass of most natural DOM molecules. Al-Reasi et al. (2011) provide a recent summary of the optical and chemical features of DOM that help to quantify their protective features against metal toxicity.

In seawater, the importance of metal speciation has been less well studied from a toxicological standpoint. This is partly because there has been much less metal toxicity research in the marine environment, and partly seawater composition is less variable than freshwater composition. The most obvious variable is salinity, with all major ions covarying when salinity changes. For many metals, complexation by the high levels of Cl^- present dominates speciation in seawater, and this, combined with the greater availability of other anions (some of which create insoluble salts) plus the protective effect of competition by high concentrations of Na^+, Mg^{2+}, and Ca^{2+}, means that most metals are far less toxic in seawater than in freshwater. Chromium (Reid, Chapter 8) and As (McIntyre and Linton, Chapter 6, Vol. 31B) are notable exceptions. The most interesting and as yet poorly studied aspects of metal speciation occur in the brackish waters of estuaries, locations where metals are often discharged by sewage treatment plants and industries. Here DOM levels may be high, while major ion concentrations, alkalinity, and pH may all be highly variable depending on tide and season, and salinity-dependent ionoregulatory physiology may also play an important role in metal toxicity (Grosell et al., 2007).

4. SOURCES OF METALS AND ECONOMIC IMPORTANCE

The authors of each of the metal-specific chapters were asked to briefly address both this and the following topic (Section 5), which at first glance may seem to have little to do with the theme of the volumes. The reason for this is simple. In the author's experience, many physiologists and toxicologists working on metal effects in fish know that the metal of interest can be weighed out of an analytical grade bottle, but have little real-world experience as to where metals come from, their numerous applications, and their environmental impacts. This is unfortunate on several levels.

Firstly, this sort of background information is essential in making experimental approaches environmentally relevant. For example, with Al, most of the dispersion into natural waters comes not from mining, but from the effects of acidified water on local geology, and this is almost exclusive to soft waters such as those in Scandinavia and the Canadian Shield.

Therefore, it makes little sense to test Al in hard water with a low pH. Yet it is very relevant for the fate of migrating salmon smolts to evaluate what happens when soft, Al-enriched acidified water runs into the sea (see Wilson, Chapter 2, Vol. 31B). Secondly, because metal toxicity and physiology are so critically dependent on speciation (Section 3), understanding the source is critically dependent. Until recently, the major anthropogenic discharge of Ag was by the photographic industry, and the form discharged, silver thiosulfate ($Ag(S_2O_3)_n^-$), had negligible toxicity and very low bioavailability, yet the discharges were paradoxically regulated as though they mainly comprised the highly toxic free silver ion, Ag^+ (see Wood, Chapter 1, Vol. 31B). Finally, a very practical consideration is that most aquatic metals research is funded by the industries that discharge or market the specific metals, or by their research associations, or by government agencies interested in improving environmental regulations. Hopeful applicants for research funding would be well advised to inform themselves of the needs of the industries and agencies from a socioeconomic viewpoint.

Another important aspect of this topic is that natural background levels of metals in lakes and rivers may vary widely because of differences in local geology, and the aquatic organisms that live there tend to be genetically adapted to the local levels of metals (the "metalloregion concept") (Chapman and Wang, 2000; Chapman et al., 2003; Fairbrother et al., 2007). A metal concentration that is benign to fish in one region where background levels are high may be toxic to fish historically living in a region where background concentrations are low. Indeed, if the metal is essential (see Section 8), deficiency symptoms may result if organisms from a high background are tested in control water lacking the metal. In addition, if fish from a high background are held for a prolonged period in control water before toxicity testing, they may upregulate their uptake mechanisms for an essential metal so as to counteract deficiency. The outcome could be greater toxicity when these fish are exposed to elevated levels of that same metal in a subsequent toxicity test.

5. ENVIRONMENTAL SITUATIONS OF CONCERN

Almost invariably, research on the aquatic toxicology and physiology of metals has been driven by environmental situations of concern, and these in turn have often helped to illuminate mechanisms of toxicity. For example, the collapse of most fish populations in Belews Lake, North Carolina, USA, after contamination with Se-rich water from a fly-ash settling pond led to a discovery of fundamental importance: Se induces reproductive failure through teratogenic effects on early life stages via maternal transmission of trophically acquired Se (see Janz, Chapter 7). Another example relates to Al. In poorly buffered soft-water catchments, observations of dead and dying fish at only moderately acidic pH's led to the discovery that the real culprit was Al, mobilized from bedrock by acidic precipitation, rather than the acidity itself. Furthermore, the fact that dying fish were suffering from both respiratory and ionoregulatory distress led to the discovery of the two main mechanisms of acute Al toxicity (see Wilson, Chapter 2, Vol. 31B). The environmental devastation associated with smelter emissions in the Sudbury and Rouyn-Noranda areas of Canada has elicited a vast amount of invaluable laboratory and field studies of metal impacts. Most importantly, it has provided evidence that metal exposure leads to both direct toxicity and indirect damage (via food web effects) in wild fish (see Couture and Pyle, Chapter 9). The other chapters of these two volumes illustrate many comparable situations.

6. ACUTE AND CHRONIC AMBIENT WATER QUALITY CRITERIA

This is where science behind the chapters in these volumes meets public policy: a vast amount of research may be boiled down to just a few numbers, a formula, or a computer program, which then has immense socioeconomic consequences. In virtually all jurisdictions, the goal is to provide realistic environmental protection while not overly impeding economic development. However, the balance between these concerns and how they are put into regulatory practice varies dramatically among different jurisdictions (Chapman et al., 1996b). On a global basis, or among different metals, there is certainly no level playing field. In practical terms it is often difficult for even an informed scientist to find out exactly what a regulation is and how it is applied, because the information is often only available in the grey literature or on constantly changing websites. This is particularly true in Europe, where a panoply of national and overlapping EU regulations, guidelines, and jurisdictions are in a present state of flux, though a concerted effort is now being made to develop EU-wide standards. In Canada and the USA, individual provinces and states may choose to adopt national standards or develop their own guidelines. Nevertheless, the authors of each of the metal-specific chapters were asked to survey current **ambient water quality criteria (AWQC)** in major jurisdictions. The careful reader will notice a heterogeneity among chapters in these summaries, which reflects both the present situation and the difficulty in obtaining exact information. Regardless, the summarized data are useful because future toxicological research will be most useful if it is carried out at metal exposure levels that are relevant to AWQC levels, rather than at orders of magnitude higher concentrations as is often done. The following explanation of concepts may be helpful in interpreting these AWQC. Further background on AWQC is provided by the excellent reviews of Chapman et al. (1998, 2003).

The **precautionary principle** was originally developed in the late 1980s as a principle for regulating discharge of hazardous material into the North Sea, and more recently has been used as a basis for aquatic metal regulations, particularly in the EU (reviewed by Fairbrother and Bennet, 1999). Many definitions exist, but a widely accepted one arose from the Rio Earth Summit of 1992: "Where there are threats of serious or irreversible damage, lack of full scientific certainty shall not be used as a reason for postponing cost-effective measures to prevent environmental degradation." Therefore, the result is often extremely low AWQC values that are derived not from scientific study, but rather as very cautious "best guesses" to protect the environment in the face of uncertainty. An important long-term objective of toxicological research should be to replace AWQC based on the precautionary principle with those based on rigorously collected scientific data.

Application or **safety factors** are another tool for dealing with uncertainty (Chapman et al., 1998). In this widely used approach, a threshold concentration value based on scientific study is divided by an application or safety factor to produce a lower number for the AWQC. Depending on the degree of uncertainty and the jurisdiction, the factor is anywhere from 2 to 1000, with larger factors being used when data quantity or quality is low, i.e. when uncertainty is high. In essence, the precautionary principle can be viewed as application of a safety factor that may approach infinity (i.e. the impractical goal of zero tolerance).

In some jurisdictions, **PBT criteria (persistence, bioaccumulation, toxicity)** have been applied to metals. This approach originated in the 1970s as a very useful tool to classify organic chemicals (e.g. DDT, dioxins, PCBs) based on their persistence in the environment (i.e. time to break down), inherent toxicity, and their potential to bioaccumulate in organisms (often estimated by octanol–water partition coefficients). However, their application to metals is inappropriate for numerous reasons (Adams and Chapman, 2005): (1) as elements, metals can never break down, i.e. they persist forever (Skeaff et al., 2002); (2) their potential for bioaccumulation cannot be estimated from octanol–water partition coefficients, and indeed meta-analyses of experimental and field data indicate that bioconcentration and bioaccumulation factors (BCFs and BAFs) are inversely related to exposure concentration, i.e. lowest when the hazard is highest (McGeer et al., 2003; DeForest et al., 2007; see Section 9); and (3) toxicity is not intrinsic, but entirely a function of chemical speciation and solubility (see Section 3).

Similar to PBT, attempts to develop criteria based on **critical body or tissue residues** for metals in field-collected fish have proven generally unsuccessful. Although valuable tools for regulating organic chemicals, they do not work for most metals because of the ability of aquatic organisms to regulate metals (i.e. minimize bioaccumulation at sites of internal toxicity), and to store them in inert forms. Metal burdens do not necessarily relate to toxicity, though their measurement may be a useful procedure for diagnosing the cause of ecosystem disturbance. Two important exceptions are Se (Janz, Chapter 7) and Hg (Kidd and Batchelar, Chapter 5, Vol. 31B), where organic-bound forms of Hg and Se do bioaccumulate internally in a manner predictive of chronic toxicity. Some AWQC are now based on critical tissue residue thresholds for these two metals. Recently, Adams et al. (2010) have reviewed the field, and proposed new approaches to make the critical tissue residue approach more useful for other metals.

Drinking water criteria (DWC) are designed to protect human health, and should not be confused with **AWQC**. The general public often believes that criteria stringent enough to protect human health should also protect aquatic ecosystem health, but this is certainly not the case for metals. The human digestive tract is far more resistant to most metals than the gills of fish or aquatic invertebrates. Fish would die at most of the DWC values listed in Table 1.1, which illustrates that the DWC for eight metals are on average 195-fold higher (range = 8–897 ×) than the AWQC!

Table 1.1
Comparison of drinking water criteria (DWC) versus ambient water quality criteria (AWQC) for selected metals

Metal	DWC (μg L^{-1})	AWQC (μg L^{-1})	Ratio
Cu	1300	1.45[c]	897
Cd	5.0	0.16	31
Pb	50[a]	1.2	42
Hg	6.0[b]	0.77	8
Ni	610	29	21
Se	170	5.0	34
Zn	7400	65.7	113
Ag	50[a]	0.12[d]	417

All values are those proposed or implemented by the US Environmental Protection Agency (EPA), except where otherwise noted when EPA data were lacking. AWQC values are designed for protection against chronic toxicity to aquatic organisms [i.e. criterion continuous concentrations (CCCs)] at a hardness of 50 mg L^{-1} CaCO$_3$.
[a]Environment Canada.
[b]World Health Organization.
[c]Calculated by the biotic ligand model for EPA moderately hard reference water.
[d]Proposed but not implemented by the EPA.

Water quality criteria are usually but not always based on dissolved, rather than total metal concentrations (see Section 3). **Acute AWQC** are generally derived from data based on short-term toxicity tests (typically 96 h for fish, 48 h for small invertebrates such as daphnia) in which the animals are not fed and the endpoint is death. For metals, daphnia and other cladocerans in freshwater, and mollusk and echinoderm larvae in seawater often prove more sensitive than fish and, therefore, essentially drive acute AWQC. **Chronic AWQC** are usually based on long-term tests (typically > 30 days for fish, 7–21 days for invertebrates) where the animals are fed and the endpoints (generally termed EC or effective concentration values in chronic tests) may be death, growth inhibition, or reduction in reproductive output. The presence of food, by binding metals and providing energy, often protects the invertebrates to a greater extent than fish. Early life stages of fish often prove to be very sensitive in these tests, and thereby influence chronic AWQC. Acute AWQC are designed to protect aquatic life against short-term surges in pollutant concentrations, whereas chronic criteria are designed to provide lifetime protection. Most jurisdictions rely solely on chronic criteria, but some use only acute or both. For example, the US EPA has both criteria for many metals. The acute EPA **criterion maximum concentration (CMC)** specifies the highest average concentration of a material in ambient water to which an aquatic community can be exposed briefly without resulting in an unacceptable adverse effect. In practice, the CMC is the concentration that cannot be exceeded over a 1 h averaging period more than once every 3 years. The chronic EPA **criterion continuous concentration (CCC)** specifies the highest average concentration of a material in ambient water to which an aquatic community can be exposed indefinitely without resulting in an unacceptable adverse effect. In practice, the CCC is the concentration that cannot be exceeded over a 4 day averaging period more than once every 3 years.

Data requirements among jurisdictions vary greatly, but acute and chronic AWQC are usually derived from **species** or **genera sensitivity distributions (SSD)**, in which acceptable mean test data (e.g. LC50 values) for different species or genera are plotted on the y-axis against the percentage rank in order from most sensitive (i.e. the lowest LC50) to least sensitive (i.e. the highest LC50) on the x-axis. Then various statistical techniques are used to derive the desired endpoint. For example, the EPA either interpolates the LC50 value at the 5th percentile, or more usually extrapolates it using a regression based on the four most sensitive LC50 values in the percentage rank distribution. This 5th percentile LC50 is termed the **final acute value (FAV)** and is then divided by a factor of 2 to yield the **CMC** (acute AWQC). A parallel approach may be used to derive the **CCC** (chronic AWQC), but here the endpoints used are generally lower values than EC50. For example, the EPA uses either an EC_{20} or the geometric mean of the **no observed (adverse) effect concentration (NOEC)** and the **lowest observed (adverse) effect concentration (LOEC)**. This geometric mean is often termed the **maximum acceptable toxicant concentration (MATC)**. Rarely are there sufficient chronic test data which are available and acceptable to the EPA to generate a reliable chronic SSD. Therefore, an alternative procedure used by the US EPA and some other jurisdictions is to extrapolate the chronic AWQC from the acute AWQC. This is generally done by first dividing the acute LC50 by the chronic value (e.g. EC_{20} or MATC) on a species-specific basis to yield the **acute-to-chronic ratio (ACR)**, then taking the geometric mean of all the ACR values. The acute AWQC is then divided by the mean ACR to yield the chronic AWQC, though a variant in EPA procedure is that the FAV, not the CMC, is divided by the ACR to yield the CCC (chronic AWQC).

In the EU, chronic AWQC may be stated as **predicted no observed effect concentrations (PNEC)**, which are derived either from statistical analyses of SSDs of true chronic test endpoints (taken as the NOEC or the EC_{10}), or by dividing the lowest acute LC50 by a large application factor (e.g. 1000). The use of the SSD approach for deriving chronic AWQC is more common in the EU because standards for acceptance of chronic data are more lenient than those of the US EPA.

In many jurisdictions, AWQC for metals are adjusted for one measured water chemistry characteristic, the **hardness** of the site-specific water, harder waters having higher AWQC. Hardness is traditionally expressed as the sum of calcium plus magnesium concentrations, quantified as $CaCO_3$ equivalents in $mg\,L^{-1}$. For example, if water contained $40\,mg\,L^{-1}$ calcium (1 mmol L^{-1}) and $12\,mg\,L^{-1}$ magnesium (0.5 mmol L^{-1}), then the hardness would be 1.5 mmol L^{-1} or $150\,mg\,L^{-1}$ expressed as $CaCO_3$ equivalents. The EPA, for example, expresses many metal AWQC as hardness-based equations of the following form:

$$\ln\,[AWQC] = A \times \ln\,[hardness] + B$$

where A = slope (generally close to 1.0), B = ordinate intercept, LC50 is in units of $\mu g\,L^{-1}$, and hardness is in $mg\,L^{-1}$. Such equations may also be used to adjust LC and EC values before they are put into SSDs. The rationale is that hardness appears to be generally protective for most metals; Fe (Bury et al., Chapter 4), Se (Janz, Chapter 7), and As (McIntyre and Linton, Chapter 6, Vol. 31B) are notable exceptions. The mechanistic explanations are explored in the metal-specific chapters, and include direct protection by the competition of Ca^{2+} and Mg^{2+} with cationic metals for binding and uptake sites on organisms (Section 3), stabilization of tight junctions and membrane integrity of the gills by these ions (Hunn, 1985), and natural covariation with hardness of other protective features of water chemistry such as complexing agents, e.g. alkalinity (bicarbonate and carbonate), chloride, and sulfate (Meyer, 1999).

A natural evolution from modifying AWQC according to one water chemistry characteristic is to take all relevant water chemistry characteristics into account (e.g. pH, alkalinity, DOC, all major cations and anions), and this is what the **biotic ligand model (BLM)** is designed to accomplish. The BLM, described in Section 3, is now approved to generate site-specific AWQC for Cu in the USA (see Grosell, Chapter 2, and Paquin et al., Chapter 9, Vol. 31B), for decision-making and for modifying LC and EC values prior to entry into SSDs for several metals in numerous jurisdictions, and for in-depth investigations by some regulatory authorities using the **tiered approach**. This enlightened approach is where the level of protection is geared to the use and value of an ecosystem. Thus, standards would be lower (a higher AWQC) in an industrial harbor than in a pristine salmonid lake in a national park, but intermediate in water bodies in a mixed farmland. The approach is often coupled to the use of **trigger values** as AWQC, such that violation of the AWQC for a certain class of ecosystem triggers a more in-depth investigation to see whether, and to what degree, the ecosystem is impaired. The outcome may be a site-specific AWQC, perhaps based on the BLM, for that particular water body.

7. MECHANISMS OF TOXICITY

In accord with the usage of terms for AWQC, **acute toxicity** for fish refers to mechanisms that are operative in causing lethality at concentrations effective in 96 h tests, whereas **chronic toxicity** refers to mechanisms causing pathology or performance decrements in trials lasting 21–30 days (or longer, i.e. up to lifetime).

7.1. Acute Toxicity

The gills, which generally comprise over 50% of the surface area of the fish and are in intimate and continuous contact with the external water, are the primary target. At high enough concentrations, virtually all toxicants elicit profound morphological changes in the gills caused by an acute, generalized inflammatory response. This results in rapid death by suffocation due to edematous swelling, cellular lifting and necrosis, lamellar fusion, greatly increased water-to-blood diffusion distance, and impeded blood and water flow through and across the respiratory lamellae. Mallatt

(1985) provided a comprehensive review of these responses, which he divided into 14 elements, most of which were non-specific to any particular toxicant.

Of greater interest are the metal-specific mechanisms of toxicity that are operative at concentrations around the 96 h LC50 levels. In seawater fish these remain poorly understood, but in freshwater fish they have been well described for most metals. Wood (2001) provided a detailed summary of these mechanisms, and the conclusions of that review, together with more recent additional mechanistic details, are reinforced by each of the metal-specific chapters of this two-volume book. In particular, several metals appear to specifically target the active ionic uptake pathways on the gills, probably by using the pathways as a route of entry through "ionic mimicry" (Clarkson, 1993; Busselburg, 1995; Bury et al., 2003), as described in Section 9. For example, As (McIntyre and Linton, Chapter 6, Vol. 31B) mimics phosphate, Cu (Grosell, Chapter 2) and Ag (Wood, Chapter 1, Vol. 31B) mimic sodium; Zn (Hogstrand, Chapter 3), Co (Blust, Chapter 6), Cd (McGeer et al., Chapter 3, Vol. 31B), Pb (Mager, Chapter 4, Vol. 31B), and Sr (Chowdhury and Blust, Chapter 7, Vol. 31B) mimic calcium; Mo and Cr mimic sulfate (Reid, Chapter 8), and Ni (Pyle and Couture, Chapter 5) mimics magnesium. Through this mimicry, metals may actually reduce the uptake of an essential nutrient ion at the gills by more than just simple competition, sometimes resulting in death from the associated deficiency (e.g. hyponatremia, hypocalcemia).

For example, relevant chapters describe how Cu and Ag not only compete with Na^+ for apical uptake exchangers and channels, but also inhibit the basolateral Na^+/K^+-ATPase (Fig. 1.3). This key enzyme energizes active Na^+ and Cl^- uptake, and also contributes to basolateral Na^+ extrusion from the cytoplasm into the bloodstream. These two metals also potently inhibit intracellular carbonic anhydrase in the ionocytes; this enzyme hydrates CO_2 so as to produce the H^+ ions that are normally exchanged against Na^+ and the HCO_3^- ions that are normally exchanged against Cl^- at the apical membrane. As a result, Na^+ and Cl^-, the two major extracellular electrolytes, decline in concentration in the blood plasma (hyponatremia and hypochloremia) until death results owing to a circulatory collapse associated with fluid shifts along the resulting osmotic gradient from extracellular to intracellular compartments throughout the organism. This often occurs in concert with disturbances of acid–base balance and ammonia excretion, processes that depend on apical H^+ extrusion. Similarly, Zn, Pb, Co, and Cd compete with Ca^{2+} for entry through apical Ca^{2+} channels (ECac), then later inhibit the basolateral high-affinity Ca^{2+}-ATPase that powers active Ca^{2+} uptake (Fig. 1.3). Some of these also appear to interfere with intracellular carbonic anhydrase.

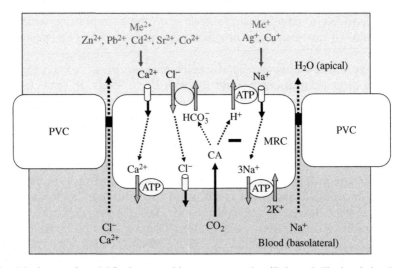

Fig. 1.3. A general model for how metal ions may enter the gill through "ionic mimicry" and thereby compete with nutritive ions for uptake, and, if in high enough concentration, eventually block nutritive ion uptake by inhibiting the ATP-dependent basolateral enzymes that normally power these processes. These nutritive ion uptake processes are critical to life in freshwater fish, because they must occur continuously to offset the passive losses of Na^+, Cl^-, and Ca^{2+}, which are shown as occurring by diffusion through the paracellular channels between the pavement cells (PVCs) and mitochondria-rich cells (MRCs). Monovalent Ag^+ and Cu^+ (after hypothesized reduction of Cu^{2+} to Cu^+ by a surface-bound reductase, not shown) compete with Na^+ for entry through a putative apical sodium channel (shown) and/or an Na^+/H^+ exchanger (not shown). Eventually they inhibit basolateral Na^+/K^+-ATPase. Divalent Zn^{2+}, Pb^{2+}, Cd^{2+}, Sr^{2+}, and Co^{2+} compete with Ca^{2+} for entry through an apical voltage-independent calcium channel (probably ECac), and eventually inhibit basolateral high-affinity Ca^{2+}-ATPase (shown) or an Na^+/Ca^{2+} exchanger (not shown). Some of these metals may also inhibit the intracellular carbonic anhydrase enzyme (CA), which provides the acid–base equivalents (H^+ and HCO_3^- ions) needed for exchange by the apical processes. For convenience, all processes are shown in a single MRC, but they may actually occur in different types of MRCs, or even in PVCs. Modified from an unpublished diagram by Fernando Galvez. **SEE COLOR PLATE SECTION.**

Eventually, a fatal hypocalcemia occurs owing to a failure of Ca^{2+}-dependent nerve and muscle function, sometimes complicated by acid–base disturbance. In addition, a common additive toxic effect of many of these same metals is to increase the efflux rates of Na^+, Cl^-, Ca^{2+}, and other nutrient ions from the gills by opening up the paracellular leakage pathway, either by displacing external Ca^{2+} ions (which maintain the integrity of the junctions) or by causing inflammation and associated changes in cell volume which weaken the tight junctions.

However these "mimicry effects" are not always the cause of death. For example, while Ni may serve as a Mg^{2+} antagonist, the branchial inflammatory effect leading to an inhibition of respiratory gas exchange appears to be the key mechanism of Ni lethality (Pyle and Couture, Chapter 5). Molybdenum mimics sulfate, but causes death by a similar respiratory mechanism (Reid, Chapter 8). The same mechanism also appears to apply to Al, but only at mildly acidic pH; at lower pH's, ionoregulatory dysfunction predominates (Wilson, Chapter 2, Vol. 31B). Iron also appears to be a respiratory toxicant (Bury et al., Chapter 4), and in the case of both Al and Fe, flocculent precipitation of metal complexes on the gill surface may compound the basic inflammatory response. Mercury appears to kill fish by a potent blockade of neural function and inhibition of key metabolic enzymes (Kidd and Batchelar, Chapter 5, Vol. 31B). However, for several metals in freshwater fish (Se, Janz, Chapter 7; Cr, Reid, Chapter 8; As, McIntyre and Linton, Chapter 6, Vol. 31B), and for most metals in seawater fish, there is as yet no clear "smoking gun" as to the cause of acute toxicity. Two clear exceptions are acute Cu (Grosell, Chapter 2) and acute Ag toxicity (Wood, Chapter 1, Vol. 31B) in marine teleosts; as in freshwater, these metals appear to target Na^+ and Cl^- regulation, with toxic actions exerted on transport functions in both gills and gut.

7.2. Chronic Toxicity

For regulatory criteria, the accepted endpoints can be only death, reduced growth, or reduced reproductive output, but from a scientific viewpoint, this is an unnecessarily narrow definition, and authors were asked to address any relevant pathophysiological mechanisms. Of particular concern is the practice in many jurisdictions of extrapolating chronic AWQC from acute AWQC based on the acute-to-chronic ratio, with the application of the same hardness-equation or BLM-derived correction factors (see Section 6). Two unwritten assumptions of this procedure are that: (1) the mechanisms of chronic toxicity are the same as those of acute toxicity, and (2) water chemistry properties (e.g. hardness, DOC, alkalinity, pH) that protect against acute toxicity provide the same relative protection against chronic toxicity. To the author's knowledge, there is little direct evidence that the latter is true, and some evidence that it is not. For example, DOC, which is very protective against the acute toxicity of Ag, has only a slight protective effect against chronic Ag toxicity (Wood, Chapter 1, Vol. 31B). With respect to the former assumption of common acute and chronic mechanisms of toxicity, the only metals for which there is clear evidence that this may be true are Ni (Pyle and Couture, Chapter 5), Ag (Wood, Chapter 1, Vol. 31B), and Al (Wilson, Chapter 2, Vol. 31B). Ni and Al, both acutely and chronically, cause diffusive limitations on respiratory gas transfer at the gills, and Ag causes both acute and chronic disruptions in Na^+ and Cl^- regulation in freshwater fish. Indeed, the opposite is clearly true for some metals. For example, acute Pb exposure appears to kill by inducing hypocalcemia, but sensitive endpoints of growth and reproductive inhibition during chronic Pb exposure are associated with scoliosis and neural and hematological disturbances rather than hypocalcemia (Mager, Chapter 4, Vol. 31B). Similarly, acute Cu exposure appears to kill by causing a failure of Na^+ and Cl^- regulation, but the most sensitive chronic endpoint is reproductive impairment, occurring at concentrations far below those required to induce mortality or reduce growth, and where there is no chronic ionoregulatory dysfunction (Grosell, Chapter 2).

Unlike acute toxicity, where lethality can often be attributed to only one or two mechanisms (ionoregulatory and respiratory disturbances), there is often a plethora of chronic toxic mechanisms. It is probably more realistic to assume that the fish's health gradually "runs down" owing to the combined load of many disturbances, with the eventual result of one or more of decreased survival, growth, or reproductive output. These disturbances include costs of acclimation ("damage repair"; McDonald and Wood, 1993), detoxification (metallothionein, glutathione synthesis; see Section 11.5), immune suppression (e.g. Mushiake et al., 1984), and the "burning out" of an ability to mount a corticosteroid stress response (e.g. Hontela, 1997), impacts that are common to many metals.

Two additional impacts deserve particular comment: oxidative stress and disruption of sensory function. A rapidly increasing body of literature (e.g. Payne et al., 1998; Craig et al., 2007, 2009) has reinforced early concerns that many metals induce oxidative stress in aquatic animals by catalyzing the Fenton reaction within cells (Di Giulio et al., 1989). This results in the generation of free hydroxyl radicals and hydrogen peroxide (H_2O_2) [reactive oxygen species (ROS)] that cause lipid peroxidation, protein carbonylation, DNA damage, and general damage to cellular functions (Lushchak, 2011). The effects occur rapidly but are unlikely to cause acute mortality; rather, they accelerate aging, leading to general deterioration of physical condition during chronic exposure. There is also a growing realization that many metals can impair sensory function (olfaction and mechanoreception) at exposure levels much lower than those causing other typical chronic endpoints (Scott and Sloman, 2004; Pyle and Wood, 2007). Again, the effects are rapid but persistent, and the ecological consequences are potentially devastating, as discussed below in Section 13.

Finally, there is now general acceptance that in the field, diet-borne metals may play an important role in chronic toxicity (e.g. Dallinger et al., 1987; Clearwater et al., 2002; Mathews and Fisher, 2009; Couture and Pyle, Chapter 9); this is particularly true for Se (Janz, Chapter 7). However, this route of exposure is not considered in most laboratory-based chronic testing, a troubling oversight. In future, it is hoped that chronic tests will be carried out with food items that have been equilibrated to the same concentration of metal as is being used in the waterborne exposure (Meyer et al., 2005).

8. ESSENTIALITY OR NON-ESSENTIALITY OF METALS

An understanding of why and how some metals and metalloids are essential for life illuminates their mechanistic physiology. These include the "macronutrients" (Na, K, Ca, Mg), as well as many of the "micronutrients" (Cu, Zn, Fe, Mn, Mo, Ni, Co, Se, Cr, V), which are the focus of this volume. One-third of all proteins are believed to require a metal cofactor for normal function (Rosenzweig, 2002), and of these about 3000 proteins require Zn, representing about 10% of the entire genome in humans (Andreini et al., 2005; Passerini et al., 2007)! Cu and Fe, through their participation in the electron transport chain, lie at the very core of aerobic life, and Se is a component of a unique amino acid, selenocysteine, with its own codon. Most scientists favor the view that the chemistry of the prebiotic environment in which life first began determined whether or not certain metals were incorporated as essential catalysts or cofactors for life processes, while subsequent environmental changes dictated that some became more important (e.g. Zn, Cu, Fe) and others less important (e.g. Ni, Mo) (Dupont et al., 2010). However, the nature of the original environment and the subsequent changes remains controversial, and makes for interesting reading beyond the scope of this chapter (see e.g. Williams, 1997; Nielsen, 2000; Mulkidjanian, 2009; Dupont et al., 2010).

As time passes, more and more metals are being proven essential in various organisms, yet outside the "big five" (Cu, Zn, Fe, Co, and Se), there have been very few tests for essentiality in fish. Therefore, the placement of metals such as Ni, Cr, and Mo in Volume 31A ("essential"), but As, Cd, and Sr in Volume 31B ("non-essential") is somewhat arbitrary based on a weight of evidence approach for other vertebrates. Indeed, all of these elements are now considered essential in at least some other life form, but none has been rigorously evaluated in fish. The question is an important one because it should influence the way we think about their regulation both in the organisms (i.e. homeostasis or detoxification?) and in the environment (i.e. deficiency or excess?) (Chapman and Wang, 2000). If a metal is essential, there should be a bell-shaped concentration–response (or dose–effect) curve for health, with symptoms of deficiency occurring at low concentrations and toxicity at high concentrations, with a plateau inbetween where the fish's

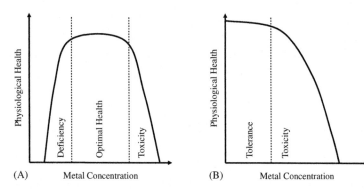

Fig. 1.4. Conceptual diagrams illustrating the differences in concentration–response relationships with respect to organism health between (A) essential metals and (B) non-essential metals. Modeled after Chapman and Wang (2000).

physiology performs optimally (Fig. 1.4A). However, if a metal is truly non-essential, there will only be a plateau of tolerance where physiology is normal in the range where excretion and/or detoxification mechanisms can keep up with entry rate. Beyond this range, toxicity will occur (Fig. 1.4B). Environmental regulations such as AWQC should be cognizant of these differences; for example, they should not dictate metal concentrations below the natural background levels that fish or their prey organisms require.

9. POTENTIAL FOR BIOCONCENTRATION AND/OR BIOMAGNIFICATION OF METALS

Bioconcentration refers to the fold extent to which the concentration of a chemical (i.e. a specific metal in these volumes) in an aquatic organism exceeds that in its aqueous environment, and may be expressed as the **bioconcentration factor (BCF)** in liters per kilogram (L kg^{-1}) (Chapman et al., 1996a; McGeer et al., 2003). BCFs are usually calculated as the ratio of the metal concentration in the whole body of the fish to the total metal concentration in the water, but variants do exist. For example, the calculation may be done on an organ-specific basis (e.g. liver concentration; see Hogstrand, Chapter 3) or a chemical species-specific basis (e.g. dissolved or ionic metal concentration in the water). There is an implicit assumption that the exposure is long enough for an equilibrium to be reached. Technically, the BCF is based on bioaccumulation from water only, so can only be measured in laboratory studies where the fish are either fasted or fed clean food. When the measurement is made on the basis of field studies, the calculation is done the same way, but is termed the **bioaccumulation factor (BAF)** as the metal can be bioaccumulated from both the water and the food, i.e. ingested prey items and sediment in equilibrium with the water.

BCFs and BAFs were originally derived for classifying the hazard associated with various organic chemicals (Chapman et al., 1996a). However, they are not useful for hazard assessment or environmental regulation of metals, as convincingly shown by the meta-analyses of McGeer et al. (2003) and DeForest et al. (2007) for many different categories of aquatic organisms. These data-mining exercises demonstrated for a vast range of metals, regardless of essentiality, that the BCF/BAF values are inversely related to exposure concentration (Fig. 1.5), i.e. highest when hazard is lowest, and lowest when hazard is highest, which is intuitively opposed to the original concept. Interestingly, the one exception in these analyses was Hg, probably because of its lipophilicity in the methyl mercury form (Kidd and Batchelar, Chapter 5, Vol. 31B). A primary assumption of the original BCF/BAF approach is that these indices are independent of the exposure concentration. This assumption is generally true for organics (Fig. 1.5) because they are lipophilic, favoring the lipid-rich environment of the fish relative to the water, and entering across gill and gut membranes by

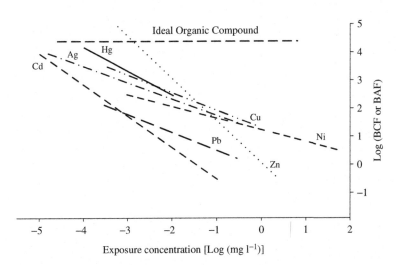

Fig. 1.5. Relationships between bioconcentration factors (BCFs) or bioaccumulation factors (BAFs) versus the logarithm (base 10) of metal concentration in the respective exposures for Zn, Cd, Pb, inorganic Hg, Ni, and Ag. Note the inverse relationships for all metals. The hypothetical flat-line relationship for an ideal organic compound is also shown. The lines have been taken from the meta-analyses of literature data for salmonids (or in their absence "all fish") performed by McGeer et al. (2003).

simple diffusion. Therefore, they bioaccumulate in organisms in a manner consistent with their octanol–water partition coefficients. However, the potential of most metals for bioaccumulation cannot be estimated from octanol–water partition coefficients. Metals are generally hydrophilic, not lipophilic, and therefore, in order to cross the barrier of lipid-rich cell membranes, they must be taken up by specific transporters or channels. These are often saturable and physiologically regulated, especially those serving the uptake of essential metals or their mimics.

Furthermore, the link between metal BCF/BAF and toxicity is tenuous because aquatic organisms are able to regulate metals internally (i.e. minimize bioaccumulation at key sites of toxicity) and to store them in inert forms. Tissue metal burdens include metals that are serving an essential role (e.g. as enzyme cofactors), metals that are stored in a non-toxic form (e.g. on metallothioneins or in granules), and metals that are actually causing toxicity (Section 11.4).

Nevertheless, the BCF/BAF concept does represent a useful index for comparing how different metals are handled physiologically by different organisms at different exposure concentrations. For example, most fish are able to regulate the essential metal Zn at very constant internal concentrations (Hogstrand, Chapter 3); as a result, BCFs are very high at low waterborne levels, and very low at high waterborne levels (Fig. 1.5). However, most fish are less effective in regulating non-essential, highly toxic metals such as Ag (Wood, Chapter 1, Vol. 31B) and Pb (Mager, Chapter 4, Vol. 31B). Thus, the internal metal burdens increase to a much greater extent with waterborne concentrations, and the BCF values exhibit less change with exposure concentration (Fig. 1.5). However, BCF/BAF values for Cu and Ni (essential) and Cd and Hg (non-essential) do not conform to this same rule of thumb (Fig. 1.5).

Biomagnification refers to the fold extent to which the concentration of a chemical increases across trophic levels (usually three or more). Again, it is an indicator of hazard for organic compounds, where high **biomagnification factors (BMFs)** indicate that secondary poisoning may occur in higher trophic level consumers. In contrast to BCFs, BMFs are less than 1.0 for most metals (i.e. biodilution occurs), because trophic transfer efficiency is low. Important exceptions are the organometallic compounds such as methyl mercury (Kidd and Batchelar, Chapter 5, Vol. 31B) and various organo-selenium compounds (Janz, Chapter 7). These can exhibit BMFs substantially above 1.0 because their high lipid solubility ensures high assimilation efficiency in the digestive tract, high retention efficiency in the consumers, and therefore high trophic transfer efficiencies. The potential consequence here is secondary poisoning in organisms higher up the food chain, which for Se may be the fish themselves (reproductive impairment), whereas for Hg, concern focuses mainly on the birds and mammals that may eat the fish (neural, biochemical, and reproductive impairment).

10. CHARACTERIZATION OF UPTAKE ROUTES

In freshwater fish, the gills are the dominant route of uptake for most waterborne metals because of their large surface area, thin water-to-blood diffusion distance, and abundance of active transport pumps designed to acquire nutrient ions from the external water. Furthermore, freshwater fish exhibit very little drinking, in distinct contrast to seawater fish, which exhibit obligatory drinking as part of their overall osmoregulatory strategy to keep internal body fluids substantially hypotonic to the external seawater (Evans et al., 2005; Grosell, 2006). Therefore, uptake through the gut is important only for foodborne metals in freshwater fish, whereas in seawater fish, both waterborne and foodborne metals may be taken up through the gut. There are several reports that drinking in marine teleosts may be inhibited by the presence of metals, perhaps as a result of taste aversion (Grosell et al., 1999, 2004; Hogstrand et al., 1999). While reducing metal uptake, this may compound ionoregulatory disturbances.

10.1. Gills

Metal uptake through the gills may occur by three different routes.

Firstly, for some of the essential metals, there appear to be metal-specific carriers designed to take up metals from low concentrations in the external water. Virtually all of the evidence for such carriers has come to light in the last decade, and remains circumstantial, as it is based on molecular expression and competitive inhibition studies. For Cu, Ctr1 (the high-affinity copper transporter) and DMT1 (the promiscuous divalent metal transporter) are the likely candidates on the apical membranes, while a Cu-ATPase appears to be present on the basolateral membranes of gill ionocytes (Grosell, Chapter 2). For Zn, apical ZIP (zinc importer) and basolateral ZnT (zinc exporter) transporters may carry out comparable functions (Hogstrand, Chapter 3), while for Fe, apical DMT1 and basolateral ferroportin do the job once trivalent Fe^{3+} has been reduced to Fe^{2+} by a hypothesized epithelial reductase (Bury et al., Chapter 4). Non-essential metals may also be taken up by such pathways. For example, Cd is transported by one isoform of trout DMT1 expressed in *Xenopus* oocytes (Cooper et al., 2007).

Secondly, there is stronger evidence that metals may masquerade as other ions on active transport pathways designed to take up nutrient metals ("ionic mimicry") (Clarkson, 1993; Busselburg, 1995; Bury et al., 2003). Thus, Ag and Cu (probably after reduction of Cu^{2+} to Cu^{+}) are taken up through the Na^{+} pathway, while divalent Zn, Co, Cd, Pb, and Sr are taken up through the Ca^{2+} pathway (see Fig. 1.3). The evidence for these routes is laid out in the relevant metal-specific chapters and includes the use of pharmacological blocking agents, competition studies, and experimental manipulation of uptake rates of the nutrient ion. For example, branchial Cu uptake in trout was reduced by high water $[Na^{+}]$, by phenamil (which blocks Na^{+} channels), by bafilomycin (which blocks the v-type H^{+}-ATPase that energizes Na^{+} channels), and by experimental downregulation of active Na^{+} uptake (Grosell and Wood, 2002; Pyle et al., 2003). Similarly, Cd uptake in trout was reduced by high water $[Ca^{2+}]$ (Niyogi and Wood, 2004b), by lanthanum (Verbost et al., 1989), which blocks voltage-insensitive Ca^{2+} channels (ECac), and by experimental downregulation of active Ca^{2+} uptake (Baldisserotto et al., 2004).

Thirdly, metals may simply diffuse in across the gill epithelium, driven by the electrochemical gradient from water to blood. Free metal levels are negligible in the bloodstream because of the presence of numerous binding molecules (Section 11.2), and the gradient will be particularly favorable for entry of positively charged metal ions by this route in freshwater fish, where the transepithelial potential is such that the blood-side of the gills is negative (Evans et al., 2005; Marshall and Grosell, 2006). It seems probable that the diffusive route should be the paracellular pathways between the gill cells, but this has never been proven.

A powerful technique for the analysis of branchial uptake (or indeed uptake through any pathway) is to describe its concentration dependence (often mislabeled as "kinetic analysis") using radiolabeled metal (Wood, 1992), as has been done, for example, with Zn (Spry and Wood, 1989), Ag (Bury and Wood, 1999), Cu (Grosell and Wood, 2002), and Cd (Niyogi and Wood, 2004b). The rate of metal influx (J_{in}) can usually be described by the equation:

$$J_{in} = \frac{J_{max}[X]}{K_m + [X]} + m[X]$$

where $[X]$ is the dissolved or ionic metal concentration in the water, J_{max} is the maximum transport rate, K_m is the affinity constant, and m is the slope of a linear component that passes through the origin. In this formulation, the second term is a lumped Fick term for diffusive uptake. The first term is the Michaelis–Menten equation for saturable carrier-mediated transport of a substrate which yields a hyperbolic relationship between J_{in} and $[X]$.

An important caveat is that the K_m and J_{max} values recorded may well be blended values reflecting the simultaneous action of several transport systems. When the flux period is short, the negative logarithm of the K_m may provide the log K_D values of gill metal-binding (see Fig. 1.2A) described for the geochemical models (GSIM, FIAM, and BLM) in Section 3. This analysis also allows direct comparison of the concentration dependence of uptake for different metals, as well as quantitative analysis of competitive (altered K_m) versus non-competitive (altered J_{max}) interactions among different metals and nutrient ions, as well as identification of the quantitative importance of the diffusive component. The reader should note that the $m[X]$ term is analogous but not homologous to the $k_u C_W$ term used in kinetic models (Paquin et al., Chapter 9, Vol. 31B), where k_u is the dissolved metal uptake rate coefficient determined by radiotracer flux measurements conducted at extremely low concentrations (C_W) of the dissolved metal in the exposure water. In both cases, uptake is linearly proportional to concentration, but in the $k_u C_W$ formulation, the usually unstated assumption is that substrate (metal) level is so low that carrier-mediated uptake occurs on the almost linear part of the Michaelis–Menten curve close to the origin, and is not distinguishable from diffusive uptake.

10.2. Gut

The same three general mechanisms as for gills (metal-specific carriers, substitution on nutrient ion transporters, and simple diffusion) apply to metal uptake via the gastrointestinal tract. However, metals may additionally bind to amino acids (e.g. L-histidine, L-cysteine) in the chyme and undergo "piggy-back" transport on amino acid transporters (Glover and Hogstrand, 2002; Glover and Wood, 2008). Some of the metal-specific carriers appear to be members of the same families as those in the gills, e.g. Ctr1, DMT1, ZIP, ferroportin, although the exact isoforms and their transport characteristics may differ. The gastrointestinal nutrient ion transporters are not the same as those at the gills, so the mimicry scheme shown in Fig. 1.3 for the gills does not necessarily apply to the gut. For example, elevated chyme [Na$^+$] stimulates rather than inhibits Cu uptake through the trout gut, both *in vivo* (Kjoss et al., 2005) and *in vitro* (Nadella et al., 2007), which is very different from the situation at the gills (Grosell and Wood, 2002). Moreover, the Ca^{2+} channels in the teleost gut appear to be of L-type voltage-gated Ca^{2+} channels (Larsson et al., 1998), which are very different from the Cd-sensitive voltage-insensitive Ca^{2+} channels (ECac) in the gills (Verbost et al., 1989; Shahsavarani et al., 2006). Nevertheless, since high chyme [Ca^{2+}] reduces gut Cd uptake both *in vivo* (Franklin et al., 2005; Wood et al., 2006) and *in vitro* (Klinck and Wood,

2011), the gut Ca^{2+} uptake system does seem to play some role in gastrointestinal Cd uptake.

In general, gut transporters are designed to function at the much higher substrate levels present in the chyme and/or ingested seawater. Thus, nutrient ion transporters have much higher K_m values (i.e. lower affinities), generally in the mmol L^{-1} rather than µmol L^{-1} range. Similarly, the metal-specific transporters also have much higher K_m values, reflecting the fact that metal levels in food or chyme are often in the mg kg^{-1} range, in contrast to the ng–µg L^{-1} range in water. Ojo and Wood (2008) compared the uptake rates of six metals (Cu, Zn, Cd, Ag, Pb, Ni) across the gut with those across the gills of rainbow trout at luminal concentrations one to four orders of magnitude higher than in typical waterborne exposures, and concluded that uptake rates were similar across the two surfaces.

As in the gills, this Michaelis–Menten approach for saturable carrier-mediated transport is very useful for physiological characterization of gut uptake mechanisms, but actual relationships are sometimes closer to linear than hyperbolic both *in vitro* (e.g. Klinck and Wood, 2011 for Cd) and *in vivo* (e.g. Kamunde and Wood, 2003 for Cu). Standard kinetic models for food-route uptake assume a linear relationship between concentration in the food and uptake. Uptake is described by $\alpha_f I_f C_f$, where α_f is metal absorption efficiency, I_f is food ingestion rate, and C_f is metal concentration in the food (Paquin et al., Chapter 9, Vol. 31B). Making α_f concentration-dependent will convert a linear relationship into a hyperbolic one.

At least for the three major essential elements (Cu, Grosell, Chapter 2; Zn, Hogstrand, Chapter 3; Fe, Bury et al., Chapter 4), the bulk of normal uptake appears to take place from the food in the gut, while the gills serve as a dynamic fine-tuning mechanism for homeostasis. The gills can greatly increase or decrease their uptake rates at times of dietary deficiency or excess, respectively, but gut uptake rates do not appear to be modified in a reciprocal manner by waterborne deficiency or excess. Different portions of the gut appear to vary in their quantitative importance for the uptake of the various metals, but an interesting recent development is the emerging role of the stomach as an important site for uptake of at least three metals in freshwater trout: Cd (Wood et al., 2006), Cu (Nadella et al., 2006, 2010), and Ni (Leonard et al., 2009). Perhaps this is not too surprising in light of the fact that metal levels are highest and pH is lowest, yielding more free metal ions in the chyme in this compartment.

10.3. Other Routes

In general, other potential routes of metal uptake have received little attention. However, at least for one electrolyte (Ca^{2+}), significant uptake from water does occur via the skin (Perry and Wood, 1985), presumably facilitated by the presence of ionocytes and a secondary circulation close to the surface. Therefore, it is not surprising that small amounts of Cd also may be taken up by the skin (Wicklund-Glynn, 2001). Consequently, the potential cutaneous uptake of other metals that are calcium analogues (Zn, Co, Pb, Sr) is worthy of future study. Some metals are taken up by the olfactory route (Tjälve and Henriksson, 1999). Quantitatively, this pathway is probably very small, but toxicological impacts may be disproportionately larger. For example, inorganic Hg, Cd, Ni, and Mn are taken up from waterborne exposures by the olfactory rosettes, transported along olfactory nerves via axonal transport, and accumulated in the olfactory bulbs, thereby providing a potential direct route of access to the brain. At least in the case of Cd (McGeer et al., Chapter 3, Vol. 31B) and Hg (Kidd and Bachelar, Chapter 5, Vol. 31B), there is direct evidence of resulting behavioral deficits (Section 13). However, the degree to which the various metals actually penetrate into the brain, and thereby cause impairment, varies (Tjälve et al., 1986; Borg-Neczak and Tjalve, 1996; Rouleau et al., 1999; Scott et al., 2003).

11. CHARACTERIZATION OF INTERNAL HANDLING

Once taken up at gills, gut, or skin, metals are transported through the bloodstream and delivered to target organs where essential metals may contribute to normal metabolic functions, and both essential and non-essential metals may be stored, detoxified, or transformed, exert toxic effects, and/or be directed to excretory processes.

11.1. Biotransformation

Metals are fundamentally different from organic contaminants because as elements, metals can never break down, i.e. they persist forever (Skeaff et al., 2002). Nevertheless, they undergo biotransformation processes that may increase or decrease their toxicity. For example, an apparent reduction of Fe^{3+} to Fe^{2+} (Bury et al., Chapter 4) and Cu^{2+} to Cu^+ (Grosell, Chapter 2) occurs at the surfaces of gills and gut. These transformations render these two metals more bioavailable, although as yet there is only one demonstration that such enzymes occur in fish (ferric reductase activity in the gut of trout; Carriquiriborde et al., 2004). In the case of some metals, biological transformations may have occurred even before the metals are taken up. For example, Hg (Kidd and Batchelar, Chapter 5, Vol. 31B) and As (McIntyre and Linton, Chapter 6, Vol. 31B) may be methylated, Co may be fixed as cobalamin (Blust, Chapter 6), and Se may be incorporated into amino acids (selenomethionine, selenocysteine; Janz, Chapter 7), all by the metabolism of microorganisms (bacteria, algae). In general, methylation greatly increases uptake and bioaccumulation of metals. These same biotransformations may also occur in the tissues of the fish. In addition, a variety of detoxifying processes occurs in fish tissues (e.g. biologically mediated complexation by glutathione, metallothionein, and reduced sulfur compounds to form granules). Essential elements undergo biotransformation into critical enzymes and transport proteins (e.g. Cu into ceruloplasmin and cytochromes, Zn into carbonic anhydrase, Fe into transferrin, ferritin, and hemoglobin).

11.2. Transport through the Bloodstream

Metals may be transported through the bloodstream via red blood cells (RBCs) or via a variety of complexes and compounds in the bloodstream. The partitioning between the two components is very metal specific: most Pb is found in the RBCs after entry via the band 3 protein (Mager, Chapter 4, Vol. 31B), most Cr in the plasma (Reid, Chapter 8), but Zn is carried in both compartments (Hogstrand, Chapter 3). Ag is also found in both plasma and RBCs under control conditions, but the RBCs appear to be protected from additional uptake when plasma levels surge (Wood, Chapter 1, Vol. 31B). Inorganic Hg is carried mainly in the plasma, while methyl Hg is carried mainly in the RBCs (Kidd and Batchelar, Chapter 5, Vol. 31B). While the author is aware of no direct electrode measurements, it seems likely that at least for cationic trace metals, free ion levels are vanishing low in plasma and erythrocytic cytoplasm. This is because there are so many potential carriers for cationic metals, many of which are promiscuous in their affinities, e.g. transferrin (principally but not exclusively for Fe), ceruloplasmin (principally but not exclusively for Cu), transcobalamin (for Co), selenoprotein P (for Se), vitellogenin (for Ca but accepting other metals which are calcium analogues), and metallothioneins, albumins, globulins, glycoproteins, cysteine (and other amino acids), and glutathione, all of which accept many metals. In addition, many metals are expected to complex with small anions normally present in blood plasma; for example U, which is mainly present as the uranyl ion (UO_2^{2+}) is largely bound by bicarbonate and citrate (Goulet et al., Chapter 8, Vol. 31B). One notable exception is Mo, which occurs as the anion molybdate (MoO_4^{2-}) in the plasma (Reid, Chapter 8).

11.3. Accumulation in Specific Organs

Accumulation is one of the best-studied areas in piscine metals physiology, likely because of the ease of measurement, and a vast amount of information has been summarized in the various chapters. Metal accumulation patterns certainly differ among metals, but a few general conclusions may be drawn. The brain appears to be preferentially protected against the accumulation of many metals, probably by the blood–brain barrier. In contrast, the liver and kidney serve as scavenging and clearance organs, usually accumulating the highest concentration. The gut and the gills follow, while concentrations in the white muscle are usually much lower, which is important because the latter is the tissue that is mainly used for human consumption. Nevertheless, since liver and kidney normally constitute less than 5% of body weight, whereas white muscle may represent over 50%, the highest absolute metal burdens may actually be in white muscle. This may be exacerbated by the common practice where white muscle is lumped together with skin, scales, and bone as the "carcass". Calcium-analogue metals tend to accumulate preferentially in bone and scales, especially during chronic exposures.

Both exposure time and exposure route substantially alter tissue-specific accumulation patterns. This is in part a result of the different rates of tissue perfusion by blood among organs. During chronic exposure and depuration, there is often a progressive trend over time for levels to stabilize or decrease in most organs while increasing in liver, kidney, and/or skeleton. Not surprisingly, gill levels, as a percentage of whole body burden, tend to be higher in waterborne exposures, whereas gut levels tend to be higher in dietary exposures in the laboratory, though as yet it remains unclear whether this can be used as a diagnostic tool in the field.

11.4. Subcellular Partitioning of Metals

This is an extremely active area of research. For essential metals such as Cu (Grosell, Chapter 2), Zn (Hogstrand, Chapter 3), and Fe (Bury et al., Chapter 4), attention is mainly directed at cell-level homeostasis, i.e. how the metal is sequentially moved from one cellular pool to another via a series of chaperones, and the techniques involved are highly sophisticated. These include molecular characterization of the chaperones, autoradiography, and fluorescent imaging using metal-specific fluorophores. For non-essential metals (Cd in particular has been well studied in this regard; McGeer et al., Chapter 3, Vol. 31B), the focus is on whether the metal is partitioned into subcellular pools in which it will exert a toxic effect or in which it will be detoxified. Methodology usually involves mechanical techniques such as homogenization, differential centrifugation, heat treatment, and size-exclusion chromatography to separate the various fractions (e.g. Wallace and Lopez, 1996). In laboratory studies, this approach is often aided by the use of radiotracers to more easily quantify the metal levels in the various fractions (e.g. Galvez et al., 2002; Ng and Wood, 2008), but it has also been applied very successfully in field studies using "cold" analytical techniques (e.g. Kraemer et al., 2006; Goto and Wallace, 2010).

In general, the distribution of metals has been partitioned into five separate pools: (1) cellular debris (CD, membranes), (2) metallothionein-like (heat stable) proteins (MTLPs), (3) heat-sensitive proteins or enzymes (HSPs or ENZs), (4) metal-rich granules (MRGs), and (5) organelle fractions (ORGs; nucleus, mitochondria, microsomes). Different combinations of the subcellular fractions have been proposed to represent a metabolically active and metal-sensitive fraction (MSF: ORG and HSP) and a metabolically detoxified metal fraction (MDF: MTLP and MRG).

Some general conclusions may be drawn. Firstly, metals appear in all cellular compartments irrespective of the exposure concentration, time, and tissue burden, which partially contradicts the original "spillover hypothesis" (Winge et al., 1974; Hamilton and Mehrle, 1986) postulating that low cellular loads of toxic metals would be entirely sequestered in the MDF until a threshold breakthrough occurs. Nevertheless, some evidence exists that there may be a critical threshold concentration in the MSF at which overt toxicity starts to occur. There is also evidence that over time, metals are moved from the MSF to the MDF pools, and that long-term surviving fish in chronic laboratory or field exposures have partitioned the bulk of their tissue metal burdens into the MDF, though MSF levels may still be substantially elevated. Indeed, there is a growing belief that the metal burden in the MSF, or some subfraction thereof, could serve as a more useful replacement for the critical tissue residue (see Section 6) in environmental risk assessment (Vijver et al., 2004; Luoma and Rainbow, 2008; Adams et al., 2010).

11.5. Detoxification and Storage Mechanisms

For most metals, two molecules (glutathione and metallothionein) predominate for both detoxification and temporary storage, while a third mechanism (formation of MRGs) serves for permanent detoxification and storage. Originally, MRGs were thought to be restricted to invertebrates. However, Goto and Wallace (2010) recently demonstrated that killifish naturally inhabiting metal-polluted sites store large amount of metals in MRGs. These may form as a result of lysosomal processing of MT-bound metals, and contain metals complexed to phosphate and sulfide as insoluble precipitates. Calcium-analogue metals are also stored in bone, but it is not clear whether this deposition is benign, and whether such storage is reversible. Iron has its own intracellular storage molecule, ferritin (Bury et al., Chapter 4), but comparable reservoirs have not been identified for the other essential metals.

Glutathione (GSH) is the major non-protein reservoir of reduced thiol groups in most cells; this tripeptide chelates many different metal cations with a 1:1 or 1:2 stoichiometry, as soon as they enter the cell (Połeć-Pawlak et al., 2007). GSH is always present at quite high levels (several mmol L^{-1}) in cells, but its synthesis may be increased in metal-exposed fish as a result of upregulation of glutathione synthetase and glutathione reductase. However, these enzymes also may be directly inhibited by metals, such that GSH levels may either decrease (e.g. Cu, Grosell, Chapter 2) or increase (e.g. Cd, McGeer et al., Chapter 3, Vol. 31B) in fish that are chronically exposed to metals. An added benefit of GSH is that it also serves to scavenge damaging ROS (see Section 7.2) produced by the metal-catalyzed Fenton reaction.

Metallothionein (MT) is a low molecular weight protein in which about 30% of the amino acid residues are cysteine. This molecule will complex seven atoms of divalent metal cations (e.g. Cd^{2+}, Zn^{2+}, Hg^{2+}) or 12 atoms of monovalent metal cations (e.g. Ag^+, Cu^+). Though normally present at much lower concentrations than GSH, its affinities for metals are generally much higher. Two of the four isoforms (MT-1 and MT-2; it is unclear whether the other two are present in fish) are induced in response to elevated intracellular levels of many different metals. This then means that the effects of metals that induce MT are necessarily time-dependent. The regulation of metallothionein synthesis occurs in response to the binding of free metal ions to a transcription factor, *mtf1*, which in turn binds to metal responsive elements (MREs) in the 5′ regulatory region of the metallothionein genes (Kling and Olsson, 1995). MTs also respond to oxidative stress by releasing Zn^{2+} and contribute to the detoxification of ROS (Chiaverini and De Ley, 2010). It was originally believed that all binding sites on MTs were always occupied, such that new metal-binding occurred only by new MT synthesis or by displacement of lower binding strength metals (e.g. Zn) by stronger binding metals (e.g. Ag, Hg, Cu). However, recent evidence indicates that different Zn binding sites within the same MT molecule may differ by four orders of magnitude in their affinities for Zn, and that unsaturated MT with up to three available metal binding sites exists in the cell (Krezel and Maret, 2007). It is easy, therefore, to see how MT could immediately participate in metal detoxification, as well as serve as a storage reservoir for essential metals.

11.6. Homeostatic Controls

In one sense, it could be argued that both essential and non-essential metals are homeostatically regulated in the extracellular compartment of fish, inasmuch as virtually all are rapidly cleared from the blood plasma after injection, thanks to the scavenging functions of organs, cells, and molecules described in subsections 11.3, 11.4, and 11.5. For example, bolus injections of radiolabeled metal salts of Sr (Boroughs and Reid, 1958), Cu (Grosell et al., 2001), Cd and Zn (Chowdhury et al., 2003), and Ag (Wood et al., 2010) were cleared from the blood of diverse species with half-times of only 0.5–2.0 h. Furthermore, in several of these studies, there was evidence of faster clearance in fish that had been pre-exposed to the metal for several weeks, suggesting that acclimation had occurred. It is also very clear that in chronically exposed fish, the relative increases in plasma metal levels are far less than those in the whole body burden. There is also evidence that increases in plasma metal concentrations may be less in dietary exposures than in waterborne exposures, perhaps because the hepatic portal system carries the absorbed metal directly to the liver, where it is scavenged (e.g. Grosell, Chapter 2).

While adaptive, these phenomena do not necessarily indicate true homeostasis, i.e. negative feedback regulation around a set-point involving sensors, afferent and efferent pathways, and an integrating centre. Nevertheless, there is evidence that this must occur, at least for the essential metals, because plasma Cu, Zn, and Fe levels are tightly regulated, and dietary deficiencies or excesses of these metals are counteracted by reciprocal changes in gill uptake, as elaborated in the metal-specific chapters. However, essentially nothing is known about the extracellular metal sensors, the integrating centres, or the afferent and efferent pathways involved. One exception is hepcidin, a 20–26 amino acid polypeptide, which serves as a Fe-regulating hormone, as discussed by Bury et al. (Chapter 4). Hepcidin is synthesized mainly in the liver, the major Fe-sequestering organ. Multiple isoforms occur in fish (Martin-Antonio et al., 2009). Their basic role is to control Fe levels by regulating the absorption of dietary Fe from the intestine and perhaps the gills, the release of recycled Fe from macrophages, and the mobilization of stored Fe from hepatocytes. Increased circulatory Fe is taken up by hepatocytes which then stimulate hepcidin synthesis and release, resulting in a reduction in ferroportin activity, and therefore in Fe transport into the bloodstream at uptake epithelia (De Domenico et al., 2008). Despite extensive research, this is the only hormone yet identified which is dedicated to trace metal homeostasis.

12. CHARACTERIZATION OF EXCRETION ROUTES

This is an area that has been only sparsely studied. Relative to our understanding of uptake mechanisms, knowledge is minimal. Whole body excretion rates of most metals are slow, with half-times of weeks to months.

12.1. Gills

Given their large surface area, thin blood-to-water diffusion distance, high blood perfusion rate, and dominant contribution to the diffusive loss rates of major cations and anions (Na^+, K^+, Ca^{2+}, Mg^{2+}, Cl^-; Evans et al., 2005), we might expect the gills to be a major site of metal excretion, especially since excretion through other routes appears to be very low. However, there is little evidence on this topic. The best evidence may be for Zn (Hardy et al., 1987) and As (Oladimeji et al., 1984), where substantial portions of administered doses of radiolabeled metal were eliminated by the head region of rainbow trout. The gills also depurate their burden of flocculated aluminum hydroxides very rapidly upon return to clean water, but here the mechanism is probably sloughing of mucus-bound metal rather than true excretion from the blood or cells (Playle and Wood, 1991), i.e. removal of metal that never entered the fish tissues. In other instances where gill metal burdens have been rapidly cleared (e.g. Cu: Grosell et al., 1997; Ag: Wood et al., 2002), evidence points to clearance into the blood rather than clearance into the water, though this may not be true of Cr (Van der Putte et al., 1981). Nevertheless, Grosell et al. (2001) reported a substantial loss of ^{64}Cu radioactivity to the external water, apparently not attributable to urinary or biliary excretion, in trout infused with a very high dose of radiolabeled Cu. There is a need for much more work on the potential excretory role of the gills.

12.2. Kidney

The few measurements of metal excretion via the urine of freshwater fish (by bladder catheterization, e.g. Cu, Grosell et al., 1998; Zn, Spry and Wood, 1985; Ni, Pane et al., 2005; Cd, Chowdhury and Wood, 2007; Pb, Patel et al., 2006) indicate that it usually increases as a result of exposure, but that absolute metal excretion rates through this route remain generally low. This probably reflects the fact that most metals are tightly bound to macromolecules in the blood plasma, many of which are too large to pass through the glomerular filtration sieve (see Section 11.2). Urinary excretion rates of most metals are likely even lower in marine teleosts, where glomerular filtration rates are reduced and urine flow rates are only a fraction of those in freshwater fish (Marshall and Grosell, 2006). An interesting exception is the Mg-analogue Ni, where substitution on Mg-secretory mechanisms in the marine teleost kidney may make this an important excretory pathway (Pane et al., 2006).

12.3. Liver/Bile

For those metals which have been assayed in the gall bladder bile of exposed fish, the data suggest that this excretory route may be quite important for Cu, Ag, Cr, Sr, and Cd, but less so for Zn, Fe, As, Hg, or Ni, as summarized in the various metal-specific chapters. These conclusions are based simply on whether or not biliary concentrations are substantial relative to plasma levels or administered doses. However, an important word of caution is that bile stored in the gall bladder does not necessarily represent bile secreted into the digestive tract (Grosell et al., 2000), nor can we assume that the metals present even in secreted bile will be excreted, because reuptake by the intestine may well occur. Indeed, the high levels of metals measured in the intestinal tissue of freshwater teleosts exposed to waterborne metals (i.e. much higher than explicable by the negligible drinking rates), as reported in many chapters, may be due to such reuptake.

To the author's knowledge, only one study has measured the flow rate and composition of secreted bile in metal-exposed fish. Grosell et al. (2001) chronically cannulated the cystic bile duct of rainbow trout, and demonstrated that the biliary secretion rate of Cu increased greatly in fish that had been preacclimated to waterborne Cu and then further loaded with a radiolabeled Cu infusion. However, the Cu secretion rate remained a very small fraction of the Cu infusion rate. There is now some information on the mechanisms by which Cu enters the bile in fish (Grosell, Chapter 2) but very little for other metals. The pathways of metal entry into the bile and the actual rates of biliary excretion are important topics for future investigation. An important consideration is that gall bladder discharge occurs more frequently when fish are fed, but the chemistry of the bile changes, so feeding/fasting may greatly alter biliary excretion rates.

12.4. Gut

There is much confusion on this topic in the literature. This is in part because metal excretion by the gut cannot be quantified by simple measurement of metal excretion rate in the feces. Fecal metals may simply have not been assimilated from the food, or they may have been added via the bile. True metal excretion by the gut represents metal secreted with digestive juices, transported from blood-to-lumen, or sloughed inside detached enterocytes and then excreted via the feces or rectal fluid. In mammals, this can be an important excretion route, and for a few metals (Fe, Zn, Ni, Co, Cd, Cr, and methyl Hg) there is indirect evidence that the same may be true in fish, as outlined in the various metal-specific chapters. Again this is a subject where much more research is needed.

13. BEHAVIORAL EFFECTS OF METALS

Rapid progress in the past decade has reinforced earlier research (reviewed by Atchison et al., 1987; Scott and Sloman, 2004) indicating that behavior of aquatic animals, including fish, is extremely sensitive to many metals, often at levels that are close to or even below AWQC. In many cases, the mechanism appears to involve attraction or avoidance at very low levels, followed by interference with chemosensory, mechanosensory, and/or cognitive functions at slightly higher levels, often associated with changes in activity patterns. Unfortunately, this information has been ignored or discounted by most regulatory authorities, such that behavioral disturbance cannot be used as an endpoint in deriving AWQCs, and such information is usually overlooked in ecological risk assessments.

This is troubling because attraction to unfavorable areas, or displacement from otherwise favorable areas could have considerable ecological cost in wild fish. Furthermore, reductions in the ability to detect and avoid predators, locate prey, maintain social hierarchies, find suitable mates and spawning grounds, or undertake directional migrations could result in "ecological death". Mirza et al. (2009) demonstrated that wild yellow perch from metal-contaminated lakes detected olfactory signals at the electrophysiological level, but failed to respond to them at the behavioral level owing to a deficit in processing. However, such behavioral impairment would not have been detected in typical laboratory tests where food, mates, and spawning substrate are provided, so it remains controversial whether AWQCs derived from classic toxicity testing in the laboratory are protective in the field. Furthermore, water chemistry characteristics that are protective against metal toxicity to the gills (e.g. hardness) may not exhibit the same degree of protectiveness against olfactory toxicity. This is certainly true for Cu (e.g. Saucier and Astic, 1995; Pyle and Mirza, 2007; McIntyre et al., 2008; Green et al., 2010; Meyer and Adams, 2010; Grosell, Chapter 2). There are similar concerns for Zn (Hogstrand, Chapter 3), Ni (Pyle and Couture, Chapter 5), Cd (McGeer et al., Chapter 3, Vol. 31B, McGeer et al.), Pb (Mager, Chapter 4, Vol. 31B), Hg (Kidd and Batchelar, Chapter 5, Vol. 31B), and As (McIntyre and Linton, Chapter 6, Vol. 31B), where behavioral effects have been reported at waterborne levels below AWQCs. It is hoped that the focus on these endpoints in the current volumes will help to inform future decision making by regulatory authorities.

14. MOLECULAR CHARACTERIZATION OF METAL TRANSPORTERS, STORAGE PROTEINS, AND CHAPERONES

With the advent of new molecular techniques, there has been an explosion of information in this area in the last few years with respect to the molecules that import, shuttle, store, export, and detoxify essential metals, as amply illustrated by the chapters on Cu (Grosell, Chapter 2), Zn (Hogstrand, Chapter 3),

Fe (Bury et al., Chapter 4), and Se (Janz, Chapter 7). As these same molecules often also handle other metals, authors were asked to address this topic with respect to each metal; many chose to incorporate the information into other sections to avoid redundancy, as in the present chapter (Sections 10 and 11).

15. GENOMIC AND PROTEOMIC STUDIES

To the skeptic, these new techniques are similar to fishing with dynamite: costly, indiscriminate, almost invariably successful at catching large numbers, but the catch may be largely unrecognizable, because expressed sequence tags (ESTs), peptide sequences, and protein coordinates are often unidentified, and databases are poorly annotated. Nevertheless, when applied thoughtfully, these approaches can yield valuable insights for further hypothesis testing, especially with respect to patterns of gene response and the interconnections of responses, which can be elucidated with modern data handling tools. Figure 1.6 provides a typical example.

One of the main attractions of these approaches is that they often yield unexpected findings. For example, in a recent time-course examination of gill responses to moderately elevated waterborne Zn in zebrafish, Zheng et al. (2010) concluded that this treatment reactivated developmental pathways and stimulated stem cell differentiation in the gills. In another microarray study with chronic exposures of zebrafish to much lower, environmentally relevant levels of Cu, Craig et al. (2009) found that the directional pattern of gene response was strongly dependent on exposure concentration, and that the indirect effects of Cu exposure, through activation of a corticosteroid response, regulated gene expression to a much greater degree than did the direct effects. Fully 30% of the responding genes had a glucocorticoid-response element in their promoter region, while only 2.5% had a metal-response element. This was in distinct contrast with a study using the same microarray to investigate the genomic responses of a zebrafish cell line to high levels of Zn. Hogstrand et al. (2008) reported that over 44% of genes significantly regulated by Zn in this exposure contained one or more putative metal-response elements. This is just a small sample of the exciting data that are now emerging from genomic and proteomic investigations.

16. INTERACTIONS WITH OTHER METALS

It is probably safe to assume that more than 90% of the studies reviewed in these two volumes were performed using exposures to only one metal at a

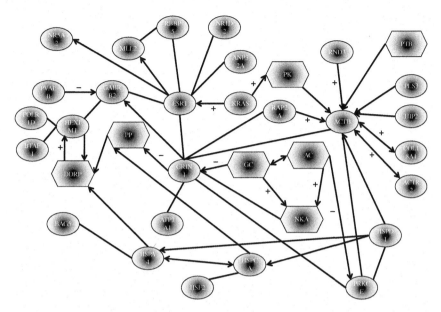

Fig. 1.6. A complex, and by no means complete, schematic network of the interactions between significantly regulated transcripts (directionally independent) from the liver tissue of zebrafish exposed to waterborne copper. Pathway analysis provides a useful tool to the scientist to allow identification of specific targets and further interaction by cross-referencing citation databases, such as PubMed, with statistically regulated transcripts of known gene ontology from a given microarray. In this instance, a custom-designed 16,730 65-mer oligonucleotide microarray chip was used to assess the transcriptomic response of the zebrafish liver to an insult of either 8 or 15 μg L^{-1} waterborne Cu, and this representation is a modified version of the pathway determined by GeneSpring GX software (Agilent Technologies). Ovals represent significantly regulated proteins, whereas hexagons represent regulated enzymes, as determined through statistical analysis of microarray data. The subsequent interaction is determined through analysis of citation databases (KEGG pathway analysis using homologous human interactions, as zebrafish databases are poorly annotated), and interactions are represented by either a line indicating binding or an arrow indicating directional regulation between proteins. Where data were available, directional regulation is indicated by either a + (positive regulation) or a − (negative regulation). Abbreviations are as follows. *Proteins*: ACTB: beta-actin; ACTR2: actin-related protein-2; ANP32A: acidic nuclear phosphoprotein 32A; ATP1A1, Na$^+$/K$^+$-ATPase, alpha-1a; BAG5, BCL2-associated athanogene 5; BTAF1: BTAF1 transcription factor associated 170 kDa subunit; COL181A1: collagen type 18a1; ESR1: estrogen receptor 1; GABRB2: GABA$_A$ receptor beta 2; GRIN1: ionotropic glutamate receptor 1; HEXIM1: hexamethylene bis-acetamide inducible 1; HIP2: huntingtin interacting protein 2; HSF2: heat shock transcription factor 2; HSPA4: heat shock 70 kDa protein 4; HSPCA: heat shock 90 kDa protein 1, alpha; HSPD1: heat shock 60 kDa protein 1 (chaperonin); KRAS: v-Ki-ras2 Kirsten rat sarcoma viral oncogene homolog; MLL2: mixed-lineage leukemia 2; NR1D2: nuclear receptor subfamily 1, group D, member; NR5A2: nuclear receptor subfamily 5, group A, member 2; PLS3: plastin 3 (T isoform); POLR1D: polymerase (RNA) I polypeptide D, 16 kDa; PRKCE: protein kinase C, epsilon; PVALB: parvalbumin; RAP2A: member of RAS oncogene family; RBBP5: retinoblastoma binding protein 5; RND3: Rho family GTPase 3. *Enzymes*: AC: adenylate cyclase; DDRP: DNA-directed RNA polymerase; GC: guanylate cyclase; NKA: sodium/potassium-exchanging ATPase; PK: protein kinase; PP: phosphoprotein phosphatase; PTB: protein-tyrosine phosphatase. Drawn by Paul Craig based on data presented in Craig et al. (2009).

time. This is entirely logical as it is much easier to elucidate basic principles in laboratory exposures when confounding variables are kept to a minimum. Indeed, sophisticated modeling approaches such as the BLM, although now calibrated for a number of different metals, deal with each metal in isolation (Paquin et al., Chapter 9, Vol. 31B). Yet in the real world, metals never occur in isolation, and aquatic organisms are often exposed to a cocktail of different metals, especially in water bodies near major industrial operations (Pyle et al., 2005; Couture and Pyle, Chapter 9).

As first elaborated by Playle (2004), the BLM is an ideal approach for predicting the effects of such mixtures. While there were some innovative multiple metal investigations at the end of the twentieth century (e.g. Hickie et al., 1993; Mount et al., 1994; Farag et al., 1994; Ribeyre et al., 1995; Richards et al., 2001) before the advent of the BLM approach, there appears to have been rather less activity in the last few years in this area. However, two fairly recent studies have examined combined metal effects, and while the gill-binding findings were predictable, the physiological/toxicological responses were not. For example, long-term Cd- and Cu-acclimated rainbow trout exhibited different changes in gill log K values for Cd (expected), but Cu-acclimated trout exhibited cross-acclimation to Cd whereas Zn-acclimated trout did not (McGeer et al., 2007), which is counter-intuitive considering that Cd and Zn are both Ca^{2+} antagonists while Cu is an Na^+ antagonist. Also in trout, Cd and Pb, both of which are Ca^{2+} antagonists, exhibited less than additive gill binding (expected), yet more than additive ionoregulatory toxicity (unexpected) (Birceanu et al., 2008). By asking authors to highlight what little is currently known about the effects of combined metal exposures, the goal is to foster more research into this very important area.

ACKNOWLEDGMENTS

The author is grateful to the many funding bodies that have supported research on metal effects on fish in his laboratory: industries, industrial research organizations, environmental agencies, and NSERC. Thanks to Peter Chapman and Kevin Brix for advice on regulatory issues, Christer Hogstrand for advice on metallothionein, Fernando Galvez for an early version of Fig. 1.3, Paul Craig for Fig. 1.6, and Sunita Nadella for drawing the other figures. This chapter is dedicated to the memory of Bernie Simons, who was instrumental in the author's early investigations into aluminum effects on fish.

REFERENCES

Adams, W. J., and Chapman, P. M. (2005). *Assessing the Hazard of Metals and Inorganic Metal Substances in Aquatic and Terrestrial Systems: Summary of a SETAC Pellston Workshop.* SETAC Press, Pensacola, FL.

Adams, W. J., Blust, R., Borgmann, U., Brix, K. V., DeForest, D. K., Green, A. S., Meyer, J., McGeer, J. C., Paquin, P., Rainbow, P., and Wood, C. (2010). Utility of tissue residues for predicting effects of metals on aquatic organisms. *Integr. Environ. Assess. Manage.* **7**, 75–98.

Allison, J. D., Brown, D. S., and Novo-Gradac, K. J. (1991). *MINTEQA2/PRODEFA2, A Geochemical Assessment Model for Environmental Systems. Version 3.0, User's Manual.* US Environmental Protection Agency, Washington, DC.

Al-Reasi, H. A., Wood, C. M. and Smith, D. S. (2011). Physicochemical and spectroscopic properties of natural organic matter (NOM) from various sources and implications for ameliorative effects on metal toxicity to aquatic biota. *Aquat. Toxicol.* doi: 10.1016/j.aquatox.2011.02.015.

Andreini, C., Banci, L., Bertini, I., and Rosato, A. (2005). Counting the zinc-proteins encoded in the human genome. *J. Proteome Res.* **5**, 196–201.

Atchison, G. J., Henry, M. G., and Sandheinrich, M. B. (1987). Effects of metals on fish behavior: a review. *Environ. Biol. Fish.* **18**, 11–25.

Baldisserotto, B., Kamunde, C., Matsuo, A., and Wood, C. M. (2004). A protective effect of dietary calcium against acute waterborne cadmium uptake in rainbow trout. *Aquat. Toxicol.* **67**, 57–73.

Birceanu, O., McGeer, J. C., Chowdury, M. J., Gillis, P., Wood, C. M., and Wilkie, M. P. (2008). Modes of metal toxicity and impaired branchial ionoregulation in rainbow trout exposed to mixtures of Pb and Cd in soft water. *Aquat. Toxicol.* **89**, 221–231.

Boroughs, H., and Reid, D. F. (1958). The role of the blood in the transportation of strontium^{90}–yittrium^{90} in teleost fish. *Biol. Bull.* **115**, 64–73.

Borg-Neczak, K., and Tjalve, H. (1996). Uptake of Hg-203($^{2+}$) in the olfactory system in pike. *Toxicol. Lett.* **84**, 107–112.

Brown, V. M. (1968). The calculation of the acute toxicity of mixtures of poisons to rainbow trout. *Water Res.* **2**, 723–733.

Brown, V. M., Shaw, T. L., and Shurben, D. G. (1974). Aspects of water quality and the toxicity of copper to rainbow trout. *Water Res.* **8**, 797–803.

Bury, N. R., and Wood, C. M. (1999). Mechanism of branchial apical silver uptake by rainbow trout is via the proton-coupled Na$^+$ channel. *Am. J. Physiol.* **277**, R1385–R1391.

Bury, N. R., Walker, P. A., and Glover, C. N. (2003). Nutritive metal uptake in teleost fish. *J. Exp. Biol.* **206**, 11–23.

Busselburg, D. (1995). Calcium channels as targets sites of heavy metals. *Toxicol. Lett.* **82/83**, 255–261.

Carriquiriborde, P., Handy, R. D., and Davies, S. J. (2004). Physiological modulation of iron metabolism in rainbow trout (*Oncorhynchus mykiss*) fed low and high iron diets. *J. Exp. Biol.* **207**, 75–86.

Carson, R. L. (1962). *Silent Spring.* Houghton Mifflin, New York.

Chakoumakos, C., Russo, R. C., and Thurston, R. V. (1979). Toxicity of copper to cutthroat trout (*Salmo clarki*) under different conditions of alkalinity, pH, and hardness. *Environ. Sci. Technol.* **13**, 213–219.

Chapman, P. M., and Wang, F. (2000). Issues in ecological risk assessment of inorganic metals and metalloids. *Hum. Ecol. Risk Assess.* **6**, 965–988.

Chapman, P. M., Allen, H. E., Godtfredsen, K., and Z'Graggen, M. N. (1996a). Evaluation of BCFs as measures for classifying and regulating metals. *Environ. Sci. Technol.* **30**, 448–452.

Chapman, P. M., Thornton, I., Persoone, G., Janssen, G., Godtfredsen, G., and Z'Graggen, M. N. (1996b). International harmonization related to persistence and bioavailability. *Hum. Ecol. Risk Assess.* **2**, 393–404.

Chapman, P. M., Fairbrother, A., and Brown, D. (1998). A critical evaluation of safety (uncertainty) factors for ecological risk assessment. *Environ. Toxicol. Chem.* **17**, 99–108.

Chapman, P. M., Wang, F., Janssen, C. R., Goulet, R. R., and Kamunde, C. M. (2003). Conducting ecological risk assessments of inorganic metals and metalloids: current status. *Hum. Ecol. Risk Assess.* **9**, 641–697.

Chiaverini, N., and De Ley, M. (2010). Protective effects of metallothionein on oxidative stress-induced DNA damage. *Free Radic. Res.* **44**, 605–613.

Chowdhury, M. J., and Wood, C. M. (2007). Renal function in the freshwater rainbow trout after dietary cadmium acclimation and waterborne cadmium challenge. *Comp. Biochem. Physiol. C* **145**, 321–332.

Chowdhury, M. J., Grosell, M., McDonald, D. G., and Wood, C. M. (2003). Plasma clearance of cadmium and zinc in non-acclimated and metal-acclimated trout. *Aquat. Toxicol.* **64**, 259–275.

Clarkson, T. W. (1993). Molecular and ionic mimicry of toxic metals. *Annu. Rev. Pharmacol. Toxicol.* **32**, 545–571.

Clearwater, S. J., Farag, A. M., and Meyer, J. S. (2002). Bioavailability and toxicity of dietborne copper and zinc to fish. *Comp. Biochem. Physiol. C* **132**, 269–313.

Cooper, C. A., Shayeghi, M., Techau, M. E., Capdevila, D. M., MacKenzie, S., Durrant, C., and Bury, N. R. (2007). Analysis of the rainbow trout solute carrier 11 family reveals iron import at $\geq$ pH 7. 4 and a functional isoform lacking transmembrane domains 11 and 12. *FEBS Lett.* **581**, 2599–2604.

Craig, P. M, Wood, C. M., and McClelland, G. B. (2007). Oxidative stress response and gene expression with acute copper exposure in zebrafish (*Danio rerio*). *Am. J. Physiol.* **293**, R1882–1892.

Craig, P. M., Hogstrand, C., Wood, C. M., and McClelland, G. B. (2009). Gene expression endpoints following chronic waterborne copper exposure in a genomic model organism, the zebrafish, *Danio rerio. Physiol. Genom.* **40**, 23–33.

Dallinger, R., and Kautzky, H. (1985). The importance of contaminated food for the uptake of heavy metals by rainbow trout (*Salmo gairdneri*): a field study. *Oecologia* **67**, 82–89.

Dallinger, R., Prosi, F., Segner, H., and Back, H. (1987). Contaminated food and uptake of heavy metals by fish: a review and a proposal for further research. *Oecologia* **73**, 91–98.

De Domenico, I., Nemeth, E., Nelson, J. M., Phillips, J. D., Ajioka, R. S., Kay, M. S., Kuschner, J. P., Ganz, T., Ward, D. M., and Kaplan, J. (2008). The hepcidin-binding site on ferroportin is evolutionarily conserved. *Cell Metab.* **8**, 146–156.

DeForest, D. K., Brix, K. V., and Adams, W. J. (2007). Assessing metal bioaccumulation in aquatic environments: the inverse relationship between bioaccumulation factors, trophic transfer factors and exposure concentration. *Aquat. Toxicol.* **84**, 236–246.

Di Giulio, R., Washburn., P., Wenning., R., Winston, G., and Jewell, C. (1989). Biochemical responses in aquatic animals: a review of determinants of oxidative stress. *Environ. Toxicol. Chem.* **8**, 1103–1123.

Di Toro, D. M., Allen, H. E., Bergman, H. L., Meyer, J. S., Paquin, P. R., and Santore, R. C. (2001). Biotic ligand model of the acute toxicity of metals. 1. Technical basis. *Environ. Toxicol. Chem.* **20**, 2383–2396.

Duffus, J. H. (2002). "Heavy metals" – a meaningless term. *Pure Appl. Chem.* **74**, 793–807.

Dupont, C., Butcher, A., Valas, R., Bourne, P., and Caetano-Anolles, G. (2010). History of biological metal utilization inferred through phylogenomic analysis of protein structures. *Proc. Natl. Acad. Sci. U.S.A.* **107**, 10567–10572.

Evans, D. H., Piermarini, P. M., and Choe, K. P. (2005). The multifunctional fish gill: dominant site of gas exchange, osmoregulation, acid–base regulation, and excretion of nitrogenous waste. *Physiol. Rev.* **85**, 97–177.

Fairbrother, A., and Bennet, R. S. (1999). Ecological risk assessment and the precautionary principle. *Hum. Ecol. Risk Assess.* **5**, 943–949.

Fairbrother, A., Wenstel, R., Sappington, K., and Wood, W. (2007). Framework for metals risk assessment. *Ecotoxicol. Environ. Saf.* **68**, 145–227.

Farag, A. M., Boese, C. J., Bergman, H. L., and Woodward, D. F. (1994). Physiological changes and tissue metal accumulation in rainbow trout exposed to foodborne and waterborne metals. *Environ. Toxicol. Chem.* **13**, 2021–2029.

Franklin, N. M., Glover, C. N., Nicol, J. A., and Wood, C. M. (2005). Calcium/cadmium interactions at uptake surfaces in rainbow trout: waterborne versus dietary routes of exposure. *Environ. Toxicol. Chem.* **24**, 2954–2964.

Galvez, F. H., Mayer, G. D., Wood, C. M., and Hogstrand, H. (2002). The distribution kinetics of waterborne silver-110m in juvenile rainbow trout. *Comp. Biochem. Physiol.* **131C**, 367–378.

Glover, C. N., and Hogstrand, C. (2002). Amino acid modulation of *in vivo* intestinal zinc absorption in freshwater rainbow trout. *J. Exp. Biol.* **205**, 151–158.

Glover, C. N., and Wood, C. M. (2008). Absorption of copper and copper–histidine complexes across the apical surface of freshwater rainbow trout intestine. *J. Comp. Physiol. B* **78**, 101–109.

Goto, D., and Wallace, W. G. (2010). Metal intracellular partitioning as a detoxification mechanism for mummichogs (*Fundulus heteroclitus*) living in metal-polluted salt marshes. *Mar. Environ. Res.* **69**, 163–171.

Green, W. W., Mirza, R. S., Wood, C. M., and Pyle., G. G. (2010). Copper binding dynamics and olfactory impairment in fathead minnows *Pimephales promelas*. *Environ. Sci. Technol.* **44**, 1431–1437.

Grosell, M. (2006). Intestinal anion exchange in marine fish osmoregulation. *J. Exp. Biol.* **209**, 2813–2827.

Grosell, M., and Wood, C. M. (2002). Copper uptake across rainbow trout gills: mechanisms of apical entry. *J. Exp. Biol.* **205**, 1179–1188.

Grosell, M. H., Hogstrand, C., and Wood, C. M. (1997). Cu uptake and turnover in both Cu-acclimated and non-acclimated rainbow trout (*Oncorhynchus mykiss*). *Aquat. Toxicol* **38**, 257–276.

Grosell, M. H., Hogstrand, C., and Wood, C. M. (1998). Renal Cu and Na excretion and hepatic Cu metabolism in both Cu-acclimated and non-acclimated rainbow trout (*Oncorhynchus mykiss*). *Aquat. Toxicol.* **40**, 275–291.

Grosell, M., DeBoeck, G., Johannsson, O., and Wood, C. M. (1999). The effects of silver on intestinal ion and acid–base regulation in the marine teleost fish *Parophrys vetulus*. *Comp. Biochem. Physiol. C* **124**, 259–270.

Grosell, M., O'Donnell, M. J., and Wood, C. M. (2000). Hepatic versus gallbladder bile composition – *in vivo* transport physiology of the gallbladder in the rainbow trout. *Am. J. Physiol.* **278**, R1674–R1684.

Grosell, M., McGeer, J. C., and Wood, C. M. (2001). Plasma copper clearance and biliary copper excretion are stimulated in copper-acclimated trout. *Am. J. Physiol.* **280**, R796–R806.

Grosell, M., McDonald, M. D., Walsh, P. J., and Wood, C. M. (2004). Effects of prolonged copper exposure on the marine gulf toadfish (*Opsanus beta*). II. Copper accumulation, drinking rate, and Na^+/K^+-ATPase activity. *Aquat. Toxicol.* **68**, 263–275.

Grosell, M., Blanchard, J., Brix, K. V., and Gerdes, R. (2007). Physiology is pivotal for interactions between salinity and acute copper toxicity to fish and invertebrates. *Aquat. Toxicol.* **84**, 162–172.

Hamilton, S. J., and Mehrle, P. M (1986). Metallothionein in fish: review of its importance in assessing stress from metal contaminants. *Trans. Am. Fish. Soc.* **115**, 596–609.

Han, F. X. X., Banin, A., Su, Y., Monts, D. L., Plodinec, M. J., Kingery, W. L., and Triplett, G. E. (2002). Industrial age anthropogenic inputs of heavy metals into the pedosphere. *Naturwissenschaften* **89**, 497–504.

Handy, R. D., Henry, T. B., Scown, T. M., Johnston, B. D., and Tyler, C. R. (2008a). Manufactured nanoparticles: their uptake, and effects on fish – a mechanistic analysis. *Ecotoxicology* **17**, 396–409.

Handy, R. D., von der Kammer, F., Lead, J. R., Hassellov, M., Owen, R., and Crane, M. (2008b). The ecotoxicology and chemistry of manufactured nanoparticles. *Ecotoxicology* **17**, 287–314.

Hardy, R. W., Sullivan, C. V., and Koziol, A. M. (1987). Absorption, body distribution, and excretion of dietary zinc by rainbow trout (*Salmo gairdneri*). *Fish Physiol. Biochem.* **3**, 133–143.

Hickie, B. E., Hutchinson, N. J., Dixon, D. G., and Hodson, P. V. (1993). Toxicity of trace metal mixtures to alevin rainbow trout (*Oncorhynchus mykiss*) and larval fathead minnow (*Pimephales promelas*) in soft, acidic water. *Can. J. Fish. Aquat. Sci.* **50**, 1348–1355.

Hodson, M. E. (2004). Heavy metals – geochemical bogey men? *Environ. Poll.* **129**, 341–343.

Hogstrand, C., and Wood, C. M. (1998). Towards a better understanding of the bioavailability, physiology, and toxicology of silver in fish: implications for water quality criteria. *Environ. Toxicol. Chem.* **17**, 547–561.

Hogstrand, C., Galvez, F., and Wood, C. M. (1996). Toxicity, silver accumulation and metallothionein induction in freshwater rainbow trout during exposure to different silver salts. *Environ. Toxicol. Chem.* **15**, 1102–1108.

Hogstrand, C., Ferguson, E. A., Galvez, F., Shaw, J. R., Webb, N. A., and Wood, C. M. (1999). Physiology of acute silver toxicity in the starry flounder (*Platichthys stellatus*) in seawater. *J. Comp. Physiol. B* **169**, 461–473.

Hogstrand, C., Zheng, D., Feeney, G., Cunningham, P., and Kille, P. (2008). Zinc-controlled gene expression by metal-regulatory transcription factor 1 (MTF1) in a model vertebrate, the zebrafish. *Biochem. Soc. Trans.* **36**, 1252–1257.

Holm-Jensen, I. (1948). Osmotic regulation in *Daphnia magna* under physiological conditions and in the presence of heavy metals. *Det KGL. Danske Videnskabers Selvskab, Biologiske Meddelser* **20**, 4–64.

Hontela, A. (1997). Interrenal dysfunction in fish from contaminated sites: *in vivo* and *in vitro* assessment. *Environ. Toxicol. Chem.* **17**, 44–48.

Hunn, J. B. (1985). Role of calcium in gill function in freshwater fishes. *Comp. Biochem. Physiol.* **A82**, 543–547.

Jeffery, W. G. (2001). *A World of Metals: Finding, Making, and Using Metals* (2nd edn.). International Council on Metals in the Environment, Ottawa.

Jones, J. R. E. (1939). The relation between the electrolytic solution pressures of the metals and their toxicity to the stickleback (*Gasterosteus aculeatus* L.). *J. Exp. Biol.* **16**, 425–437.

Kamunde, C. N., and Wood, C. M. (2003). The influence of ration size on copper homeostasis during sublethal dietary copper exposure in juvenile rainbow trout, *Oncorhynchus mykiss*. *Aquat. Toxicol.* **62**, 235–254.

Kjoss, V. A., Kamunde, C. N., Niyogi, S., Grosell, M., and Wood, C. M. (2005). Dietary Na does not reduce dietary Cu uptake by juvenile rainbow trout. *J. Fish. Biol.* **66**, 468–484.

Klaine, S. J., Alvarez, P. J. J., Batley, G. E., Fernandes, T. F., Handy, R. D., Lyon, D. Y., Mahendra, S., McLaughlin, M. J., and Lead, J. R. (2008). Nanomaterials in the environment: behavior, fate, bioavailability, and effects. *Environ. Toxicol. Chem.* **27**, 1825–1851.

Klinck, J. and Wood, C. M. (2011). *In vitro* characterization of cadmium transport along the gastro-intestinal tract of freshwater rainbow trout (*Oncorhynchus mykiss*). *Aquat. Toxicol.* (in press).

Kling, P., and Olsson, P. E. (1995). Regulation of the rainbow trout metallothionein-A gene. *Mar. Environ. Res.* **39**, 117–120.

Kraemer, L. D., Campbell, P. G. C., and Hare, L. (2006). Seasonal variations in hepatic Cd and Cu concentrations and in the sub-cellular distribution of these metals in juvenile yellow perch (*Perca flavescens*). *Environ. Pollut.* **142**, 313–325.

Krezel, A., and Maret, W. (2007). Dual nanomolar and picomolar Zn(II) binding properties of metallothionein. *J. Am. Chem. Soc.* **129**, 10911–10921.

Larsson, D., Lundgren, T., and Sundell, K. (1998). Ca^{2+} uptake through voltage-gated L-type Ca^{2+} channels by polarized enterocytes from Atlantic cod, *Gadus morhua. J. Membr. Biol.* **164**, 229–237.

Leonard, E. M., Nadella, S. R., Bucking, C., and Wood, C. M. (2009). Characterization of dietary Ni uptake in the rainbow trout, *Oncorhynchus mykiss. Aquat. Toxicol.* **95**, 205–216.

Lewis, G. N. (1923). *Valence and the Structure of Atoms and Molecules.* Chemical Catalogue Company, New York.

Lloyd, R., and Herbert, D. W. M. (1962). The effect of the environment on the toxicity of poisons to fish. *J. Inst. Public Health Eng.* **61**, 132–145.

Luoma, S. N., and Rainbow, P. S. (2008). *Metal Contamination in Aquatic Environments: Science and Lateral Management.* Cambridge University Press, Cambridge.

Lushchak, V. (2011). Environmentally induced oxidative stress in aquatic animals. *Aquat. Toxicol.* **101**, 13–30.

MacRae, R. K., Smith, D. E., Swoboda-Colberg, N., Meyer, J. S., and Bergman, H. L. (1999). Copper binding affinity of rainbow trout (*Oncorhynchus mykiss*) and brook trout (*Salvelinus fontinalis*) gills: implications for assessing bioavailable metal. *Environ. Toxicol. Chem.* **18**, 1180–1189.

Mallatt, J. (1985). Fish gill structural changes induces by toxicants and other irritants: a statistical review. *Can. J. Fish. Aquat. Sci.* **42**, 630–648.

Marshall, W. S., and Grosell, M. (2006). Ion transport, osmoregulation and acid–base balance. In: *Physiology of Fishes* (D. Evans and J.B. Claiborne, eds.), 3rd edn, pp. 179–230. CRC Press, Boca Raton, FL.

Martin-Antonio, B., Jimenez-Cantizano, R. M., Salas-Leiton, E., Infante, C., and Manchado, M. (2009). Genomic characterization and gene expression analysis of four hepcidin genes in the redbanded seabream (*Pagrus auriga*). *Fish Shellfish Immun.* **26**, 483–491.

Mathews, T., and Fisher, N. S. (2009). Dominance of dietary intake of metals in marine elasmobranch and teleost fish. *Sci. Tot. Environ* **407**, 5156–5161.

McDonald, D. G., and Wood, C. M. (1993). Branchial acclimation to metals. In: *Fish Ecophysiology* (J.C. Rankin and F.B. Jensen, eds.), pp. 297–321. Chapman, London.

McGeer, J. C., Playle, R. C., Wood, C. M., and Galvez, F. (2000). A physiologically based biotic ligand model for predicting the acute toxicity of waterborne silver to rainbow trout in fresh waters. *Environ. Sci. Technol.* **34**, 4199–4207.

McGeer, J. C., Brix, K. V., Skeaff, J. M., Deforest, D. K., Brigham, S. I., Adams, W. J., and Green, A. (2003). Inverse relationship between bioconcentration factor and exposure concentration for metals: implications for hazard assessment of metals in the aquatic environment. *Environ. Toxicol. Chem.* **22**, 1017–1037.

McGeer, J. C., Nadella, S. R., Alsop, D. H., Hollis, L., Taylor, L. N., McDonald, D. G., and Wood, C. M. (2007). Influence of exposure to Cd or Cu on acclimation and the uptake of Cd in rainbow trout (*Oncorhynchus mykiss*). *Aquat. Toxicol.* **84**, 190–197.

McIntyre, J. K., Baldwin, D. H., Meador, J. P., and Scholz, N. L. (2008). Chemosensory deprivation in juvenile coho salmon under varying water chemistry conditions. *Environ. Sci. Technol.* **42**, 1352–1358.

Meyer, J. S. (1999). A mechanistic explanation for the ln(LC50) vs ln(hardness) adjustment equation for metals. *Environ. Sci. Technol.* **33**, 908–912.

Meyer, J. S., and Adams, W. J. (2010). Relationship between biotic ligand model-based water quality criteria and avoidance and olfactory responses to copper by fish. *Environ. Toxicol. Chem.* **29**, 2096–2103.

Meyer, J. S., Adams, W. J., Brix, K. V., Luoma, S. N., Mount, D. R., Stubblefield, W. A., and Wood, C. M. (eds). *Toxicity of Dietborne Metals to Aquatic Organisms*. SETAC Press, Pensacola, FL.

Mirza, R. S., Green, W. R., Connor, S., Weeks, A. C., Wood, C. M., and Pyle, G. G. (2009). Do you smell what I smell? Olfactory impairment in wild yellow perch from metal-contaminated waters. *Ecotox. Environ. Saf.* **72**, 677–683.

Morel, F. (1983). *Principles of Aquatic Chemistry*. John Wiley and Sons, New York.

Mount, D. R., Barth, A. K., Garrison, T. D., Barten, K. A., and Hockett, J. R. (1994). Dietary and waterborne exposure of rainbow trout (*Oncorhynchus mykiss*) to copper, cadmium, lead and zinc using a live diet. *Environ. Toxicol. Chem.* **13**, 2031–2041.

Morgan, T. P., and Wood, C. M. (2004). A relationship between gill silver accumulation and acute silver toxicity in the freshwater rainbow trout: support for the acute silver biotic ligand model. *Environ. Toxicol. Chem.* **23**, 1261–1267.

Mulkidjanian, A. Y. (2009). On the origin of life in the zinc world: I. Photosynthesizing, porous edifices built of hydrothermally precipitated zinc sulfide as cradles of life on Earth. *Biol. Direct* **4**, 26.

Mushiake, K., Muroga, K., and Nakai, T. (1984). Increased susceptibility of Japanese eel *Anguilla japonica* to *Edwardsielle tarda* and *Pseudomonas anguilliseptica* following exposure to copper. *Bull. Jpn. Soc. Sci. Fish.* **50**, 1797–1801.

Nadella, S. R., Bucking, C., Grosell, M., and Wood., C. M. (2006). Gastrointestinal assimilation of Cu during digestion of a single meal in the freshwater rainbow trout (*Oncorhynchus mykiss*). *Comp. Biochem. Physiol. A* **143**, 394–401.

Nadella, S. R., Grosell, M., and Wood, C. M. (2007). Mechanisms of dietary Cu uptake in freshwater rainbow trout: evidence for Na-assisted Cu transport and a specific metal carrier in the intestine. *J. Comp. Physiol. B* **177**, 433–446.

Nadella, S. R., Hung, C. Y., and Wood, C. M. (2010). Mechanistic characterization of gastric copper transport in rainbow trout. *J. Comp. Physiol. B* **181**, 27–41.

Ng, T. Y. T., and Wood, C. M. (2008). Trophic transfer and dietary toxicity of Cd from the oligochaete to the rainbow trout. *Aquat. Toxicol.* **87**, 47–59.

Nielsen, F. H. (2000). Evolutionary events culminating in specific minerals becoming essential for life. *Eur. J. Nutr.* **39**, 62–66.

Nieboer, E., and Richardson, D. H. S. (1980). The replacement of the nondescript term "heavy metals" by a biologically and chemically significant classification of metal ions. *Environ. Pollut. (Series B)* **1**, 3–26.

Niyogi, S., and Wood, C. M. (2004a). Biotic ligand model, a flexible tool for developing site-specific water quality guidelines for metals. *Environ. Sci. Technol.* **38**, 6177–6192.

Niyogi, S., and Wood, C. M. (2004b). Kinetic analyses of waterborne Ca and Cd transport and their interactions in the gills of rainbow trout (*Oncorhynchus mykiss*) and yellow perch (*Perca flavescens*), two species differing greatly in acute waterborne Cd sensitivity. *J. Comp. Physiol. B* **174**, 243–253.

Ojo, A. A., and Wood, C. M. (2008). *In vitro* characterization of cadmium and zinc uptake via the gastro-intestinal tract of the rainbow trout (*Oncorhynchus mykiss*): interactive effects and the influence of calcium. *Aquat. Toxicol.* **89**, 55–64.

Oladimeji, A. A., Quadri, S. U., and deFreitas, A. S. W. (1984). Measuring the elimination of arsenic by the gills of rainbow trout (*Salmo gairdneri*) by using a two compartment respirometer. *Bull. Environ. Contam. Toxicol.* **32**, 661–668.

Pagenkopf, G. K. (1983). Gill surface interaction model for trace-metal toxicity to fishes: role of complexation, pH, and water hardness. *Environ. Sci. Technol.* **17**, 342–347.

Pagenkopf, G. K., Russo, R. C., and Thurston, R. V. (1974). Effect of complexation on toxicity of copper to fishes. *J. Fish. Res. Bd Can.* **31**, 462–465.

Pane, E. F., Bucking, C., Patel, M., and Wood, C. M. (2005). Renal function in the freshwater rainbow trout (*Oncorhynchus mykiss*) following acute and prolonged exposure to waterborne nickel. *Aquat. Toxicol.* **72**, 119–133.

Pane, E. F., McDonald, M. D., Curry, H. N., Blanchard, J., Wood, C. M., and Grosell, M. (2006). Hydromineral balance in the marine gulf toadfish (*Opsanus beta*) exposed to waterborne or infused nickel. *Aquat. Toxicol.* **80**, 70–81.

Paquin, P. R., Santore, R. C., Wu, K. B., Kavvadas, C. D., and Di Toro, D. M. (2000). The biotic ligand model: a model of the acute toxicity of metals to aquatic life. *Environ. Sci. Pol.* **3**, 175–182.

Paquin, P. R., Gorsuch, J. W., Apte, S., Batley, G. E., Bowles, K. C., Campbell, P. G. C., Delos, C. G., Di Toro, D. M., Dwyer, R. L., Galvez, F., Gensemer, R. W., Goss, G. G., Hogstrand, C., Janssen, C. R., McGeer, J. M., Naddy, R. B., Playle, R. C., Santore, R. C., Schneider, U., Stubblefield, W. A., Wood, C. M., and Wu, K. B. (2002). The biotic ligand model: a historical overview. *Comp. Biochem. Physiol. C* **133**, 3–35.

Passerini, A., Andreini, C., Menchetti, S., Rosato, A., and Frasconi, P. (2007). Predicting zinc binding at the proteome level. *BMC Bioinform.* **8**, 39.

Patel, M., Rogers, J. G., Pane, E. F., and Wood, C. M. (2006). Renal responses to acute lead waterborne exposure in the freshwater rainbow trout (*Oncorhynchus mykiss*). *Aquat. Toxicol.* **80**, 362–371.

Payne, J. F., Malins, D. C., Gunselman, S., Rahimtula, A., and Yeats, P. A. (1998). DNA oxidative damage and vitamin A reduction in fish from a large lake system in Labrador, Newfoundland, contaminated with iron-ore mine tailings. *Mar. Environ. Res.* **46**, 289–294.

Perry, S. F., and Wood, C. M. (1985). Kinetics of branchial calcium uptake in the rainbow trout: effects of acclimation to various external calcium levels. *J. Exp. Biol.* **116**, 411–433.

Playle, R. C. (1998). Modelling metal interactions at fish gills. *Sci. Tot. Environ.* **219**, 147–163.

Playle, R. C. (2004). Using multiple metal–gill binding models and the toxic unit concept to help reconcile multiple-metal toxicity results. *Aquat. Toxicol.* **67**, 359–370.

Playle, R. C., and Wood, C. M. (1991). Mechanisms of aluminium extraction and accumulation at the gills of rainbow trout, *Oncorhynchus mykiss* (Walbaum), in acidic soft water. *J. Fish Biol.* **38**, 791–805.

Playle, R. C., Dixon, D. G., and Burnison, K. (1993a). Copper and cadmium binding to fish gills: modification by dissolved organic carbon and synthetic ligands. *Can. J. Fish. Aquat. Sci.* **50**, 2667–2677.

Playle, R. C., Dixon, D. G., and Burnison, K. (1993b). Copper and cadmium binding to fish gills: estimates of metal-gill stability constants and modelling of metal accumulation. *Can. J. Fish. Aquat. Sci.* **50**, 2678–2687.

Połeć-Pawlak, K, Ruzik, R., and Lipiec, E. (2007). Investigation of Cd(II), Pb(II) and Cu(I) complexation by glutathione and its component amino acids by ESI-MS and size exclusion chromatography coupled to ICP-MS and ESI-MS. *Talanta* **2**, 1564–1572.

Pyle, G. G., and Mirza, R. S. (2007). Copper-impaired chemosensory function and behavior in aquatic animals. *Hum. Ecol. Risk Assess.* **13**, 492–505.

Pyle, G., and Wood, C. M. (2007). Predicting "non-scents": rationale for a chemosensory-based biotic ligand model. *Austral. J. Ecotoxicol.* **13**, 47–51.

Pyle, G. G., Kamunde, C. N., McDonald, D. G., and Wood, C. M. (2003). Dietary sodium inhibits aqueous copper uptake in rainbow trout (*Oncorhynchus mykiss*). *J. Exp. Biol.* **206**, 609–618.

Pyle, G. G., Rajotte, J. W., and Couture, P. (2005). Effects of industrial metals on wild fish populations along a metal contamination gradient. *Ecotoxicol. Environ. Saf.* **61**, 287–312.

Rauch, J. N. and Graedel, T. E. (2007). Earth's anthrobiogeochemical copper cycle. *Global Biogeochem. Cycles* **21**, GB2010, 13 pp., doi: 10.1029/2006GB002850.

Rauch, J. N. and Pacyna, J. M. (2009). Earth's global Ag, Al, Cr, Cu, Fe, Ni, Pb, and Zn cycles. *Global Biogeochem. Cycles* **23**, GB2001, 16 pp, doi: 10.1029/2008GB003376.

Ribeyre, F., Amiard-Triquet, C., Boudou, A., and Amiard, J.-C. (1995). Experimental study of interactions between five trace elements – Cu, Ag, Se, Zn and Hg – toward their bioaccumulation by fish (*Brachydanio rerio*) from the direct route. *Ecotoxicol. Environ. Saf.* **32**, 1–11.

Richards, J. G., Curtis, P. K., Burnison, B. K., and Playle, R. C. (2001). Effects of natural organic matter source on reducing metal toxicity to rainbow trout (*Oncorhynchus mykiss*) and on metal binding to their gills. *Environ. Toxicol. Chem.* **20**, 1159–1166.

Rosenzweig, A. (2002). Metallochaperones: bind and deliver. *Chem. Biol.* **9**, 673–677.

Rouleau, C., Borg-Neczak, K., Gottofrey, J., and Tjälve, H. (1999). Accumulation of waterborne mercury (II) in specific areas of fish brain. *Environ. Sci. Technol.* **33**, 3384–3389.

Santore, R. C., Di Toro, D. M., Paquin, P. R., Allen, H. E., and Meyer, J. S. (2001). Biotic ligand model of the acute toxicity of metals. 2. Application to acute copper toxicity in freshwater fish and daphnia. *Environ. Toxicol. Chem.* **20**, 2397–2402.

Saucier, D., and Astic, L. (1995). Morpho-functional alterations in the olfactory system of rainbow trout (*Oncorhynchus mykiss*) and possible acclimation in response to long-lasting exposure to low copper levels. *Comp. Biochem. Physiol. A* **112**, 273–284.

Schecher, W. D., and McAvoy, D. C. (1992). MINEQL+: a software environment for chemical equilibrium modelling. *Comput. Environ. Urban Syst.* **16**, 65–76.

Scott, G. R., and Sloman, K. A. (2004). The effects of environmental pollutants on complex fish behaviour: integrating behavioural and physiological indicators of toxicity. *Aquat. Toxicol.* **68**, 369–392.

Scott, G. R., Sloman, K. A., Rouleau, C., and Wood, C. M. (2003). Cadmium disrupts behavioural and physiological responses to alarm substance in juvenile rainbow trout (*Oncorhynchus mykiss*). *J. Exp. Biol.* **206**, 1779–1790.

Shahsavarani, A., McNeill, B., Galvez, F., Wood, C. M., Goss, G. G., Hwang, P. P., and Perry, S. F. (2006). Characterization of the branchial epithelial calcium channel (ECaC) in freshwater rainbow trout (*Oncorhynchus mykiss*). *J. Exp. Biol.* **209**, 1928–1943.

Shaw, B. J. and Handy, R. D. (2011). Physiological effects of nanoparticles on fish: a comparison of nanometals versus metal ions. *Environ. Int.* doi: 10.1016/j.envint.2011.03.009.

Skeaff, J. M., Durbeuil, A. A., and Brigham, S. I. (2002). The concept of persistence as applied to metals for aquatic hazard identification. *Environ. Toxicol. Chem.* **21**, 2581–2590.

Spry, D. J., and Wood, C. M. (1985). Ion flux rates, acid–base status, and blood gases in rainbow trout exposed to toxic zinc in natural soft water. *Can. J. Fish. Aquat. Sci.* **42**, 1332–1341.

Spry, D. J., and Wood, C. M. (1989). A kinetic method for the measurement of zinc influx in the rainbow trout and the effects of waterborne calcium. *J. Exp. Biol.* **142**, 425–446.

Thurman, E. M. (1985). *Geochemistry of Natural Waters*. Martinus Nijhof; Dordrecht: Dr. W. Junk Publishers, Boston, MA.

Tipping, E. (1994). WHAM – a chemical–equilibrium model and computer code for waters, sediments, and soils incorporating a discrete site electrostatic model of ion-binding by humic acid. *Comput. Geosci.* **20**, 973–1023.

Tjälve, H., and Henriksson, J. (1999). Uptake of metals in the brain via olfactory pathways. *Neurotoxicology* **20**, 181–196.

Tjälve, H., Gottofrey, J., and Bjorklund, I. (1986). Tissue disposition of ^{109}Cd $^{2+}$ in the brown trout (*Salmo trutta*) studied by autoradiography and impulse counting. *Toxicol. Environ. Chem.* **12**, 31–45.

Van der Putte, I., Lubbers, J., and Kolar, K. (1981). Effect of pH on uptake, tissue distribution and retention of hexavalent chromium in rainbow trout (*Salmo gairdneri*). *Aquat. Toxicol.* **1**, 3–18.

Verbost, P. M., Van Rooij, J., Flik, G., Lock, R. A. C., and Wendelaar Bonga, S. E. (1989). The movement of cadmium through freshwater trout branchial epithelium and its interference with calcium transport. *J. Exp. Biol.* **145**, 185–197.

Vijver, M. G., van Gestel, C. A. M., Lanno, R. P., van Straalen, N. M., and Peijnenberg, W. J. G. M. (2004). Internal metal sequestration and its ecotoxicological relevance: a review. *Environ. Sci. Technol.* **38**, 4705–4712.

Wallace, W. G., and Lopez, G. L. (1996). Relationship between the subcellular cadmium distribution in prey and cadmium trophic transfer to a predator. *Estuaries* **19**, 923–930.

Wicklund-Glynn, A. (2001). The influence of zinc on apical uptake of cadmium in the gills and cadmium influx to the circulatory system in zebrafish (*Danio rerio*). *Comp. Biochem. Physiol. C* **128**, 165–172.

Williams, R. (1997). The natural selection of the chemical elements. *Cell. Mol. Life Sci.* **53**, 816–829.

Winge, D., Krasno, J., and Colucci, A. V. (1974). Cadmium accumulation in rat liver: correlation between bound metal and pathology. In: *Trace Element Metabolism in Animals – 2* (W.G. Hoekstra, J.W. Suttie, H.E. Ganther and W. Mertz, eds.), pp. 500–502. Baltimore, MD: University Park.

Wood, C. M. (1992). Flux measurements as indices of H^+ and metal effects on freshwater fish. *Aquat. Toxicol.* **22**, 239–264.

Wood, C. M. (2001). Toxic responses of the gill. In: *Target Organ Toxicity in Marine and Freshwater Teleosts*, Vol. 1, Organs (D. Schlenk and W.H. Benson, eds.), pp. 1–89. Taylor and Francis, New York.

Wood, C. M., Grosell, M., Hogstrand, C., and Hansen, H. (2002). Kinetics of radiolabeled silver uptake and depuration in the gills of rainbow trout (*Oncorhynchus mykiss*) and European eel (*Anguilla anguilla*): the influence of silver speciation. *Aquat. Toxicol.* **56**, 197–213.

Wood, C. M., Franklin, N., and Niyogi, S. (2006). The protective role of dietary calcium against cadmium uptake and toxicity in freshwater fish: an important role for the stomach. *Environ. Chem.* **3**, 389–394.

Wood, C. M., Grosell, M., McDonald, D. M., Playle, R. C., and Walsh, P. J. (2010). Effects of waterborne silver in a marine teleost, the gulf toadfish (*Opsanus beta*): effects of feeding and chronic exposure on bioaccumulation and physiological responses. *Aquat. Toxicol.* **99**, 138–148.

Zheng, D., Kille, P., Feeney, G. P., Cunningham, P., Handy, R. D., and Hogstrand, C. (2010). Dynamic transcriptomic profiles of zebrafish gills in response to zinc supplementation. *BMC Genom.* **11**, 553.

Zitko, V., and Carson, W. G. (1976). A mechanism of the effects of water hardness on the lethality of heavy metals to fish. *Chemosphere* **5**, 299–303.

Zitko, P., Carson, W. V., and Carson, W. G. (1973). Prediction of incipient lethal levels of copper to juvenile Atlantic salmon in the presence of humic acid by cupric electrode. *Bull. Environ. Contam. Toxicol.* **10**, 265–271.

2

COPPER

MARTIN GROSELL

Homeostasis and Toxicology of Essential Metals: Volume 31A
FISH PHYSIOLOGY

The present text provides a review of waterborne as well as dietary copper (Cu) toxicity and Cu homeostasis in fish, and leads to suggestions for further research in this area. Copper, although essential for life, is a potent toxicant and as such, delicate homeostatic controls have evolved at the organismal and cellular level. During exposure to elevated levels of Cu in the water or the diet, homeostatic systems may become overwhelmed such that cellular Cu levels increase to a point where protein function becomes impaired. A range of potential cellular targets for Cu manifest in altered physiology and toxicity, at the organ and organismal level during Cu exposure; these targets and organismal responses are discussed. Copper toxicity is not simply a matter of ambient concentrations as water chemistry greatly influences not only the bioavailability of this metal but also the physiology and thus susceptibility of fish. This complexity has recently been realized and is being considered in current environmental regulations for Cu. Prolonged exposure to Cu elicits an acclimation response which includes a compensatory response of the functions impaired by Cu and adjustments in the homeostatic control of the metal, enabling fish to survive despite continued exposure and to tolerate subsequent exposures to higher concentrations. Although not quantified, these acclimation responses must occur at a cost to the organism, possibly explaining observations of reduced growth, reproductive output, and swimming performance. The literature review forming the basis for this chapter was completed by December 2010.

1. INTRODUCTION

The redox potential of copper (Cu) is utilized by a number of enzymes, including mitochondrial cytochrome *c* oxidase, which makes Cu an essential element for all aerobic organisms (Solomon and Lowery, 1993). However, the redox properties of Cu can also lead to formation of reactive oxygen species (ROS) when cellular Cu levels are elevated, and Cu readily binds to histidine, cysteine, and methionine sites in proteins, potentially leading to their dysfunction (Harris and Gitlin, 1996). As a consequence of Cu's essentiality and potency as a toxicant, organisms must balance between Cu deficiency and excess, a feat accomplished by sophisticated homeostatic control systems. Copper deficiency can be induced under laboratory conditions as evident from reduced growth, but documented cases of deficiency in natural populations are lacking. In contrast to terrestrial vertebrates which only obtain Cu from dietary sources, fish can take up Cu from their diet and across the gills from the ambient water (Miller et al., 1993; Kamunde et al., 2002b). It appears that interactions exist between the two uptake pathways such that branchial Cu uptake is regulated in response to dietary Cu availability and overall Cu status (Kamunde et al., 2002b).

In addition to providing for nutritional uptake of Cu, the gill in freshwater and marine fish is also the target for acute and to some extent chronic Cu toxicity as Cu impairs a number of the exchange functions of this multifunctional organ (Grosell et al., 2002, 2007a). Furthermore, in marine teleosts, waterborne Cu ingested with seawater can interact with the intestinal epithelium to impair osmoregulation (Marshall and Grosell, 2005).

Regardless of the target organ, Cu toxicity is not simply a function of the environmental Cu concentration, as factors such as complexation with organic and inorganic negatively charged molecules, and competition between Cu and cations for uptake exert a strong influence on toxicity (Taylor et al., 2000; McGeer et al., 2000a, 2002; Paquin et al., 2002).

2. CHEMICAL SPECIATION AND OTHER FACTORS AFFECTING TOXICITY IN FRESHWATER AND SEAWATER

2.1. Inorganic Speciation Across Salinities

Inorganic speciation of Cu in seawater is dominated mainly by $CuCO_3$ and $Cu(CO_3)_2^{2-}$, with only a small fraction of Cu being present as ionic Cu^{2+}, $CuOH^-$ and $Cu(OH)_2$. Similar inorganic Cu speciation applies to freshwater of high alkalinity and pH (Blanchard and Grosell, 2005), while $CuOH^-$ and $Cu(OH)_2$ complexes dominate in high pH water with low alkalinity (Chakoumakos et al., 1979; Erickson et al., 1996). However, in freshwaters with intermediate or low alkalinity and lower pH, ionic Cu^{2+} is more prevalent, and becomes the dominant form at neutral and acidic pH (Chakoumakos et al., 1979; Cusimano et al., 1986) (Fig. 2.1).

2.2. Toxic Inorganic Forms of Copper

There is general consensus that ionic Cu^{2+} is the main toxic form of Cu, although $CuOH^-$ and $Cu(OH)_2$ complexes have also been demonstrated to be bioavailable and exert toxicity (Chakoumakos et al., 1979; Erickson et al., 1996; Paquin et al., 2002). In contrast, Cu carbonate complexes are generally assumed not to be available for exerting toxicity, although it appears that these carbonate complexes may be available for accumulation to some extent (Grosell et al., 2004a; Blanchard and Grosell, 2005).

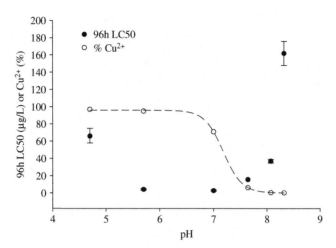

Fig. 2.1. Fraction of Cu present as ionic Cu^{2+} as a function of water pH and corresponding 96 h LC50 for juvenile salmonids. Data obtained from cutthroat trout (3–10 g) for alkaline pH at water hardness of 18–31 mg $CaCO_3$ L^{-1} (Chakoumakos et al., 1979) and rainbow trout (2.60–2.88 g) for neutral and acidic waters at a hardness of 9.2 mg L^{-1} (Cusimano et al., 1986).

2.3. Influence of Cations

In addition to Cu speciation, dissolved cations impact Cu uptake and toxicity. It is generally assumed that the ameliorating effects of cations on the toxicity of Cu as Cu^{2+} (and several other metals) to freshwater organisms is due to competition for uptake by the gill. Among the main cations of relevance (Na^+, H^+, Ca^{2+}, Mg^{2+}, and K^+), Mg^{2+} has been demonstrated to have no impact on acute Cu toxicity while K^+ has been demonstrated to enhance Cu toxicity without affecting uptake by gill tissue (Grosell and Wood, 2002). Thus, for K^+ and Mg^{2+} there is nothing to suggest competition with Cu for uptake at the gill. In contrast, H^+, Na^+, and Ca^{2+} all act to reduce the toxicity of Cu. For protons, the protection against toxicity becomes evident when considering waters of pH < 6 for which more than 95% of Cu is present as Cu^{2+}. Figure 2.1 illustrates the markedly reduced toxicity of approximately one order of magnitude observed when comparing pH 5.7 to 4.7. The protective effect of $[H^+]$ is assumed to be due to competitive interactions between H^+ and Cu^{2+} for uptake into the gill tissue (Di Toro et al., 2001; Paquin et al., 2002). However, classic studies by Playle and co-workers demonstrated no effect of pH 4.8 compared to 6.3 on gill Cu uptake by fathead minnows in ion-poor waters (Playle et al., 1992), suggesting that the protective effects against toxicity may not be related simply to reduced Cu uptake, but may be a physiologically based protection (see Section 6). When considering interactions between cations and Cu as it pertains to gill Cu accumulation and related effects, it is important to recognize that similar interactions occur at dissolved organic carbon (DOC) binding sites and that caution must be taken when interpreting experimental data without reports of DOC concentrations.

Similarly, Ca^{2+}, which offers clear protection against Cu toxicity assessed as acute mortality (Chakoumakos et al., 1979; Erickson et al., 1997; Meyer et al., 1999; Welsh et al., 2000; Naddy et al., 2002), is assumed to compete with Cu^{2+} for uptake by the gills of freshwater fish. However, short-term exposures of rainbow trout to Cu showed no protective effect of Ca^{2+} on gill Cu accumulation (Grosell and Wood, 2002). Similarly, longer exposures of zebrafish to Cu revealed no effect of ambient Ca^{2+} on hepatic or gut Cu accumulation (Craig et al., 2010), although reduced branchial accumulation was observed. The reduced gill Cu accumulation observed in zebrafish in the presence of Ca^{2+} was not accompanied by a reduction in oxidative damage. Overall, these observations suggest that protection from Ca^{2+} against acute Cu toxicity is likely not related to reduced uptake but rather related to a physiological protection against Cu toxicity (see Section 6.1.7).

In contrast to H^+ and Ca^{2+}, Na^+ clearly protects against short-term Cu uptake by rainbow trout gills (Grosell and Wood, 2002), which likely explains, at least in part, the protection offered against acute Cu toxicity by ambient Na^+ seen in fathead minnows (Erickson et al., 1996). Furthermore, ambient Na^+ offers strong protection against oxidative stress and Cu-induced gene expression in zebrafish during prolonged exposures (Craig et al., 2010). However, although reduced Cu uptake in the presence of Na^+ demonstrates competitive interactions, reduced toxicity in the presence of Na^+ is likely the product of both reduced Cu uptake and other physiological protection (see Section 6.1.7).

2.4. Influence of Dissolved Organic Carbon

Possibly the strongest effect on Cu toxicity is exerted by dissolved organic matter (DOM), typically referred to as dissolved organic carbon (DOC), since organic carbon is often what is measured to determine the concentration of DOM. Copper binds to DOC with a high affinity and this complexation prevents or reduces Cu binding and uptake, which acts to lower or prevent toxicity (Playle et al., 1993; Playle and Dixon, 1993; Richards et al., 2001; Sciera et al., 2004). A large number of studies have confirmed the relationship between DOC concentrations and Cu toxicity in freshwater fish and invertebrates during acute exposures, and studies of prolonged Cu exposure have revealed similar relationships. During 30 day exposures to Cu, rainbow trout showed reduced Cu accumulation in gills and liver as well as full protection against Cu-induced osmoregulatory disturbances in the presence of low levels of DOC added as humic acid (McGeer et al., 2002).

The impact of DOC on Cu toxicity to marine fish has yet to be determined, in part owing to the higher Cu tolerance of marine fish (Grosell et al., 2007a) and subsequent need for high Cu concentrations to induce toxicity. However, DOC has been demonstrated to reduce Cu toxicity to marine invertebrates (Arnold, 2005; Arnold et al., 2006), confirming the relationship between DOC and Cu toxicity also for marine waters.

2.4.1. INFLUENCE OF DOC QUALITY

Natural organic matter (NOM) is comprised of humic acids, fulvic acids, carbohydrates, proteins, and lipids (Thurman, 1985). In general, humic acids are of high molecular weight and darkly colored, while fulvic acids are lightly colored and of low molecular weight (Morel and Hering, 1993; Peuravouri and Pihlaja, 2010). Highly colored allochthonous (terrestrially derived) NOM has been reported to decrease metal accumulation by fish gills and decrease toxicity to a greater degree than less colored autochthonous-like (algae-derived) NOM (Richards et al., 2001; Schwartz et al., 2004), likely due to more negatively charged aromatic binding sites for metals in allochthonous NOM (Richards et al., 2001; Luider et al., 2004). The relationship between the quality of NOM (degree of aromaticity) and degree of protectiveness has been approximated using various specific absorption coefficients (Richards et al., 2001; De Schamphelaere and Janssen, 2004) and fluorescence indices (Schwartz et al., 2004) and it is possible to incorporate this parameter in the biotic ligand model (BLM; see Section 5.1).

However, a recent study demonstrated that different sources of DOM with varying aromaticity had no impact on Cu–gill binding in rainbow trout at low Cu concentrations ($<126\,\mu g\,L^{-1}$), but an expected effect at higher ($>126\,\mu g\,L^{-1}$) concentrations (Gheorghiu et al., 2010). Thus, protection offered by different sources of NOM may not be related to different impacts on Cu–gill binding but rather to effects of NOM on the gill. NOM is thought to play a role in promoting ion uptake and preventing diffusive ion loss (Gonzalez et al., 2002; Matsuo et al., 2004; Galvez et al., 2008). Specifically, NOM with greater aromaticity acts to hyperpolarize the gill epithelium of freshwater fish to a greater extent than low aromaticity sources (Galvez et al., 2008). Gill epithelial hyperpolarization would counteract increased diffusive Na^+ loss and reduced Na^+ uptake in the presence of Cu, a key component to acute toxicity (see Section 6.1.7), and thus explain the observed higher protection of aromatic NOM sources.

2.4.2. INFLUENCE OF EQUILIBRATION TIME ON DOC–Cu INTERACTIONS: CONCERN FOR TOXICITY TESTING

Equilibration time between Cu and DOC/NOM has been demonstrated to impact the degree of protection offered by DOC/NOM, at least for freshwater cladocerans and fish (Kim et al., 1999; Santore et al., 2001). Caution must be exercised in future studies to ensure sufficient time (several hours) for stable Cu–DOC complexes to form.

3. SOURCES OF COPPER IN THE ENVIRONMENT AND ITS ECONOMIC IMPORTANCE

Copper is present naturally in the Earth's crust and as such is generally found in surface waters, with naturally occurring Cu concentrations reported to range from 0.2 to 30 µg L^{-1} in freshwater systems (USEPA, 2007), 0.06 to 17 µg L^{-1} in costal systems (Klinkhammer and Bender, 1981; van Geen and Luoma, 1993; Kozelka and Bruland, 1998), and 0.001 to 0.1 µg L^{-1} in oceanic waters (Bruland, 1980; Coale and Bruland, 1988; Sherrell and Boyle, 1992). However, anthropogenic input can raise ambient Cu concentrations to 100 µg L^{-1} or more and Cu can reach levels as high as 200 mg L^{-1} in mining areas (USEPA, 2007). In addition to mining activities, the leather industry, fabricated metal producers, and electric equipment contribute Cu to surface waters (Patterson et al., 1998). Combustion of fossil fuel, municipal waste waters, manure, fertilizers, and antifouling measures (paint and wood preservatives) further contribute Cu to the aquatic environment.

The world mine production of Cu in 2009 was estimated at 15.8 million tonnes which, when combined with the average price for 2009 of US $2.30 per pound, amounts to a global value of Cu mining of approximately $73 billion in 2009 alone (US Geological Survey, 2009).

4. ENVIRONMENTAL SITUATIONS OF CONCERN

After mercury, Cu is the metal most frequently reported to result in impaired water quality in the USA (Reiley, 2007). When combined with its economic importance (see above) this justifies the impressive effort towards understanding Cu toxicity to aquatic organisms and the establishment of site-specific water quality criteria in the USA and European Union (EU) (see below).

It follows from the above discussion that in waters of low ionic strength, low concentrations of NOM, and low pH, free ionic Cu^{2+}, and thus potential toxicity, will be high. Such water chemistry characteristics apply to freshwater environments in the Canadian Shield and northern Scandinavia, with lakes in the Canadian Shield having received considerable attention with respect to environmental impacts of Cu on fish health and fish communities (Taylor et al., 2004; Klinck et al., 2007; Bourret et al., 2008; Gauthier et al., 2009; Pierron et al., 2009). In addition to chemical factors, extremely dilute and acid environments present a challenge to most fish with respect to ionoregulation, which may act synergistically with Cu (and other metals) toxicity.

Recent studies have revealed that extreme salinities, low as well as high (33 ppt), render euryhaline fish more sensitive to Cu than intermediate salinities, although fish in seawater show greater tolerance to Cu than those in freshwater (Grosell et al., 2007a). It is likely that salinities higher than seawater (33 ppt), as found periodically in tide pools and in many lakes in arid environments, may render fish even more sensitive to Cu than seen in normal seawater owing to the increased ionic gradient and thus osmoregulatory demand. Similarly, estuarine environments with fluctuating salinities may render fish more sensitive to Cu exposures; however, these conditions remain to be investigated.

5. ACUTE AND CHRONIC AMBIENT WATER QUALITY CRITERIA

Water quality criteria for Canada, the USA, Australia/New Zealand, and South Africa, as well as criteria under development for member countries of the EU, are summarized in Table 2.1. Criteria for the USA, Europe, and Australia/New Zealand are based principally on species sensitivity distribution diagrams and are designed to protect 95% of species in a given ecosystem with, at least in the case of the US Environmental Protection Agency (EPA), consideration of taxa diversity of the system in question. For South Africa, Canada, and Australia/New Zealand, water quality criteria for Cu in freshwater are hardness based, which was also the case for the USA until the BLM was adopted by the EPA in 2007. While positive correlations between water hardness and Cu tolerance of freshwater organisms, at least for acute exposures, support hardness-based criteria, such criteria fail to incorporate DOC, which is arguably the strongest modifier of Cu toxicity (see above). Protection against Cu toxicity from Ca^{2+}, one of the hardness ions, and from alkalinity, which often covaries with Ca^{2+}, is well documented as discussed above, but hardness-based

Table 2.1
Overview of water quality criteria or guidelines for select countries/regions

Region/country	Hardness (mg CaCO$_3$ L^{-1})	Freshwater (μg L^{-1})		Marine (μg L^{-1})	
		Acute	Chronic	Acute	Chronic
South Africa[a]	<60	1.6	0.53	None	
	60–119	4.6	1.5		
	120–180	7.5	2.4		
	>180	12	2.8		
Canada[b]	<120	2		None. For British	
	120–180	3		Columbia: 2 μg L^{-1}	
	>180	4		(30 day average) or max. 3 μg L^{-1}	
USA[c]		BLM adjusted		3.1	
Europe (EU)[d,e]		BLM adjusted		2.6 or 4.7 (DOC corrected at 2 mg C L^{-1})	
Australia/ New Zealand[f]	50	2.2 (hardness adjusted, sliding scale)		1.3	

Information obtained from: [a]Department of Water Affairs and Forestry (1996), [b]CCME (Canadian Council of Ministers of the Environment) (2007), [c]USEPA (2007), [d,e]ECB (European Chemicals Bureau) (2008a,b), [f]ANZECC/ARMCANZ (2000). See text for further details. BLM: biotic ligand model; DOC: dissolved organic carbon.

criteria are founded purely on correlations rather than a mechanistic understanding.

Of the five countries/regions listed in Table 2.1, only three (USA, EU, and Australia/New Zealand) have established marine criteria that do not differentiate between acute and chronic exposures. Although Canada has yet to establish marine criteria, the province of British Columbia has established both acute and chronic marine criteria (Table 2.1).

5.1. The Biotic Ligand Model

The BLM was first developed and implemented by the US EPA for Cu in 2007 based on approximately two decades of experimental support culminating in two synthesizing papers in 2001 (Di Toro et al., 2001; Santore et al., 2001). The historical as well as technical aspects of the BLM have been reviewed extensively (Paquin et al., 2002) and a detailed discussion of the approach is beyond the scope of the present chapter. In brief, Cu accumulation on/by the gill is predicted by the BLM from cation–gill binding constants and Cu–anion as well as Cu–DOC binding constants, and assuming that ionic Cu^{2+} and $CuOH^-$ are the Cu species resulting in toxicity. In general, the BLM predicts acute toxicity from water chemistry with high accuracy for individual species and can be calibrated to account for differences in sensitivity among species. The BLM is employed to establish chronic water quality criteria in the USA, an extension based largely on acute-to-chronic ratios since limited information is available about the influence of water chemistry on chronic Cu toxicity.

So far, the BLM has not been employed for marine and estuarine environments in any region or country, although the EU considers the influence of DOC to establish higher criteria values in the presence of DOC based on recent work demonstrating protection against Cu toxicity for marine organisms (Arnold, 2005; Arnold et al., 2006).

Estuarine environments often display elevated Cu concentrations and hardly any research has examined the influence of water chemistry on toxicity at intermediate salinities. In addition and in contrast to freshwater organisms, osmoregulatory strategies among estuarine and marine organisms vary considerably such that similar mechanisms of toxicity in different taxa cannot be assumed, as discussed recently (Grosell et al., 2007a; Bielmyer and Grosell, 2010). However, from a perspective of providing protection for fish, water quality criteria are likely to be protective since fish rarely are among the most sensitive organisms tested, although a few sublethal endpoints indicating toxicity are reported for concentrations approaching criteria values (discussed in the following).

6. MECHANISMS OF TOXICITY

6.1. Acute Waterborne Exposure in Freshwater

Several endpoints in freshwater fish are affected by Cu exposure and will be discussed in the following. Although differences in water chemistry among studies, if reported, make it challenging to evaluate which of the following endpoints are more sensitive, they are discussed in order of perceived sensitivity from highest to lowest.

6.1.1. OLFACTION AND MECHANORECEPTION

The ability to respond to olfactory cues is critical for predator avoidance, prey localization, social interactions, homing, and successful reproduction, and as such is critical for survival and population health. The past decade has seen a sharp increase in studies demonstrating Cu-induced olfactory impairment, in many cases at low and environmentally relevant concentrations. Copper-induced impairment has been illustrated by behavioral assays and by direct recordings of electro-olfactograms from the olfactory epithelia. Impacts of Cu on olfaction have been recorded for fright responses (Beyers and Farmer, 2001; Carreau and Pyle, 2005; Sandahl et al., 2007), feeding stimulants (amino acids) (Steele et al., 1990; Baldwin et al., 2003; Sandahl et al., 2006; Green et al., 2010), bile salts (presumed homing stimuli), catecholamines, and steroid as well as non-steroid pheromones (Sandahl et al., 2007; Kolmakov et al., 2009). In other words, Cu has the potential to affect most, if not all, behavioral aspects of fish biology through interactions with the olfactory epithelia (Pyle and Mirza, 2007).

The interaction between Cu and the olfactory epithelium stems from Cu accumulation directly in the olfactory epithelium and appears to be of general nature since multiple receptor pathways are impacted by Cu exposure, as indicated by the broad range of stimuli affected (see above) and a wide range of genes, including some coding for ion channels, G-proteins, and olfactory receptors, being downregulated in the olfactory epithelium of zebrafish upon Cu exposure (Tilton et al., 2008).

The onset of olfactory inhibition is fast (minutes) and seems to persist for weeks or longer, even though some signs of recovery despite continued exposure have been reported (Saucier and Astic, 1995; Beyers and Farmer, 2001).

The persistence of the olfactory inhibition by Cu combined with high sensitivity, as illustrated by effect levels in the low $\mu g\ L^{-1}$ range (Baldwin et al., 2003; Carreau and Pyle, 2005; Sandahl et al., 2007; Green et al., 2010), makes olfactory inhibition one of the most pressing environmental concerns regarding fish exposed to Cu. Furthermore, it appears that water hardness (adjusted as $CaCl_2$) has variable effects on Cu-induced inhibition of electro-olfactograms, ranging from no effect (Baldwin et al., 2003) to a modest protective effect (Bjerselius et al., 1993; McIntyre et al., 2008). This lack of uniform protection from Ca^{2+} is alarming since water quality criteria effectively are adjusted based on water hardness throughout Canada and the USA, and since even BLM-adjusted guidelines for Europe employ a Ca^{2+} protection term (see Section 5). The lack of protection from Ca^{2+} against effects on the olfactory epithelium is consistent with recent reports of minor effects of ambient Ca^{2+} on short-term Cu accumulation by the olfactory epithelium (Green et al., 2010). Specifically, a, 20-fold increase in ambient Ca^{2+} results in only a modest 50% reduction in olfactory epithelial Cu binding.

The apparent lack of competition between Cu and Ca^{2+} for uptake by the gill epithelium (see Section 2.3) is similar to what has been reported for the olfactory epithelium. However, in contrast to the olfactory epithelium, where there is good agreement between the lack of (or modest) impact of Ca^{2+} on epithelial Cu uptake and effect, ambient Ca^{2+} does offer protection against acute toxicity (observed as mortality). This disparity offers additional support for the conclusion that protection offered by Ca^{2+} against Cu-induced mortality is of a physiological nature rather than simply the product of cation competition (see Section 3.3). The mechanisms of Ca^{2+} protection against Cu toxicity as evaluated by ion flux measurements or mortality are discussed in Section 6.1.7.

However, in agreement with predictions from the BLM on Cu-induced mortality, ambient HCO_3^-, at least to some extent (Winberg et al., 1992; McIntyre et al., 2008), and NOM levels (McIntyre et al., 2008) ameliorate Cu-induced inhibition of olfactory responses in salmonids in a way that seems consistent with predicted Cu speciation, with Cu^{2+} as the main toxic form of Cu. Indeed, a recent evaluation of olfactory impairment by Cu in freshwater fish suggests that the current BLM offers protection against this endpoint for fish examined to date (Meyer and Adams, 2010).

In addition to affecting olfactory neurons, mechanoreception by lateral line hair cells or neuromasts has been reported to be impacted by Cu exposure at relatively low ambient Cu concentrations, an impact that is in part related to oxidative stress (see Section 6.1.6 for a discussion of oxidative stress) (Johnson et al., 2007; Olivari et al., 2008; Linbo et al., 2009). The neuromasts are important peripheral specialized neurons allowing for the detection of water movement relative to the body surface of the fish. While the sensory input provided by these neurons is important for orientation relative to water currents and sensing of swim speed, it is also important for detecting water movement caused by approaching predators, prey, or conspecifics.

In agreement with the impact of water chemistry on Cu toxicity to the olfaction sensory system, it appears that water chemistry parameters, with the exception of DOC, offer minor protection against Cu toxicity to neuromasts (Linbo et al., 2009). While DOC offers strong protection against Cu-induced damage to hair cells, it appears that the impact of water chemistry on Cu toxicity to peripheral neurons, olfactory or mechanosensory, is less and/or different than the impact on toxicity assessed by mortality and predicted by the BLM (discussed above). Although the impact of Cu on mechanoreception can be inferred from damage to neuromasts, it should be noted that functional indications of such impacts have yet to be presented in the peer-reviewed literature.

6.1.2. BEHAVIORAL RESPONSES

6.1.2.1. Avoidance. Copper has been demonstrated to elicit avoidance behavior in freshwater fish, with response concentrations in the low $\mu g\ L^{-1}$ range (Hansen et al., 1999; Svecevicius, 2001; Moreira-Santos et al., 2008). Implicit to an avoidance response is the ability to detect the substance being avoided, but the mechanism of Cu detection is unknown. In addition, it is unknown to what extent Cu avoidance is displayed by marine fish.

6.1.2.2. Interactions between Copper Exposures and Social Behavior. Salmonids and many other fish establish social hierarchies by agonistic encounters or interactions in laboratory settings as well as in their natural environment. For rainbow trout, exposure at 15% of the 96 h LC50 is of no consequence for the establishment of dominance hierarchies, whether groups of fish are exposed together or whether exposed individuals are brought to interact with unexposed fish (Sloman et al., 2003a). However, social status affects Cu uptake and accumulation. Subordinate fish display higher branchial and hepatic Cu accumulation during acute exposures than dominant fish, a difference that is attributable to differences in Na^+ uptake rates (Sloman et al., 2002, 2003b). Copper uptake in freshwater rainbow trout, and possibly other freshwater fish, occurs at least in part via Na^+ uptake pathways (Grosell and Wood, 2002) (see Section 9.1). Subordinate fish display higher Na^+ uptake than dominant fish, likely accounting for the higher Cu accumulation rates.

6.1.3. AMMONIA EXCRETION

Perhaps the most consistently observed response to sublethal Cu exposure in freshwater and seawater fish is elevated plasma ammonia/ ammonium (Lauren and McDonald, 1985; Wilson and Taylor, 1993a,b; Beaumont et al., 1995; Wang et al., 1998; Grosell et al., 2003, 2004b). Elevated plasma cortisol, which has been reported from Cu-exposed fish (Donaldson and Dye, 1975; De Boeck et al., 2001), stimulates protein catabolism and thereby ammonia production and could thus explain hyperammonemia.

Absolute ammonia excretion rates by common carp (De Boeck et al., 1995a) and rainbow trout (Lauren and McDonald, 1985) do not appear to change in response to Cu exposure despite situations where elevated plasma: water gradients would likely favor higher ammonia excretion rates. A similar lack of impact on ammonia excretion was observed in freshwater killifish, although seawater-acclimated fish displayed reduced ammonia excretion during Cu exposure (Blanchard and Grosell, 2006). Overall, these results suggest that ammonia excretion may be impaired by Cu exposure which, coupled with increased metabolic ammonia production, results in elevated plasma ammonia concentrations during Cu exposure.

Recent studies have reported that fed fish are more sensitive to acute Cu exposure than fasted individuals, despite lower Cu accumulation (Hashemi et al., 2008a,b), which appears to correlate with higher ammonia accumulation in fed individuals (Kunwar et al., 2009).

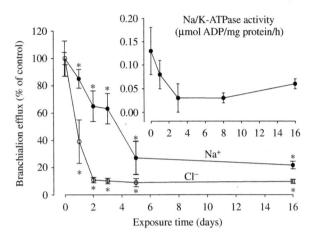

Fig. 2.2. Gill Na^+/K^+-ATPase activity, and Na^+ and Cl^- excretion by gulf toadfish during 16 days of exposure to 3.48 mg Cu L^{-1}. Na^+/K^+-ATPase data showed a trend towards reduced activity during Cu exposure although no significant differences were observed (Grosell et al., 2004a). Na^+ and Cl^- efflux rates are expressed as a percentage of initial controls and were significantly lower (*) at all times during Cu exposure. Details regarding experimental procedures as well as absolute flux values are reported elsewhere (Grosell et al., 2011).

The mechanism of ammonia excretion by aquatic organisms has long been a controversial subject, although it is clear that the primary route of excretion is the gill and that NH_3, rather than NH_4^+, likely is transported across the epithelium. In addition, the importance of boundary layer acidification and thus diffusion trapping by NH_3 to NH_4^+ conversion in the boundary layer is generally recognized, as is the importance of branchial carbonic anhydrase (CA) (Wright et al., 1989; Randall and Wright, 1989; Wilson et al., 1994). Considering that Cu is a potent inhibitor of CA, historically used to distinguish between CA isoforms (Magid, 1967), and has been demonstrated to inhibit CA at least in crustaceans (Vitale et al., 1999), Cu-induced CA inhibition seems a likely explanation for reduced ammonia excretion. However, to date, no studies have demonstrated Cu-induced inhibition of CA activity in fish. Considering the consistent response of ammonia excretion and acid–base balance during Cu exposure, the lack of CA effects is surprising and ought to be further examined. Possibly, the high dilution volumes necessitated by the delta pH CA assay (Henry, 1991) dilute Cu to the point where there is no longer inhibition. To a lesser extent, similar dilution may also cause underestimation of Cu inhibition of Na^+/K^+-ATPase assays. Na^+/K^+-ATPase assays have been successful in demonstrating Cu-induced inhibition but the degree of inhibition is never complete and there are examples of clearly reduced Na^+ transport (Fig. 2.2) without apparent reductions in Na^+/K^+-ATPase activity (Grosell et al., 2004a).

There is a clear need to investigate the effect of Cu on rhesus (Rh) proteins, which have recently been shown to be involved in ammonia excretion (Nawata et al., 2007, 2010b; Nakada et al., 2007a,b; Hung et al., 2008; Braun et al., 2009; Tsui et al., 2009; Wright and Wood, 2009).

6.1.4. ACID–BASE BALANCE

Sublethal Cu exposure may result in acid–base balance disturbances even at concentrations that fail to induce or only result in modest osmoregulatory disturbances (Pilgaard et al., 1994; Wang et al., 1998). Sublethal Cu exposure, if concentrations are sufficiently high, results in elevation of extracellular pH despite no change in partial pressure of carbon dioxide (PCO_2), and therefore appears to be the result of elevated plasma HCO_3^- (Pilgaard et al., 1994; Wang et al., 1998). The reason for this metabolic alkalosis induced by Cu exposure remains unknown, but could include effects on branchial Cl^-/HCO_3^- exchange and/or interactions with CA. At higher Cu concentrations (or lower hardness) a respiratory acidosis may occur in combination with a metabolic alkalosis without impact on arterial oxygen (O_2) levels (Pilgaard et al., 1994). Such selective inhibition of CO_2 excretion without impact on O_2 strongly implies that CA is inhibited by Cu exposure, although direct evidence for such inhibition in fish exposed to Cu is lacking (see Section 6.1.3). The ability of fish in freshwater as well as seawater to compensate for hypercapnia (elevated ambient CO_2) is strongly impaired by Cu exposure, which acts to reduce the compensatory increase in extracellular HCO_3^- concentrations (Larsen et al., 1997; Wang et al., 1998). The impact of combined Cu and hypercapnia exposure on acid–base balance is similar to the impact of combined exposure to hypercapnia and CA inhibitors (Georgalis et al., 2006), again pointing to CA as a likely target for Cu toxicity.

High, often lethal, Cu exposure concentrations result in severe acidosis owing to impaired branchial gas exchange (see Section 6.1.8).

6.1.5. SWIMMING PERFORMANCE

Exposure to sublethal Cu concentrations may impair maximum sustainable swimming speeds (Ucrit; Brett, 1964) even at concentrations of 12–35% of the 96 h LC50 (Waiwood and Beamish, 1978; Beaumont et al., 1995, 2000). At such low Cu concentrations (relative to lethal concentrations), the loss of swimming performance is apparently not due to reduced oxygen transfer, although that may be the case at higher exposure concentrations (see Section 6.1.8), since arterial O_2 and CO_2 levels remained unaffected in brown trout displaying reduced Ucrit (Beaumont et al., 1995; Waiwood and Beamish, 1978). Thus, reduced swimming performance during Cu exposure may occur at Cu concentrations insufficient to result in severe gill damage. Furthermore, it appears that reduced swimming performance occurs despite a lack of increased blood viscosity and local hypoxia resulting from osmoregulatory disturbances (see Section 6.1.7) (Waiwood and Beamish, 1978). Possible explanations for reduced aerobic capacity during sublethal Cu exposure are related to the commonly observed hyperammonemia (see Section 6.1.3). Ammonia regulates a number of metabolic pathways and, in addition, can alter membrane potentials via displacement of K^+ in ion-exchange mechanisms, resulting in depolarization of neurons and muscle cells (Beaumont et al., 1995, 2000; Taylor et al., 1996).

Impacts of Cu on swimming performance are evident early during exposure (hours to days) (Beaumont et al., 1995, 2000) and may in some cases be transient (Waiwood and Beamish, 1978). More prolonged exposures, however, have also been reported to result in reduced swimming performance, although this effect occurs in interaction with food intake. Increased feeding rates alone result in an increased metabolic demand and a decreased aerobic swimming performance. Higher feeding rates render rainbow trout more susceptible to sublethal Cu exposure as evaluated by swimming performance (McGeer et al., 2000b). This observation suggests that the impairment of swimming performance is related to an increased O_2 demand during prolonged Cu exposure, which has been observed for both brown trout and rainbow trout (Waiwood and Beamish, 1978; Beaumont et al., 1995). Such an increased O_2 demand is likely due to an increased metabolic cost of compensating for Cu-induced physiological impairments rendering fewer resources for aerobic performance and reducing aerobic scope. The impacts of Cu can thus be considered a loading stress rather than a limiting stress (Brett, 1964).

6.1.6. Oxidative Stress

Oxidative stress in response to Cu exposure is initiated early during exposure, although it may be met with enzymatic and non-enzymatic antioxidant defense mechanisms. Prolonged exposures (days to weeks), however, may result in oxidation of DNA, protein, and lipids when cellular antioxidant capacity is exceeded. Some of these same effects occur as a result of chronic Cu exposure, and both the phenomena and the mechanisms involved appear to be similar in freshwater and seawater fish. Consequently all these areas are discussed in the present section. Copper levels required to induce oxidative stress likely vary between freshwater and seawater fish, as is the case for other endpoints, but this remains to be addressed experimentally in a systematic manner.

Metals, including Cu, can result in oxidative stress by (1) inhibition of antioxidant enzymes, (2) alterations in the mitochondrial electron-transfer chain, (3) the formation of ROS, a process referred to as the Fenton reaction, and (4) depletion of cellular glutathione (Pruell and Engelhardt, 1980; Shukla et al., 1987; Freedman et al., 1989; Stohs and Bagchi, 1995; Rau et al., 2004; Wang et al., 2004). Protection against ROS, which are generated continuously during aerobic metabolism, is normally achieved by antioxidant enzymes such as catalase, glutathione peroxidase, and Cu/Zn superoxide dismutase (SOD). In addition, glutathione reductase, metal-lothionein, and heat shock protein 70 (HSP70) are involved in protection against oxidative stress (Sato and Bremner, 1993; McDuffee et al., 1997; Evans and Halliwell, 2001). Oxidation of DNA, proteins, and lipids, leading to DNA adduct formation, protein carbonyls, and lipid peroxidation, respectively, may occur if the combined cellular antioxidant capacity comprised by the above enzymes and proteins is exceeded by the rate of ROS formation.

6.1.6.1. Impact of Copper on Antioxidant Enzymes. ROS formation is clearly inducible by Cu exposure in fish gill cells (Bopp et al., 2008) and hepatocytes (Krumschnabel et al., 2005). In accord with these findings, catalase gene expression and enzymatic activity has been reported to increase within a few days of exposure in gills, liver, and kidney of freshwater fish (Hansen et al., 2006; Craig et al., 2007). However, it appears that catalase gene expression and enzymatic activity are not tightly correlated and that catalase activity is controlled in part by enzyme activation rather than transcription (Hansen et al., 2006; Craig et al., 2007). It is also clear that Cu can contribute to oxidative stress by inhibiting antioxidant enzymes, as illustrated by observations of decreased catalase activity in gill, hepatic, and renal tissue in Cu-exposed fish (Ahmad et al., 2005; Hansen et al., 2007; Sampaio et al., 2008).

Similarly diverse responses are reported for glutathione peroxidase, which may show increased messenger RNA (mRNA) expression (Hansen et al., 2007) or lack of expression change despite changes in other antioxidant enzymes (Hansen et al., 2006). In agreement with the diverse mRNA transcription responses, both increased and decreased glutathione perox-idase enzyme activity levels have been reported from fish during short-term Cu exposure (Radi and Matkovics, 1988; Ahmad et al., 2005).

SOD mRNA expression may be increased within days in response to Cu exposure in gill, liver, and kidney (Cho et al., 2006; Hansen et al., 2006), and often results in increased SOD enzymatic activity (Sanchez et al., 2005; Hansen et al., 2006; Sampaio et al., 2008), although decreased SOD activity in response to Cu exposure has also been reported (Vutukuru et al., 2006). One possible reason for the variable reported responses in SOD (as well as catalase and glutathione peroxidase) activity may be that both transcrip-tional and enzymatic responses appear to be transient, even during continued Cu exposure (Sanchez et al., 2005). This transient response can reasonably be interpreted as a complex interaction between a need to defend against accumulation of ROS on one hand and the direct inhibitory action of Cu on these antioxidant enzymes on the other.

6.1.6.2. Other Protection Against Oxidative Stress. Appropriate levels of reduced glutathione are important for homeostatic redox balance and are the product of total glutathione present and the ratio of reduced (GSH) versus oxidized glutathione (GSSH) (Carlberg and Mannervik, 1985). A first line of cellular defense against metals is chelation and detoxification as well as scavenging of oxyradicals by reduced glutathione (Sies, 1999). The ratio of GSH to GSSH is determined by glutathione peroxidase (see above), which facilitates the oxidation of glutathione and glutathione reductase that reduces GSSH back to GSH (Carlberg and Mannervik, 1985; Hansen et al., 2006). In addition to altering the GSH/GSSH ratio, Cu has been reported to inhibit glutathione synthetase (Canesi et al., 1999) and furthermore forms stable complexes with GSH, hence decreasing GSH levels in the cytosol (Brouwer et al., 1993). Indeed, reduced glutathione levels have been observed in gills, liver, and kidney from freshwater and seawater fish exposed to waterborne Cu (Ahmad et al., 2005; Sanchez et al., 2005; Almroth et al., 2008). Glutathione reductase mRNA expression may be increased in gill and hepatic tissue during waterborne Cu exposure (Hansen et al., 2006; Minghetti et al., 2008). The shared function of glutathione reductase and glutathione peroxidase in glutathione turnover is illustrated by the positive correlation in expression of these enzymes in the gills and liver of Cu-exposed brown trout (Hansen et al., 2006).

Metallothionein is generally accepted as a metal scavenger, with two of the four isoforms (MT-1 and MT-2) being inducible in response to, among other stimuli, metals (Kling and Olsson, 1995; Chiaverini and De Ley, 2010). The regulation of metallothionein synthesis occurs in response to the binding of free metal ions to transcription factors, the metal-responsive elements (MREs) in the 5′ regulatory region of the metallothionein genes (Kling and Olsson, 1995). In addition to metals, free oxygen radicals are known to increase metallothionein mRNA, at least in mammals, a response mediated by a combination of antioxidant response elements and MREs, suggesting a direct role for metallothionein in antioxidant defense (Kling and Olsson, 2000; Chiaverini and De Ley, 2010). Indeed, metallothionein, compared to SOD or glutathione, is highly efficient in quenching superoxide radicals (Chiaverini and De Ley, 2010). A large number of Cu exposure studies has revealed elevated metallothionein expression and protein levels in target organs for Cu accumulation (McCarter and Roch, 1984; Roch and McCarter, 1984; Grosell et al., 1997, 1998b; Hogstrand et al., 1989, 1991;

Kling and Olsson, 1995; Cheung et al., 2004; Wu et al., 2007; Minghetti et al., 2008), but mass balance considerations point to a minor role of metallothioneins as Cu storage proteins (Hogstrand et al., 1991; Grosell et al., 1997, 1998b). The relatively minor contribution of metallothioneins to binding of the total cell Cu content may suggest that the antioxidant properties of metallothioneins are a more important function of these proteins in the response to Cu exposure.

HSP70 acts as a molecular chaperone and forms an important part of the cellular response to oxidative stress by protecting the protein machinery. Constitutive forms of HSP70 are present in unstressed cells, whereas the inducible form is synthesized by fish in response to stressors, including Cu (Sanders et al., 1995; Boone and Vijayan, 2002; Feng et al., 2003). Copper-induced expression of HSP70 has been reported from isolated hepatocytes, freshwater fish gills (Hansen et al., 2007), freshwater fish hepatocytes, and marine fish kidneys (Boone and Vijayan, 2002; Feng et al., 2003). Oxidative stress and associated damage in fish livers during Cu exposure is a likely outcome of hepatic Cu accumulation since the liver, in general, accumulates the highest tissue Cu concentrations of any tissue (see Section 10). Oxidative stress-induced damage as evident from elevated HSP70 levels often corresponds with the proposed target tissues; the gills in freshwater fish and the kidneys in marine fish (Stagg and Shuttleworth, 1982a; Grosell et al., 2003).

6.1.6.3. Damage from Oxidative Stress. DNA damage, lipid peroxidation, and protein carbonyls have been reported from Cu-exposed fish although rarely, if ever, from the same species in the same study. Consequently, it is not possible to evaluate which of these endpoints are most sensitive as indicators of oxidative stress induced by Cu exposure. However, a single study demonstrating DNA damage (by COMET assay) found no evidence for lipid peroxidation [by thiobarbituric acid reactive substances (TBARS) assay] in rainbow trout gill cells (Bopp et al., 2008), suggesting that DNA damage is a more sensitive indicator than lipid peroxidation. In agreement with this observation are reports of DNA damage in red blood cells isolated from fish exposed to Cu concentrations in the low $\mu g\,L^{-1}$ range in freshwater as well as seawater (Gabbianelli et al., 2003; Santos et al., 2010). Nevertheless, lipid peroxidation has been reported frequently for Cu-exposed fish in seawater and freshwater (Roméo et al., 2000; Ahmad et al., 2005; Vutukuru et al., 2006; Hoyle et al., 2007), while Cu-induced protein carbonyl formation has been reported in at least one case (Craig et al., 2007; Almroth et al., 2008). The DNA damage response to Cu exposure is modulated by pH in a way that suggests that this endpoint is also most sensitive to Cu as ionic Cu^{2+} (Bopp et al., 2008).

6.1.7. NaCl Uptake

Acute toxicity leading to mortality arising from Cu exposure is associated with an osmoregulatory disturbance in freshwater fish (Grosell et al., 2002), where Na^+ and Cl^- balance may be impaired, even during exposure to sublethal Cu concentrations (McKim et al., 1970; Lewis and Lewis, 1971; Christensen et al., 1972; Schreck and Lorz, 1978; Stagg and Shuttleworth, 1982a). The reduction of plasma Na^+ and Cl^- can be due to both reduced uptake and increased loss, at least at higher Cu concentrations. In early and elegant work on rainbow trout it was demonstrated that the reason for impaired Na^+ homeostasis was reduced Na^+ uptake at lower Cu concentrations and a combination of reduced uptake and also increased Na^+ loss at higher concentrations. The increased Na^+ loss at higher concentrations was attributed to displacement of calcium by Cu in tight junction proteins, thereby increasing the paracellular permeability (Lauren and McDonald, 1985). This interaction between calcium and acute Cu toxicity at the physiological level likely explains, at least in part, the protective effect of calcium against acute Cu toxicity (see Section 2.3).

A relatively strong correlation between Na^+ uptake rates, or Na^+ turnover, and sensitivity to acute Cu exposure expressed as 96 h LC50 demonstrates that disruption of Na^+ (and Cl^-) uptake is the cause of Cu-induced mortality in freshwater fish (reviewed by Grosell et al., 2002). The resulting net loss of Na^+ and Cl^- during Cu exposure leads to reduced blood plasma osmolality and ultimately a fluid shift from plasma to tissues including red blood cells. The resulting reduction in plasma volume and swelling of blood cells combined with splenic release of blood cells elevates hematocrit and thus blood viscosity which, combined with increased vascular resistance due to catecholamine-induced systemic vasoconstriction, leads to cardiovascular collapse. Thus, although acute Cu-induced mortality in freshwater fish is due to a cascade of events leading to cardiovascular collapse, the onset of the response is inhibition of branchial Na^+ and Cl^- uptake and possibly increased diffusive loss (Lauren and McDonald, 1985; Wilson and Taylor, 1993b).

6.1.7.1. Inhibition of Na^+/K^+-ATPase. A detailed kinetic analysis of branchial Na^+ uptake by freshwater-acclimated rainbow trout in the presence and absence of Cu revealed that reduced Na^+ uptake was the product of a mixed non-competitive (decreased transport capacity) and competitive inhibition (decreased affinity) after 24 h of Cu exposure and that these effects co-occurred with inhibition of maximal Na^+/K^+-ATPase activity in gill homogenates (Lauren and McDonald, 1987a,b). These observations led to the conclusion that inhibition of the branchial

Na^+/K^+-ATPase enzyme led to a reduction in branchial Na^+ uptake. In support of this conclusion, later studies reported parallel inhibition of Na^+ uptake and gill Na^+/K^+-ATPase enzyme activity (Pelgrom et al., 1995; Sola et al., 1995) and the degree of Na^+/K^+-ATPase inhibition has been reported to correlate positively with gill Cu accumulation (Li et al., 1998). The mechanism by which Cu inhibits Na^+/K^+-ATPase appears to be through interference with the Mg^{2+} binding (Li et al., 1996) which is critical for phosphorylation and thus transport by the enzyme (Skou, 1990; Skou and Esmann, 1992).

6.1.7.2. Inhibition of Apical Na^+ Entry. Although no detailed information exists on the time course of Cu-induced inhibition, gill Cu levels do not reach maximum levels until 6 h of exposure (Grosell et al., 1997). With this in mind, observations of Cu-induced inhibition of Na^+ uptake during only 2 h of exposure (Grosell and Wood, 2002) are likely attributable to inhibition of an apical Na^+ entry step rather than the basolateral Na^+/K^+-ATPase enzyme. This interpretation is supported by a strictly competitive inhibition (reduced Na^+ affinity) between Cu and Na^+ uptake during 2 h exposures, which is distinct from the mixed competitive and non-competitive inhibition observed in fish exposed for 24 h. The observations of mixed competitive and non-competitive inhibition likely reflect action of Cu at the apical Na^+ entry step (competitive) and the basolateral Na^+/K^+-ATPase (non-competitive) (Fig. 2.3).

At least for rainbow trout, it is clear that Cu interacts with a Na^+ channel in the apical membrane, which could possibly explain the competitive inhibition of Na^+ uptake during short-term Cu exposures. However, additional Na^+ uptake pathways in the form of apical Na^+/H^+ exchanger isoforms (mainly NHE2 and NHE3) are also present and could be the target for Cu-induced inhibition of Na^+ uptake, either directly or indirectly (Fig. 2.3).

In addition to the proposed direct effects of Cu on apical Na^+ entry steps, an indirect effect cannot be dismissed. For both Na^+ uptake via an apical Na^+ channel and NHEs, H^+ availability for excretion across the apical membrane is central. Na^+ uptake via Na^+ channels is fueled in part by the hyperpolarizing effect of an electrogenic apical H^+ pump, while the importance of cellular substrate in the form of H^+ for NHEs is obvious (Fig. 2.3). Carbonic anhydrase-facilitated hydration of CO_2 provides H^+ and HCO_3^- ions for exchange during Na^+ and Cl^- uptake, respectively. As discussed above (Section 6.1.3), the possibility that Cu exposure inhibits branchial CA *in situ* cannot be dismissed despite the lack of Cu-induced CA inhibition as measured in tissue homogenates. A Cu-induced inhibition of CA could possibly explain the early onset of Na^+ uptake inhibition by

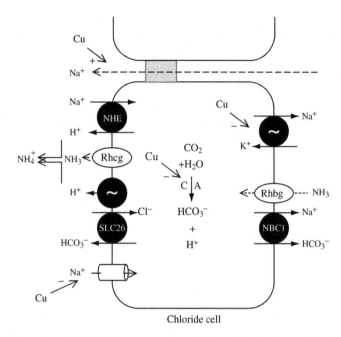

Fig. 2.3. Schematic and simplified representation of Cu-sensitive freshwater fish gill transport processes relevant for salt balance, acid–base balance and ammonia excretion (see text for details) (Evans et al., 2005). Apical Na^+/H^+ exchange (NHE), Cl^-/HCO_3^- exchange (SLC26), H^+ extrusion via the proton pump ($\sim$) and apical Na^+ channels are included, but the likely involvement of multiple isoforms of NHEs and SLC26 members as well as anion exchangers of the SLC4 family, Ca^{2+} transport pathways, and basolateral ion channels have been omitted for clarity. Furthermore, the depicted transport processes are likely occurring in different chloride cell types and the diagram should therefore be considered as overall branchial transport regardless of cell types. Copper-induced inhibition of Na^+/K^+-ATPase ($\sim$) has been documented repeatedly (Grosell et al., 2002) and competitive inhibition of Na^+ uptake across the apical membrane has been suggested by altered Na^+-uptake kinetics following only a few hours of Cu exposure (Grosell and Wood, 2002). In addition, Na^+ efflux via paracellular pathways has been reported to be increased by exposure to high Cu concentrations (Lauren and McDonald, 1985, 1986). Although no studies have demonstrated inhibition of branchial carbonic anhydrase (CA) during or following Cu exposure, reports of acid–base balance disturbance are abundant, and ammonia excretion is the parameter most consistently impacted by Cu. These observations could all be related to reduced CA activity since CA provides acidic and basic equivalents for exchange with the environment and allows for ammonia excretion. Ammonia excretion occurs via basolateral and apical Rhbg and Rhcg transporters, respectively (Wright and Wood, 2009) and relies on apical acid excretion and thus cellular CA activity.

depleting apical Na^+ entry steps from cellular substrate in the form of H^+. Indeed, pharmacological inhibition of CA can in some cases result in inhibited Na^+ uptake, illustrating the link between CA and Na^+ uptake (Boisen et al., 2003).

The possibility of direct interaction between Cu and the H^+ pump has yet to be examined and it therefore cannot be ruled out that direct inhibition of the H^+ pump can explain the inhibition of Na^+ uptake immediately after onset of exposure.

6.1.7.3. Inhibition of Cl^- Transport. No direct measurements of Cl^- uptake in freshwater fish during Cu exposure have been reported; however, observations of parallel reductions in plasma Cl^- and Na^+ concentrations during Cu exposure (Lauren and McDonald, 1985; Wilson and Taylor, 1993b) strongly imply that gill Cl^- uptake, like Na^+ uptake, is potently impaired by Cu, as is the case for Ag (Morgan et al., 1997, 2004). In addition, Cu may increase diffusive Cl^- loss via paracellular pathways, as seen for Na^+ (Lauren and McDonald, 1985). Indeed, Cu inhibits Cl^- secretion by the opercular epithelium (Crespo and Balasch, 1980), a model for the gill of the killifish, and has been documented to greatly inhibit Cl^- extrusion by the marine teleost, the gulf toadfish (Fig. 2.2). Since no direct link between Na^+/K^+-ATPase activity and Cl^- uptake has been established, the parallel reduction in plasma Na^+ and Cl^- during acute Cu exposure may suggest that CA, which links apical Na^+ and Cl^- entry (Fig. 2.2), is a likely target for acute Cu exposure and that inhibition of this enzyme can explain the parallel effects on Na^+ and Cl^- homeostasis.

6.1.8. RESPIRATORY DISTRESS

Copper exposure at sublethal levels may result in reduced oxygen consumption, a transient effect at lower concentrations and persistent effect at higher concentrations (De Boeck et al., 1995a). In addition, the critical oxygen concentration was greatly increased in response to Cu exposure, indicating that Cu-exposed fish lose the ability to sense reduced oxygen, the ability to regulate oxygen consumption, or both (De Boeck et al., 1995a). The environmental significance of these findings is obvious, as sublethal Cu exposure may result in increased susceptibility to environmental hypoxia. Conversely, combined exposure to Cu and hypoxia has been reported to prevent the recovery of plasma ions seen in fish exposed to Cu only (Pilgaard et al., 1994).

High Cu concentrations resulting in rapid mortality induced hypoxic hypercapnia in the arterial blood, likely due in part to gill histopathologies such as cell swelling and thickening of lamellae, effectively increasing the blood–water diffusive distance, leading to impaired gas transfer across the gill (Wilson and Taylor, 1993b). The PCO_2 accumulation resulting from the impaired gas exchange at the gill leads to a pronounced respiratory acidosis as well as a substantial metabolic acidosis, as indicated by increased blood lactate. Reduced O_2 uptake at the gill combined with increased blood viscosity results in impaired O_2 delivery during exposure to high Cu concentrations. However, mortality likely occurs as a result of ionoregulatory disturbances that induce elevated blood pressure, leading to cardiovascular collapse rather than hypoxemia (Wilson and Taylor, 1993b) (see Section 6.1.7), although clear differences exist among species, with cyprinids apparently responding differently than salmonids (De Boeck et al., 2007b).

6.2. Acute Effects of Waterborne Copper Exposure in Seawater

Copper-induced alterations of sensory systems have not been examined in seawater fish. Although greater competitive interactions among cations and Cu for interactions with ion channels and membrane proteins occur in seawater compared to freshwater, impacts of Cu on olfaction and mechanoreception in seawater are also likely and clearly worthy of examination. Similarly, with the exception of demonstrations of altered schooling behavior (Koltes, 1985), there are no observations of Cu-induced alterations in fish behavior or social interactions among marine fishes. As discussed above (Section 6.1.6), there appear to be no mechanistic differences between seawater and freshwater with respect to oxidative stress and damage induced by Cu exposure, although the specific tissues displaying such effects likely will differ between seawater and freshwater. However, excretion of nitrogenous waste, acid–base balance, and salt and water balance are all physiological functions that may also be impaired during Cu exposure in seawater, as discussed below.

6.2.1. EFFECTS OF COPPER ON DRINKING

Marine fish continuously lose water to their hyperosmotic environment and rehydrate by ingesting seawater along with any contaminants that may be dissolved in the water. In a marine teleost exposed to high Cu levels ($mg\ L^{-1}$ range) a dramatic reduction in drinking rate was observed within 3 and 24 h of exposure (Grosell et al., 2004a), before any other observed physiological effects, indicating that the response may have been a taste aversion. Reduced drinking rate in the presence of Cu can be viewed as an adaptive response as it would act to limit ingestion of Cu, but can only serve as a short-term response as it would in itself result in dehydration, and a threshold environmental concentration required to elicit the taste aversion remains to be addressed.

6.2.2. NITROGENOUS (WASTE) EXCRETION

6.2.2.1. Teleosts. Both acid–base balance and osmoregulation were impacted by Cu exposure in Atlantic cod (Larsen et al., 1997), while studies on seawater-acclimated rainbow trout revealed elevated plasma ammonia and possibly a modest acid–base balance disturbance during exposure to similar Cu concentrations (400–500 μg L^{-1}) (Wilson and Taylor, 1993a). In seawater-acclimated killifish impaired ammonia excretion has been observed with no evidence of an osmoregulatory disturbance during exposure to Cu at 120 μg L^{-1} (Blanchard and Grosell, 2006). Furthermore, in gulf toadfish, plasma urea and ammonia were elevated at Cu levels where osmoregulatory indicators such as plasma osmolality and ion concentrations were unaltered (Grosell et al., 2004b). These observations point to nitrogenous waste excretion as a sensitive physiological function targeted by Cu exposure; however, the effect of Cu on nitrogenous waste excretion was not dose dependent as it was with osmoregulatory disturbances, implicating osmoregulatory failure as the cause of mortality induced by acute Cu exposure (Grosell et al., 2004b).

Unlike freshwater fish, very little is known about the mechanisms of ammonia excretion in marine fish, although rhesus proteins are likely involved (Hung et al., 2007; Nakada et al., 2007b). It is unclear, for example, whether CA is important for ammonia excretion in marine species as it appears to be for freshwater fish (see above), and therefore hard to speculate about the possible mechanisms of Cu-induced impairment of ammonia excretion. For at least a single species, the mudskipper, amiloride and ouabain are effective in blocking or reducing ammonia excretion, implying that Na$^+$/H$^+$ exchangers (NHEs) and Na$^+$/K$^+$-ATPase may be directly involved in transporting ammonia (Randall et al., 1999). However, caution is warranted for extrapolating these observations to marine fish in general as the mudskipper is unusual with respect to ammonia handling and highly tolerant of elevated ambient ammonia.

6.2.2.2. Elasmobranchs. Unlike teleost fish, which are constantly voiding ammonia as a waste product, marine elasmobranchs retain nitrogenous compounds for osmoregulatory purposes (Marshall and Grosell, 2005). As such, plasma urea concentrations in elasmobranchs exceed 200 mM owing to high production rates (Mommsen and Walsh, 1989, 1991) and efficient retention by the gill (Fines et al., 2001) and kidney (Schmidt-Nielsen et al., 1972); plasma ammonia concentrations are comparable to those of teleost fish (Wood et al., 1995, 2005; Grosell et al., 2003).

At relatively high Cu concentrations (~ 1 mg L^{-1}), increased branchial permeability results in elevated urea and trimethylamine oxide (TMAO) loss, leading to reduced concentrations of these two important osmolytes in the plasma of Cu-exposed elasmobranchs (De Boeck et al., 2007a). Even at lower Cu concentrations (~ 30–100 μg L^{-1}) plasma ammonia is elevated, as often reported for freshwater teleost fish. As in teleosts, elevated plasma ammonia is likely due to increased protein breakdown arising from elevated cortisol (Vanderboon et al., 1991). However, the situation in elasmobranchs is a little more complicated. In addition to the two components mentioned above, the conversion of ammonia to glutamine for urea production, a reaction mediated by glutamine synthetase, is important for setting circulating levels of ammonia (Mommsen and Walsh, 1989, 1991). Glutamine synthetase in the gill tissue has been proposed to be responsible for the low apparent branchial ammonia permeability in elasmobranchs (Wood et al., 1995), but could also be important for setting circulating levels of ammonia. Conversion of ammonia to glutamine in the gill tissue would act to suppress gill tissue ammonia levels and thus increase the clearance of plasma ammonia. Considering the very high gill Cu accumulation levels in elasmobranchs (40–50-fold increase over control levels) (Grosell et al., 2003; De Boeck et al., 2007a), it is possible that Cu acts to inhibit branchial glutamine synthetase and thus the clearance of plasma ammonia.

6.2.3. Acid–base Balance

Copper exposure appears to have varied effects on acid–base status in fish, ranging from no pH change despite increased plasma HCO_3^- in rainbow trout (Wilson and Taylor, 1993a) to a metabolic acidosis in Atlantic cod exposed to the same Cu concentrations (0.4 mg L^{-1}) (Larsen et al., 1997). A time-course study performed on the gulf toadfish exposed to approximately 3 mg Cu L^{-1} revealed an initial (day 2–3) compensated respiratory acidosis followed by a clear metabolic acidosis at 16 days of exposure (Grosell et al., 2011). A single study reports what appears to be a mixed metabolic and respiratory acidosis in the elasmobranch spiny dogfish exposed to approximately 1 mg Cu L^{-1} (De Boeck et al., 2007a).

Evidence of much more pronounced impacts on acid–base physiology comes from combined exposures to hypercapnia and Cu. Teleost fish compensate effectively for hypercapnia-induced acidosis by net retention of HCO_3^- and/or excretion of acid across the gills (Heisler and Neumann, 1977; Heisler, 1993). Retention of HCO_3^- and excretion of protons occur via exchange and cotransport proteins ultimately relying on cellular hydration of CO_2, mediated by CA (Fig. 2.4) (Claiborne et al., 1999; Edwards et al., 2005; Deigweiher et al., 2008). Fish exposed to combined hypercapnia and Cu display a greatly impaired ability to compensate for hypercapnia-induced respiratory acidosis, suggesting that one or more of the exchangers, cotransporters, or/and CA are targeted by Cu (Larsen et al., 1997). In addition, it cannot be dismissed that inhibition of Na^+/K^+-ATPase, which establishes and maintains ionic gradients to fuel transport of acid–base equivalents via exchangers and cotransporters, occurs in response to Cu exposure (see below).

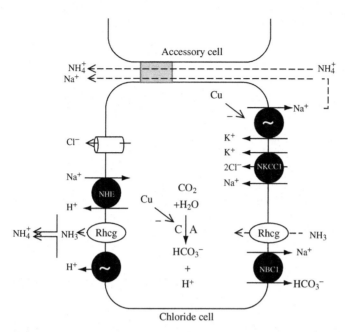

Fig. 2.4. Schematic and simplified representation of Cu-sensitive marine teleost gill transport processes relevant for salt balance, acid–base balance and ammonia excretion (see text for details) (Evans et al., 2005). Apical transporters relevant for salt and acid–base balance include the proton pump ($\sim$), Na^+/H^+ exchangers (NHEs), and Cl^- channels, while the secretory Na^+:K^+:2Cl^- cotransporter (NKCC1), the Na^+/K^+-ATPase ($\sim$; NKA) and the Na^+:HCO_3^- cotransporter contribute in the basolateral membrane. Multiple isoforms of NHEs, including NHE1, are likely present but have been omitted for clarity. In addition, putative ammonia excretion pathways are included. Strong evidence supports the role of basolateral Rhbg and apical Rhcg in ammonia excretion by freshwater teleosts (Wright and Wood, 2009) and some evidence suggests involvement of Rh proteins in marine teleosts (Nawata et al., 2010a). Paracellular efflux of ammonium has long been accepted as a pathway for excretion (Wilkie, 2002). In addition to inhibition of Na^+ and Cl^- excretion, presumably by inhibition of NKA, acid–base balance disturbance and inhibition of ammonia excretion is commonly reported for marine fish during Cu exposure. Inhibition of cellular carbonic anhydrase (CA) is a likely explanation for impaired acid–base balance and could explain reduced ammonia excretion assuming that transcellular ammonia excretion is occurring in marine fish. Although inhibition of CA provides a convenient explanation for responses to Cu, no studies to date have demonstrated inhibited CA activity as a result of Cu exposure.

In addition to exchange of acid–base equivalents by the gill, the marine teleost intestine contributes to overall acid–base status of the fish due to high base secretion rates involved in osmoregulatory processes (Grosell, 2006; Grosell et al., 2009b). Intestinal base secretion appears to be involved in dynamic acid–base regulation, at least during postfeeding events

(Taylor and Grosell, 2006, 2009; Taylor et al., 2007). and is impaired by Cu exposure (see below).

6.2.4. SALT AND WATER BALANCE: TELEOSTS

Marine teleost fish maintain roughly 300 mOsm in the extracellular fluids and are thus forced to drink seawater to replace water lost by diffusion to their concentrated environment. Water absorption by the intestine is coupled to active uptake of Na^+ and Cl^-, which is subsequently actively excreted by the gill (Marshall and Grosell, 2005). A consequence of this osmoregulatory strategy is that two organs, the gill and the gastrointestinal tract, are exposed directly to waterborne Cu. Copper concentration dependency of osmoregulatory disturbance and ultimate failure strongly suggest that the cause of mortality in marine fish exposure to toxic Cu levels is osmoregulatory failure.

6.2.4.1. Impact on Gills. Early attempts to identify the mechanism of Cu toxicity in marine fish demonstrated disruption of salt balance and that gill Na^+/K^+-ATPase of the European flounder is sensitive to Cu (Stagg and Shuttleworth, 1982b). A similar elevation of plasma Na^+ and Cl^- concentrations was observed for Atlantic cod and is consistent with inhibition of branchial Na^+/K^+-ATPase, which drives the active excretion of Na^+ and Cl^- across the gill (Fig. 2.4). Elevated plasma Na^+ and Cl^- concentrations were also observed in gulf toadfish exposed to Cu despite a lack of inhibition of branchial Na^+/K^+-ATPase activity in tissue homogenates (Grosell et al., 2011). However, as discussed above, these enzymatic assays, and assays for CA activity, are performed on homogenized tissue preparations, which involves dilution of the Cu originally present in the tissue. In addition, assay conditions are often optimized for maximal enzymatic activity, which may not truly reflect *in situ* conditions. It is therefore possible that *in situ* Na^+/K^+-ATPase activity was indeed inhibited to account for the elevated plasma Na^+ and Cl^- concentrations during Cu exposure. In any case, a recent report (Grosell et al., 2011) demonstrates unequivocally that Cu exposure results in reduced unidirectional Na^+ as well as Cl^- excretion by the gulf toadfish gill, with 90% inhibition of Cl^- excretion within 2 days of exposure and around 70% inhibition of Na^+ excretion within 5 days of exposure to 3 mg Cu L^{-1} (Fig. 2.2).

6.2.4.2. Impact on the Gastrointestinal Tract. In addition to the gills, the gastrointestinal tract plays a critical role in marine teleost osmoregulation, and owing to seawater ingestion the intestinal epithelium is exposed directly to Cu during waterborne exposures. As for the gills, toadfish intestinal tissue Na^+/K^+-ATPase activity is not affected by Cu exposure during 30 day

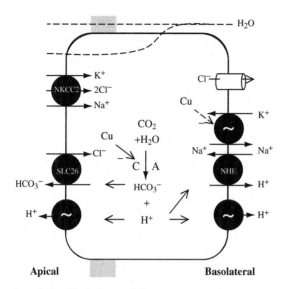

Fig. 2.5. Schematic and simplified diagram of transport processes relevant to salt and water absorption by the marine teleost intestine as well as likely targets for Cu exposure (see text for further details) (Grosell et al., 2009b). Transporters in the apical membrane includes two Cl^- uptake pathways, the absorptive $Na^+{:}K^+{:}2Cl^-$ (NKCC2), apical anion exchange via SCL26 family members as well as the proton pump. Basolateral transporters include Na^+/K^+-ATPase ($\sim$; NKA), Na^+/H^+ exchange (NHE), Cl^- channels, and the proton pump. Cytosolic carbonic anhydrase (CA) is critical for the net excretion of base and acid across the apical and basolateral membranes, respectively. Observations of reduced luminal HCO_3^- *in vivo* as well as reduced HCO_3^- secretion and Na^+ absorption by isolated intestinal tissue following Cu exposure point to inhibition of carbonic anhydrase and NKA, although Cu-induced inhibition of these enzymes has yet to be demonstrated.

exposures to as much as 3 mg Cu L^{-1} (Grosell et al., 2004b). Nevertheless, intestinal fluid ionic composition indicates that intestinal Cl^- absorption, which is central to water absorption, is reduced following 8 or more days of Cu exposure (Grosell et al., 2011). Based on intestinal fluid chemical composition, it appears that Na^+ absorption is not impacted by Cu exposure.

Absorption of Cl^- by the marine teleost fish intestine occurs in part via Cl^-/HCO_3^- exchange, which leads to the accumulation of high HCO_3^- concentrations in intestinal fluids (Grosell, 2006; Grosell et al., 2009b) (Fig. 2.5). With this and observations of apparently reduced intestinal Cl^- absorption in mind, it is not surprising that recent studies revealed reduced concentrations of HCO_3^- in intestinal fluids of toadfish exposed to Cu (Grosell et al., 2011), suggesting that Cl^-/HCO_3^- exchange in the intestinal tissue is inhibited by Cu. Indeed, direct measurements of HCO_3^- transport across isolated intestinal epithelia reveal that 5 and 16 days of exposure to 3 mg Cu L^{-1} results in complete inhibition of intestinal HCO$_3^-$ secretion (Grosell et al., 2011). In addition to serving solute-coupled water absorption, Cl$^-$/HCO$_3^-$ exchange is important for CaCO$_3$ precipitate formation, which acts to reduce luminal osmotic pressure and thereby facilitate water absorption (Wilson et al., 2002). Thus, Cu-induced inhibition of intestinal Cl$^-$/HCO$_3^-$ exchange impairs solute-coupled water absorption as well as water absorption facilitated by CaCO$_3$ precipitation.

The mechanism by which Cu inhibits intestinal Cl$^-$/HCO$_3^-$ exchange is unknown but the exchange process is fueled in part by Na$^+$/K$^+$-ATPase coupled to basolateral Na$^+$/H$^+$ exchange (Grosell and Genz, 2006), Na$^+$: HCO$_3^-$ cotransporters (Taylor et al., 2010) and the H$^+$ pump, and relies in part on cellular substrate provided from CA-mediated hydration of endogenous CO$_2$ (Grosell and Genz, 2006; Grosell et al., 2007b, 2009a) (Fig. 2.5). This topic clearly warrants further investigation.

6.2.4.3. Impact on the Kidney.   As for freshwater fish, there is no evidence for Cu-induced impairment of renal salt and water handling in marine teleosts or elasmobranchs. However, in contrast to freshwater teleosts, marine species seem to accumulate Cu in renal tissue, making it possible that renal salt and water handling could be compromised during Cu exposure (Grosell et al., 2003, 2004a).

6.2.5. SALT AND WATER BALANCE: ELASMOBRANCHS

One elasmobranch study reports no effects of Cu exposure ($<$100 μg L^{-1}) on plasma electrolytes (Grosell et al., 2003), while another study employing Cu concentrations from 500 to 1500 μg L^{-1} shows increased plasma Na$^+$ and Cl$^-$ concentrations, as has been reported for teleosts (De Boeck et al., 2007a). The likely explanation for this observation is increased gill permeability of Cu-exposed elasmobranchs, as also evident from loss of urea and TMAO. There is at present no evidence to suggest that the main salt-secreting organ of elasmobranchs, the rectal gland, is impacted by Cu exposure.

6.3. Acute Effects of Waterborne Copper Exposure at Intermediate Salinities

Urbanization is concentrated near the coasts and often near estuaries, and results in sources for Cu release into environments of intermediate and often fluctuating salinities. At the same time, estuarine areas are often nursing grounds for larval and juvenile fish. On this background there is a surprising paucity in the Cu literature dealing with toxicity at intermediate salinities. Overall acute Cu toxicity is less in seawater compared to freshwater, as would be expected simply from a cation competition perspective, and expectations of gradually increased tolerance to acute Cu exposure with increasing salinity seem justified. Indeed, Cu accumulation in killifish gill and liver decreased with increasing salinity from freshwater to 28 ppt as would be expected owing to cation competition (Blanchard and Grosell, 2005). However, for juvenile killifish, acute toxicity is highest at the extreme salinities (freshwater and seawater), with the greatest tolerance observed at intermediate salinities (Grosell et al., 2007a), illustrating a clear disconnect between cation competition and accumulation on one hand and acute toxicity on the other (Grosell et al., 2007a). Assuming that acute Cu toxicity leading to mortality is the consequence of osmoregulatory failure across all salinities, it is possible to explain a large part of the variation ($>90\%$) in Cu tolerance simply by considering the absolute Na^+ gradients (plasma [Na^+]/water [Na^+]) between the blood plasma and the surrounding water (Grosell et al., 2007a). Thus, the greater the Na^+ gradient, the greater the diffusive Na^+ loss (at low salinities) or gain (at high salinities) and thereby the greater the need for active Na^+ transport driven by the Na^+/K^+-ATPase. Copper-induced inhibition of the Na^+/K^+-ATPase thus presents less of a challenge to osmoregulation at intermediate salinities where the need for active Na^+ transport is minimal.

The driving force for Na^+ diffusion is not strictly a function of concentration gradients as the electrical potential difference across the gill epithelium also acts on the diffusion of Na^+. While the chemical Na^+ gradients were known at the time of the above analyses, only more recently were the electrical potential differences across the gill epithelium of the euryhaline killifish documented (Wood and Grosell, 2008). With both chemical and electrical gradients available for the killifish gill epithelium at a full range of ambient salinities, the electrochemical gradient (ECp) for the diffusive movement of Na^+, and thus the predicted sensitivity to Cu, can be calculated as described recently (Bielmyer and Grosell, 2010):

$$ECp = TEP - \left(\left(\frac{RT}{zF} \right) 2.303 \, \log \left(\frac{[Na_i^+]}{[Na_o^+]} \right) \right)$$

See Fig. 2.6 for details. Figure 2.6 illustrates the strong relationship between ECp and acute Cu toxicity and demonstrates that the driving force for diffusive Na^+ movement across fish gills is the main factor determining relative sensitivity to Cu across salinities. Indeed, physiology (as represented by ECp) rather than Cu speciation seems pivotal for Cu sensitivity when considering a full range of ambient salinities (Bielmyer and Grosell, 2010).

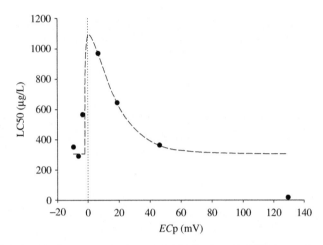

Fig. 2.6. Relationship between acute Cu 96 h LC50 for juvenile killifish tested at a range of salinities as a function of the calculated electrochemical potential (ECp) for Na^+ (see Section 6.3 for details). The ECp was calculated from measured ambient Na^+ concentrations during acute toxicity testing assuming plasma $[Na^+]$ of 150 mM at all salinities (Grosell et al., 2007a) and recently reported measured transepithelial potentials for adult killifish across a range of salinities (Wood and Grosell, 2008) using the following parameters: TEP, the transepithelial potential difference; $[Na_i^+]$ and $[Na_o^+]$), the Na^+ concentrations in the blood and water, respectively; z, the valence of the ion in question (1 for Na^+); and R, T, and F: the gas constant, absolute temperature, and Faraday's constant, respectively. Values for TEP used in the ECp calculations were 3.0, 2.0, −6.3, 0, 2.7, 10.1, and 19.6 for freshwater, and 2.5, 5, 10, 15, 22, and 35 ppt salinity, respectively. For freshwater and 2.5 ppt, TEP values from freshwater-acclimated fish were used, while TEP values from seawater-acclimated fish were used at all higher salinities as fish acclimated to intermediate salinities resemble seawater acclimated fish greatly with respect to TEP (Wood and Grosell, 2009). The curve was fitted to the data points using the Weibull, five-parameter non-linear regression function in SIGMAPLOT 8.0 and peaks at $ECp = 0$, where there is maximal Cu tolerance and no driving force for Na^+ diffusion. Reproduced from Bielmyer and Grosell (2010) with permission.

Although it seems clear that osmoregulatory disturbance is the mechanism of acute Cu toxicity leading to mortality across the full range of salinities, other physiological parameters discussed above are also likely to be affected in fish exposed to Cu at intermediate salinities. For example, ammonia excretion has been documented to be affected by Cu exposure at intermediate salinities (Blanchard and Grosell, 2006). Other physiological functions, including acid–base balance and olfaction as well as mechanoreception, may also be impacted by Cu exposure at intermediate salinities, areas clearly worthy of study.

Last, but not least, it is worth noting that estuarine environments are characterized not only by intermediate salinities but also by often dramatic salinity fluctuations. Undoubtedly, osmoregulatory adjustments associated with such salinity fluctuations are demanding for piscine inhabitants of estuaries and it is completely unknown how Cu exposure affects the ability to tolerate salinity change and fluctuations.

6.4. Chronic Waterborne Copper Exposure

The number of studies of truly chronic Cu toxicity including full life-cycle exposures or at least exposures of several sensitive (early) life stages is relatively low, whereas reports of exposure effects on juvenile or adult fish for 4–6 weeks are more abundant. Based on a comprehensive evaluation of early life-stage testing as indicative of truly chronic tests (McKim, 1977), the US EPA accepts 30 day toxicity testing on early life stages as a predictor of chronic toxicity in the absence of full life-cycle exposure data. The 4–6 week studies are often referred to as chronic but should be considered as "prolonged" exposures. Truly chronic as well as "prolonged" studies are discussed below.

6.4.1. ACCLIMATION

In the toxicology literature "acclimation" typically denotes a higher tolerance (i.e. increased LT50 or LC50) to an elevated, normally lethal, concentration of a toxicant arising from previous prolonged sublethal exposure to the toxicant (McDonald and Wood, 1993). Evidence of Cu acclimation was first reported in 1981 (Dixon and Sprague, 1981) and has since been confirmed repeatedly for freshwater fish (Buckley et al., 1982; McCarter and Roch, 1984). Acclimation to Cu has also been reported using a definition of "true acclimation" (Prosser, 1973), meaning a return to normal physiological conditions despite continued exposure. Examples include recovered appetite and growth (Lett et al., 1976; Buckley et al., 1982), and restoration of Na^+ homeostasis (Lauren and McDonald, 1987a,b). Restoration of Na^+ homeostasis in freshwater-acclimated rainbow trout during continued exposure involves increased synthesis of Na^+/K^+-ATPase to compensate for the Cu-induced inhibition of this enzyme (Lauren and McDonald, 1987a; McGeer et al., 2000b). In addition to an increased amount of Na^+/K^+-ATPase and recovered Na^+ uptake parameters, fresh-water rainbow trout appear to reduce diffusive Na^+ loss, presumably across the gills (Lauren and McDonald, 1987b). Branchial morphological adjust-ments during Cu exposure include an increased mitochondria-rich cell number and area of apically exposed mitochondria-rich cell membrane (Pelgrom et al., 1995). Such morphological changes may in part account for the changes in Na^+/K^+-ATPase activity discussed above. In addition to branchial modifications, a role for the kidney in restoration of Na^+ balance during continued exposure was demonstrated by increased renal Na^+ reabsorption following 3 and 30 days of Cu exposure contributing to fully recovered plasma Na^+ levels by day 30 (Grosell et al., 1998b).

Evidence for an acclimation response in seawater-acclimated European flounder has been reported (Stagg and Shuttleworth, 1982b), although Na^+ (and Cl^-) excretion by the gulf toadfish does not appear to recover during continued 30 days of exposure (Grosell et al., 2011). Although acclimation according to any definition has yet to be conclusively demonstrated for marine teleosts exposed to Cu, some evidence of compensatory responses is present. Following an initial reduction (see above), a compensatory increase in drinking rates is seen in gulf toadfish from 8 to 30 days of Cu exposure when disturbances of branchial Na^+ and Cl^- excretion and osmoregulation are evident (Grosell et al., 2004a). In addition, the ionic composition of intestinal fluids strongly suggests an enhanced uptake of Na^+ and water by the distal portions of the intestine during a 30 day exposure, which again would act to compensate for osmoregulatory distress caused initially by Cu exposure (Grosell et al., 2004b).

Despite acclimation which involves restoration of osmoregulatory balance and adjustments of homeostatic control of Cu (see below), chronic exposure results in a range of late effects discussed in the following.

6.4.2. IMPACT ON REPRODUCTIVE OUTPUT

Among the truly chronic studies, two stand out to identify reproductive output as being a highly sensitive and obviously ecologically relevant endpoint. An early study on brook trout revealed that reproductive output was impaired at 17 μg Cu L^{-1} (McKim and Benoit, 1971), while a later study on bluntnose minnow found reduced reproductive output at 18 μg Cu L^{-1} (Horning and Neiheisel, 1979) in soft and intermediate hardness waters, respectively. In a follow-up study, 9.4 μg Cu L^{-1} was found to be of no consequence to reproductive output of brook trout, suggesting that the effect threshold for chronic toxicity for brook trout lies between 9.4 and 17 μg Cu L^{-1} (McKim and Benoit, 1974).

Reproductive impairment in response to Cu exposure occurs at concentrations far below those required to induce mortality or reduce growth. Indeed, the 96 h LC50 for adult bluntnose minnow was reported to be 220–270 μg Cu L^{-1} and thus more than 10-fold higher than the effect threshold for reproductive impairment (Horning and Neiheisel, 1979). This difference in effect thresholds is not reflected by the acute versus chronic water quality criteria (see above) and although caution has to be employed when comparing across species, it appears that water chemistry (hardness) has little, if any, effect on chronic toxicity as assessed by reproductive output.

The mechanism by which Cu impairs reproduction is unknown but could be a simple outcome of reduced resource availability arising when energy must be devoted to dealing with Cu-induced physiological challenges rather than a direct endocrine effect. However, various endocrine systems appear to be influenced by Cu exposure (see below).

6.4.3. Impact on Sensory Systems

Olfactory impairment occurs almost immediately after onset of exposure and the effects seem to be highly persistent (see above). Rainbow trout hatchlings and 1-year-old fish exposed to Cu (20–40 µg L^{-1}) for 40 weeks showed no or only partial recovery of olfactory discrimination performance based on behavioral observations, and several weeks were required for recovery after termination of exposure (Saucier et al., 1991; Saucier and Astic, 1995). These observations are in agreement with similar studies on embryonic fathead minnow, which show persistent impairment of chemo-sensory function following termination of exposure. There is some evidence for partial recovery of olfactory discrimination ability despite continued exposure, at least at lower exposure concentration (Saucier and Astic, 1995). However, long delays in recovery following termination of exposure in developing fish (Saucier et al., 1991; Carreau and Pyle, 2005) suggest that this life stage is more vulnerable to impairment of olfaction induced by long-term exposure.

Intuitively, it seems clear that impaired sensory function (olfactory as well as mechanoreception) may translate into reduced ability to detect and avoid predators, and to locate prey and suitable spawning grounds, and result in problems with orientation and control of swimming speed and direction. However, at present there have been no reports of links between impaired sensory function, as demonstrated by laboratory studies on sensory neurons, and fish behavior or higher order effects such as altered predator–prey interactions or impacted reproductive success.

6.4.4. Impact on Immune Function

Copper is a frequently used fungicide in aquaculture, but Cu exposure has also been reported to increase susceptibility to viral and bacterial disease. Even short-term Cu exposure at low concentrations (9% of 96 h LC50 or 3–4 µg L^{-1}) has been reported to induce greater mortality due to *Vibrio anguillarum* infections in Chinook salmon and rainbow trout when exposed experimentally to the pathogen in holding water (Baker et al., 1983). A study using wild-caught European eel carrying *V. anguillarum* revealed that unexposed control fish remained healthy for up to 12 months whereas fish exposed to 30–60 µg Cu L^{-1} showed mortality within 50–120 days of exposure and symptoms consistent with *V. anguillarum* infection.

Indeed, blood collected from infected, Cu-exposed individuals and eels collected immediately after death contained *V. anguillarum*, while blood from unexposed eels contained no bacteria (Rødsæther et al., 1977). Similar responses have been reported from fish subjected to combined Cu exposure and pathogen infection via injection or brief exposures to high pathogen concentrations in water (Hetrick et al., 1979; Mushiake et al., 1984).

Macrophage function such as phagocytosis is affected by Cu exposure (Mushiake et al., 1985; Khangarot and Tripathi, 1991; Rougier et al., 1994), as are the blastogenic and antibody production responses (O'Neill, 1981; Anderson et al., 1989), decreasing the magnitude of both specific humoral and cellular immune responses (Dethloff et al., 1998). Consistent with this range of Cu-induced effects on the immune response of fish are observations of Cu-induced alterations in transcription of immune-system-related genes following either waterborne Cu exposure (Geist et al., 2007) or Cu injections (Osuna-Jimenez et al., 2009; Prieto-Alamo et al., 2009). In contrast to the majority of the above studies, which were all conducted on freshwater fish, the gene expression studies were performed on marine species and suggest that Cu affects the immune response of freshwater as well as seawater fish.

The exact mechanism by which Cu interacts with the immune system is unknown but an elegant experiment performed over 25 years ago may shed some light on this question: corticosteroid injection enhanced susceptibility to pathogens in a similar manner to Cu exposure, suggesting that the effects of Cu on immune responses by fish may be secondary to a more general stress response (Mushiake et al., 1984) often reported for Cu-exposed fish.

6.4.5. IMPACT ON STRESS RESPONSE

The hypothalamic–pituitary–interrenal (HPI) axis of freshwater fish is activated even during short-term Cu exposure at sublethal concentrations, resulting in elevated plasma cortisol levels (Dethloff et al., 1999; De Boeck et al., 2001; Teles et al., 2005) and, at least in some cases, hyperglycemia (Pelgrom et al., 1995). However, this elevation in plasma cortisol is transient in nature, with levels returning to control values within a few weeks despite continuous exposure (Dethloff et al., 1999; De Boeck et al., 2001). The return to control cortisol levels during continuous exposure, however, does not mean that the HPI axis is not impacted during prolonged Cu exposure. An elegant laboratory study of Cu impacts on the ability to mobilize a full stress response demonstrated that 30 days of exposure to relatively low Cu concentrations (30–80 µg L^{-1}) greatly reduced cortisol release in response to a subsequent acute and severe stress (Gagnon et al., 2006). Similar effects have been reported for fish collected from contaminated sites (Gravel et al., 2005). The ability of adrenocortical cells, isolated from Cu-exposed fish, to release cortisol in response to adrenocorticotropic hormone (ACTH) was enhanced rather than decreased compared to cells isolated from control fish, illustrating that reduced cortisol synthesis is not the reason for reduced stress responses *in vivo* (Gagnon et al., 2006). These observations led to the suggestion of reduced pituitary ACTH production and release, and/or altered ACTH receptor function in adrenocortical cells of Cu-exposed fish (Gravel et al., 2005). Finally it cannot be dismissed that Cu acts to alter cortisol turnover by affecting clearance of plasma cortisol. Regardless of the mechanism(s), it is clear that Cu potentially affects the ability of freshwater fish to mobilize a full stress response to additional stressors and that this loss of ability is not related directly to a reduced steroidogenic capacity. Whether Cu affects the HPI axis in marine fish during Cu exposure is unknown, although it seems likely.

6.4.6. IMPACT ON DEVELOPMENT, GROWTH, AND SURVIVAL

Growth inhibition appears to occur at concentrations somewhat below or at mortality threshold concentrations (Buckley et al., 1982; Hansen et al., 2002; Kamunde et al., 2005; Niyogi et al., 2005; Besser et al., 2007). However, at least in some cases, growth inhibition occurs at concentrations of less than 20% of the corresponding 96 h LC50 (Hansen et al., 2002), although other studies report no growth inhibition even at concentrations as high as 80% of 96 h LC50 values (McGeer et al., 2002). Increased metabolic load (see Section 6.1.5) and/or possibly reduced food conversion efficiency likely contribute to the observed Cu-induced growth inhibition, but transiently reduced appetite appears to also be a contributing factor (Buckley et al., 1982). Larvae and early juveniles are more sensitive to Cu exposure than embryos, as assessed by growth and survival for a relatively high number of species tested (Sauter et al., 1976; McKim et al., 1978), and can reasonably be expected to be more sensitive than adults. Indeed, early life-stage testing with a duration of 60 days has been demonstrated to provide a fair surrogate for truly chronic exposures (McKim and Benoit, 1974; McKim, 1977; McKim et al., 1978), although it does not take into account the possibility of reduced reproductive output, which is of obvious ecological relevance and appears to be highly sensitive (McKim and Benoit, 1971; Horning and Neiheisel, 1979).

6.4.7. OTHER EFFECTS

Dose-dependent depletion of the two major monoamine neurotransmitters, serotonin [5-hydroxytryptamine (5-HT)] and dopamine (DA), in the telencephalon has been reported for carp exposed to sublethal Cu concentrations. In addition, Cu caused a depletion of DA in the hypothalamus and brain stem (De Boeck et al., 1995b). The serotonergic system is integrally linked with the HPI axis (discussed above) and changes in the ratios of 5-hydroxyindoleacetic acid (5-HIAA), the main 5-HT metabolite, to 5-HT are often associated with general stressors. Increases in the 5-HIAA:5-HT ratio are generally taken as evidence for increased activities in the brain 5-HT system (Winberg and Nilsson, 1993) caused by activation of the HPI axis. While it is therefore appealing to hypothesize that the impacts of Cu on central nervous system 5-HT metabolism are secondary to the stress response often induced by Cu (see Section 6.4.2), experiments on carp showing reduced 5-HT levels rather than increased 5-HIAA concentrations suggest a reduced 5-HT synthesis rate rather than increased turnover (De Boeck et al., 1995b). At least in mammals, 5-HT synthesis is decreased during even mild hypoxia (Katz, 1981; Prioux-Guyonneau et al., 1982; Freeman et al., 1986) which, based on blood lactate concentrations, likely occurred in the Cu-exposed carp showing reduced 5-HT synthesis (De Boeck et al., 1995b). Regardless of the reason for reduced 5-HT levels in the brain of Cu-exposed fish, depletion of brain 5-HT could possibly explain the reduced appetite of Cu-exposed fish (Johnston et al., 1992).

6.5. Chronic Dietary Exposure

Elevated environmental Cu levels not only result in direct waterborne exposure to fish, but may lead to accumulation of Cu in invertebrates and other fish prey items, ultimately resulting in potential dietary exposure. Indeed, several reports of elevated prey Cu (and other metals) concentrations in contaminated environments illustrate that dietary exposure and thus toxicity may occur.

In most cases, these studies were performed with natural diets contaminated with Cu as well as other metals, and are thus challenging to interpret from a Cu effect perspective. However, the above studies with field-collected invertebrate diets, in most cases, revealed significant growth inhibition typically at exposure concentrations lower than those observed to cause effects in studies using artificially formulated diets (Hansen et al., 2004). The interpretation from this observation is that naturally incorporated metal is more available for uptake, and thus has more deleterious effects than metals spiked into artificial diets, an interpretation that is in agreement with a number of studies on invertebrates (Hook and Fisher, 2001, 2002; Bielmyer et al., 2006). Nevertheless, the majority of dietary Cu toxicity studies on fish have been and continue to be performed using artificial diets spiked with Cu salts.

Dietary Cu toxicity to fish has been discussed extensively in an excellent review by Clearwater et al. (2002). They found that the relationship between dietary Cu exposure and induced effects was much better predicted when dietary exposure is expressed as dietary Cu dose rather than dietary Cu concentration. For example, the effect threshold for rainbow trout was estimated by Clearwater and co-workers to be approximately 44 mg Cu kg^{-1} body weight per day, while the effect threshold for Atlantic salmon fry and small parr smolts was estimated at 15 mg Cu kg^{-1} body weight per day. Channel catfish appeared to be the most sensitive fish tested by the year 2002, with sublethal effects at 0.4–0.9 mg Cu kg^{-1} body weight per day. However, similar studies did not demonstrate toxicity when channel catfish were given a dietary dose of 1 mg kg^{-1} body weight per day, illustrating that additional factors to species differences and dietary dose influence susceptibility to dietary Cu. Such factors could include diet composition, feeding regime, and water chemistry.

Since the 2002 review (Clearwater et al., 2002), additional studies have shown effects of Cu on rainbow trout at doses of 42 mg Cu kg^{-1} day^{-1} (Hansen et al., 2004) and 15 mg Cu kg^{-1} day^{-1} (Campbell et al., 2002, 2005), possibly suggesting that rainbow trout and Atlantic salmon show similar sensitivity to dietary Cu exposure. Nevertheless, species differences in sensitivity as suggested by Clearwater et al. (2002) likely exist since recent reports of a marine rockfish (Kang et al., 2005) and zebrafish (Alsop et al., 2007) show sublethal effects during exposure to doses as low as 2 and 2.7 mg Cu kg^{-1} day^{-1}, respectively.

6.5.1. GROWTH AND MORTALITY

The majority of studies investigating dietary Cu toxicity report reduced growth (Lanno et al., 1985; Baker, 1998; Berntssen et al., 1999a,b; Clearwater et al., 2002; Hansen et al., 2004; Kim and Kang, 2004; Kang et al., 2005; Shaw and Handy, 2006). Reduced growth rates during dietary Cu exposure, in cases of high dietary Cu concentrations, may be attributed to food refusal (Lanno et al., 1985) or at least decreased food intake (Baker, 1998; Shaw and Handy, 2006). However, reduced growth is often reported despite lack of effects on food intake or in parallel to reduced feeding, and seems to be caused by a reduction in conversion of food energy to biomass (Lanno et al., 1985; Hansen et al., 2004; Kang et al., 2005). The basolateral membrane of the intestinal epithelium appears to be rate limiting for Cu uptake (Clearwater et al., 2002) and Cu consequently accumulates in the intestinal tissue during dietary exposures. With this in mind, it is possible that reduced food conversion is caused by effects of Cu on digestive enzymes (Li et al., 2007) and/or reduced nutrient absorption.

6.5.2. METABOLIC EFFECTS

At least two studies suggest that dietary Cu exposure, albeit at high concentrations (500–730 mg kg^{-1}), may impart metabolic costs. Although the resting metabolic rate of rainbow trout exposed to Cu was not elevated above control values, Cu-exposed fish swam less and thus covered less distance relative to oxygen consumed (Handy et al., 1999). Copper-fed fish that swam at the same velocity as control fish had higher oxygen consumption rates (Campbell et al., 2002), and generally swam at lower speeds, indicating a metabolic effect of dietary Cu. In both the above studies, feeding and growth rates were not affected by Cu and it thus appears that the reduced swimming activity in the laboratory compensates, at least in part, for the increased metabolic demand, but this may have severe consequences on a fish's natural environment. Indeed, the impacts of dietary Cu exposure on subtle but likely significant behavioral parameters illustrate that such effects may affect natural populations. A recent study revealed that social interactions among Cu-exposed rainbow trout were altered compared to controls and those social interactions between control fish and Cu-exposed fish were heavily biased in favor of the unexposed controls (Campbell et al., 2005).

Based on the above observations, it appears that dietary Cu acts as a metabolic loading stressor much as is the case for waterborne exposures, although a systematic analysis of loading versus limiting stress during dietary Cu exposure remains to be performed.

6.5.3. OXIDATIVE STRESS

Mechanisms of Cu-induced oxidative stress are discussed in some detail above (Section 6.1.6). Evidence of oxidative stress in intestinal and hepatic tissues, the two tissues showing most pronounced Cu accumulation during dietary exposure, has been reported for freshwater as well as marine species (Overnell and McIntosh, 1988; Berntssen et al., 1999a; Hoyle et al., 2007). In addition to lipid peroxidation, elevated intestinal and hepatic metallothionein levels are commonly observed during dietary exposure (Overnell and McIntosh, 1988; Handy et al., 1999; Berntssen et al., 1999a). While metallothionein is generally accepted as a metal binding protein, it also confers protection against ROS and thus protects against oxidative stress (see Section 6.1.6.2).

6.5.4. OTHER EFFECTS

Dietary Cu exposure has been reported to deplete hepatic glycogen and lipid reserves (Hoyle et al., 2007), which is in agreement with observations of reduced food intake and an apparent increased metabolic cost of exposure. However, marked hepatic lipidosis (increased cellular fat stores) has also been observed in fish exposed to dietary Cu (Shaw and Handy, 2006), suggesting hepatotoxicity, as also evident from observations of direct hepatic necrosis (Hansen et al., 2004). In addition, morphological changes in the gut are often observed during and following dietary Cu exposure (Woodward et al., 1995; Berntssen et al., 1999a; Kamunde et al., 2001).

A recent and very interesting study reports effects of dietary Cu on the retinoid system in zebrafish (Alsop et al., 2007). Retinal acts as the chromophore of rhodopsin and, in fish, is deposited into eggs for embryonic development (Irie and Seki, 2002). In addition, retinoids have antioxidant properties (Caiccio et al., 1993) and can thus be depleted by toxicants, such as Cu, that may lead to ROS. Alsop et al. (2007) revealed that dietary Cu exposure resulted in significant depletion of retinoids, but this was without impact on reproductive output.

One obvious void in the literature on dietary Cu toxicity, with the exception of the study by Alsop et al. (2007), is a lack of systematic assessment of reproductive effects, and no full life-cycle tests have been performed to date. Considering the apparent potential for dietary Cu exposure to cause metabolic loading stress, it seems likely that altered energy allocation leading to reduced reproductive output may occur, especially in fish fed a fixed ration and thereby being calorie restricted.

7. ESSENTIALITY OF COPPER

Copper is an essential element for all aerobic organisms since its redox potential is utilized by mitochondrial cytochrome c oxidase, and Cu acts as a cofactor for a high number of other enzymes (Solomon and Lowery, 1993). Evidence for an important role for Cu as a micronutrient in teleost fish comes from observations of reduced growth under conditions of low ambient and dietary Cu (Ogino and Yang, 1980; Gatlin and Wilson, 1986; Kamunde et al., 2002b). More recently, the use of the zebrafish model system has allowed for unequivocal demonstrations of Cu essentiality in developing teleost fish embryos. The high-affinity Cu transporter 1 (Ctr1), which is a highly conserved transmembrane protein mediating internalization specifically of Cu, is maternally loaded in zebrafish and expressed abundantly in brain and the developing intestine. Antisense morpholino knockdown of Ctr1 in developing zebrafish results in early larval mortality, illustrating the importance of Cu, and specifically uptake via Ctr1, for central nervous system development (Mackenzie et al., 2004).

A recently described zebrafish mutant, *calamity*, is defective in the zebrafish orthologue of the Menkes disease gene (atp7a). The human ATP7A gene product is a Cu-transporting P-type ATPase required for Cu absorption and homeostasis (Lutsenko and Petris, 2002; Lutsenko et al., 2007). Human patients with Menkes disease develop a series of severe dysfunctions related to Cu deficiency normally resulting in death during early infancy (Kaler, 1998). The *calamity* zebrafish mutant displays a phenotype similar to human Menkes patients, but Cu metabolism and normal development in *calamity* embryos can be restored by injection of human ATP7A RNA (Mendelsohn et al., 2006). Furthermore, a complete rescue of the Cu-deficient defects of *calamity* was seen with the generation of a wild-type zebrafish atp7a protein (Madsen et al., 2008). These observations demonstrate a role for atp7a in zebrafish Cu acquisition and homeostasis as well as the essentiality of Cu for embryonic development of teleost fish.

8. POTENTIAL FOR BIOCONCENTRATION AND BIOMAGNIFICATION OF COPPER

In general, biomagnification, defined as increasing tissue burdens over three trophic levels, is not considered a major factor for Cu (Lewis and Cave, 1982; Suedel et al., 1994), presumably owing to the relatively strong homeostatic control of this essential element (see below). Bioconcentration factors (accumulation from dissolved sources) and bioaccumulation factors (accumulation from dissolved and dietary sources combined) for Cu in freshwater organisms in general are inversely related to exposure concentrations (McGeer et al., 2003; DeForest et al., 2007). These observed general patterns most likely stem from carrier-mediated Cu uptake, which is saturable, and thus explain disproportional uptake as a function of concentration in the water and diet. In addition, continuous Cu excretion and stimulated Cu excretion at higher exposure concentrations may contribute to the observed pattern, as discussed below.

9. CHARACTERIZATION OF UPTAKE ROUTES

In contrast to higher vertebrates which rely exclusively on dietary Cu, or at least intestinal uptake, to meet requirements for this element, fish have an additional route of Cu uptake directly from the water across the gill epithelium. The majority of studies of Cu homeostasis in fish are performed in freshwater and the following discussion is therefore focused on freshwater fish unless otherwise stated.

9.1. Gills

Under normal dietary and waterborne Cu concentrations the gill plays a minor, yet significant, role in Cu acquisition, providing approximately 10% of the required Cu. However, when dietary Cu concentrations are reduced, the gill contributes more than 60% of whole-body Cu uptake and is thus very important for Cu homeostasis. Conversely, during exposure to elevated dietary Cu, uptake from the water across the gills accounts for less than 1% of total Cu uptake (Miller et al., 1993; Kamunde et al., 2002b). Reduced dietary Cu concentrations result in an elevated gill Cu uptake capacity, suggesting that Cu uptake pathways across the gill are somehow regulated based on organismal Cu status (Kamunde et al., 2002b). Similarly, exposure to elevated waterborne Cu results in reduced branchial Cu uptake (Grosell et al., 1997, 1998a; Kamunde et al., 2002a) contributing to maintaining Cu homeostasis and protecting the gill epithelium against toxic effects of Cu. In contrast, prolonged exposure to elevated waterborne Cu does not seem to reduce dietary Cu uptake (Kamunde et al., 2002a). Thus, it appears that while intestinal uptake plays a quantitatively important role in Cu acquisition, regulated branchial Cu uptake plays a role in dynamic regulation of Cu homeostasis.

The affinity constants (K_m) for gill Cu uptake are low (meaning high affinity) and thus difficult to measure, and require the use of radioisotopes owing to the significant background levels of Cu in gill tissue. The few measurements to date suggest that freshwater fish gill Cu affinity is less than $1 \mu g \, L^{-1}$ (i.e. $<20 \, nmol \, L^{-1}$) and certainly within environmentally realistic Cu concentrations (Grosell and Wood, 2002; Taylor et al., 2003), although this is likely to be influenced by water chemistry. Furthermore, it appears that gill Cu uptake affinities are similar among species; however, they are sensitive to general water chemistry, with low ionic strength water conferring a higher Cu affinity (Taylor et al., 2003).

Although Cu uptake from the water clearly occurs in marine fish (Stagg and Shuttleworth, 1982a,b; Grosell et al., 2003, 2004a; Blanchard and Grosell, 2005), nothing is known about gill Cu uptake affinity and branchial contributions to overall Cu homeostasis, and only little is known about how this may be regulated based on Cu status.

In general, the onset of waterborne Cu exposure results in a rapid accumulation of Cu in gill tissue (within a few hours) after which a second steady state is often reached, with elevated, yet stable gill Cu levels (Grosell et al., 1996, 1997, 1998a). Copper uptake continues at reduced rates, even when this second steady state is reached, in situations where entry of Cu from the water across the apical membrane into the gill cells is balanced by transfer of Cu across the basolateral membrane to the circulatory system.

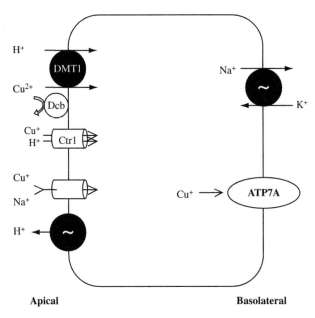

Fig. 2.7. Schematic presentation of suggested Cu uptake pathways across (freshwater) fish gills (see text for further details). The divalent cation transporter (DMT1) and the high-affinity copper transporter (Ctr1) are both favored by the acidic boundary layer provided by H^+ extrusion via the proton pump or other acid-excreting transporters. In addition, Cu uptake occurs via apical Na^+ channels when ambient Na^+ concentrations are low. Ctr1 and likely the Na^+-sensitive pathway favor uptake of Cu^+ rather than Cu^{2+}. Copper reductase activity remains to be demonstrated for the fish gill but the duodenal cytochrome b protein (Dcb) acts to reduce Cu in the mammalian intestine and a similar protein is hypothesized to facilitate Cu^+ uptake by fish gills. Export of Cu across the basolateral membrane is likely via the fish orthologue to the human Menkes protein ATP7A.

It thus appears that the basolateral membrane is rate-limiting for gill Cu uptake, at least early during exposure, but that regulation of the apical and possibly basolateral transport steps is initiated relatively rapidly (Grosell et al., 1996, 1997, 1998a). Apical and basolateral transport events involved in gill Cu uptake are discussed in the following and illustrated in Fig. 2.7.

At least two and possibly three apical Cu uptake pathways are functional in the freshwater fish gill. Nothing is known about branchial apical Cu uptake pathways in marine fish. On the basis of competitive interactions between Cu and Na^+, inhibition of Cu uptake by the proton pump inhibitor, bafilomycin, and the Na^+-channel blocker, phenamil, it appears that Cu uptake occurs via an apical Na^+ channel in rainbow trout (Grosell and Wood, 2002). The nature of the Na^+ channel involved in Na^+ (and Cu) uptake by freshwater fish remains elusive. Based on residual Cu uptake even in the presence of high Na^+ concentrations, and differential inhibition of Cu and Na^+ uptake by phenamil, it appears that both a Na^+-sensitive and a Na^+-insensitive uptake pathway exist (Grosell and Wood, 2002). Both Cu uptake pathways exhibit high affinity, with half maximal transport (K_m) occurring at 0.5–1 $\mu g\ L^{-1}$ (i.e. <20 nmol L^{-1}), well within environmentally realistic levels even for uncontaminated waters. At low Na^+ concentrations the Na^+-sensitive pathway dominates Cu uptake, but with an IC50 of 104 $\mu M\ Na^+$, the Na^+-insensitive uptake pathway dominates in waters of higher ionic strength (Grosell and Wood, 2002). The presence of Na^+-sensitive and Na^+-insensitive branchial Cu uptake pathways has yet to be demonstrated for other species, but has been confirmed for rainbow trout in several studies (Kamunde et al., 2003, 2005; Pyle et al., 2003; Sloman et al., 2003b).

The nature of the carrier(s) responsible for Na^+-insensitive Cu uptake is unknown, but at least two candidates can be identified. The high-affinity, high-specificity Cu transporter Ctr1 is an obvious candidate for Cu uptake, and is expressed in fish gill tissue (Mackenzie et al., 2004; Minghetti et al., 2008; Craig et al., 2009, 2010) However, waterborne exposure appears to have no impact on Ctr1 expression in gill tissue (Minghetti et al., 2008; Craig et al., 2010), suggesting that Ctr1 is not involved in regulation of Cu uptake across the gill tissue, or that regulation by Ctr1 is not at the transcriptional level. An additional potential carrier mediating Na^+-insensitive Cu uptake is the divalent metal transporter (DMT1), which is also expressed in fish gills (Bury et al., 2003). DMT1 is relatively promiscuous with respect to substrate and has been demonstrated to transport Cu (Gunshin et al., 1997; Knopfel et al., 2005), although this has yet to be verified for fish DMT1s.

Of these three putative Cu uptake carriers in the apical membrane, Ctr1 and presumably the Na^+-channel transport reduced Cu^+ rather than Cu^{2+}, whereas DMT1 would facilitate Cu^{2+} uptake. Free ionic Cu is present predominantly as Cu^{2+} in natural waters and uptake via the Na^+-sensitive pathway and part of the Na^+-insensitive pathway mediated by Ctr1 may therefore require the presence of a metal reductase in the apical membrane. A metal reductase from fish gills has yet to be described, but the mammalian duodenal cytochrome b protein has been reported to act as a ferric and cupric reductase in mammalian cells and acts to facilitate uptake of Fe and possibly Cu via DMT1 and Ctr1, respectively (Wyman et al., 2008; Scheiber et al., 2010). Ferric reductase activity has been demonstrated in intestinal tissue of rainbow trout (Carriquiriborde et al., 2004) and examining the presence of similar reductase activity in the apical membrane of fish gills and its role in Fe and Cu uptake offers an exciting area for future study.

In mammals, two Cu P-type ATPases, ATP7A and ATP7B, serve delivery of Cu into the Golgi for incorporation into cuproenzymes, but are also involved in secretion of Cu from cells (Mercer and Llanos, 2003; Mercer et al., 2003). Under normal cellular Cu load, the Cu-ATPases reside in the trans-Golgi network, but excess Cu induces trafficking of ATP7A and ATP7B to the plasma membrane and the vesicular secretory compartment, respectively (Mercer et al., 2003). In mammals, ATP7A is expressed in most tissues but is absent or expressed at low levels in the liver where ATP7B is abundant. Excess cellular Cu in polarized cells, like the intestinal epithelium, causes ATP7A to target the basolateral membrane and thereby facilitate Cu export from the enterocytes to the blood. Fish Cu ATPase orthologues, atp7a and atp7b, have recently been identified from a marine fish and share similar tissue distribution with mammals and appear to serve similar functions (Minghetti et al., 2010). Notably, atp7a is expressed in gill tissue and could contribute the basolateral step in branchial Cu uptake. Earlier evidence indicated that transport of Cu across the basolateral membrane of trout gills was carrier mediated, exhibited saturation kinetics, and was sensitive to the P-type ATPase inhibitor, vanadate (Campbell et al., 1999). In addition, silver-stimulated ATPase activity has been observed in isolated basolateral membrane vesicles (Bury et al., 1999), and Ag uptake by such vesicles is inhibited by Cu but not other metals in a dose-dependent manner (Bury et al., 2003). These latter observations have been interpreted as silver being transported by a basolateral Cu ATPase in the basolateral membrane of rainbow trout gills (Bury et al., 2003).

While excess Cu can alter expression of Cu ATPase in some tissues of the sea bream, branchial atp7a transcription does not appear to be altered by waterborne or dietary exposure. While this may suggest that atp7a is not responsible for regulated branchial Cu uptake in fish, it should be noted that trafficking of existing proteins between the trans-Golgi and the plasma membrane rather than transcriptional changes is the mode of regulation of this protein, at least in mammals. The possibility of similar regulation in fish gills offers an exciting area for further study.

9.2. Gastrointestinal Tract

As discussed in Section 9.1, above, the gastrointestinal tract dominates whole-body Cu uptake under normal conditions and certainly under conditions of elevated dietary Cu. Overall, Cu assimilation efficiency by the gastrointestinal tract is relatively low ($<50\%$) (Clearwater et al., 2000; Kamunde and Wood, 2003; Kjoss et al., 2005b; Nadella et al., 2006a) and occurs primarily in the distal intestinal segments (Handy et al., 2000; Nadella et al., 2006a), although considerable uptake by the gastric mucosa has also been reported (Nadella et al., 2006a, 2011). A study comparing uptake in all regions of the gastrointestinal tract found that the anterior region, the pyloric ceca, accounted for the majority of uptake. Similar conclusions were reached after experiments with isolated intestinal segments showing highest transepithelial Cu transport rates across the anterior intestine (Ojo and Wood, 2007; Ojo et al., 2009). These conflicting results likely are attributable to the fact that the study by Clearwater and co-workers employed a protocol of injecting Cu in a saline solution without food into the stomach, whereas the studies by Ojo and co-workers employed intestinal sac preparations with saline and no food present in the lumen. Under more physiological conditions, the presence of food in the intestinal lumen will trigger the release of bile with very high Cu concentrations from the gall bladder into the anterior region of the intestine, masking any uptake in this intestinal region to yield no or limited net uptake of Cu (Nadella et al., 2006a). Gall bladder bile contains up to 20 μg Cu mL^{-1} (Grosell et al., 1998a,b, 2001) and even release of low volumes of bile into the intestinal lumen can therefore influence net flux rates considerably. Nevertheless, it is clear from studies on isolated intestinal tissue that the anterior intestinal region is capable of unidirectional Cu uptake at high rates (Nadella et al., 2006b), rates that may be balanced by biliary secretion in intact, postprandial fish.

9.2.1. INTESTINAL UPTAKE

A diagram of putative intestinal Cu transport pathways is displayed in Fig. 2.8. As for Cu uptake by the gill, the rate-limiting step for intestinal uptake appears to be the basolateral membrane, as evident by fast and substantial Cu accumulation on intestinal tissues during dietary exposure preceding accumulation in internal tissues (Berntssen et al., 1999a; Handy et al., 2000; Clearwater et al., 2000, 2002; Kamunde et al., 2001). Both atp7a and atp7b are highly expressed in the intestinal tissue of the sea bream, but only atp7a responds to Cu exposure at the transcriptional level (Minghetti et al., 2010). Dietary exposure results in a substantial reduced atp7a expression while waterborne exposure leads to a robust expression increase. As for dietary exposure, waterborne Cu exposure results in Cu accumulation in intestinal tissues of marine fish due to the intake of waterborne Cu with ingestion of seawater for osmoregulatory purposes (Grosell et al., 2003, 2004a) (see Section 6.2.4.2). Thus, increased atp7a transcription during waterborne exposure is consistent with the role of atp7a (and atp7b) in cellular Cu excretion during Cu excess, whereas the reduced expression observed during dietary exposure seems counter-intuitive. However, trafficking of the atp7a protein in addition to the transcriptional responses is possible considering that this is the primary mode of regulation in mammals (Mercer et al., 2003), and a role for atp7a in intestinal Cu uptake seems likely given the dynamic transcriptional response to Cu exposure.

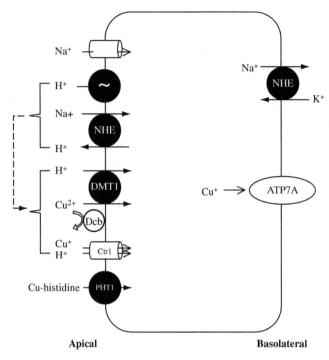

Fig. 2.8. Schematic presentation of suggested Cu uptake pathways across the intestinal epithelium of teleost fish (see text for further details). Three distinct transporters allow for uptake across the apical membrane. The divalent metal transporter (DMT1) as well as the high-affinity copper transporter (Ctr1) appear to contribute to Cu uptake across the apical membrane. The apparent sodium-dependent Cu uptake by fish intestine is likely due to increased acid excretion across the apical membrane via the Na$^+$/H$^+$ exchanger (NHE) and the proton pump in the presence of high luminal sodium since the activity of both these transporters is stimulated by reduced extracellular pH (Nadella et al., 2007). In addition to these two Cu transporters, a peptide/histidine transporter (PHT1) has been suggested to be responsible for high-affinity uptake of Cu–histidine complexes (Glover and Wood, 2008a,b). Export of Cu across the basolateral membrane is likely via the fish orthologue to the human Menkes protein ATP7A.

In addition to Cu ATPases, a role for a putative Cu:Cl$^-$ symporter has been proposed by Handy and co-workers based on observations of DIDS (anion transport inhibitor) sensitivity and correlation between the rate of Cu appearance in serosal solutions and mucosal Cl$^-$ concentrations (Handy et al., 2000). While metal ion:Cl$^-$ symporters have been observed in other cell types (Bury et al., 2003), a range of alternative interpretations for the observations presented by Handy and co-workers is possible since the experimental manipulations must have resulted in altered cellular acid–base status and associated changes. The possibility of a $Cu:Cl^-$ symporter in the basolateral membrane of teleost fish intestinal tissue requires further examination.

Apical entry of Cu into the intestinal epithelial cells appears to be complicated and involves several parallel pathways. Interactions between sodium and intestinal Cu uptake were examined to demonstrate a sodium-dependent system (Kjoss et al., 2005a; Nadella et al., 2007) rather than the sodium-sensitive system known for the gill. For the trout intestine, higher luminal sodium stimulates Cu uptake and similar observations have been reported for the intestine of mammals (Wapnir, 1991). A comprehensive examination of the linkage between sodium and Cu absorption across the apical membrane of the intestinal epithelium revealed that coupling is likely indirect (Nadella et al., 2007). The most parsimonious interpretation of the studies by Nadella and co-workers is that elevated luminal sodium facilitates H^+ extrusion across the apical membrane by NHE or by an apical H^+-ATPase, which may fuel Na^+ uptake via an apical Na^+ channel. Elevated H^+ concentrations near the apical surface possibly stimulate Cu uptake via Ctr1 and/or DMT1, both of which are stimulated by protons. While there is no evidence to date for the presence of NHEs or Na^+ channels in the apical membrane of freshwater or marine teleost intestines, thermodynamically both could be involved in intestinal Na^+ uptake (Grosell, 2010) and NHEs are certainly involved in this capacity in many vertebrates, including mammals (Kiela et al., 2006). Recently, apical H^+-pump activity has been demonstrated in marine fish (Grosell et al., 2007b, 2009a,b; Wood et al., 2010) and could thus contribute to acid-stimulated Cu absorption by Ctr1 and/or DMT1.

Evidence suggests that Ctr1 as well as DMT1 operate in parallel to provide apical Cu entry, but no effects were observed by addition of the reducing agent ascorbate (Nadella et al., 2006b). These observations may be explained by switching between Ctr1, which utilizes Cu^+ as substrate, and DMT1, which transports Cu^{2+}, as the oxidation state of Cu changes such that DMT1 dominates under oxidizing conditions while Ctr1 would dominate under reducing conditions. An alternative interpretation is that endogenous reductase activity in the apical membrane controls the oxidation state of Cu regardless of luminal reducing (or oxidizing) agents. The presence of a duodenal cytochrome b protein, which has been reported to act as a ferric and cupric reductase in mammalian cells and acts to facilitate iron and possibly Cu uptake via DMT1 and Ctr1, respectively, has been demonstrated (Wyman et al., 2008; Scheiber et al., 2010) and could contribute to Cu uptake by fish. Indeed, ferric reductase activity has been reported for fish intestinal tissue (Carriquiriborde et al., 2004).

An additional Cu uptake pathway, likely involving absorption of a histidine–Cu complex, has been described recently (Nadella et al., 2006b; Glover and Wood, 2008a,b). Isolated intestinal segments in sac preparations as well as isolated brush border membrane vesicles from freshwater-acclimated rainbow trout show L-histidine, but not D-histidine-stimulated high-affinity Cu uptake (Nadella et al., 2006b; Glover and Wood, 2008a). Although intact intestinal tissue in sac preparations suggests interaction between L-histidine, sodium and Cu uptake, experiments with isolated brush border membranes have revealed that this L-histidine-stimulated pathway is distinct from the sodium-dependent uptake pathways discussed above (Glover and Wood, 2008a). The amino acid transporter involved in the L-histidine-stimulated Cu uptake shows some amino acid substrate selectivity, and results indicate that Cu uptake in the presence of L-histidine occurs by a transport system distinct from that responsible for Cu uptake alone, as well as distinct from the transport system mediating uptake of histidine alone. The amino acid substrate selectivity is unusual and although the nature of the histidine/Cu transporter remains elusive, observations suggest that a fish orthologue to the peptide/histidine transporter PHT1 (SLC15A4) is likely responsible for the uptake of histidine/Cu complexes. While PHT1 is sodium independent, it displays pH sensitivity (Glover and Wood, 2008b). The pH dependence of the histidine–Cu complex uptake pathway was not examined for the trout intestine. Nevertheless, the pH dependence of PHT1 suggests that all identified Cu uptake pathways in the trout intestine are pH sensitive. This apparently ubiquitous pH dependence raises an interesting question about mechanisms of Cu uptake by the marine teleost intestinal epithelium since the intestinal lumen pH of marine species is considerably higher than in freshwater teleosts (Wilson, 1999; Grosell, 2006; Grosell and Taylor, 2007; Grosell et al., 2009b).

9.2.2. GASTRIC UPTAKE

Observations of processing a single meal containing Cu revealed the stomach as a potential site for Cu absorption in freshwater trout (Nadella et al., 2006a). Subsequent studies recently confirmed Cu uptake by the gastric mucosa and demonstrated it to be stimulated by low pH, and to be insensitive to phenamil, silver, and other divalent essential metals. These observations rule out a role for DMT1 in gastric Cu uptake and point to a possible role for Ctr1, which is expressed in the gastric mucosa (Nadella et al., 2011). This characterization of gastric Cu uptake was performed using salines to which Cu salts were added as a reasonable first approach, but results should be interpreted with caution. The presence of organic material and certainly naturally incorporated Cu may greatly influence the availability for uptake by the gastric mucosa and calls for further study.

10. CHARACTERIZATION OF INTERNAL HANDLING

10.1. Transport through the Bloodstream

Waterborne Cu exposure may result in elevated blood Cu concentrations (Buckley et al., 1982; Stagg and Shuttleworth, 1982a; Pelgrom et al., 1995; Grosell et al., 1997, 1998a; Kamunde and MacPhail, 2008), although this is not necessarily the case for marine fish (Grosell et al., 2003, 2004a). Dietary Cu exposure appears to have no influence on blood Cu levels (Kamunde and Wood, 2003) despite generally contributing more significantly to whole-body uptake and accumulation. The reason for the lack of increase in blood Cu levels during dietary exposure could be that Cu entering the circulation from the gut flows directly into the hepatic portal vein and that the liver, the main homeostatic organ for Cu, clears any excess Cu efficiently. The majority of blood Cu is associated with the plasma fraction, which accounts for 90% or more of whole-blood Cu levels (Grosell et al., 1997).

The observed increase in plasma Cu following waterborne exposure is often transient, with normalization of plasma Cu concentrations despite continuous exposure (Buckley et al., 1982; Grosell et al., 1997). In some cases a second modestly elevated steady state is reached within days of exposure (Stagg and Shuttleworth, 1982a; Pelgrom et al., 1995; Grosell et al., 1998a; Kamunde and MacPhail, 2008). These observations suggest that plasma Cu is under tight homeostatic control, as confirmed by observations of very fast clearance of plasma Cu following single bolus injections (Carbonell and Tarazona, 1994; Grosell et al., 2001). Acclimation to prolonged exposure leads to an increased clearance of plasma Cu in rainbow trout during continued Cu infusion experiments, as evident from lower steady-state concentrations in acclimated versus non-acclimated fish, with the majority of plasma Cu being cleared by the liver (Grosell et al., 2001).

10.1.1. PLASMA PROTEIN ASSOCIATION

In mammals, plasma Cu is found in two distinct pools. One pool is tightly bound to ceruloplasmin, a plasma protein of approximately 134 kDa (Cousins, 1985; Harris, 1991; Linder et al., 1998), while the other pool is associated with albumin and amino acids and is believed to be derived from recent uptake (Marceau and Aspin, 1973; Frieden, 1980; Weiner and Cousins, 1983; Cousins, 1985). Copper-containing ceruloplasmin is synthesized in the mammalian liver and is released to the blood for delivery of Cu to extrahepatic tissues (Linder et al., 1998). Ceruloplasmin is present in fish (Siwiki and Studnicka, 1986; Syed and Coombs, 1986; Cogoni et al., 1990; Pelgrom et al., 1995) and likely plays a similar role.

Copper derived from recent uptake by the European eel during waterborne exposures was associated with a 70 kDa plasma protein and also with low molecular weight substances, likely to be albumin and amino acids, respectively (Grosell, 1996), suggesting very similar plasma protein distribution of Cu in fish compared to mammals.

10.2. Accumulation in Specific Organs

In mammals, the liver is the main homeostatic organ in Cu metabolism. Copper derived from dietary uptake is effectively cleared from the blood by the liver, where it is incorporated into ceruloplasmin for transport to extrahepatic organs, stored in Cu–protein complexes, or excreted via the bile (Cousins, 1985). While the liver in fish represents a small fraction of the total body mass, it typically accounts for 25–60% of whole-body Cu, with tissue concentrations often exceeding those of other tissues by one to two orders of magnitude, implying a role for the liver in Cu metabolism in fish (Miller et al., 1993; Grosell et al., 1997, 1998a,b; Kamunde et al., 2002b; Kamunde and Wood, 2003; Kjoss et al., 2005b). The high hepatic Cu accumulation in fish is explained by a very efficient clearance of plasma Cu by the liver. A study employing injections or infusions of radiolabeled Cu found that plasma Cu is cleared rapidly by the liver such that half concentrations in plasma are reached within 32–40 min after injection, and that a total of 80% of the injected dose is found in the hepatic tissues 72 h postinjection (Grosell et al., 2001). The use of radiotracers allowed for the assessment of substrate selectivity for various tissues and demonstrated that the liver readily takes up newly accumulated Cu and Cu present before injections, but that skeletal muscle, for example, preferentially accumulates "old" Cu (Grosell et al., 2001). These observations agree perfectly with what is known from mammalian systems in which Cu derived from recent uptake is accumulated by the liver, incorporated into ceruloplasmin, and then released to the circulation for uptake by extrahepatic tissues (see Section 10.2).

The high rate of liver-specific Cu accumulation suggests a high tissue-specific expression of known Cu-carrier proteins and/or the presence of unique carriers in the hepatic tissue. While Ctr1 is expressed in hepatic tissues, expression levels in other tissues are as high or higher, suggesting that high hepatic Cu uptake rates cannot be accounted for by Ctr1 alone (Minghetti et al., 2008). Information from the mammalian literature suggests the presence of additional hepatic Cu import pathways. Deletion of hepatic Ctr1 in mice results in only mild reductions in hepatic Cu concentrations, cuproenzyme activities, and growth rates, suggesting that other Cu import pathways can compensate for the absence of Ctr1 (Kim et al., 2009). In addition, Cu uptake from Cu–albumin complexes by hepatocytes shows no sensitivity to silver (eliminating Ctr1) and only modest sensitivity to Zn and Fe, eliminating a significant role for DMT1 in hepatic Cu uptake from albumin and strongly implicating as yet unidentified Cu transport pathways in hepatic tissue (Moriya et al., 2008).

10.3. Fate of Cellular Copper

Regardless of the cell type and mode of Cu uptake, cellular free Cu concentrations must be controlled tightly to avoid toxicity while ensuring sufficient levels of Cu to preserve function of individual proteins/enzymes. Distribution of hepatic cellular Cu in control and Cu-exposed fish was recently demonstrated to be similar, and Cu was found mainly in the "metabolically active" fraction, with only 32–40% being associated with heat-stable proteins [metallothioneins (MTs)] and NaOH-resistant gran-uoles (Kamunde and MacPhail, 2008). These results agree with a number of earlier studies showing that MTs account for a minor fraction of cellular Cu under control as well as excess Cu conditions (Hogstrand et al., 1989; Hogstrand and Haux, 1991; Grosell et al., 1997, 1998b; Kraemer et al., 2005; Giguére et al., 2006). Relatively little is known about cellular partitioning of Cu among proteins in fish, but great strides have been made in understanding mammalian cellular Cu metabolism owing to two genetically linked and fatal Cu disorders, Wilson's disease (defect in ATP7B) and Menkes (defect in ATP7A) disease. What is known about cellular mechanisms of Cu homeostasis in fish (Ctr1, atp7a and atp7b, and MT) however, suggests that fish greatly resemble mammals and a brief discussion of mammalian cellular Cu metabolism (summarized in Fig. 2.9) therefore seems warranted.

Cellular free Cu concentrations are extremely low, estimated to be less than 10^{-18} M (O'Halloran and Culotta, 2000), and can be assumed to be present as Cu^+ rather than Cu^{2+} owing to the reducing intracellular environment (Banci et al., 2010a). Copper is believed to be associated with glutathione immediately after entering a human cell. Glutathione (GSH) is the most abundant intracellular Cu ligand (mM levels) but has a relatively low affinity for Cu when compared to Cu chaperone proteins. From GSH, Cu can enter one of four principal pathways (Banci et al., 2010a): (1) Cu can bind to the Cu chaperone for Cu-ATPases (ATP7A and ATP7B), HAH1, which effectively transfers Cu to the cytosolic metal-binding domains of Cu-ATPases for transport into the Golgi for cuproprotein synthesis, or across the plasma membrane in cases of excess cellular Cu; (2) Cu can bind to the chaperone for Cu,Zn-SOD, CCS, before being bound to either cytosolic or mitochondrial Cu,Zn-SOD; (3) Cu can bind to the chaperone Cox-17 before being transferred to its protein partners (Sco1/Sco2), involved in the

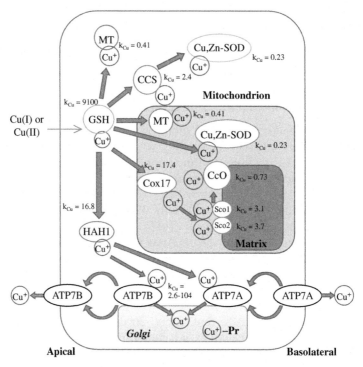

Fig. 2.9. Copper binding proteins and ligands, apparent Cu binding affinities (K_{Cu}; 10^{-15} M) and Cu transport pathways in eukaryotic cells (redrawn from Banci et al., 2010a). The lower the K_{Cu} number, the higher the affinity (see text for further details). Cellular Cu uptake can occur by several different transporters depending on cell and tissue type (see Figs. 2.7 and 2.8) but cellular Cu is present as Cu^+ due to the reducing intracellular milieu. GSH: glutathione; MT: metallothionein; CCS: Cu chaperone for Cu,Zn superoxide dismutase (SOD1); HAH1: Cu chaperone for the Cu ATPases (ATP7A and ATP7B). The Cu ATPases have a total of six Cu-binding domains with a range of K_{Cu} values (2.6–104); Cox 17, Sco1 and Sco2: Cu-chaperones and co-chaperones of cytochrome c oxidase. The Cu ATPases deliver Cu for incorporation into cupro-proteins (Cu^+-Pr) in the Golgi under normal Cu conditions. During periods of elevated cellular Cu, the Cu ATPases traffics to the plasma membranes for excretion of Cu. ATP7A targets basolateral membranes in polarized cells while ATP7B targets the cannalicular membrane (apical) in hepatocytes and facilitates biliary Cu excretion. **SEE COLOR PLATE SECTION.**

assembly of the Cu_A site of cytochrome c oxidase (CcO) within the mitochondria; or (4) Cu can bind directly to MT isoform 2 (MT-2), the inducible MT isoform in fish (see Section 6.1.6.2), located in both the cytoplasmic and intermembrane space compartment (Fig. 2.9).

Although all chaperones are present at much lower concentrations (micromolar range) than GSH, their higher affinity for Cu facilitates exchange of Cu from GSH to HAH1, Cox-17, and CCS. The protein partners for each of these chaperones all have higher affinities than the respective chaperones, which drives Cu transfer from HAH1, Cox-17, and CCS chaperones to Cu-ATPases, CcO, and Cu,Zn-SOD, respectively (Fig. 2.9) (Banci et al., 2010a). The Cu-binding affinity of MT-2 is only superseded by that of Cu,Zn-SOD, yet Cu delivery to Cu-ATPases and CcO is ensured. This apparent paradox is explained by fast Cu transfer kinetics observed in all Cu-handling proteins and specific protein–protein recognition (Banci et al., 2010a,b). In addition, simple mass-balance considerations for fish, specifically for hepatic tissue, suggest that MTs combined can account for no more than 30–40% of the total cellular Cu pool (see beginning of this section).

Although the above considerations are based solely on observations made on mammalian systems, and only two cellular chaperone components of Cu handling (Cox-17 and HAH1) have been examined in fish (Craig et al., 2007; Minghetti et al., 2010), the high conservation of Cu transporters examined to date suggests that similar pathways are likely to be present in teleosts. Despite these apparently tightly controlled Cu delivery mechanisms, Cu-induced inhibition of Na^+/K^+-ATPase and likely cytosolic CA occurs (see above), begging the determination of affinity and abundance of Cu-binding sites on these enzymes to better understand cellular Cu delivery pathways resulting in enzyme inhibition.

11. CHARACTERIZATION OF EXCRETION ROUTES

Not surprisingly considering the liver's role in Cu homeostasis, Cu excretion is hepatobiliary in mammals and this appears to be an important route in freshwater and seawater teleosts (Grosell et al., 1997, 1998a,b, 2001, 2004a). Very high gall bladder bile Cu concentrations (up to 20 μg mL^{-1}) are the products of hepatic excretion with hepatic bile containing 0.6–3.0 μg mL^{-1} (Grosell et al., 2001) and differential absorption of NaCl and water across the gall bladder epithelium leaving constituents such as Cu highly concentrated (Grosell et al., 2000). Biliary Cu excretion is stimulated by waterborne (Grosell et al., 1998a,b, 2001) as well as dietary Cu exposure (Lanno et al., 1987; Andreasson and Dave, 1995; Kamunde et al., 2001), illustrating a dynamic role of hepatobiliary excretion in Cu homeostasis.

Mechanisms of hepatobiliary Cu excretion in fish are not thoroughly understood, but the identification of atp7b and the demonstration of hepatic transcriptional responses of this protein to Cu status strongly imply a role for Cu excretion (Minghetti et al., 2010). In mammals, regulation of Cu excretion via ATP7B appears to be via protein trafficking between the trans-Golgi and excretory vesicular compartments rather than at the transcriptional level, suggesting some differences between mammals and fish. However, although atp7b is regulated at the transcriptional level, this fish orthologue of the human ATP7B could also be subject to trafficking as part of the hepatic homeostatic response.

In addition to Cu excretion by ATP7B in mammals, Cu excretion via the canalicular multiorganic anion transporter (cMOAT) (Houwen et al., 1990; Dijkstra et al., 1996) and by lysosomal Cu excretion has been documented (Gross et al., 1989). In addition to evidence for atp7b involvement in piscine hepatobiliary Cu excretion, evidence suggests that lysosomal Cu excretion may occur in fish during Cu excess (Lanno et al., 1987; Segner, 1987).

In accord with very low renal Cu accumulation in most freshwater teleosts, regardless of the route of Cu exposure, urinary Cu excretion is extremely low, and plays no appreciable role in freshwater teleost Cu homeostasis as it is unaffected by Cu exposure duration, at least during waterborne exposure (Grosell et al., 1998b). These observations are in good agreement with the tight association of plasma Cu with relatively large plasma proteins (albumin and ceruloplasmin, see above) rendering Cu unavailable for glomerular filtration. A few studies of marine teleosts show elevated renal Cu concentrations following waterborne exposures (Grosell et al., 2003, 2004a), which may suggest that urinary Cu excretion is more prominent in marine than in freshwater teleosts.

In trout continuously infused with radiolabeled Cu (Grosell et al., 2001), there was considerable appearance of radiolabeled Cu in the water that could not be accounted for by hepatobiliary or renal excretion, indicating branchial excretion of Cu. The latter exceeded that of hepatobiliary excretion rates but showed no signs of being regulated based on Cu status of the fish (Grosell et al., 2001). To the author's knowledge, this study represents the only attempt to quantify Cu excretion by the gill. However, the presence of atp7b in gill tissue and its apparent regulation based on Cu status support a possible role of the gill in Cu excretion since atp7b targets the apical region under situations of excess Cu (Minghetti et al., 2010).

12. BEHAVIORAL EFFECTS OF COPPER

Copper exerts a multitude of behavioral effects on fish, effects which have been most extensively documented in freshwater. Observations of altered behavior in response to Cu exposure are discussed in detail in

Sections 6.1.2 (Direct behavioral observations), 6.2.1 (Drinking), and 6.4.7 (Serotonergic systems).

13. MOLECULAR CHARACTERIZATION OF COPPER TRANSPORTERS, STORAGE PROTEINS, AND CHAPERONES

Copper is essential yet a potent toxicant and as such delicate systems have evolved to ensure adequate uptake and cellular handling and storage of the metal. These systems are discussed in detail in Sections 9 (Uptake), 10 (Internal Handling), and 11 (Excretion) above.

14. GENOMIC AND PROTEOMIC STUDIES

The use of especially genomic tools to study Cu toxicity and homeostasis has seen a rise during the past 5 years and has revealed a plethora of pathways being influenced by Cu exposure (Craig et al., 2009, 2010). Many of the gene expression responses appear to be associated with general stress response pathways or energy metabolism, while others indicate more specific responses to Cu exposures. Indeed, genomic studies have confirmed the induction of oxidative stress and/or compensatory responses during Cu exposure (see Section 6.1.6) and have increased our understanding of cellular Cu handling (see Sections 9–11, above).

15. INTERACTIONS WITH OTHER METALS

While studies of interactions between waterborne Cu and other metals exist, studies of exposure to dietary Cu and other metals have all been conducted using mixed metal loads in the diet (Dallinger and Kautzky, 1985; Miller et al., 1992; Farag et al., 1994, 1999; Woodward et al., 1994, 1995; Mount et al., 1994; Hansen et al., 2004), with few attempts (Mount et al., 1994) to directly characterize the possible antagonistic or synergistic interactions between different metals. Combined and individual waterborne Cu and Cd exposures have demonstrated interactions between these two metals with respect to both metal accumulation and induced effects on salt and water balance (Pelgrom et al., 1994, 1995). Prolonged exposure to sublethal Cu concentrations appears to enhance Cd uptake during subsequent exposures (McGeer et al., 2007). Considering that many metals share toxic actions, including formation of ROS, synergistic interactions are likely and clearly worthy of examination, especially since metal contamination often involves more than a single metal.

Perhaps of particular relevance from a mechanistic perspective is the possible interactions between Cu and Ag (see Wood, Chapter 10, Vol. 31B), which appear to share similar mechanisms of toxicity, including competitive interactions with apical sodium entry, potent inhibition of the Na^+/K^+-ATPase and possibly inhibition of CA (Grosell et al., 2002). Considering these shared modes of action, synergistic interactions between Cu and Ag seem likely. In addition, silver appears to stimulate sodium-dependent Cu uptake by rainbow trout intestine (Nadella et al., 2007) and Cu uptake by rainbow trout gills (M. Grosell and C. M. Wood, unpublished).

Ag has been demonstrated to stimulate sodium uptake by a number of vertebrate epithelia but is also a potent inhibitor of sodium transport across gill epithelia (Morgan et al., 1997; Bury and Wood, 1999). While these discrepancies could reflect species differences, they might also be related to differences in Ag concentration or chemical speciation. It seems conceivable that Ag at low concentrations may stimulate sodium uptake pathways and thereby Cu uptake by sodium-sensitive and sodium-dependent Cu uptake pathways. Furthermore, Ag employed at higher concentrations with Cu is likely to result in toxic responses.

16. KNOWLEDGE GAPS AND FUTURE DIRECTIONS

Suggestions for areas in need of study are interspersed throughout this chapter. In the following, areas of particular concerns are highlighted to provide inspiration for further research in the area of Cu toxicity and homeostasis in fish.

While it is clear that Cu exposure induces oxidative stress, the relationship between exposure concentrations and the cellular response cascade needs to be addressed. Determination of Cu threshold concentrations for ROS formation, enzymatic and non-enzymatic antioxidant responses, and direct effects on antioxidant enzymes and the oxidative damage occurring when antioxidant defenses are exceeded is desirable. As such, integrative studies quantifying the response to a range of Cu concentrations of individual components of the oxidative stress response are called for.

Common to the impairment of sensory physiology, avoidance, and behavioral modifications elicited by Cu exposure is that it is largely restricted to acute exposure in freshwater environments. Furthermore, such observations are rarely related to performance or fitness-based endpoints.

Clearly, there is a need for additional studies of marine species, the potential for recovery or acclimation of sensory systems during continued exposure, or possibly the enhanced sensitivity following chronic exposures. Such studies would benefit from examining links between effects observed at the cellular level and whole-animal behavioral responses, and better yet, population-level impacts.

While Cu exposure has been shown to directly inhibit CA activity in invertebrates (Vitale et al., 1999), the same has not been demonstrated in fish, although a range of observations points to CA as a likely and sensitive target for Cu toxicity, including reduced ammonia excretion, acid–base balance disturbance, and impaired intestinal anion exchange during Cu exposure. Alternatives to the conventional delta pH method applied to tissue homogenates must be applied to more thoroughly test the possibility that CA is a sensitive target for Cu exposure, as has been demonstrated for silver.

Many areas of great ecological importance are subject to fluctuations in salinity and other physical/chemical parameters, conditions which in themselves are challenging. Very little is known about Cu exposure possibly interfering with the ability of fish to cope with such naturally and frequently occurring environmental challenges. However, at least one study has revealed that Cu exposure in freshwater impairs downstream migration and survival of Coho salmon in seawater (Lorz and McPherson, 1976). Along similar lines of thinking, it is unknown how Cu exposure might impact the ability of fish to tolerate extremely high salinities exceeding those commonly found in seawater.

Acclimation to Cu exposure with respect to osmoregulation and Cu homeostasis is relatively well documented. In contrast, some doubt exists with respect to acclimation of the olfactory system and the mechanosensory systems. Little if anything is known about the potential for acclimation with respect to acid–base balance disturbances and handling of nitrogenous waste during Cu exposure. It is likely that studies of compensatory mechanisms involved in restoring acid–base balance and ammonia excretion (if such adjustments indeed occur) would shed light on the mechanisms by which Cu affects these physiological processes and as such, acclimation to Cu forms a very fruitful area for further research.

Full life-cycle dietary Cu exposure studies are desperately needed and should include a direct comparison of diets with naturally incorporated Cu and diets spiked with Cu salts. Such studies are logistically challenging but feasible with smaller model species with relatively short generation times, such as the zebrafish and the fathead minnow in freshwater, or pupfish in seawater and intermediate salinities.

The esophagus in seawater teleosts is responsible for 50% or more of the NaCl uptake from the gastrointestinal tract and is in intimate contact with contaminants in the ingested seawater. As such, studies of potential Cu impacts on transport by this very important gastrointestinal segment seem timely.

While great strides have been made in predicting acute toxicity, attempts to predict chronic toxicity for environmental regulatory purposes are limited to extrapolations from knowledge of acute toxicity. This practice relies on the assumption that modes of Cu action are the same during short-term and long-term exposures, an assumption that is largely untested. There is a demand for information about how water chemistry and dietary composition may affect fish during long-term Cu exposure. Studies addressing this void ideally should include full life-cycle exposures and naturally incorporated Cu during dietary exposures.

Studies of interactions among metals are scarce and studies of interactions between Cu and other environmental stressors, including hypoxia, elevated temperature, and ambient CO_2 levels associated with global climate change, are an important and fruitful area for research. Global climate change in itself will impose stress on aquatic organisms and will also tend to shift Cu from less toxic carbonate/bicarbonate and hydroxide complexes towards ionic Cu, increasing the potency of the Cu exposure alone.

Last, but not least, many questions remain to be answered about Cu homeostasis in fish as well as mammals. Fish appear to resemble mammals in many ways with respect to Cu homeostasis but offer interesting differences that could be exploited to gain insight into vertebrate Cu homeostasis. Recent studies using fish as model organisms (Mackenzie et al., 2004; Mendelsohn et al., 2006; Madsen et al., 2008; Minghetti et al., 2008, 2010) are excellent examples of the potential of fish for comparative Cu physiology studies of vertebrates.

ACKNOWLEDGMENTS

Enjoyable discussions with Kevin V. Brix and David Deforest about environmental regulation of Cu are acknowledged. Dr Andrew Esbaugh provided insightful comments on an early version of this text.

REFERENCES

Ahmad, I., Oliveira, M., Pacheco, M., and Santos, M. A. (2005). *Anguilla anguilla* L. oxidative stress biomarkers responses to copper exposure with or without beta-naphthoflavone pre-exposure. *Chemosphere* **61**, 267–275.

Almroth, B. C., Sturve, J., Stephensen, E., Holth, T. F., and Förlin, L. (2008). Protein carbonyls and antioxidant defenses in corkwing wrasse (*Symphodus melops*) from a heavy metal polluted and PAH polluted site. *Mar. Environ. Res.* **66**, 271–277.

Alsop, D., Brown, S., and Van Der Kraak, G. (2007). The effects of copper and benzo[a]pyrene on retinoids and reproduction in zebrafish. *Aquat. Toxicol.* **82**, 281–295.

Anderson, D. P., Dixon, O. W., Bodammer, J. E., and Lizzio, E. F. (1989). Suppression of antibody-producing cells in rainbow trout spleen sections exposed to copper *in vitro*. *J. Aquat. Anim. Health* **1**, 57–61.

Andreasson, M., and Dave, G. (1995). Transfer of heavy metals from sediments to fish, and their biliary excretion. *J. Aquat. Ecosyst. Health* **4**, 221–230.

ARMCANZ/ARMCANZ (2000). *Australian and New Zealand Guidelines for Fresh and Marine Water Quality*. http://www.mincos.gov.au/publications/australian_and_new_zealand_guidelines_for_fresh_and_marine_water_quality

Arnold, W. R. (2005). Effects of dissolved organic carbon on copper toxicity: implications for saltwater copper criteria. *Int. Environ. Assess. Manag.* **1**, 34–39.

Arnold, W. R., Cotsifas, J. S., and Corneillie, K. M. (2006). Validation and update of a model used to predict copper toxicity to the marine bivalve *Mytilus* sp. *Environ. Toxicol.* **21**, 65–70.

Baker, R. J., Knittel, M. D., and Fryer, J. L. (1983). Susceptibility of chinook salmon, *Oncorhynchus tshawytscha* (Walbaum), and rainbow trout, *Salmo gairdneri*, Richardson, to infection with *Vibrio anguillarum* following sublethal copper exposure. *J. Fish Biol.* **6**, 267–275.

Baker, R. T. M. (1998). Chronic dietary exposure to copper affects growth, tissue lipid peroxidation, and metal composition of the grey mullet. *Chelon labrosus. Mar. Environ. Res.* **45**, 357–365.

Baldwin, D. H., Sandahl, J. F., Labenia, J. S., and Scholz, N. L. (2003). Sublethal effects of copper on coho salmon: impacts on nonoverlapping receptor pathways in the peripheral olfactory nervous system. *Environ. Toxicol. Chem.* **22**, 2266–2274.

Banci, L., Bertini, I., Ciofi-Baffoni, S., Kozyreva, T., Zovo, K., and Palumaa, P. (2010a). Affinity gradients drive copper to cellular destinations. *Nature* **465**, 645–650.

Banci, L., Bertini, I., McGreevy, K. S., and Rosato, A. (2010b). Molecular recognition in copper trafficking. *Nat. Prod. Rep.* **27**, 695–710.

Beaumont, M. W., Butler, P. J., and Taylor, E. W. (1995). Exposure of brown trout, *Salmo trutta*, to sub-lethal copper concentrations in soft acidic water and its effect upon sustained swimming performance. *Aquat. Toxicol.* **33**, 45–63.

Beaumont, M. W., Butler, P. J., and Taylor, E. W. (2000). Exposure of brown trout, *Salmo trutta*, to a sub-lethal concentration of copper in soft acidic water: effects upon muscle metabolism and membrane potential. *Aquat. Toxicol.* **51**, 259–272.

Berntssen, M. H. G., Hylland, K., Wendelaar Bonga, S. E., and Maage, A. (1999a). Toxic levels of dietary copper in Atlantic salmon (*Salmo salar* L.) parr. *Aquat. Toxicol.* **46**, 87–99.

Berntssen, M. H. G., Lundebye, A. K., and Maage, A. (1999b). Effects of elevated dietary copper concentrations on growth, feed utilization and nutritional status of Atlantic salmon (*Salmo salar* L.) fry. *Aquaculture* **174**, 167–181.

Besser, J. M., Mebane, C. A., Mount, D. R., Ivey, C. D., Kunz, J. L., Greer, I. E., May, T. W., and Ingersoll, C. G. (2007). Sensitivity of mottled sculpins (*Cottus bairdi*) and rainbow trout *(Onchorhynchus mykiss)* to acute and chronic toxicity of cadmium, copper, and zinc. *Environ. Toxicol. Chem.* **26**, 1657–1665.

Beyers, D. W., and Farmer, M. S. (2001). Effects of copper on olfaction of Colorado pikeminnow. *Environ. Toxicol. Chem.* **20**, 907–912.

Bielmyer, G. K., and Grosell, M. (2010). Emerging issues in marine metal toxicity. *Essential Rev. Exp. Biol.* **2**, 129–158.

Bielmyer, G. K., Grosell, M., and Brix, K. V. (2006). Toxicity of silver, zinc, copper, and nickel to the copepod, *Acartia tonsa*, exposed via a phytoplankton diet. *Environ. Sci. Technol.* **40**, 2063–2068.

Bjerselius, R., Winberg, S., Winberg, Y., and Zeipel, K. (1993). Ca^{2+} protects olfactory receptor function against acute Cu(II) toxicity in Atlantic salmon. *Aquat. Toxicol.* **25**, 125–138.

Blanchard, J., and Grosell, M. (2005). Effects of salinity on copper accumulation in the common killifish (*Fundulus heteroclitus*). *Environ. Toxicol. Chem.* **24**, 1403–1413.

Blanchard, J., and Grosell, M. (2006). Copper toxicity across salinities from freshwater to seawater in the euryhaline fish, *Fundulus heteroclitus*: is copper an ionoregulatory toxicant in high salinites?. *Aquat. Toxicol.* **80**, 131–139.

Boisen, A. M. Z., Amstrup., J., Novak, I., and Grosell, M. (2003). Sodium and chloride transport in soft water and hard water acclimated zebrafish (*Danie rerio*). *Biochim. Biophys. Acta* **1618**, 207–218.

Boone, A. N., and Vijayan, M. M. (2002). Constitutive heat shock protein 70 (HSP70) expression in rainbow trout hepatocytes. *Comp. Biochem. Physiol. C* **132**, 223–233.

Bopp, S. K., Abicht, H. K., and Knauer, K. (2008). Copper-induced oxidative stress in rainbow trout gill cells. *Aquat. Toxicol.* **86**, 197–204.

Bourret, V., Couture, P., Campbell, P. G., and Bernatchez, L. (2008). Evolutionary ecotoxicology of wild yellow perch (*Perca flavescens*) populations chronically exposed to a polymetallic gradient. *Aquat. Toxicol.* **86**, 76–90.

Braun, M. H., Steele, S. L., Ekker, M., and Perry, S. F. (2009). Nitrogen excretion in developing zebrafish (*Danio rerio*): a role for Rh proteins and urea transporters. *Am. J. Physiol.* **296**, F994–F1005.

Brett, J. R. (1964). The respiratory metabolism and swimming performance of young sockeye salmon. *J. Fish. Res. Bd Can.* **21**, 1183–1226.

Brouwer, M., Hoexum-Brouwer, T., and Cashon, R. E. (1993). A putative glutathione-binding binding site in Cd/Zn-metallothionein identified by equilibrium binding and molecular-modelling studies. *Biochem. J.* **294**, 219–225.

Bruland, K. W. (1980). Oceanographic distributions of cadmium, zinc, nickel, and copper in the North Pacific. *Earth Planet Sci. Lett.* **47**, 176–198.

Buckley, J. T., Roch, M., McCarter, J. A., Rendell, C. A., and Matheson, A. T. (1982). Chronic exposure of coho salmon to sublethal concentrations of copper – I. Effect on growth, on accumulation and distribution of copper, and on copper tolerance. *Comp. Biochem. Physiol. C* **72**, 15–19.

Bury, N. R., and Wood, C. M. (1999). Mechanisms of branchial apical silver uptake by rainbow trout is via the proton coupled Na^+ channel. *Am. J. Physiol.* **277**, R1385–R1391.

Bury, N. R., Grosell, M., Grover, A. K., and Wood, C. M. (1999). ATP-dependent silver transport across the basolateral membrane of rainbow trout gills. *Toxicol. Appl. Pharmacol.* **159**, 1–8.

Bury, N. R., Walker, P. A., and Glover, C. N. (2003). Nutritive metal uptake in teleost fish. *J. Exp. Biol.* **206**, 11–23.

Caiccio, M., Valenza, M., Tesoriere, L., Bongiorno, A., Albiero, R., and Livrea, M. A. (1993). Vitamin A inhibits doxorubicin-induced membrane lipid peroxidation in rat tissues *in vivo*. *Arch. Biochem. Biophys.* **302**, 103–108.

Campbell, H. A., Handy, R. D., and Nimmo, M. (1999). Copper uptake kinetics across the gill of rainbow trout (*Oncorhynchus mykiss*) measured using an improved isolated perfused head technique. *Aquat. Toxicol.* **46**, 177–190.

Campbell, H. A., Handy, R. D., and Sims, D. W. (2002). Increased metabolic cost of swimming and consequent alterations of circadian activity in rainbow trout (*Oncorhynchus mykiss*) exposed to dietary copper. *Can. J. Fish. Aquat. Sci.* **59**, 768–777.

Campbell, H. A., Handy, R. D., and Sims, D. W. (2005). Shifts in a fish's resource holding power during a contact paired interaction: the influence of a copper-contaminated diet in rainbow trout. *Physiol. Biochem. Zool.* **78**, 706–714.

Canesi, L., Viarengo, A., Leonzio, C., Filippelli, M., and Gallo, G. (1999). Heavy metals and glutathione metabolism in mussel tissues. *Aquat. Toxicol.* **46**, 67–76.

Carbonell, G., and Tarazona, J. V. (1994). Toxicokinetics of copper in rainbow trout (*Oncorhynchus mykiss*). *Aquat. Toxicol.* **29**, 213–221.

Carlberg, I., and Mannervik, B. (1985). Glutathione-reductase. *Methods Enzymol.* **113**, 484–490.

Carreau, N. D., and Pyle, G. G. (2005). Effect of copper exposure during embryonic development on chemosensory function of juvenile fathead minnows (*Pimephales promelas*). *Ecotoxicol. Environ. Saf.* **61**, 1–6.

Carriquiriborde, P., Handy, R. D., and Davies, S. J. (2004). Physiological modulation of iron metabolism in rainbow trout (*Oncorhynchus mykiss*) fed low and high iron diets. *J. Exp. Biol.* **207**, 75–86.

CCME (Canadian Council of Ministers of the Environment). (2007). Canadian Water Quality Guidelines for the Protection of Aquatic Life: summary table. In: *Canadian Environmental Quality Guidelines, 1999*. Winnipeg: Canadian Council of Ministers of the Environment.

Chakoumakos, C., Russo, R. C., and Thurston, R. V. (1979). Toxicity of copper to cutthroat trout (*Salmo clarki*) under different conditions of alkalinity, pH and hardness. *Environ. Sci. Technol.* **13**, 219.

Cheung, A. P. L., Lam, T. H. J., and Chan, K. M. (2004). Regulation of tilapia metallothionein gene expression by heavy metal ions. *Mar. Environ. Res.* **58**, 389–394.

Chiaverini, N., and De Ley, M. (2010). Protective effects of metallothionein on oxidative stress-induced DNA damage. *Free Radic. Res.* **44**, 605–613.

Cho, Y. S., Choi, B. N., Kim, K. H., Kim, S. K., Kim, D. S., Bang, I. C., and Nam, Y. K. (2006). Differential expression of Cu/Zn superoxide dismutase mRNA during exposures to heavy metals in rockbream (*Oplegnathus faciatus*). *Aquaculture* **253**, 667–679.

Christensen, G. M., McKim, J. M., Brungs, W. A., and Hunt, E. P. (1972). Changes in the blood of the brown bullhead (*Ictalurus nebulosus* Lesueur) following short and long term exposure to copper(II). *Toxicol. Appl. Pharmacol.* **23**, 417–427.

Claiborne, J. B., Blackston, C. R., Choe, K. P., Dawson, D. C., Harris, S. P., Mackenzie, L. A., and Morrison-Shetlar, A. I. (1999). A mechanism for branchial acid excretion in marine fish: identification of multiple Na^+/H^+ antiporter (NHE) isoforms in gills of two seawater teleosts. *J. Exp. Biol.* **202**, 315–324.

Clearwater, S. J., Baskin, S. J., Wood, C. M., and McDonald, D. G. (2000). Gastrointestinal uptake and distribution of copper in rainbow trout. *J. Exp. Biol.* **203**, 2455–2466.

Clearwater, S. J., Farag, A. M., and Meyer, J. S. (2002). Bioavailability and toxicity of dietborne copper and zinc to fish. *Comp. Biochem. Physiol. C* **132**, 269–313.

Coale, K. H., and Bruland, K. W. (1988). Copper complexation in the Northeast Pacific. *Limnol. Oceanogr.* **33**, 1084–1101.

Cogoni, A., Farci, R., Cau, A., Melis, A., Medda, R., and Floris, G. (1990). Mullet plasma: serum amine oxidase and ceruloplasmin: purification and properties. *Comp. Biochem. Physiol. C* **95**, 297–300.

Cousins, R. J. (1985). Absorption, transport, and hepatic metabolism of copper and zinc: special reference to metallothionein and ceruloplasmin. *Physiol. Rev.* **65**, 238–309.

Craig, P. M., Wood, C. M., and McClelland, G. B. (2007). Oxidative stress response and gene expression with acute copper exposure in zebrafish (*Danio rerio*). *Am. J. Physiol. Regul. Integr. Comp. Physiol.* **293**, R1882–R1892.

Craig, P. M., Hogstrand, C., Wood, C. M., and McClelland, G. B. (2009). Gene expression endpoints following chronic waterborne copper exposure in a genomic model organism, the zebrafish. *Danio rerio. Physiol. Genomics* **40**, 23–33.

Craig, P. M., Wood, C. M., and McClelland, G. B. (2010). Water chemistry alters gene expression and physiological end points of chronic waterborne copper exposure in zebrafish. *Danio rerio. Environ. Sci. Technol.* **44**, 2156–2162.

Crespo, S., and Balasch, J. (1980). Mortality, accumulation, and distribution of zinc in the gill system of the dogfish following zinc treatment. *Bull. Environ. Contam. Toxicol.* **24**, 940–944.

Cusimano, R. F., Brakke, D. F., and Chapman, G. A. (1986). Effects of pH on the toxicities of cadmium, copper, and zinc to steelhead trout (*Salmo gairdneri*). *Can. J. Fish. Aquat. Sci.* **43**, 1497–1503.

Dallinger, R., and Kautzky, H. (1985). The importance of contaminated food for the uptake of heavy metals by rainbow trout (*Salmo gairdneri*): a field study. *Oecologia* **67**, 82–89.

De Boeck, G., Desmet, H., and Blust, R. (1995a). The effect of sublethal levels of copper on oxygen consumption and ammonia excretion in the common carp. *Cyprinus carpio. Aquat. Toxicol.* **32**, 127–141.

De Boeck, G., Nilsson, G. E., Elofsson, U., Vlaeminck, A., and Blust, R. (1995b). Brain monoamine levels and energy status in common carp (*Cyprinus carpio*) after exposure to sublethal levels of copper. *Aquat. Toxicol.* **33**, 265–277.

De Boeck, G., Vlaeminck, A., Balm, P. H. M., Lock, R. A. C., De Wachter, B., and Blust, R. (2001). Morphological and metabolic changes in common carp, *Cyrpinus carpio*, during short-term copper exposure: interactions between Cu^{2+} and cortisol elevation. *Environ. Toxicol. Chem.* **20**, 374–381.

De Boeck, G., Hattink, J., Franklin, N. M., Bucking, C. P., Wood, S., Walsh, P. J., and Wood, C. M. (2007a). Copper toxicity in the spiny dogfish (*Squalus acanthias*): urea loss contributes to the osmoregulatory disturbance. *Aquat. Toxicol.* **84**, 133–141.

De Boeck, G., van der ven, K., Meeus, W., and Blust, R. (2007b). Sublethal copper exposure induces respiratory distress in common carp and gibel carp but not in rainbow trout. *Comp. Biochem. Physiol. C* **144**, 380–390.

DeForest, D. K., Brix, K. V., and Adams, W. J. (2007). Assessing metal bioaccumulation in aquatic environments: the inverse relationship between bioaccumulation factors, trophic transfer factors and exposure concentrations. *Aquat. Toxicol.* **84**, 236–246.

De Schamphelaere, K. A. C., and Janssen, C. R. (2004). Effects of dissolved organic carbon concentration and source, pH, and water hardness on chronic toxicity of copper to *Daphnia magna. Environ. Toxicol. Chem.* **23**, 1115–1122.

Deigweiher, K., Koschnick, N., Pörtner, H. O., and Lucassen, M. (2008). Acclimation of ion regulatory capacities in gills of marine fish under environmental hypercapnia. *Am. J. Physiol.* **295**, R1660–R1670.

and Department of Water Affairs and Forestry. (2008). *South Africa Water Quality Criteria Guidelines.* Vol. 7. *Aquatic Ecosystems.* Department of Water Affairs and Forestry, Pretoria.

Dethloff, G. M., Schlenk, D., Khan, S., and Bailey, H. C. (1998). The effects of copper on blood and biochemical parameters of rainbow trout (*Oncorhynchus mykiss*). *Arch. Environ. Contam. Toxicol.* **36**, 415–423.

Dethloff, G. M., Schlenk, D., Hamm, J. T., and Bailey, H. C. (1999). Alterations in physiological parameters of rainbow trout (*Oncorhynchus mykiss*) with exposure to copper and copper/zinc mixtures. *Ecotoxicol. Environ. Saf.* **42**, 253–264.

Dijkstra, M., Havinga, R., Vonk, R. J., and Kuipers, F. (1996). Bile secretion of cadmium, silver, zinc and copper in the rat. Involvement of various transport systems. *Life Sci.* **59**, 1237–1246.

Di Toro, D. M., Allen, H. E., Bergman, H. L., Meyer, G., Paquin, P. R., and Santore, R. C. (2001). A biotic ligand model of the acute toxicity of metals. 1. Technical basis. *Environ. Toxicol. Chem.* **20**, 2383–2396.

Dixon, D. G., and Sprague, J. B. (1981). Acclimation to copper by rainbow trout (*Slamo gairdneri*) – a modifying factor in toxicity. *Can. J. Fish. Aquat. Sci.* **38**, 880–888.

Donaldson, E. M., and Dye, H. M. (1975). Corticosteroid concentrations in the sockeye salmon (*Oncorhynchus nerka*) exposed to low concentrations of copper. *J. Fish. Res. Bd Can.* **32**, 533–539.

ECB (European Chemicals Bureau) (2008a). *Copper, copper II sulphate pentahydrate, copper(I) oxide, copper(II)oxide, dicopper chloride trihydroxide.* European Union risk assessment report. Brussels: European Copper Institute.

ECB (European Chemicals Bureau) (2008b). *Copper, copper II sulphate pentahydrate, copper(I) oxide, copper(II)oxide, dicopper chloride trihydroxide. Marine effects assessment.* European Union risk assessment report. Brussels: European Copper Institute.

Edwards, S. L., Wall, B. P., Morrison-Shetlar, A., Sligh, S., Weakley, J. C., and Claiborne, J. B. (2005). The effect of environmental hypercapnia and salinity on the expression of NHE-like isoforms in the gills of a euryhaline fish (*Fundulus heteroclitus*). *J. Exp. Zool. A Comp. Exp. Biol.* **303**, 464–475.

Erickson, R. J., Benoit, D. A., Mattson, V. R., Nelson, H. P., Jr., and Leonard, E. N. (1996). The effects of water chemistry on the toxicity of copper to fathead minnows. *Environ. Toxicol. Chem.* **15**, 181–193.

Erickson, R. J., Kleiner, C. F., Fiandt, J. T., and Highland, T. L. (1997). Effect of acclimation period on the relationship of acute copper toxicity to water hardness for fathead minnows. *Environ. Toxicol. Chem.* **16**, 813–815.

Evans, D. H., Piermarini, P. M., and Choe, K. P. (2005). The multifunctional fish gill: dominant site of gas exchange, osmoregulation, acid–base regulation, and excretion of nitrogenous waste. *Physiol. Rev.* **85**, 97–177.

Evans, P., and Halliwell, B. (2001). Micronutrients: oxidant/antioxidant status. *Br. J. Nutr.* **85** (Suppl. 2), S67–S74.

Farag, A. M., Boese, C. J., Woodward, D. F., and Bergman, H. L. (1994). Physiological changes and tissue accumulation in rainbow trout exposed to foodborne and waterborne metals. *Environ. Toxicol. Chem.* **13**, 2021–2029.

Farag, A. M., Woodward, D. F., Brumbaugh, W., Goldstein, J. N., MacConnell, E., Hogstrand, C., and Barrows, F. T. (1999). Dietary effects of metal-contaminated invertebrates in the Coeur d'Alene River, Idaho, on cutthroat trout. *Trans. Am. Fish. Soc.* **128**, 578–592.

Feng, Q., Boone, A. N., and Vijayan, M. M. (2003). Copper impact on heat shock protein 70 expression and apoptosis in rainbow trout hepatocytes. *Comp. Biochem. Physiol. C* **135**, 345–355.

Fines, G. A., Ballantyne, J. S., and Wright, P. A. (2001). Active urea transport and an unusual basolateral membrane composition in the gills of a marine elasmobranch. *Am. J. Physiol.* **280**, R16–R24.

Freedman, J. H., Ciriolo, M. R., and Peisach, J. (1989). The role of glutathione in copper metabolism and toxicity. *J. Biol. Chem.* **264**, 5598–5605.

Freeman, G. P., Nielsen, P., and Gibson, G. E. (1986). Monoamine neurotransmitter metabolism and locomotor activity during chemical hypoxia. *J. Neurochem.* **46**, 733–738.

Frieden, E. (1980). Ceruloplasmin: a multi-functional metalloprotein of vertebrate plasma. In: *Biological Roles of Copper*, pp. 98–124. Elsevier/North-Holland, New York.

Gabbianelli, R., Lupidi, G., Villarini, M., and Falcioni, G. (2003). DNA damage induced by copper on erythrocytes of gilthead sea bream, *Sparus aurata* and mollusk *Scapharca inaequivalvis*. *Arch. Environ. Contam. Toxicol.* **45**, 350–356.

Gagnon, A., Jumarie, C., and Hontela, A. (2006). Effects of Cu on plasma cortisol and cortisol secretion by adrenocortical cells of rainbow trout (*Oncorhynchus mykiss*). *Aquat. Toxicol.* **78**, 59–65.

Galvez, F., Donini, A., Playle, R. C., Smith, D. S., O'Donnell, M. J., and Wood, C. M. (2008). A matter of potential concern: natural organic matter alters the electrical properties of fish gills. *Environ. Sci. Tech.* **42**, 9385–9390.

Gatlin, D. M., and Wilson, R. P. (1986). Dietary copper requirements of fingerling channel catfish. *Aquaculture* **54**, 277–285.

Gauthier, C., Campbell, P. G., and Couture, P. (2009). Condition and pyloric caeca as indicators of food web effects in fish living in metal-contaminated lakes. *Ecotoxicol. Environ. Saf.* **72**, 2066–2074.

van Geen, A., and Luoma, S. N. (1993). Trace metals (Cd, Cu, Ni and Zn) and nutrients in coastal waters adjacent to San Francisco Bay, California. *Estuaries* **16**, 559–566.

Geist, J., Werner, I., Eder, K. J., and Leutenegger, C. M. (2007). Comparisons of tissue-specific transcription of stress response genes with whole animal endpoints of adverse effect in striped bass (*Morone saxatilis*) following treatment with copper and esfenvalerate. *Aquat. Toxicol.* **85**, 28–39.

Georgalis, T., Perry, S. F., and Gilmour, K. M. (2006). The role of branchial carbonic anhydrase in acid–base regulation in rainbow trout (*Oncorhynchus mykiss*). *J. Exp. Biol.* **209**, 518–530.

Gheorghiu, C., Smith, D. S., Al-Reasi, H. W., McGeer, J. C., and Wilkie, M. P. (2010). Influence of natural organic matter (NOM) quality on Cu–gill binding in the rainbow trout (*Oncorhynchus mykiss*). *Aquat. Toxicol.* **97**, 343–352.

Giguére, A., Campbell, P. G. C., Hare, L., and Couture, P. (2006). Sub-cellular partitioning of cadmium, copper, nickel and zinc in indigenous yellow perch (*Perca flavescens*) sampled along a polymetallic gradient. *Aquat. Toxicol.* **77**, 178–189.

Glover, C. N., and Wood, C. M. (2008a). Absorption of copper and copper–histidine complexes across the apical surface of freshwater rainbow trout intestine. *J. Comp. Physiol. B* **178**, 101–109.

Glover, C. N., and Wood, C. M. (2008b). Histidine absorption across apical surfaces of freshwater rainbow trout intestine: mechanistic characterization and the influence of copper. *J. Membr. Biol.* **221**, 87–95.

Gonzalez, R. J., Wilson, R. W., Wood, C. M., Patrick, M. L., and Val, A. L. (2002). Diverse strategies for ion regulation in fish collected from the ion-poor, acidic Rio Negro. *Physiol. Biochem. Zool.* **75**, 37–47.

Gravel, A., Campbell, P. G. C., and Hontela, A. (2005). Disruption of the hypothalamo-pituitary-interrenal axis in 1+ yellow perch (*Perca flavescens*) chronically exposed to metals in the environment. *Can. J. Fish. Aquat. Sci.* **62**, 982–990.

Green, W. W., Mirza, R. S., Wood, C. M., and Pyle, G. G. (2010). Copper binding dynamics and olfactory impairment in fathead minnows (*Pimephales promelas*). *Environ. Sci. Technol.* **44**, 1431–1437.

Grosell, M. (1996). Acclimation to copper by freshwater teleost fish: Cu uptake, metabolism and elimination during exposure to elevated ambient Cu concentrations. PhD thesis, The August Krogh Institute, University of Copenhagen.

Grosell, M. (2006). Intestinal anion exchange in marine fish osmoregulation. *J. Exp. Biol.* **209**, 2813–2827.

Grosell, M. (2010). The role of the gastrointestinal tract in salt and water balance. In: *Fish Physiology* (M Grosell, A.P Farrell and C.J Brauner, eds), Vol. 30, pp. 135–164. Elsevier, New York.

Grosell, M., and Genz, J. (2006). Ouabain sensitive bicarbonate secretion and acid absorption by the marine fish intestine play a role in osmoregulation. *Am. J. Physiol.* **291**, R1145–R1156.

Grosell, M., and Taylor, J. R. (2007). Intestinal anion exchange in teleost water balance. *Comp. Biochem. Physiol. A* **148**, 14–22.

Grosell, M., and Wood, C. M. (2002). Copper uptake across rainbow trout gills: mechanisms of apical entry. *J. Exp. Biol.* **205**, 1179–1188.

Grosell, M., Boetius, I., Hansen, H. J. M., and Rosenkilde, P. (1996). Influence of preexposure to sublethal levels of copper on Cu-64 uptake and distribution among tissues of the European eel (*Anguilla anguilla*). *Comp. Biochem. Physiol. C* **114**, 229–235.

Grosell, M. H., Hogstrand, C., and Wood, C. M. (1997). Cu uptake and turnover in both Cu-acclimated and non-acclimated rainbow trout (*Oncorhynchus mykiss*). *Aquat. Toxicol.* **38**, 257–276.

Grosell, M., Hansen, H. J. M., and Rosenkilde, P. (1998a). Cu uptake, metabolism and elimination in fed and starved European eels (*Anguilla anguilla*) during adaptation to water-borne Cu exposure. *Comp. Biochem. Physiol. C* **120**, 295–305.

Grosell, M. H., Hogstrand, C., and Wood, C. M. (1998b). Renal Cu and Na excretion and hepatic Cu metabolism in both Cu acclimated and non acclimated rainbow trout (*Oncorhynchus mykiss*). *Aquat. Toxicol.* **40**, 275–291.

Grosell, M., O'Donnell, M. J., and Wood, C. M. (2000). Hepatic versus gallbladder bile composition: in vivo transport physiology of the gallbladder in rainbow trout. *Am. J. Physiol.* **278**, R1674–R1684.

Grosell, M., McGeer, J. C., and Wood, C. M. (2001). Plasma copper clearance and biliary copper excretion are stimulated in copper-acclimated trout. *Am. J. Physiol.* **280**, R796–R806.

Grosell, M., Nielsen, C., and Bianchini, A. (2002). Sodium turnover rate determines sensitivity to acute copper and silver exposure in freshwater animals. *Comp. Biochem. Physiol. C* **133**, 287–303.

Grosell, M., Wood, C. M., and Walsh, P. J. (2003). Copper homeostasis and toxicity in the elasmobranch *Raja erinacea* and the teleost *Myoxocephalus octodecemspinosus* during exposure to elevated water-borne copper. *Comp. Biochem. Physiol. C* **135**, 179–190.

Grosell, M., McDonald, M. D., Walsh, P. J., and Wood, C. M. (2004a). Effects of prolonged copper exposure in the marine gulf toadfish (*Opsanus beta*). II. Drinking rate, copper accumulation and Na^+/K^+-ATPase activity in osmoregulatory tissues. *Aquat. Toxicol.* **68**, 263–275.

Grosell, M., McDonald, M. D., Wood, C. M., and Walsh, P. J. (2004b). Effects of prolonged copper exposure in the marine gulf toadfish (*Opsanus beta*). I. Hydromineral balance and plasma nitrogenous waste products. *Aquat. Toxicol.* **68**, 249–262.

Grosell, M., Blanchard, J., Brix, K. V., and Gerdes, R. (2007a). Physiology is pivotal for interactions between salinity and acute copper toxicity to fish and invertebrates. *Aquat. Toxicol.* **84**, 162–172.

Grosell, M., Gilmour, K. M., and Perry, S. F. (2007b). Intestinal carbonic anhydrase, bicarbonate, and proton carriers play a role in the acclimation of rainbow trout to seawater. *Am. J. Physiol.* **293**, R2099–R2111.

Grosell, M., Genz, J., Taylor, J. R., Perry, S. F., and Gilmour, K. M. (2009a). The involvement of H^+-ATPase and carbonic anhydrase in intestinal HCO_3^- secretion of seawater-acclimated rainbow trout. *J. Exp. Biol.* **212**, 1940–1948.

Grosell, M., Mager, E. M., Williams, C., and Taylor, J. R. (2009b). High rates of HCO_3^- secretion and Cl^- absorption against adverse gradients in the marine teleost intestine: the involvement of an electrogenic anion exchanger and H^+-pump metabolon? *J. Exp. Biol.* **212**, 1684–1696.

Grosell, M., McDonald, M. D., Wood, C. M. and Walsh, P. J. (2011). Prolonged copper exposure impairs acid–base balance, branchial Na^+ and Cl^- extrusion, nitrogenous waste excretion, intestinal transport physiology and affects ^{110m}Ag toxicokinetics in the gulf toadfish, *Opsanus beta. Aquat. Toxicol.* (in preparation).

Gross, J. B., Jr., Myers, B. M., Kost, L. J., Kuntz, S. M., and LaRusso, N. F. (1989). Biliary copper excretion by hepatocyte lysosomes in the rat. Major excretory pathway in experimental copper overload. *J. Clin. Invest.* **83**, 30–39.

Gunshin, H., Mackenzie, B., Berger, U. V., Gunshin, Y., Romero, M. F., Nussberger, S., Gollan, J. L., and Hediger, M. A. (1997). Cloning and characterization of a mammalian proton-coupled metal-ion transporter. *Nature* **388**, 482–488.

Handy, R. D., Sims, D. W., Giles, A., Cambell, H. A., and Musonda, M. M. (1999). Metabolic trade-off between locomotion and detoxification for maintenance of blood chemistry and growth parameters by rainbow trout (*Oncorhynchus mykiss*) during chronic dietary exposure to copper. *Aquat. Toxicol.* **47**, 23–41.

Handy, R. D., Musonda, M. M., and Phillips, C. (2000). Mechanisms of gastrointestinal copper absorption in the African walking catfish: copper dose-effects and a novel anion-dependent pathway in the intestine. *J. Exp. Biol.* **203**, 2365–2377.

Hansen, B. H., Romma, S., Softeland, L. I., Olsvik, P. A., and Andersen, R. A. (2006). Induction and activity of oxidative stress-related proteins during waterborne Cu-exposure in brown trout (*Salmo trutta*). *Chemosphere* **65**, 1707–1714.

Hansen, B. H., Garmo, O. A., Olsvik, P. A., and Andersen, R. A. (2007). Gill metal binding and stress gene transcription in brown trout (*Salmo trutta*) exposed to metal environments: the effect of pre-exposure in natural populations. *Environ. Toxicol. Chem.* **26**, 944–953.

Hansen, J. A., Marr, J. C. A., Lipton, J., Cacela, D., and Bergman, H. L. (1999). Differences in neurobehavioral responses of chinook salmon (*Oncorhynchus tshawytscha*) and rainbow trout (*Oncorhynchus mykiss*) exposed to copper and cobalt: behavioral responses. *Environ. Toxicol. Chem.* **18**, 1972–1978.

Hansen, J. A., Lipton, J., Welsh, P. G., Morris, J., Cacela, D., and Suedkamp, M. J. (2002). Relationship between exposure duration, tissue residues, growth, and mortality in rainbow trout (*Oncorhynchus mykiss*) juveniles sub-chronically exposed to copper. *Aquat. Toxicol.* **58**, 175–188.

Hansen, J. A., Lipton, J., Welsh, P. G., Cacela, D., and MacConnell, B. (2004). Reduced growth of rainbow trout (*Oncorhynchus mykiss*) fed a live invertebrate diet pre-exposed to metal-contaminated sediments. *Environ. Toxicol. Chem.* **23**, 1902–1911.

Harris, E. D. (1991). Copper transport: an overview. *Proc. Soc. Exp. Biol. Med.* **192**, 130–140.

Harris, Z. L., and Gitlin, J. D. (1996). Genetic and molecular basis for copper toxicity. *Am. J. Clin. Nutr.* **63**, 836S–841S.

Hashemi, S., Blust, R., and De Boeck, G. (2008a). Combined effects of different food rations and sublethal copper exposure on growth and energy metabolism in common carp. *Arch. Environ. Contam. Toxicol.* **54**, 318–324.

Hashemi, S., Blust, R., and De Boeck, G. (2008b). The effect of starving and feeding on copper toxicity and uptake in Cu acclimated and non-acclimated carp. *Aquat. Toxicol.* **86**, 142–147.

Heisler, N. (1993). Acid–base regulation. In: *The Physiology of Fishes* (D.H. Evans, ed.), pp. 343–377. CRC Press, Boca Raton, FL.

Heisler, N., and Neumann, P. (1977). Influence of sea-water pH upon bicarbonate uptake induced by hypercapnia in an elasmobranch fish (*Scyliorhinus stellaris*). *Pflügers Arch. – Eur. J. Physiol.* **368**,pp. 19–19

Henry, R. P. (1991). Techniques for measuring carbonic anhydrase activity *in vitro*: the electrometric delta pH and pH stat assays. In: *The Carbonic Anhydrases: Cellular Physiology and Molecular Genetics* (S.J. Dodgson, R.E. Tashian, G. Gros and N.D. Carter, eds), pp. 119–126. Plenum Press, New York.

Hetrick, F. M., Knittel, M. D., and Fryer, J. L. (1979). Increased susceptibility of rainbow trout to infectious hematopoietic necrosis virus after exposure to copper. *Appl. Environ. Microbiol.* **37**, 198–201.

Hogstrand, C., and Haux, C. (1991). Binding and detoxification of heavy-metals in lower-vertebrates with reference to metallothionein. *Comp. Biochem. Physiol. C* **100**, 137–141.

Hogstrand, C., Lithner, G., and Haux, C. (1989). Relationship between metallothionein, copper and zinc in perch (*Perca fluviatilis*) environmentally exposed to heavy metals. *Mar. Environ. Res.* **28**, 179–182.

Hogstrand, C., Lithner, G., and Haux, C. (1991). The importance of metallothionein for the accumulation of copper, zinc and cadmium in environmentally exposed perch, *Perca fluviatilis*. *Pharmacol. Toxicol.* **68**, 492–501.

Hook, S. E., and Fisher, N. S. (2001). Reproductive toxicity of metals in calanoid copepods. *Mar. Biol.* **138**, 1131–1140.

Hook, S. E., and Fisher, N. S. (2002). Relating the reproductive toxicity of five ingested metals in calanoid copepods with sulfur affinity. *Mar. Environ. Res.* **53**, 161–174.

Horning, W. B., and Neiheisel, T. W. (1979). Chronic effect of copper on the bluntnose minnow, *Pimephales notatus* (Rafinesque). *Arch. Environ. Contam. Toxicol.* **8**, 545–552.

Houwen, R., Dijkstra, M., Kuipers, F., Smit, E. P., Havinga, R., and Vonk, R. J. (1990). Two pathways for biliary copper excretion in the rat. The role of glutathione. *Biochem. Pharmacol.* **39**, 1039–1044.

Hoyle, I., Shaw, B. J., and Handy, R. D. (2007). Dietary copper exposure in the African walking catfish, *Clarias gariepinus*: transient osmoregulatory disturbances and oxidative stress. *Aquat. Toxicol.* **83**, 62–72.

Hung, C. C., Tsui, T. K. N., Wilson, J. M., Nawata, C. M., Wood, C. M., and Wright, P. A. (2007). Rhesus glycoprotein gene expression in the mangrove killifish, *Kryptolebias marmoratus*, exposed to elevated environmental ammonia levels and air. *J. Exp. Biol.* **210**, 2419–2429.

Hung, C. C., Nawata, C. M., Wood, C. M., and Wright, P. A. (2008). Rhesus glycoprotein and urea transporter genes are expressed in early stages of development of rainbow trout (*Oncorhynchus mykiss*). *J. Exp. Zool. A Ecol. Genet. Physiol.* **309**, 262–268.

Irie, T., and Seki, T. (2002). Retinoid composition and retinal localization in the eggs of teleost fishes. *Comp. Biochem. Physiol. B* **131**, 209–219.

Johnson, A., Carew, E., and Sloman, K. A. (2007). The effects of copper on the morphological and functional development of zebrafish embryos. *Aquat. Toxicol.* **84**, 431–438.

Johnston, W. L., Atkinson, J. L., and Galnville, N. T. (1992). Effect of PCPA or tryptophan on brain serotonin and on consumption of a high protein or high carbohydrate diet by rainbow trout. *Oncorhynchus mykiss*. *J. Nutr. Biochem.* **3**, 421–428.

Kaler, S. G. (1998). Metabolic and molecular bases of Menkes disease and occipital horn syndrome. *Pediatr. Dev. Pathol.* **1**, 85–98.

Kamunde, C., and MacPhail, R. (2008). Bioaccumulation and hepatic speciation of copper in rainbow trout (*Oncorhynchus mykiss*) during chronic waterborne copper exposure. *Arch. Environ. Contam. Toxicol.* **54**, 493–503.

Kamunde, C., and Wood, C. M. (2003). The influence of ration size on copper homeostasis during sublethal dietary copper exposure in juvenile rainbow trout, *Oncorhynchus mykiss*. *Aquat. Toxicol.* **62**, 235–254.

Kamunde, C. N., Grosell, M., Lott, J. N. A., and Wood, C. M. (2001). Copper metabolism and gut morphology in rainbow trout (*Oncorhynchus mykiss*) during chronic sublethal dietary copper exposure. *Can. J. Fish. Aquat. Sci.* **58**, 293–305.

Kamunde, C., Clayton, C., and Wood, C. M. (2002a). Waterborne vs. dietary copper uptake in rainbow trout and the effects of previous waterborne copper exposure. *Am. J. Physiol.* **283**, R69–R78.

Kamunde, C., Grosell, M., Higgs, D., and Wood, C. M. (2002b). Copper metabolism in actively growing rainbow trout (*Oncorhynchus mykiss*): interactions between dietary and waterborne copper uptake. *J. Exp. Biol.* **205**, 279–290.

Kamunde, C. N., Pyle, G. G., McDonald, D. G., and Wood, C. M. (2003). Influence of dietary sodium on waterborne copper toxicity in rainbow trout. *Oncorhynchus mykiss. Environ. Toxicol. Chem.* **22**, 342–350.

Kamunde, C. N., Niyogi, S., and Wood, C. M. (2005). Interaction of dietary sodium chloride and waterborne copper in rainbow trout (*Oncorhynchus mykiss*): copper toxicity and sodium and chloride homeostasis. *Can. J. Fish. Aquat. Sci.* **62**, 390–399.

Kang, J.-C., Kim, S.-G., and Jang, S.-W. (2005). Growth and hematological changes of rockfish, *Sebastes schlegeli* (Hilgendorf) exposed to dietary Cu and Cd. *J. World Aquacult. Soc.* **36**, 188–195.

Katz, I. R. (1981). Interaction between the oxygen and tryptophan dependence of synaptosomal tryptophan hydroxylase. *J. Neurochem.* **37**, 447–451.

Khangarot, B. S., and Tripathi, D. M. (1991). Changes in humoral and cell-mediated immune responses and in skin and respiratory surfaces of catfish, *Saccobranchus fossilis*, following copper exposure. *Ecotoxicol. Environ. Saf.* **22**, 291–308.

Kiela, P. R., Xu, H., and Ghishan, F. K. (2006). Apical Na^+/H^+ exchangers in the mammalian gastrointestinal tract. *J. Physiol. Pharmacol.* **57**(Suppl. 7), 51–79.

Kim, H., Son, H. Y., Bailey, S. M., and Lee, J. (2009). Deletion of hepatic Ctr1 reveals its function in copper acquisition and compensatory mechanisms for copper homeostasis. *Am. J. Physiol.* **296**, G356–G364.

Kim, S. D., Ma, H., Allen, H. E., and Cha, D. K. (1999). Influence of dissolved organic matter on the toxicity of copper to *Ceriodaphnia dubia*: effects of complexation kinetics. *Environ. Toxicol. Chem.* **18**, 2433–2437.

Kim, S. G., and Kang, J. C. (2004). Effect of dietary copper exposure on accumulation, growth and hematological parameters of the juvenile rockfish. *Sebastes schlegeli. Mar. Environ. Res.* **58**, 65–82.

Kjoss, V. A., Grosell, M., and Wood, C. M. (2005a). The influence of dietary Na on Cu accumulation in juvenile rainbow trout exposed to combined dietary and waterborne Cu in soft water. *Arch. Environ. Contam. Toxicol.* **49**, 520–527.

Kjoss, V. A., Kamunde, C. N., Niyogi, S., Grosell, M., and Wood, C. M. (2005b). Dietary Na does not reduce dietary Cu uptake by juvenile rainbow trout. *J. Fish Biol.* **66**, 468–484.

Klinck, J. S., Green, W. W., Mirza, R. S., Nadella, S. R., Chowdhury, M. J., Wood, C. M., and Pyle, G. G. (2007). Branchial cadmium and copper binding and intestinal cadmium uptake in wild yellow perch (*Perca flavescens*) from clean and metal-contaminated lakes. *Aquat. Toxicol.* **84**, 198–207.

Kling, P., and Olsson, P. E. (1995). Regulation of the rainbow trout metallothionein-A gene. *Mar. Environ. Res.* **39**, 117–120.

Kling, P. G., and Olsson, P. E. (2000). Involvement of differential metallothionein expression in free radical sensitivity of RTG-2 and CHSE-214 cells. *Free Rad. Biol. Med.* **28**, 1628–1637.

Klinkhammer, G. P., and Bender, M. L. (1981). Trace metal distribution in the Hudson river estuary. *Estuar. Coast. Shlf. Sci.* **12**, 629–643.

Knopfel, M., Smith, C., and Solioz, M. (2005). ATP-driven copper transport across the intestinal brush border membrane. *Biochem. Biophys. Res. Commun.* **330**, 645–652.

Kolmakov, N. N., Hubbard, P. C., Lopes, O., and Canario, A. V. (2009). Effect of acute copper sulfate exposure on olfactory responses to amino acids and pheromones in goldfish (*Carassius auratus*). *Environ. Sci. Technol.* **43**, 8393–8399.

Koltes, K. H. (1985). Effects of sublethal copper concentrations in the structure and activity of Atlantic silverside schools. *Trans. Am. Fish. Soc.* **114**, 413–422.

Kozelka, P. B., and Bruland, K. W. (1998). Chemical speciation of dissolved Cu, Zn, Cd, Pb in Narragansett Bay, Rhode Island. *Mar. Chem.* **60**, 267–282.

Kraemer, L. D., Campbell, P. G. C., and Landis, H. (2005). Dynamics of Cd, Cu and Zn accumulation in organs and in the sub-cellular fractions in field transplanted juvenile yellow perch (*Perca flavescens*). *Environ. Pollut.* **138**, 324–337.

Krumschnabel, G., Manzl, C., Berger, C., and Hofer, B. (2005). Oxidative stress, mitochondrial permeability transition, and cell death in Cu-exposed trout hepatocytes. *Toxicol. Appl. Pharmacol.* **209**, 62–73.

Kunwar, P. S., Tudorache, C., Eyckmans, M., Blust, R., and DeBoeck, G. (2009). Influence of food ration, copper exposure and exercise on the energy metabolism of common carp (*Cyprinus carpio*). *Comp. Biochem. Physiol. C* **149**, 113–119.

Lanno, R. P., Singer, S. J., and Hilton, J. W. (1985). Maximum tolerable and toxicity level of dietary copper in rainbow trout (*Salmo gairdneri* Richardson). *Aquaculture* **49**, 257–268.

Lanno, R. P., Hicks, B., and Hilton, J. W. (1987). Histological observations on intrahepatocytic copper-containing granules in rainbow trout reared on diets containing elevated levels of copper. *Aquat. Toxicol.* **10**, 251–263.

Larsen, B. K., Portner, H. O., and Jensen, F. B. (1997). Extra- and intracellular acid–base balance and ionic regulation in cod (*Gadus morhua*) during combined and isolated exposures to hypercapnia and copper. *Mar. Biol* **128**, 337–346.

Lauren, D. J., and McDonald, D. G. (1985). Effects of copper on branchial ionoregulation in the rainbow trout, *Salmo gairdneri* Richardson – modulation by water hardness and pH. *J. Comp. Physiol. B* **155**, 635–644.

Lauren, D. J., and McDonald, D. G. (1986). Influence of water hardness, pH, and alkalinity on the mechanisms of copper toxicity in juvenile rainbow trout, Salmo gairdneri. *Can. J. Fish. Aquat. Sci.* **43**, 1488–1496.

Lauren, D. J., and McDonald, D. G. (1987a). Acclimation to copper by rainbow trout, *Salmo gairdneri* – biochemistry. *Can. J. Fish. Aquat. Sci.* **44**, 105–111.

Lauren, D. J., and McDonald, D. G. (1987b). Acclimation to copper by rainbow trout, *Salmo gairdneri* – physiology. *Can. J. Fish. Aquat. Sci.* **44**, 99–104.

Lett, P. F., Farmer, G. J., and Beamish, F. W. (1976). Effects of copper on some aspects of the bioenergetics of rainbow trout (*Salmo gairdneri*). *J. Fish. Res. Bd Can.* **33**, 1335–1342.

Lewis, A. G., and Cave, W. R. (1982). The biological importance of copper in oceans and estuaries. *Oceanogr. Mar. Biol. Annu. Rev.* **20**, 471–692.

Lewis, S. D., and Lewis, E. M. (1971). The effects of zinc and copper on the osmolality of blood serum of the channel catfish *Ictalurus punctatus* and the golden shiner. *Notemigonus crysoleucas. Trans. Am. Fish. Soc.* **100**, 639–643.

Li, J., Lock, R. A. C., Klaren, P. H. M., Swarts, H. G. P., Stekhoven, F. M. A. H., Bonga, S. E. W., and Flik, G. (1996). Kinetics of Cu^{2+} inhibition of Na^+/K^+-ATPase. *Toxicol. Lett.* **87**, 31–38.

Li, J., Quabius, E. S., Wendelaar Bonga, S. E., Flik, G., and Lock, R. A. C. (1998). Effects of water-borne copper on branchial chloride cells and Na^+/K^+-ATPase activities in Mozambique tilapia (*Oreochromis mossambicus*). *Aquat. Toxicol.* **43**, 1–11.

Li, J.-S., Li, J.-L., and Wu, T.-T. (2007). The effects of copper, iron and zinc on digestive enzyme activity in the hybrid tilapia *Oreochromis niloticus* (L.) × *Oreochromis aureus* (Steindachner). *J. Fish Biol.* **71**, 1788–1798.

Linbo, T. L., Baldwin, D. H., McIntyre, J. K., and Scholz, N. L. (2009). Effects of water hardness, alkalinity, and dissolved organic carbon on the toxicity of copper to the lateral line of developing fish. *Environ. Toxicol. Chem.* **28**, 1455–1461.

Linder, M. C., Wooten, L., Cerveza, P., Cotton, S., Shulze, R., and Lomeli, N. (1998). Copper transport. *Am. J. Clin. Nutr.* **67**, 965S–971S.

Lorz, H. W., and McPherson, B. P. (1976). Effects of copper or zinc in freshwater on the adaptation to seawater and ATPase activity, and the effects of copper on migratory disposition of Coho salmon (*Oncorhynchus kisutch*). *J. Fish. Res. Bd Can.* **33**, 223–230.

Luider, C. D., Crusius, J., Playle, R. C., and Curtis, P. J. (2004). Influence of natural organic matter source on copper speciation as demonstrated by Cu binding to fish gills, by ion selective electrode and by DGT gel sampler. *Environ. Sci. Technol.* **38**, 2865–2872.

Lutsenko, S., and Petris, M. J. (2002). Function and regulation of the mammalian copper-transporting ATPase: insights from biochemical and cell biological approaches. *J. Membr. Biol.* **191**, 1–12.

Lutsenko, S., Barnes, N. L., Bartee, M. Y., and Dmitriev, O. Y. (2007). Function and regulation of human copper-transporting ATPases. *Physiol. Rev.* **87**, 1011–1046.

Mackenzie, N. C., Brito, M., Reyes, A. E., and Allende, M. L. (2004). Cloning, expression pattern and essentiality of the high-affinity copper transporter 1 (ctr1) in zebrafish. *Gene* **328**, 113–120.

Madsen, E. C., Morcos, P. A., Mendelsohn, B. A., and Gitlin, J. D. (2008). *In vivo* correction of a Menkes disease model using antisense oligonucleotides. *Proc. Natl. Acad. Sci. U.S.A.* **105**, 3909–3914.

Magid, E. (1967). The activity of carbonic anhydrase B and C from human erythrocytes and the inhibition of the enzymes by copper. *Scand. J. Haematol.* **2**, 257–270.

Marceau, N., and Aspin, N. (1973). The intracellular distribution of radiocopper derived from ceruloplasmin and from albumin. *Biochim. Biophys. Acta* **293**, 228–350.

Marshall, W. S., and Grosell, M. (2005). Ion transport, osmoregulation and acid–base balance. In: *Physiology of Fishes* (D. Evans and J.B. Claiborne, eds), 3rd edn, pp. 177–230. CRC Press, Boca Raton, FL.

Matsuo, A. Y. O., Playle, R. C., Val, A. L., and Wood, C. M. (2004). Physiological action of dissolved organic matter in rainbow trout in the presence and absence of copper: sodium uptake kinetics and unidirectional flux rates in hard and softwater. *Aquat. Toxicol.* **70**, 63–81.

McCarter, J. A., and Roch, M. (1984). Chronic exposure of coho salmon to sublethal concentrations of copper – III. Kinetics of metabolism of metallothionein. *Comp. Biochem. Physiol. C.* **77**, 83–87.

McDonald, D. G., and Wood, C. M. (1993). Branchial mechanisms of acclimation to metals in freshwater fish. *Fish Ecophysiology, Fish and Fisheries Series* **9**, 297–321.

McDuffee, A. T., Senisterra, G., Huntley, S., Lepock, J. R., Sekhar, K. R., Meredith, M. J., Borrelli, M. J., Morrow, J. D., and Freeman, M. L. (1997). Proteins containing non-native disulfide bonds generated by oxidative stress can act as signals for the induction of the heat shock response. *J. Cell Physiol.* **171**, 143–151.

McGeer, J. C., Playle, R. C., Wood, C. M., and Galvez, F. (2000a). A physiologically based biotic ligand model for predicting the acute toxicity of waterborne silver to rainbow trout in freshwaters. *Environ. Sci. Technol.* **34**, 4199–4207.

McGeer, J. C., Szebedinszky, C., Wood, C. M., and McDonald, D. G. (2000b). Effects of chronic sublethal exposure to waterborne Cu, Cd or Zn in rainbow trout. 1: Ionoregulatory disturbance and metabolic costs. *Aquat. Toxicol.* **50**, 231–243.

McGeer, J. C., Szebedinszky, C., McDonald, D. G., and Wood, C. M. (2002). The role of dissolved organic carbon in moderating the bioavailability and toxicity of Cu to rainbow trout during chronic waterborne exposure. *Comp. Biochem. Physiol. C* **133**, 147–160.

McGeer, J. C., Brix, K. V., Skeaff, J. M., DeForest, D. K., Brigham, S. I., Adams, W. J., and Green, A. (2003). Inverse relationship between bioconcentration factor and exposure concentration for metals: implications for hazard assessment of metals in the aquatic environment. *Environ. Toxicol. Chem.* **22**, 1017–1037.

McGeer, J. C., Nadella, S., Alsop, D. H., Hollis, L., Taylor, L. N., McDonald, D. G., and Wood, C. M. (2007). Influence of acclimation and cross-acclimation of metals on acute Cd toxicity and Cd uptake and distribution in rainbow trout (*Oncorhynchus mykiss*). *Aquat. Toxicol.* **84**, 190–197.

McIntyre, J. K., Baldwin, D. H., Meador, J. P., and Scholz, N. L. (2008). Chemosensory deprivation in juvenile coho salmon under varying water chemistry conditions. *Environ. Sci. Technol.* **42**, 1352–1358.

McKim, J. M. (1977). Evaluation of tests with early life stages of fish for predicting long-term toxicity. *J. Fish. Res Bd Can.* **34**, 1148–1154.

McKim, J. M., and Benoit, D. A. (1971). Effects of long-term exposure to copper on survival, growth, and reproduction of brook trout (*Salvelinus fontinalis*). *J. Fish. Res. Bd Can.* **28**, 655–662.

McKim, J. M., and Benoit, D. A. (1974). Duration of toxicity test for establishing "no effect" concentrations for copper with brook trout (*Salvelinus fontinalis*). *J. Fish. Res. Bd Can.* **31**, 449–452.

McKim, J. M., Christensen, G. M., and Hunt, E. P. (1970). Changes in the blood of the brook trout *Salvelinus fontinalis* after short-term and long-term exposure to copper. *J. Fish. Res. Bd Can.* **27**, 1883–1889.

McKim, J. M., Eaton, J. G., and Holcombe, G. W. (1978). Metal toxicity to embryos and larvae of eight species of freshwater fish – II: Copper. *Bull. Environ. Contam. Toxicol.* **19**, 608–616.

Mendelsohn, B. A., Yin, C., Johnson, S. L., Wilm, T. P., Solnica-Krezel, L., and Gitlin, J. D. (2006). Atp7a determines a hierarchy of copper metabolism essential for notochord development. *Cell Metab.* **4**, 155–162.

Mercer, J. F., and Llanos, R. M. (2003). Molecular and cellular aspects of copper transport in developing mammals. *J. Nutr.* **133**, 1481S–1484S.

Mercer, J. F., Barnes, N., Stevenson, J., Strausak, D., and Llanos, R. M. (2003). Copper-induced trafficking of the Cu-ATPases: a key mechanism for copper homeostasis. *Biometals* **16**, 175–184.

Meyer, J. S., and Adams, W. J. (2010). Relationship between biotic ligand model-based water quality criteria and avoidance and olfactory responses to copper by fish. *Environ. Toxicol. Chem.* **29**, 2096–2103.

Meyer, J. S., Santore, R. C., Bobbitt, J. P., Debrey, L. D., Boese, C. J., Paquin, P. R., Allen, H. E., Bergman, H. L., and Ditoro, D. M. (1999). Binding of nickel and copper to fish gills predicts toxicity when water hardness varies, but free-ion activity does not. *Environ. Sci. Technol.* **33**, 913–916.

Miller, P. A., Munkittrick, K. R., and Dixon, D. G. (1992). Relationship between concentrations of copper and zinc in water, sediment, benthic invertebrates, and tissue of white sucker (*Catostomus commersoni*) at metal-contaminated sites. *Can. J. Fish. Aquat. Sci.* **49**, 978–984.

Miller, P. A., Lanno, R. P., McMaster, M. E., and Dixon, D. G. (1993). Relative contributions of dietary and waterborne copper to tissue copper burdens and waterborne copper tolerance in rainbow trout (*Oncorhynchus mykiss*). *Can. J. Fish. Aquat. Sci.* **50**, 1683–1689.

Minghetti, M., Leaver, M. J., Carpene, E., and George, S. G. (2008). Copper transporter 1, metallothionein and glutathione reductase genes are differentially expressed in tissues of sea bream (*Sparus aurata*) after exposure to dietary or waterborne copper. *Comp. Biochem. Physiol. C* **147**, 450–459.

Minghetti, M., Leaver, M. J., and George, S. G. (2010). Multiple Cu-ATPase genes are differentially expressed and transcriptionally regulated by Cu exposure in seabream, *Sparus aurata*. *Aquat. Toxicol.* **97**, 23–33.

Mommsen, T. P., and Walsh, P. J. (1989). Evolution of urea synthesis in vertebrates – the piscine connection. *Science* **243**, 72–75.

Mommsen, T. P., and Walsh, P. J. (1991). Urea synthesis in fishes: evolutionary and biochemical perspectives. In: *Biochemistry and Molecular Biology of Fishes* Vol. 1, (P.W Hochachka and T.P Mommsen, eds), pp. 137–163. Elsevier, New York.

Moreira-Santos, M., Donato, C., Lopes, I., and Ribeiro, R. (2008). Avoidance tests with small fish: determination of the median avoidance concentration and of the lowest-observed-effect gradient. *Environ. Toxicol. Chem.* **27**, 1576–1582.

Morel, F. M. M., and Hering, J. G. (1993). *Principles and Applications of Aquatic Chemistry* (2nd edn..). John Wiley and Sons, New York.

Morgan, I. J., Henry, R. P., and Wood, C. M. (1997). The mechanism of acute silver nitrate toxicity in freshwater rainbow trout (*Oncorhynchus mykiss*) is inhibition of gill Na^+ and Cl^- transport. *Aquat. Toxicol.* **38**, 145–163.

Morgan, T. P., Grosell, M., Gilmour, K. M., Playle, R. C., and Wood, C. M. (2004). Time course analysis of the mechanism by which silver inhibits active Na^+ and Cl^- uptake in gills of rainbow trout. *Am. J. Physiol.* **287**, R234–R242.

Moriya, M., Ho, Y. H., Grana, A., Nguyen, L., Alvarez, A., Jamil, R., Ackland, M. L., Michalczyk, A., Hamer, P., Ramos, D., Kim, S., Mercer, J. F., and Linder, M. C. (2008). Copper is taken up efficiently from albumin and alpha2-macroglobulin by cultured human cells by more than one mechanism. *Am. J. Physiol.* **295**, C708–C721.

Mount, D. R., Barth, A. K., Garrison, T. D., Barten, K. A., and Hockett, J. R. (1994). Dietary and waterborne exposure of rainbow trout (*Oncorhynchus mykiss*) to copper, cadmium, lead and zinc using a live diet. *Environ. Toxicol. Chem.* **13**, 2031–2041.

Mushiake, K., Muroga, K., and Nakai, T. (1984). Increased susceptibility of Japanese eel *Anguilla japonica* to *Edwardsiella tarda* and *Pseudomonas anguilliseptica* following exposure to copper. *Bull. Jpn. Soc. Scient. Fish.* **50**, 1797–1801.

Mushiake, K., Nakai, T., and Muroga, K. (1985). Lowered phagocytosis in the blood of eels exposed to copper. *Fish Pathol.* **20**, 49–53.

Naddy, R. B., Stubblefield, W. A., May, J. R., Tucker, S. A., and Hockett, J. R. (2002). The effect of calcium and magnesium ratios on the toxicity of copper to five aquatic species in freshwater. *Environ. Toxicol. Chem.* **21**, 347–352.

Nadella, S. R., Bucking, C., Grosell, M., and Wood, C. M. (2006a). Gastrointestinal assimilation of Cu during digestion of a single meal in the freshwater rainbow trout (*Oncorhynchus mykiss*). *Comp Biochem. Physiol. C* **143**, 394–401.

Nadella, S. R., Grosell, M., and Wood, C. M. (2006b). Physical characterization of high-affinity gastrointestinal Cu transport in vitro in freshwater rainbow trout *Oncorhynchus mykiss*. *J. Comp. Physiol. B* **176**, 793–806.

Nadella, S. R., Grosell, M., and Wood, C. M. (2007). Mechanisms of dietary Cu uptake in freshwater rainbow trout: evidence for Na-assisted Cu transport and a specific metal carrier in the intestine. *J. Comp. Physiol. B* **177**, 433–446.

Nadella, S. R., Hung, C. Y., and Wood, C. M. (2011). Mechanistic characterization of gastric copper transport in rainbow trout. *J. Comp. Physiol. B* **181**, 27–41.

Nakada, T., Hoshijima, K., Esaki, M., Nagayoshi, S., Kawakami, K., and Hirose, S. (2007a). Localization of ammonia transporter Rhcg1 in mitochondrion-rich cells of yolk sac, gill, and kidney of zebrafish and its ionic strength-dependent expression. *Am. J. Physiol.* **293**, R1743–R1753.

Nakada, T., Westhoff, C. M., Kato, A., and Hirose, S. (2007b). Ammonia secretion from fish gill depends on a set of Rh glycoproteins. *FASEB J.* **21**, 1067–1074.

Nawata, C. M., Hung, C. C., Tsui, T. K., Wilson, J. M., Wright, P. A., and Wood, C. M. (2007). Ammonia excretion in rainbow trout (*Oncorhynchus mykiss*): evidence for Rh glycoprotein and H$^+$-ATPase involvement. *Physiol. Genomics* **31**, 463–474.

Nawata, C. M., Hirose, S., Nakada, T., Wood, C. M., and Kato, A. (2010a). Rh glycoprotein expression is modulated in pufferfish (*Takifugu rubripes*) during high environmental ammonia exposure. *J. Exp. Biol.* **213**, 3150–3160.

Nawata, C. M., Wood, C. M., and O'Donnell, M. J. (2010b). Functional characterization of Rhesus glycoproteins from an ammoniotelic teleost, the rainbow trout, using oocyte expression and SIET analysis. *J. Exp. Biol.* **213**, 1049–1059.

Niyogi, S., Kamunde, C. N., and Wood, C. M. (2005). Food selection, growth and physiology in relation to the dietary sodium content in rainbow trout *Oncorhynchus mykiss* under chronic waterborne Co exposure. *Aquat. Toxicol.* **77**, 210–221.

Ogino, C., and Yang, G. Y. (1980). Requirements of carp and rainbow trout for dietary manganese and copper. *Bull. Jpn. Soc. Scient. Fish.* **46**, 455–458.

O'Halloran, T. V., and Culotta, V. C. (2000). Metallochaperones: an intracellular shuttle service for metal ions. *J. Biol. Chem.* **275**, 25057–25060.

Ojo, A. A., and Wood, C. M. (2007). *In vitro* analysis of the bioavailability of six metals via the gastro-intestinal tract of the rainbow trout (*Oncorhynchus mykiss*). *Aquat. Toxicol.* **83**, 10–23.

Ojo, A. A., Nadella, S. R., and Wood, C. M. (2009). *In vitro* examination of interactions between copper and zinc uptake via the gastrointestinal tract of the rainbow trout (*Oncorhynchus mykiss*). *Arch. Environ. Contam. Toxicol.* **56**, 244–252.

Olivari, F. A., Hernández, P. P., and Allende, M. L. (2008). Acute copper exposure induces oxidative stress and cell death in lateral line hair cells of zebrafish larvae. *Brain Res.* **1244**, 1–12.

O'Neill, J. G. (1981). The humoral immune response of *Salmo trutta* L. and *Cyprinus carpio* L. exposed to heavy metals. *J. Fish Biol.* **19**, 297–306.

Osuna-Jimenez, I., Williams, T. D., Prieto-Alamo, M. J., Abril, N., Chipman, J. K., and Pueyo, C. (2009). Immune- and stress-related transcriptomic responses of *Solea senegalensis* stimulated with lipopolysaccharide and copper sulphate using heterologous cDNA microarrays. *Fish Shellfish Immunol.* **26**, 699–706.

Overnell, J., and McIntosh, R. (1988). The effect of supplementary dietary copper on copper and metallothionein levels in marine flatfish. *Mar. Environ. Res.* **26**, 237–247.

Paquin, P. R., Gorsuch, J. W., Apte, S., Batley, G. E., Bowles, K. C., Campbell, P. G. C., Delos, C. G., Di Toro, D. M., Dwyer, R. L., Galvez, F., Gensemer, R. W., Goss, G. G., Hogstrand, C., Janssen, C. R., McGeer, J. C., Naddy, R. B., Playle, R. C., Santore, R. C., Schneider, U., Stubblefield, W. A., Wood, C. M., and Wu, K. B. (2002). The biotic ligand model: a historical overview. *Comp. Biochem. Physiol. C* **133**, 3–35.

Patterson, J. W., Minear, R. A., Gasca, E., and Petropoulou, C. (1998). Industrial discharges of metals to water. In: *Metals in Surface Waters* (H.E. Allen, A.W. Garrison and G.W. Luther, eds), pp. 37–66. Ann Arbor Press, Chelsea, MI.

Pelgrom, S. M. G. J., Lamers, L. P. M., Garritsen, J. A. M., Pels, B. M., Lock, R. A. C., Balm, P. H. M., and Wendelaar Bonga, S. E. (1994). Interactions between copper and cadmium during single and combined exposure in juvenile tilapia *Oreochromis mossambicus*: influence of feeding condition on whole body metal accumulation and the effect of the metals on tissue water and ion content. *Aquat. Toxicol.* **30**, 117–135.

Pelgrom, S. M. G. J., Lock, R. A. C., Balm, P. H. M., and Bonga, S. E. W. (1995). Integrated physiological response of tilapia, *Oreochromis mossambicus*, to sublethal copper exposure. *Aquat. Toxicol.* **32**, 303–320.

Peuravouri, J., and Pihlaja, K. (2010). Structural characterization of humic substances. In: *Limnology of Humic Waters* (J. Keskitalo and P. Eloranta, eds), pp. 22–39. Backhuys, Leiden.

Pierron, F., Bourret, V., St-Cyr, J., Campbell, P. G., Bernatchez, L., and Couture, P. (2009). Transcriptional responses to environmental metal exposure in wild yellow perch (*Perca flavescens*) collected in lakes with differing environmental metal concentrations (Cd, Cu, Ni). *Ecotoxicology* **18**, 620–631.

Pilgaard, L., Malte, H., and Jensen, F. B. (1994). Physiological effects and tissue accumulation of copper in fresh water rainbow trout (*Oncorhynchus mykiss*) under normoxic and hypoxic conditions. *Aquat. Toxicol.* **29**, 197–212.

Playle, R. C., and Dixon, D. G. (1993). Copper and cadmium binding to fish gills: modification by dissolved organic carbon and synthetic ligands. *Can. J. Fish. Aquat. Sci.* **50**, 2667–2677.

Playle, R. C., Gensemer, R. W., and Dixon, D. G. (1992). Copper accumulation on gills of fathead minnows: influence of water hardness, complexation and pH of the gill micro-environment. *Environ. Toxicol. Chem.* **11**, 381–391.

Playle, R. C., Dixon, D. G., and Burnison, K. (1993). Copper and cadmiun binding to fish gills: estimates of metal–gill stability constants and modelling of metal accumulation. *Can. J. Fish. Aquat. Sci.* **50**, 2678–2687.

Prieto-Alamo, M. J., Abril, N., Osuna-Jimenez, I., and Pueyo, C. (2009). *Solea senegalensis* genes responding to lipopolysaccharide and copper sulphate challenges: large-scale identification by suppression subtractive hybridization and absolute quantification of transcriptional profiles by real-time RT-PCR. *Aquat. Toxicol.* **91**, 312–319.

Prioux-Guyonneau, M., Mocaer-Cretet, E., Redjimi-Hafsi, F., and Jacquot, C. (1982). Changes in brain 5-hydroxytryptamine metabolism induced by hypobaric hypoxia. *Gen. Pharmacol.* **13**, 251–254.

Prosser, L. (1973). *Comparative Animal Physiology*. Saunders, New York.

Pruell, R. J., and Engelhardt, F. R. (1980). Liver cadmium uptake, catalase inhibition and cadmium thionein production on killifish (*Fundulus heteroclitus*) induced by experimental cadmium exposure. *Mar. Environ. Res.* **3**, 101–111.

Pyle, G. G., and Mirza, R. S. (2007). Copper-impaired chemosensory function and behavior in aquatic animals. *Hum. Ecol. Risk Assess.* **13**, 492–505.

Pyle, G. G., Kamunde, C. N., McDonald, D. G., and Wood, C. M. (2003). Dietary sodium inhibits aqueous copper uptake in rainbow trout (*Oncorhynchus mykiss*). *J. Exp. Biol.* **206**, 609–618.

Radi, A. A., and Matkovics, B. (1988). Effects of metal ions on the antioxidant enzyme activities, protein contents and lipid peroxidation of carp tissues. *Comp. Biochem. Physiol. C* **90**, 69–72.

Randall, D. J., and Wright, P. A. (1989). The interaction between carbon dioxide and ammonia excretion and water pH in fish. *Can. J. Zool.* **67**, 2936–2942.

Randall, D. J., Wilson, J. M., Peng, K. W., Kok, T. W. K., Kuah, S. S. L., Chew, S. F., Lam, T. J., and Ip, Y. K. (1999). The mudskipper, *Periophthalmodon schlosseri*, actively transports NH_4^+ against a concentration gradient. *Am. J. Physiol.* **277**, R1562–R1567.

Rau, M. A., Whitaker, J., Freedman, J. H., and Di Giulio, R. T. (2004). Differential susceptibility of fish and rat liver cells to oxidative stress and cytotoxicity upon exposure to pro-oxidants. *Comp. Biochem. Physiol. C* **137**, 335–342.

Reiley, M. C. (2007). Science, policy, and trends of metals risk assessment at EPA: how understanding metals bioavailability has changed metals risk assessment at US EPA. *Aquat. Toxicol.* **84**, 292–298.

Richards, J. G., Curtis, P. J., Burnison, B. K., and Playle, R. C. (2001). Effects of natural organic matter source on reducing metal toxicity to rainbow trout (*Oncorhynchus mykiss*) and on metal binding to their gills. *Environ. Toxicol. Chem* **20**, 1159–1166.

Roch, M., and McCarter, J. A. (1984). Hepatic metallothionein production and resistance to heavy metals by rainbow trout (*Salmo gairdneri*) – I. Exposed to an artificial mixture of zinc, copper and cadmium. *Comp. Biochem. Physiol. C* **77**, 71–75.

Rødsæther, M., Olafsen, J., Raa, J., Myhre, K., and Steen, J. B. (1977). Copper as an initiating factor of vibriosis (*Vibrio anguillarum*) in eel (*Anguilla anguilla*). *J. Fish Biol.* **10**, 17–21.

Roméo, M., Bennani, N., Gnassia-Barelli, M., Lafaurie, M., and Girard, J. P. (2000). Cadmium and copper display different responses towards oxidative stress in the kidney of the sea bass *Dicentrarchus labrax. Aquat. Toxicol.* **48**, 185–194.

Rougier, F., Troutaud, D., Ndoye, A., and Deschaux, P. (1994). Non-specific immune response of zebrafish, *Brachydanio rerio* (Hamilton-Buchanan) following copper and zinc exposure. *Fish Shellfish Immunol.* **4**, 115–127.

Sampaio, F. G., Boijink, C. D. L., Oba, E. T., Santos, L. R. B. D., Kalinin, A. L., and Rantin, F. T. (2008). Antioxidant defenses and biochemical changes in pacu (*Piaractus mesopotamicus*) in response to single and combined copper and hypoxia exposure. *Comp. Biochem. Physiol. C* **147**, 43–51.

Sanchez, W., Palluel, O., Meunier, L., Coquery, M., Porcher, J.-M., and Aït-Aïssa, S. (2005). Copper-induced oxidative stress in the three-spined stickleback: relationship with hepatic metal levels. *Environ. Toxicol. Pharmacol.* **19**, 177–183.

Sandahl, J. F., Miyasaka, G., Koide, N., and Ueda, H. (2006). Olfactory inhibition and recovery in chum salmon (*Oncorhynchus keta*) following copper exposure. *Can. J. Fish. Aquat. Sci.* **63**, 1840–1847.

Sandahl, J. F., Baldwin, D. H., Jenkins, J. J., and Scholz, N. L. (2007). A sensory system at the interface between urban stormwater runoff and salmon survival. *Environ. Sci. Technol.* **41**, 2998–3004.

Sanders, B. M., Nguyen, J., Martin, L. S., Howe, S. R., and Coventry, S. (1995). Induction and subcellular localization of two major stress proteins in response to copper in the fathead minnow *Pimephales promelas. Comp. Biochem. Physiol. C* **112**, 335–343.

Santore, R. C., Di Toro, D. M., Paquin, P. R., Allen, H. E., and Meyer, J. S. (2001). Biotic ligand model on the acute toxicity of metals. 2. Application to acute copper toxicity in freshwater fish and daphnia. *Environ. Toxicol. Chem.* **20**, 2397–2402.

Santos, E. M., Ball, J. S., Williams, T. D., Wu, H., Ortega, F., Van Aerle, R., Katsiadaki, I., Falciani, F., Viant, M. R., Chipman, J. K., and Tyler, C. R. (2010). Identifying health impacts of exposure to copper using transcriptomics and metabolomics in a fish model. *Environ. Sci. Technol.* **44**, 820–826.

Sato, M., and Bremner, I. (1993). Oxygen free radicals and metallothionein. *Free Radic. Biol. Med.* **14**, 325–337.

Saucier, D., and Astic, L. (1995). Morpho-functional alterations in the olfactory system of rainbow trout (*Oncorhynchus mykiss*) and possible acclimation in response to long-lasting exposure to low copper levels. *Comp. Biochem. Physiol. A* **112**, 273–284.

Saucier, D., Astic, L., and Rioux, P. (1991). The effect of early chronic exposure to sublethal copper on the olfactory discrimination of rainbow trout, *Oncorhynchus mykiss*. *Environ. Biol. Fishes* **30**, 345–351.

Sauter, S., Buxton, K. S., Macek, K. J., and Petrocelli, S. R. (1976). *Effects of Exposure to Heavy Metals on Selected Freshwater Fish: Toxicity of Copper, Cadmium, Chromium and Lead to Eggs and Fry of Seven Fish Species*. US Environmental Protection Agency, Duluth, MN, EPA-600/376-105.

Scheiber, I. F., Mercer, J. F., and Dringen, R. (2010). Copper accumulation by cultured astrocytes. *Neurochem. Int.* **56**, 451–460.

Schmidt-Nielsen, B., Truninger, B., and Rabinowitz, L. (1972). Sodium-linked urea transport by the renal tubule of the spiny dogfish *Squalus acanthias*. *Comp. Biochem. Physiol. A* **42**, 13–25.

Schreck, C. B., and Lorz, H. W. (1978). Stress resistance of coho salmon (*Oncorhynchus kisutch*) elicited by cadmium and copper, and potential use of cortisol as an indicator of stress. *J. Fish. Res. Bd Can.* **35**, 1124–1129.

Schwartz, M. L., Curtis, P. J., and Playle, R. C. (2004). Influence of natural organic matter source on acute copper, lead and cadmium toxicity to rainbow trout (*Oncorhynchus mykiss*). *Environ. Toxicol. Chem.* **23**, 2889–2899.

Sciera, K. L., Isely, J. J., Tomasso, J. R., Jr., and Klaine, S. J. (2004). Influence of multiple water-quality characteristics on copper toxicity to fathead minnows (*Pimephales promelas*). *Environ. Toxicol. Chem.* **23**, 2900–2905.

Segner, H. (1987). Response of fed and starved roach, *Rutilus rutilus*, to sublethal copper contamination. *J. Fish Biol.* **30**, 423–437.

Shaw, B. J., and Handy, R. D. (2006). Dietary copper exposure and recovery in Nile tilapia. *Oreochromis niloticus*. *Aquat. Toxicol.* **76**, 111–121.

Sherrell, R. M., and Boyle, E. A. (1992). The trace metal composition of suspended particles in the oceanic water column near Bermuda. *Earth Planet. Sci. Lett.* **111**, 155–174.

Shukla, G. S., Hussain, T., and Chandra, S. V. (1987). Possible role of regional superoxide dismutase activity and lipid peroxide levels in cadmium neurotoxicity: *in vivo* and *in vitro* studies in growing rats. *Life Sci.* **41**, 2215–2221.

Sies, H. (1999). Glutathione and its role in cellular functions. *Free Radic. Biol. Med.* **27**, 916–921.

Siwiki, A., and Studnicka, M. (1986). Ceruloplasmin activity in the carp (*Cyprinus carpio* L). *Bamidgeh* **38**, 126–129.

Skou, J. C. (1990). The energy coupled exchange of Na^+ for K^+ across the cell membrane. The Na^+, K^+-pump. *FEBS Lett.* **268**, 314–324.

Skou, J. C., and Esmann, M. (1992). The Na,K-ATPase. *J. Bioenerg. Biomembr.* **24**, 249–261.

Sloman, K. A., Baker, D. W., Wood, C. M., and McDonald, G. (2002). Social interactions affect physiological consequences of sublethal copper exposure in rainbow trout, *Oncorhynchus mykiss*. *Environ. Toxicol. Chem.* **21**, 1255–1263.

Sloman, K. A., Baker, D. W., Ho, C. G., McDonald, D. G., and Wood, C. M. (2003a). The effects of trace metal exposure on agonistic encounters in juvenile rainbow trout. *Oncorhynchus mykiss*. *Aquat. Toxicol.* **63**, 187–196.

Sloman, K. A., Morgan, T. P., McDonald, D. G., and Wood, C. M. (2003b). Socially-induced changes in sodium regulation affect the uptake of water-borne copper and silver in the rainbow trout. *Oncorhynchus mykiss*. *Comp. Biochem. Physiol. C* **135**, 393–403.

Sola, F., Isaia, J., and Masoni, A. (1995). Effects of copper on gill structure and transport function in the rainbow trout. *Oncorhynchus mykiss. J. Appl. Toxicol.* **15**, 391–398.

Solomon, E. I., and Lowery, M. D. (1993). Electronic structure contributions to function in bioinorganic chemistry. *Science* **259**, 1575–1581.

Stagg, R. M., and Shuttleworth, T. J. (1982a). The accumulation of copper in *Platichthys flesus* L. and its effects on plasma electrolyte concentrations. *J. Fish Biol.* **20**, 491–500.

Stagg, R. M., and Shuttleworth, T. J. (1982b). The effects of copper on ionic regulation by the gills of the seawater-adapted flounder (*Platichthys flesus* L.). *J. Comp. Physiol.* **149**, 83–90.

Steele, C. W., Owens, D. W., and Scarfe, A. D. (1990). Attraction of zebrafish, *Brachydanio rerio*, to alanine and its suppression by copper. *J. Fish Biol.* **36**, 341–352.

Stohs, S. J., and Bagchi, D. (1995). Oxidative mechanisms in the toxicity of metal ions. *Free Radic. Biol. Med.* **18**, 321–336.

Suedel, B. C., Boraczek, J. A., Peddicord, R. K., Clifford, P. A., and Dillon, T. M. (1994). Trophic transfer and biomagnification potential of contaminants in aquatic ecosystems. *Rev. Environ. Contam. Toxicol.* **136**, 22–89.

Svecevicius, G. (2001). Avoidance response of rainbow trout *Oncorhynchus mykiss* to heavy metal model mixtures: a comparison with acute toxicity tests. *Bull. Environ. Contam. Toxicol.* **67**, 680–687.

Syed, M. A., and Coombs, T. L. (1986). Copper metabolism in the plaice, *Pleuronectes platessa* (L.); some physiological properties of ceruloplasmin. *Contemp. Themes Biochem.* **6**, 332–333.

Taylor, E. W., Beaumont, M. W., Butler, P. J., Mair, J., and Mujallid, M. S. I. (1996). Lethal and sub-lethal effects of copper upon fish: a role for ammonia toxicity? In: *Toxicology of Aquatic Pollution* (E.W. Taylor, ed.), pp. 85–113. Cambridge University Press, Cambridge.

Taylor, J. R., and Grosell, M. (2006). Feeding and osmoregulation: dual function of the marine teleost intestine. *J. Exp. Biol.* **209**, 2939–2951.

Taylor, J. R., and Grosell, M. (2009). The intestinal response to feeding in seawater gulf toadfish, *Opsanus beta*, includes elevated base secretion and increased epithelial oxygen consumption. *J. Exp. Biol.* **212**, 3873–3881.

Taylor, J. R., Whittamore, J. M., Wilson, R. W., and Grosell, M. (2007). Postprandial acid–base balance in freshwater and seawater-acclimated European flounder. *Platichthys flesus. J. Comp. Physiol.* **177**, 597–608.

Taylor, J. R., Mager, E. M., and Grosell, M. (2010). Basolateral NBC1 plays a rate-limiting role in transepithelial intestinal HCO_3^- secretion serving marine fish osmoregulation. *J. Exp. Biol.* **213**, 459–468.

Taylor, L. N., McGeer, J. C., Wood, C. M., and McDonald, D. G. (2000). Physiological effects of chronic copper exposure to rainbow trout (*Oncorhynchus mykiss*) in hard and soft water: evaluation of chronic indicators. *Environ. Toxicol. Chem.* **19**, 2298–2308.

Taylor, L. N., Wood, C. M., and McDonald, D. G. (2003). An evaluation of sodium loss and gill metal binding properties in rainbow trout and yellow perch to explain species differences in copper tolerance. *Environ. Toxicol. Chem.* **22**, 2159–2166.

Taylor, L. N., McFarlane, W. J., Pyle, G. G., Couture, P., and McDonald, D. G. (2004). Use of performance indicators in evaluating chronic metal exposure in wild yellow perch (*Perca flavescens*). *Aquat. Toxicol.* **67**, 371–385.

Teles, M, Pacheco, M., and Santos, M. A. (2005). Physiological and genetic responses of European eel (*Anguilla anguilla* L.) to short-term chromium or copper exposure – influence of preexposure to a PAH-like compound. *Environ. Toxicol.* **20**, 92–99.

Thurman, E. M. (1985). *Organic Geochemistry of Natural Waters.* Kluwer Academic, Boston, MA.

Tilton, F., Tilton, S. C., Bammler, T. K., Beyer, R., Farin, F., Stapleton, P. L., and Gallagher, E. P. (2008). Transcriptional biomarkers and mechanisms of copper-induced olfactory injury in zebrafish. *Environ. Sci. Technol.* **42**, 9404–9411.

Tsui, T. K., Hung, C. Y., Nawata, C. M., Wilson, J. M., Wright, P. A., and Wood, C. M. (2009). Ammonia transport in cultured gill epithelium of freshwater rainbow trout: the importance of Rhesus glycoproteins and the presence of an apical Na^+/NH_4^+ exchange complex. *J. Exp. Biol.* **212**, 878–892.

US Geological Survey (2009). Minerals yearbook. http://minerals.usgs.gov/minerals/pubs/commodity/myb/

USEPA (2007). *Aquatic Life Ambient Freshwater Quality Criteria – Copper.* EPA-822-R-07-001, 2007 revision. Washington, DC: US Environmental Protection Agency.

Vanderboon, J., Van Den Thillart, G. E. E. J., and Addink, A. D. F. (1991). The effects of cortisol administration on intermediary metabolism in teleost fish. *Comp. Biochem. Physiol. A* **100**, 47–53.

Vitale, A. M., Monserrat, J. M., Castilho, P., and Rodriguez, E. M. (1999). Inhibitory effects of cadmium on carbonic anhydrase activity and ionic regulation of the estuarine crab *Chasmagnathus granulata* (Decapoda, Grapsidae). *Comp. Biochem. Physiol. C* **122**, 121–129.

Vutukuru, S. S., Chintada, S., Madhavi, K. R., Rao, J. V., and Anjaneyulu, Y. (2006). Acute effects of copper on superoxide dismutase, catalase and lipid peroxidation in the freshwater teleost fish. *Esomus danricus. Fish. Physiol. Biochem.* **32**, 221–229.

Waiwood, K. G., and Beamish, F. W. H. (1978). Effects of copper, pH and hardness on the critical swimming performance of rainbow trout (*Salmo gairdneri* Richardson). *Water Res.* **12**, 611–619.

Wang, T., Knudsen, P. K., Brauner, C. J., Busk, M., Vijayan, M. M., and Jensen, F. B. (1998). Copper exposure impairs intra- and extracellular acid–base regulation during hypercapnia in the fresh water rainbow trout (*Oncorhynchus mykiss*). *J. Comp. Physiol. B* **168**, 591–599.

Wang, Y., Fang, J., Leonard, S. S., and Rao, K. M. (2004). Cadmium inhibits the electron transfer chain and induces reactive oxygen species. *Free Radic. Biol. Med.* **36**, 1434–1443.

Wapnir, R. A. (1991). Copper–sodium linkage during intestinal absorption: inhibition by amiloride. *Proc. Soc. Exp. Biol. Med.* **196**, 410–414.

Weiner, A. L., and Cousins, R. J. (1983). Hormonally produced changes in ceruloplasmin synthesis and secretion in primary cultured rat hepatocytes. *Biochem. J.* **212**, 297–304.

Welsh, P. G., Lipton, J., Chapman, G. A., and Podrabsky, T. L. (2000). Relative importance of calcium and magnesium in hardness-based modifications of copper toxicity. *Environ. Toxicol. Chem.* **19**, 1624–1631.

Wilkie, M. P. (2002). Ammonia excretion and urea handling by fish gills: Present understanding and future research challenges. *J. Exp. Zool.* **293**, 284–301.

Wilson, R. W. (1999). A novel role for the gut of seawater teleosts in acid–base balance. In: *Regulation of Acid–Base Status in Animals and Plants. SEB Seminar Series*, Vol. 68, pp. 257–274. Cambridge: Cambridge University Press.

Wilson, R. W., and Taylor, E. W. (1993a). Differential responses to copper in rainbow trout (*Oncorhynchus mykiss*) acclimated to sea water and brackish water. *J. Comp. Physiol. B* **163**, 239–246.

Wilson, R. W., and Taylor, E. W. (1993b). The physiological responses of fresh water rainbow trout, *Oncorhynchus mykiss*, during acutely lethal copper exposure. *J. Comp. Physiol. B* **163**, 38–47.

Wilson, R. W., Wright, P. M., Munger, S., and Wood, C. M. (1994). Ammonia excretion in fresh water rainbow trout (*Oncorhynchus mykiss*) and the importance of gill boundary layer acidification – lack of evidence for Na^+/NH_4^+ exchange. *J. Exp. Biol.* **191**, 37–58.

Wilson, R. W., Wilson, J. M., and Grosell, M. (2002). Intestinal bicarbonate secretion by marine teleost fish – why and how? *Biochim. Biophys. Acta* **1566**, 182–193.

Winberg, S., and Nilsson, G. E. (1993). Roles of monoamine neurotransmitters in agonistic behavior and stress reactions with particular reference to fish. *Comp. Biochem. Physiol.* **106C**, 597–614.

Winberg, S., Bjerselius, R., Baatrup, E., and Doving, K. B. (1992). The effect of Cu(II) on the electro-olfactogram (EOG) of the Atlantic salmon (*Salmo salar* L) in artificial freshwater of varying inorganic carbon concentrations. *Ecotoxicol. Environ. Saf.* **24**, 167–178.

Wood, C. M., and Grosell, M. (2008). A critical analysis of transepithelial potential in intact killifish (*Fundulus heteroclitus*) subjected to acute and chronic changes in salinity. *J. Comp. Physiol. B* **178**, 713–727.

Wood, C. M., and Grosell, M. (2009). TEP on the tide in killifish (*Fundulus heteroclitus*): effects of progressively changing salinity and prior acclimation to intermediate or cycling salinity. *J. Comp. Physiol. B* **179**, 459–467.

Wood, C. M., Part, P., and Wright, P. A. (1995). Ammonia and urea metabolism in relation to gill function and acid–base-balance in a marine elasmobranch, the Spiny dogfish (*Squalus acanthias*). *J. Exp. Biol.* **198**, 1545–1558.

Wood, C. M., Kajimura, M., Mommsen, T. P., and Walsh, P. J. (2005). Alkaline tide and nitrogen conservation after feeding in an elasmobranch (*Squalus acanthias*). *J. Exp. Biol.* **208**, 2693–2705.

Wood, C. M., Bucking, C., and Grosell, M. (2010). Acid–base responses to feeding and intestinal Cl⁻ uptake in freshwater- and seawater-acclimated killifish, *Fundulus heteroclitus*, an agastric euryhaline teleost. *J. Exp. Biol.* **213**, 2681–2692.

Woodward, D. F., Brumbaugh, W., Delonay, A. J., Little, E. E., and Smith, C. E. (1994). Effects on rainbow trout fry of a metals-contaminated diet of benthic invertebrates from the Clark Fork River, Montana. *Trans. Am. Fish. Soc.* **123**, 51–62.

Woodward, D. F., Farag, A. M., Bergman, H. L., Delonay, A. J., Little, E. E., Smith, C. E., and Barrows, F. T. (1995). Metals contaminated benthic invertebrates in the Clark Fork River, Montana – effects on age-0 brown trout and rainbow trout. *Can. J. Fish. Aquat. Sci.* **52**, 1994–2004.

Wright, P. A., and Wood, C. M. (2009). A new paradigm for ammonia excretion in aquatic animals: role of Rhesus (Rh) glycoproteins. *J. Exp. Biol.* **212**, 2303–2312.

Wright, P. A., Randall, D. J., and Perry, S. F. (1989). Fish gill water boundary layer – a site of linkage between carbon dioxide and ammonia excretion. *J. Comp. Physiol. B* **158**, 627–635.

Wu, S. M., Ho, Y.-C., and Shih, M. J. (2007). Effects of Ca^{2+} or Na^+ on metallothionein expression in tilapia larvae (*Oreochromis mossambicus*) exposed to cadmium or copper. *Arch. Environ. Contam. Toxicol.* **52**, 229–234.

Wyman, S., Simpson, R. J., McKie, A. T., and Sharp, P. A. (2008). Dcytb (Cybrd1) functions as both a ferric and a cupric reductase *in vitro*. *FEBS Lett.* **582**, 1901–1906.

3

ZINC

CHRISTER HOGSTRAND

Homeostasis and Toxicology of Essential Metals: Volume 31A
FISH PHYSIOLOGY

Speciation of zinc (Zn) in waters is modulated by pH and by dissolved organic matter (DOM), which typically binds most aqueous Zn. In most natural waters the free Zn^{2+} ion is the dominant inorganic Zn species. Total Zn concentrations in natural waters span six orders of magnitude and are heavily influenced by human activities. Lethality of waterborne Zn to fish is caused by the free Zn^{2+} ion, while DOM, calcium, and pH in the water are the principal factors modifying Zn toxicity. The ameliorating influence of water chemistry has therefore been included in some legislation governing permissible concentrations of Zn in natural waters. The principal mode of action for acute Zn toxicity to freshwater fish is inhibition of calcium uptake. Little is known about mechanisms of sublethal toxicity in fish; however, lethality is often a sensitive endpoint also in chronic exposures of freshwater fish. Although a potentially toxic element, Zn is essential for all known life forms. It is a cofactor of 10% of all proteins and functions as both a paracellular and an intracellular signaling substance. There is therefore a comprehensive set of proteins that function as transporters, chelators, and molecular sensors for Zn. This regulatory network includes two large families of transporters (Slc30; Slc39), which regulate distribution of Zn throughout the body and within cells, metallothionein and several proteins, which are either activated or inhibited by changes in concentrations of labile Zn^{2+}. These proteins are involved in regulation of Zn uptake across gut and gill by homeostatic processes that are partially understood.

1. INTRODUCTION

Zinc (Zn) is essential to all cells in all known organisms and it is the second most abundant trace element, after Fe, in most vertebrates (Vallee, 1986). While Fe is used in relatively few but abundant proteins, such as hemoglobin, the body uses Zn much more diversely and often in minute quantities. Zinc is required for a variety of basic biological processes including metabolism of proteins, nucleic acids, carbohydrates, and lipids, and is also involved in more advanced functions, such as the immune system, neurotransmission, and cell signaling (Coleman, 1992; Beyersmann, 2002; Murakami and Hirano, 2008). It has been estimated that about 10% of all proteins in eukaryotic cells bind Zn and that there are approximately 3000 Zn proteins in humans (Passerini et al., 2007). Genomic sequencing data suggest that this number is a reasonable approximation for the numbers of proteins present also in different fish species. Almost all of the Zn in cells is bound to proteins, peptides, and amino acids, but there is a minute fluctuating pool of labile cytosolic Zn^{2+} which is involved in cell signaling pathways (Murakami and Hirano, 2008; Haase and

Rink, 2009; Hogstrand et al., 2009). One of the mechanisms by which Zn^{2+} transduces intracellular signals is by inhibition of protein tyrosine phosphatases (PTPs), which for example is believed to be the molecular mechanism behind the insulin mimetic effect of Zn (Haase and Maret, 2003; Miranda and Dey, 2004; Wong et al., 2006). Compartmentalization of Zn between tissues and within cells is managed principally by two large families of Zn transporters, the ZnT (Slc30A) family and the ZIP (Slc39A) family, which between them have 21 paralogues in pufferfish (Feeney et al., 2005). Distinct distribution and activities of these transporters determine the distribution of Zn within cells and animals.

Dietary Zn has low toxicity to vertebrates, including fish (Clearwater et al., 2002). In contrast, fish and other water-breathing animals are moderately sensitive to waterborne Zn, with acute toxicity concentrations being higher than for metals such as Ag, Cd, and Cu, but lower than those for Mn and Ni (McDonald and Wood, 1993). The relatively high risk of Zn toxicity to aquatic life has led to its inclusion as a priority pollutant by the US Environmental Protection Agency (USEPA, 2002). As in the case of most borderline metals (Ahrland et al., 1958), elevated concentrations of waterborne Zn can have deleterious effects on the gill (Niyogi et al., 2008). Very high and environmentally unrealistic concentrations of Zn cause non-specific inflammation of the gill, resulting in impaired gas exchange and suffocation (Skidmore and Tovell, 1972). Lower concentrations may still be acutely toxic with a specific mechanism of action, which involves inhibition of branchial calcium uptake with consequential hypocalcemia (Spry and Wood, 1985). The most sensitive documented endpoints observed in chronic toxicity studies of different freshwater species are diverse: survival, growth, reproduction, and hatching (USEPA, 1987; De Schamphelaere et al., 2005). However, it appears that if the fish is able to survive the initial period of Zn exposure it is often able to make biochemical and physiological adjustments to restore homeostasis in a process that is known as acclimation (Hogstrand et al., 1995; De Schamphelaere and Janssen, 2004; De Schamphelaere et al., 2005). In seawater where there is a positive diffusion gradient for calcium from the water to the fish, inhibition of calcium uptake would be an unlikely mode of action and the critical mechanisms of Zn toxicity to seawater fish are unknown.

2. CHEMICAL SPECIATION OF ZINC IN FRESHWATER AND SEAWATER

The predominant inorganic species of Zn in natural waters are believed to be Zn^{2+}, $ZnCO_3^0$, $ZnSO_4^0$, $ZnOH^+$, $Zn(OH)_2^0$, $ZnCl^+$, $Zn(Cl)_2^0$, $Zn(Cl)_3^-$

$ZnHPO_4$ and $Zn(Cl)_4^{2-}$ (Rainbow et al., 1993; Vega et al., 1995; Bervoets and Blust, 2000; Evans, 2000). In addition, complexation of Zn as multinuclear sulfide clusters may be of importance even in oxic natural waters (Evans, 2000; Rozan et al., 2000). Speciation of Zn in waters is pH and chloride dependent and is also strongly influenced by the presence of dissolved organic matter (DOM), which binds Zn with relatively high affinity. Overall stability constants (log K) for Zn binding to DOM in seven European inland waters have been reported to range from 6.4 to 7.0 (Jansen et al., 1998). In the absence of DOM and at pH vlaues below 8, the free Zn^{2+} dominates speciation (Bervoets and Blust, 2000; Luoma and Rainbow, 2008). The effect of pH on Zn speciation in freshwater relates primarily to the formation of $ZnCO_3^0$ and $Zn(OH)_n^{2-n}$ species, which collectively increase incrementally in abundance up to about pH 7.5, above which there is a steep increase until they completely dominate inorganic Zn speciation above pH 8 (Bervoets and Blust, 2000; Evans, 2000; Qiu and Hogstrand, 2005). In brackish water and seawater, $ZnCl_n^{2-n}$ species become increasingly important as salinity increases (Rainbow et al., 1993). If no DOM is present, the free Zn^{2+} remains the most abundant species, contributing about 30% to the total dissolved Zn concentration in 33‰ seawater.

Most natural surface freshwaters contain appreciable concentrations of DOM, which then dominates the speciation of dissolved Zn. In a survey of Zn complexation in a number of European river waters it was found that the free Zn concentration ranged between 12 and 45% of the total Zn concentration (Jansen et al., 1998). In a study on water from a Swiss eutrophic lake it was found that the total Zn concentration was about 1.3 $\mu g \, L^{-1}$, of which 13% was the free Zn^{2+} ion (Xue and Sigg, 1994). Complexation of Zn by weak and strong organic complexes was determined to be 33.5% and 50.5%, respectively. As a comparison, in Dutch surface freshwaters, 25–30% was considered to be present as the free Zn^{2+} and 70–75% adsorbed to DOM (Cleven et al., 1993; Jansen et al., 1998). Thus, there was only a relatively small difference between 70% and 84% in these estimates of complexed dissolved Zn. The binding of Zn to DOM in natural waters is dependent on the competition for binding sites with protons (low pH) and other cations (Cheng et al., 2005).

3. SOURCES OF ZINC AND ECONOMIC IMPORTANCE

Ores containing high proportions of Zn have been used for over 2500 years for fashioning brass ornaments and for wound healing (NAS, 1979). The Babylonians may have been able to produce Zn-containing brass already in the third century BC by reduction with charcoal (Porter, 1991). However, it was not until the fourteenth century that Zn was recognized in India as a metal in its own right and a small-scale commercial production of metallic Zn and Zn oxide existed. The knowledge how to produce Zn spread to China around AD 1600, where an industry developed for the manufacturing of brass (NAS, 1979). Paracelcus is believed to have been the first European to state that Zn (or "zincum") is a unique metal with different properties from those previously discovered. Modern production of Zn usually involves an electrolytic process. Zinc oxide is leached from the roasted or calcined ore with sulfuric acid to form a zinc sulfate solution, which is electrolyzed in cells to deposit Zn on cathodes.

Zinc is the 23rd most abundant element in the Earth's crust and it is found primarily as zinc sulfide (USGS, 2010a). The identified world resources of Zn amount to 1.9×10^9 t and the global Zn mine production was 11.1×10^6 t in 2009 (USGS, 2010b). China is the largest producer (2800×10^6 t), followed by Peru (1470×10^6 t) and Australia (1300×10^6 t) (USGS, 2010b). Although Zn deposits are evenly distributed between the developed and developing worlds, as much as 25% of the in-use Zn stocks are confined to three belts, comprised of (1) the eastern seaboard of the USA from Washington DC to Boston, (2) England–Benelux–central Germany–northern Italy, and (3) South Korea and Japan (Rauch, 2009). Uses of Zn range from metal products to rubber, healthcare products, and animal feeds (USGS, 2010a). Seventy-five percent of the Zn production goes to metal applications, such as galvanization of steel, die-casting, and production of alloys (USGS, 2010b). The remaining 25% is used mainly in rubber, chemicals, paint, and agricultural products.

Zinc concentrations in aquatic environments vary immensely as a function of distance from land in the marine environment, human activities, and also natural geology (Eisler, 1993; Janssen et al., 2000; Luoma and Rainbow, 2008). Total Zn concentrations in freshwaters range from 0.02 μg L^{-1} in remote rivers to more than 1000 μg L^{-1} in areas near mining, electroplating and other metal industrial activities (Eisler, 1993; Luoma and Rainbow, 2008). However, in most freshwaters total Zn concentrations rarely exceed 50 μg L^{-1} (Eisler, 1993; Bodar et al., 2005; Luoma and Rainbow, 2008; Naito et al., 2010). Like most metals, Zn concentrations in the marine environment are typically lower than those in freshwater, with 1–60 ng L^{-1} concentrations being typical for the upper 20 m in open oceans (Eisler, 1993; Ellwood, 2004; Luoma and Rainbow, 2008). In developed estuaries and coastal areas, total dissolved Zn concentrations may reach 5 μg L^{-1}.

The combined global anthropogenic input of Zn into aquatic environments has been estimated to be 226,000 t year^{-1} (Nriagu and Pacyna, 1988).

In the European Union (EU) risk assessment on Zn and Zn compounds, detailed quantitative data on Zn discharges in the Netherlands for the year 1999 were reviewed (Bodar et al., 2005). The total Zn emission was estimated to be 2720 t year^{-1} to soil, 460 t year^{-1} to wastewater, 254 t year^{-1} to surface water, and 91 t year^{-1} to air. The highest Zn concentrations observed in waters are usually results of local mining and other metal industrial activities, but the direct contribution of mines to the global Zn input into surface waters is relatively minor compared with other sources (1.8% of total in Japan) (USEPA, 1980; Bodar et al., 2005; Naito et al., 2010; Tsushima et al., 2010). Likewise, industrial point-sources were only responsible for about 10% of the total addition of Zn into surface waters in Japan and the Netherlands (Bodar et al., 2005; Naito et al., 2010). Corrosion of galvanized products and Zn alloys is one of the largest contributors to Zn emission into surface waters, amounting to about 30% of the total in Japan and the Netherlands (Bodar et al., 2005; Naito et al., 2010). Other principal sources of Zn include atmospheric deposition of particulate matter, wastewater treatment plants, and road runoff (USEPA, 1980; Bodar et al., 2005; Naito et al., 2010). Zinc is one of the most abundant transition metals in road runoff (Legret and Pagotto, 2006; Preciado and Li, 2006) because of the use of Zn in tires and brake disks. Drainage from agricultural soils and sedimentation of excreta from fish farms can be important contributors (Eisler, 1993; Bodar et al., 2005; Dean et al., 2007; Naito et al., 2010). Thus, most of the input of Zn to aquatic environments comes from diffuse sources associated with everyday human activities and this has resulted in a general elevation of Zn concentrations of surface waters in densely populated areas.

4. ENVIRONMENTAL SITUATIONS OF CONCERN

The greatest risks for Zn toxicity most likely exist in areas of active mining or downstream of discontinued mines, where metals are continuously added to surface waters through flooding of adits and leaching from tailings. Zinc accumulation in the gills of fish and consequential toxicity are strongly dependent on water chemistry, with DOM content, calcium concentration, and pH being the principal modifiers in freshwater (Santore et al., 2002; De Schamphelaere et al., 2005; Todd et al., 2009). Santore and co-workers made a thorough meta-analysis of published acute toxicity data on Zn to rainbow trout and fathead minnows and observed that pH dependence of Zn toxicity followed a U-shaped curve with increasing LC50 below pH 6, due to competition between Zn^{2+} and H^+ ions for binding sites on the gill, and above pH 8 because of inorganic complexation (Santore et al., 2002). Thus, the potential for Zn poisoning in fish can be expected to be highest in oligotrophic lakes with very soft and non-acid waters. Such conditions can, for example, be found in Scandinavia and the Canadian Shield.

In nature, isolated exposures to Zn are highly uncommon and there are few reports on adverse effects on natural fish populations where Zn has been the only toxicant to blame. Instead fish are typically exposed to metal mixtures, although these do frequently contain elevated levels of Zn. In some areas of active or legacy mining, the Zn concentrations in the water are well within those shown to be lethal to fish in laboratory experiments using similar water chemistries. Finlayson and colleagues reported on fish kills in waters containing high concentrations of Zn and Cu downstream of mines in the Sacramento River and the Mokelumne River in California (USDoI, 1998). The water in the Mokelumne River measured 1.4 mg L^{-1} during a fish kill in 1958.

The Boulder River watershed is a region where there are well-documented effects on fish from leaching of metals from abandoned mine adits and tailing (Farag et al., 2003). The system is contaminated by several metals, including Zn, Cu, and Cd, and in some of the tributaries, close to mine sources, Zn, Cu, and Cd concentrations are extremely high (dissolved [Zn] up to 5.7 mg L^{-1}; [Cu] up to 380 µg L^{-1}; [Cd] up to 75 µg L^{-1}) and no fish are present (Farag et al., 2003). However, in one of the studied tributaries, High Ore Creek, Zn was the only metal showing seriously high levels in the water, with dissolved Zn concentrations ranging from 460 to 990 µg L^{-1} compared with Cu and Cd concentrations below 5 µg L^{-1} (hardness = 135 mg L^{-1} as $CaCO_3$). An isolated population of westslope cutthroat trout (*Oncorhynchus clarki lewisi*) existed upstream of Comet Mine in High Ore Creek, but not downstream of the mine and possibly nowhere else in the Boulder River watershed. Transplantation of hatchery-raised cutthroat trout to the High Ore Creek downstream of the mine resulted in 67–100% mortality within 96 h, suggesting that the metal concentrations, and particularly those of Zn, constituted a chemical migration barrier for the native population of westslope cutthroat trout living in the higher reaches (Farag et al., 2003).

The River Hayle in Cornwall, UK, flows through an area where mining was prolific in the sixteenth and seventeenth centuries (Brown, 1977). The last mine in the catchment closed at the beginning of the twentieth century, but concentrations of Zn and Cu remain high in the water downstream of the adits (Brown, 1977; South West Water, 1983). The River Hayle supports a small population of brown trout (*Salmo trutta*), but is void of fish and almost all invertebrates in a stretch of about 2 km (downstream of Wheal

Godolphin Adit) where dissolved Zn and Cu concentrations have exceeded 1000 and 40 μg L^{-1} (hardness ~ 100 mg L^{-1} as $CaCO_3$) (Brown, 1977; South West Water, 1983; Khan et al., 2011).

5. AMBIENT WATER QUALITY CRITERIA FOR ZINC IN VARIOUS JURISDICTIONS

Because of the extensive and diverse use of Zn by humans, Zn concentrations in water are elevated wherever there is civilization (see Section 3) and relatively frequently exceed the no observed effect concentrations (NOECs) derived in the laboratory for some aquatic species (Bodar et al., 2005; Naito et al., 2010; Tsushima et al., 2010). However, in the EU risk assessment on Zn it was concluded that, up to about 40 μg L^{-1} (called background), there was no clear relationship between Zn concentrations and ecotoxicity (Bodar et al., 2005). This creates a dilemma in setting water quality criteria [alternatively, environmental water quality standards (WQS)] aimed at protecting aquatic life as well as human welfare.

Zinc is considered a pollutant of particular concern and has been labeled a priority pollutant in the USA, a possible specific pollutant in the UK, and a list II dangerous substance (for which pollution reduction programs should be established) in the EU. Water quality guidelines/criteria for Zn in the USA, Canada, Australia/New Zealand, Japan, and South Africa are shown in Table 3.1. Most authorities recognize the ameliorating effect of hardness (calcium) on Zn toxicity to fish, but only some have implemented adjustments typically based on equations including hardness as a variable. The USA and South Africa have acute and chronic values, of which the former are intended to apply to short-term exposure episodes, such as an accidental discharge. Australia and New Zealand have a common Water Quality Guideline that does not give limit values, per se, but rather target values above which further refinement of the exposure situation is prescribed. This approach reflects the vast range of aquatic habitats represented in Australia and New Zealand and is meant to accommodate site-specific modifications. Similarly, the EU has recently undertaken a risk assessment for Zn, which includes several methodologies to implement environmental quality standards in a large and diverse region (Bodar et al., 2005). This included, among others, the use of Biotic ligand models (BLMs) (Paquin et al., 2002) to account for differences in Zn bioavailability due to water chemistry, and an added risk approach to eliminate the contributions of the natural background concentrations of Zn to the predicted environmental concentrations (PECs) as well as the predicted no observed

Table 3.1

Summary of acute and chronic ambient water quality criteria for zinc in various jurisdictions in freshwater and seawater

Jurisdiction	Reference	Acute (μg L^{-1})	Chronic (μg L^{-1})	Notes
USA	USEPA (1987)	30.6[a]	30.2[a]	Hardness 20 mg L^{-1}
		66.6[a]	65.7[a]	Hardness 50 mg L^{-1}
		215.6[a]	212.5[a]	Hardness 200 mg L^{-1}
		95.1[a]	85.6[a]	Seawater
Canada	CCME (2007)		30	No hardness adjustment
Japan	Tsushima et al. (2010)		30	No hardness adjustment
Australia/NZ	ANZECC (1992)		5.7[b]	Hardness 20 mg L^{-1} (95% protection)
			12[b]	Hardness 50 mg L^{-1} (95% protection)
			40[b]	Hardness 200 mg L^{-1} (95% protection)
			15[b]	Seawater (95% protection)
South Africa	DWAF (1996)	36[c]	3.6[c]	No hardness adjustment
	DWAF (1995)		None	Seawater
EU	EU (2000)		None	Under review
China	NEPA (1989)		100	In Wang et al. (2010)

NZ: New Zealand; EU: European Union.
[a]Total concentrations adjusted with conversion factors for total to dissolved metal.
[b]Hardness-modified "trigger values" for 95% protection, above which further refinement of assessment is required.
[c]Dissolved Zn.

effect concentrations (PNECs). Using a species sensitivity distribution plot, the 5th percentile of NOECs for freshwater species was determined to be 15.6 μg L^{-1}, a value which then would be modified by an assessment factor, background Zn concentration, and a BLM to generate site- or region-specific criteria (Bodar et al., 2005). The added risk approach and the way that BLM was applied to risk assessment were initially criticized by the EU Scientific Committee on Health and Environmental Risk (SCHER, 2007), but the use of a BLM as well as the added risk approach has recently been gaining momentum. Implicit to the added risk approach is that organisms in areas where the background Zn concentration is high have higher NOEC values for Zn than organisms from areas with less Zn, an assertion for which there is some evidence (Muyssen and Janssen, 2005). In the Japanese risk assessment of Zn the same general problem was recognized in that 15–28% of 2075 monitoring sites were exceeding Zn concentrations predicted to affect NOEC for 5% of the species (Naito et al., 2010). The approach to manage this problem was, however, fundamentally different. It was proposed to introduce a dual water quality standards system where a population-based threshold concentration (determined to be 107 μg L^{-1}

total Zn) could be used as a risk management option instead of the traditional organismal-based NOEC for 5% of the species (determined to be 26.7 µg L^{-1}). The higher population-based threshold concentration was calculated using ecological population modeling tools to determine a Zn concentration at which the population size would not decline for 95% of the species.

6. MECHANISMS OF TOXICITY

6.1. Acute Toxicity to Freshwater Fish

To humans, the risk for Zn poisoning is negligible for the general public, and health risks from Zn exposures are only of concern for workers where occupational exposure to Zn fumes or dermal exposure to Zn oxide-containing paints could lead to health effects (Bodar et al., 2005). For fish and other aquatic organisms, however, the situation is different and this is why Zn is regarded as an important pollutant. Acute toxicity values (96 h LC50) that formed the basis for the US Water Quality Criteria for Zn in freshwater ranged from 66 to 40,900 µg L^{-1}, with the lowest value recorded in a test with rainbow trout in moderately soft water (hardness = 92 mg L^{-1} as CaCO$_3$) (Cusimano et al., 1986). One of the most defining differences in toxicological effects between fish and terrestrial vertebrates is due to the presence of gills. These make fish especially prone to toxicity from waterborne chemicals and many effects of Zn exposure in fish can be derived from the interaction between Zn and molecular processes in the gill. This was recognized relatively early and it was found that high concentrations of a number of waterborne chemicals caused a stereotyped pathological response in the gill epithelium typified by mucus secretion, hypertrophy, hyperplasia, leukocyte infiltration, and epithelial lifting (Mallatt, 1985). The insult results in increased diffusional distance for respiratory gases and the fish typically dies from suffocation. In a milestone paper, Spry and Wood (1985) demonstrated that acute Zn toxicity occurs at concentrations much lower than those eliciting structural damage and hypoxia, and went on to show that acute Zn toxicity in freshwater fish is primarily caused by reduced entry of Ca^{2+} across the gill. Acute Zn exposure also leads to metabolic acidosis through stimulation of branchial net ammonia excretion and uptake of acidic equivalents from the water, but these effects alone were concluded not to be great enough to explain mortality (Spry and Wood, 1985).

Much later, details were worked out regarding the nature of Zn inhibition of calcium uptake at the rainbow trout gill and it was found that

Zn may interfere with calcium homeostasis at several levels. It was shown that Zn^{2+} competitively inhibits Ca^{2+} transfer across the apical membrane of the gill epithelial cells, indicating competition for a common uptake system (Hogstrand et al., 1995). Pharmacological intervention lent further support for this idea. Addition of the calcium channel inhibitor La^{3+} to the water decreased uptake of Ca^{2+} and Zn^{2+} into the gill by 80 and 50%, respectively (Hogstrand et al., 1996b). Furthermore, induction of hypercalcemia (through calcium injection), which caused downregulation of branchial Ca^{2+} influx through stimulation of Stanniocalcin release, also significantly reduced Zn^{2+} uptake across the gills (Hogstrand et al., 1996b). The apical Ca^{2+} channel, which is believed to be the target for this interaction, has now been identified as the epithelial calcium channel (Ecac/Trpv6) and it has been shown to be highly permeable to Zn^{2+} as well as to Ca^{2+} (Qiu and Hogstrand, 2004; Shahsavarani et al., 2006; Shahsavarani and Perry, 2006). However, Zn interacts with calcium uptake at the gill through other mechanisms as well. The high-affinity Ca^{2+}-ATPase (Pmca), which extrudes Ca^{2+} across the basolateral membrane of gill epithelial cells, is extremely sensitive to Zn^{2+}, with inhibition in rainbow trout occurring at a cytosolic free Zn^{2+} activity of 100 pM (6.5 ng L^{-1}) (Hogstrand et al., 1996b). This concentration is actually within the range of most recent estimates of the free Zn^{2+} ion concentrations in mammalian cells (Colvin et al., 2010), raising the intriguing possibility that Zn is a physiological regulator of Pmca activities in cells. As Pmca is located at the basolateral membrane of the gill epithelium, it is not in direct contact with the water and only influenced by the local free Zn^{2+} concentration within the cell.

Studies primarily on mammalian cells lend strong support for Zn^{2+} as an intracellular signaling ion that closely interacts with cellular Ca^{2+} signaling (Maret, 2001; Haase and Maret, 2003; Besser et al., 2009). Studies on mammals have shown that extracellular Zn binds to and activates the G protein-coupled receptor, GPR39, resulting in phospholipase C (PLC) activation, inositol trisphosphate (IP_3) production, and release of Ca^{2+} from the endoplasmic reticulum through the inositol trisphosphate receptor (IP3R) (Holst et al., 2007; Storjohann et al., 2008; Sharir et al., 2010). Gpr39 is present in fish and has two transcripts, gpr39-1a and gpr39-1b, in black seabream (Acanthopagrus schlegeli) (Zhang et al., 2008). Both of these are abundantly expressed in the intestine and gpr39-1b is expressed in both intestine and gill. Zinc may also cause PLC activation through inhibition of the protein tyrosine phosphatase, PTP1B (also known as PTPN1), which is extremely sensitive to fluctuations in intracellular Zn^{2+} concentrations (Haase and Maret, 2003, 2004a). Thus, there are clearly several levels at which Zn may interfere with calcium homeostasis in the gill and several of these loci may be affected by exposure of fish to toxic concentrations of Zn in the water. Although these targets might be diverse, the physiological consequence is the same, a severe inhibition of calcium uptake with lethality as the ultimate outcome.

6.2. Acute Toxicity to Seawater-living Fish

Surprisingly little is known about acute toxicity of Zn to fish in seawater and with few exceptions the data that do exist are derived from tests with nominal Zn concentrations before working with measured metal concentration became the norm. These data have been summarized in the US EPA water quality criteria for Zn – 1987 (USEPA, 1987) and in a comprehensive review of Zn hazards to the environment (Eisler, 1993). Considering the ameliorating effect of calcium against Zn toxicity in freshwater and the fact that 35‰ seawater contains 10 mM calcium, it might be expected that marine fish are well protected against Zn toxicity and that Zn toxicity should be salinity dependent. However, there was little or no difference in Zn toxicity to two tropical marine fish species at 20 and 36‰ salinity (Denton and Burdon-Jones, 1986). Furthermore, acute toxicity values to fish in seawater are not strikingly higher than those determined for freshwater fish. The available 96 h LC50 values ranged from 190 µg L^{-1} in cabezon (*Scorpaenichthys marmoratus*) to 83,000 µg L^{-1} recorded in a test with mummichog (*Fundulus heteroclitus*) (Dinnel et al., 1983; USEPA, 1987). There is even a recent publication presenting a 24 h LC50 for mudskipper (*Periophthalmus waltoni*) to Zn in filtered 38‰ seawater of 12 µg L^{-1} (Bu-Olayan and Thomas, 2008), but this value should probably be verified. So, perhaps to marine fish calcium has less of a protective role than in freshwater and perhaps it is of significance that the free Zn^{2+} ion dominates inorganic Zn speciation in seawater. Even so, it has been shown that accumulation rate of waterborne Zn by the euryhaline black sea bream (*Acanthopagrus schlegeli*) is substantially reduced as salinity increases (Zhang and Wang, 2007b). The limited data on acute Zn toxicity to marine fish do not allow many conclusions to be drawn. However, it is clear that sensitivity varies enormously between species and there is not a huge difference in sensitivities to Zn in freshwater and seawater. As with salinity, there also do not seem to be any particular differences in sensitivity with regard to life stage.

It is fair to say that almost nothing is known about the mechanism of acute Zn toxicity to seawater fish. As in freshwater fish, marine species do take up calcium across the gills, but as the calcium concentration in seawater is 10 times that of the blood plasma, hypocalcemia due to inhibition of calcium uptake seems unlikely. This is reflected in the apparent lack of salinity dependence for toxicity within the seawater range.

6.3. Chronic Toxicity

In the European Risk Assessment for Zn, NOEC values for six species of freshwater fish were used as reported in 74 chronic studies (Bodar et al., 2005). The lowest NOEC value included was for mottled sculpin (*Cottus bairdi*), which when exposed to Zn for 30 days in soft water (46 mg L^{-1} as CaCO$_3$) showed a NOEC for reduced survival of 16 µg L^{-1} (Woodling et al., 2002). As with acute Zn toxicity to freshwater fish, chronic values are heavily dependent on the water calcium concentration, and in hard water (154 mg L^{-1} as CaCO$_3$), the NOEC for survival was increased to 172 µg L^{-1} (Brinkman and Woodling, 2005). De Schamphelaere and Janssen (2004) published an insightful comparison between outcomes of different chronic studies with Zn on fish and concluded that in many cases, survival was actually the most sensitive endpoint reported. The minnow (*Phoxinus phoxinus*) is one of the exceptions, where the NOEC for growth is lower than that for survival (Bengtsson, 1974). After 30 days of exposure to Zn in relatively soft water (47 mg L^{-1} as CaCO$_3$) a reduction in growth was observed at 130 µg L^{-1} with effect on survival at 200 µg L^{-1}. Similarly, in Indian major carp (*Cirrhinus mrigala*) exposed in harder water (114 mg L^{-1} as CaCO$_3$) for 30 days there were no mortalities at the highest Zn concentration of 150 µg L^{-1}, but reduced feed intake and growth were observed at 100 µg L^{-1} and above (Mohanty et al., 2009). Reproduction may be another sensitive endpoint for chronic Zn toxicity. In a 10 month exposure of fathead minnows (hardness = 200 mg L^{-1} as CaCO$_3$), egg production was reduced by 83% at a measured Zn concentration of 180 µg L^{-1} and survival was only marginally affected (15% reduction) at the highest Zn concentration deployed, 2800 µg L^{-1} (Brungs, 1969). In a 12 week reproduction study with fathead minnows, adhesiveness and strength of eggs were affected by waterborne Zn at 147 µg L^{-1} (hardness = 46 mg L^{-1} as CaCO$_3$) with no effect on survival of breeding adults or larvae below 294 µg L^{-1} (Benoit and Holcombe, 1978). There are relatively few studies where adults have been exposed to Zn followed by analysis of reproductive output, but at least in the fathead minnow, impairment of reproduction may be the most sensitive chronic endpoint of Zn.

It has previously been noted that even in the cases where survival was not the most sensitive effect, NOEC and LOEC for survival were usually not substantially higher than the most sensitive endpoint recorded (De Schamphelaere and Janssen, 2004). From a biological perspective it makes little sense that lethality occurs before sublethal endpoints set in, but one has to bear in mind that most standardized ecotoxicity tests only include relatively crude (but admittedly ecologically relevant) outcomes, such as survival, growth, and reproductive success. Where data on the time-course of fish deaths are available, it appears that mortalities usually occur over the first 2 weeks of Zn exposure. In a study on rainbow trout exposed to a sublethal concentration of waterborne Zn it was found that growth of exposed fish was retarded over the first two weeks after which the exposed fish caught up with the controls (Hogstrand et al., 1995). This initial period was characterized by hypocalcemia, an increased protein turnover, and adjustments in Zn and calcium uptake rates, all of which were completed after 4 weeks of exposure. Similarly, the gill protein content of Zn-exposed rainbow trout was found to be reduced after 2 weeks of exposure but recovered subsequently (Sappal et al., 2009). Thus, it would appear that short-term (< 14 days) effects of Zn on survival in freshwater can be principally attributed to inhibition of calcium uptake across the gills and that fish managing to physiologically acclimate during this period may show little signs of adverse effects afterwards. The fish also acclimate to Zn in the traditional toxicological sense in that exposure to a sublethal concentration increases their tolerance (LC50) and resistance (LT50) in subsequent toxicity tests (Bradley et al., 1985).

There are few hard data available informing whether or not net loss of calcium is a major cause of chronic toxicity beyond that occurring during the initial weeks of exposure. There was no change in whole-body status of calcium in rainbow trout exposed to 150 μg L^{-1} waterborne Zn (hardness = 120 mg L^{-1} as CaCO$_3$) or a combined exposure to Zn in water and feed up to 529 μg L^{-1} (hardness = 135 mg L^{-1} as CaCO$_3$) and 590 mg kg^{-1}, respectively (Spry et al., 1988; Hogstrand et al., 1995; Alsop et al., 1999). However, except for effects on calcium regulation occurring during the first 2 weeks of exposure there were actually no signs of any further toxicity in these studies, so it could be argued that the lack of net calcium loss is consistent with absence of observed effects. Furthermore, in rainbow trout exposed to 150 μg L^{-1} waterborne Zn for 30 days in very similar conditions (hardness = 120 mg L^{-1} as CaCO$_3$) as in the experiment described above (Hogstrand et al., 1995), there was also no effect on swimming performance, although a reduced critical swimming speed was recorded for fish exposed to 250 μg L^{-1} (Alsop et al., 1999).

Effects of Zn exposure on reproduction could hypothetically be linked to reduction in calcium uptake because oocyte quality depends on the quality of vitellogenin, which normally contains large amounts of calcium, but this author could not find any information as to whether or not Zn exposure affects calcium incorporation into oocytes. Alternative explanations are feasible, as indicated by effects of Zn on reproductive function in mammals. In a recent two-generational study of Zn toxicity in rats it was found that exposure to Zn causes reduction in weight of the uterus in females and in the seminal vesicles and prostate of males (Khan et al., 2007). Such pathologies could be caused by effects of Zn on steroid hormone biosynthesis or function. For example, the human sex hormone-binding globulin (SHBG) has a regulatory Zn-binding site (Avvakumov et al., 2000). Binding of Zn to this site reduces the affinity of SHBG for estradiol, leaving its affinity for androgens unaffected. Effects on embryo development could be plausibly related to disturbance of normal Zn signaling during embryogenesis (see below), although this remains to be investigated. In humans, the most sensitive known functions to be adversely affected by high dietary Zn include hematological parameters and these are generally believed to be related to inhibition of Fe and Cu uptake in the gut (Maret and Sandstead, 2006; Stefanidou et al., 2006). Whole-body Cu and Fe status in fish are also known to be negatively influenced by high Zn levels in the diet (Knox et al., 1984; Eid and Ghonim, 1994), but this may not be a major issue for waterborne exposures because nutritional uptake of Cu and Fe occurs under most conditions chiefly from the diet (see Grosell, Chapter 2; Bury et al., Chapter 4).

7. ESSENTIALITY AND ROLES OF ZINC IN BIOLOGY

7.1. Zinc is an Essential Element

The essentiality of Zn has been quite unequivocally demonstrated. A new hypothesis originally published as two substantial back-to-back articles even puts Zn at the very center of the origin of life (Mulkidjanian, 2009; Mulkidjanian and Galperin, 2009). Once it became apparent that the atmosphere of the prebiotic Earth was probably filled by carbon dioxide (CO_2), and not by a reducing mixture of methane, hydrogen, ammonia, and water vapor, as was previously thought, earlier ideas of the original formation of life on the planet fell through. In the new hypothesis it is argued that under the high pressure of the primeval, CO_2-dominated atmosphere, ZnS could precipitate at the surface of the first continents, well within reach of solar light. ZnS and ZnO have high capacities to absorb the energy of ultraviolet (UV) light and are therefore used in many modern-day devices, such as solar panels and sunscreens. It was suggested that in the primordial Earth, solar energy captured by ZnS surfaces could drive CO_2 reduction, yielding the building blocks for the first biopolymers. ZnS could then have been a catalyst in the synthesis of longer biopolymers from simpler building blocks and preventing the first biopolymers from photodissociation. Moreover, the authors argue that the UV light may have favored the selective enrichment of photostable, RNA-like polymers (Mulkidjanian, 2009). They backed up their ideas by producing the circumstantial evidence that a disproportionately high number of the evolutionarily oldest proteins are Zn proteins (Mulkidjanian and Galperin, 2009).

The earliest recorded scientific evidence for the essentiality of Zn came in 1869 when the French botanist and chemist Jules Raulin showed that the mold *Aspergillus niger* cannot grow in its absence (Vallee, 1986). This finding was followed by the discovery that Zn was present in every plant and animal tissue that was measured. The first Zn enzyme was discovered 71 years later when Keilin and Mann isolated carbonic anhydrase. The best current estimate is that there are about 3000 Zn proteins in humans, representing 10% of the entire genome (Andreini et al., 2005; Passerini et al., 2007). Similarly, about 10% of all genes in sequenced fish genomes carry the annotation Zn binding. The majority of these proteins use Zn to create folds in the protein structure (e.g. Zn fingers) or to join different gene products together (e.g. Zn hooks) (Andreini et al., 2005; Passerini et al., 2007; Maret and Li, 2009). In addition to its structural and catalytic roles in proteins, there is a growing appreciation for Zn as a signaling substance and this includes genomic and non-genomic cell signaling and a role as a neuromodulator, or perhaps even transmitter substance (Laity and Andrews, 2007; Hirano et al., 2008; Besser et al., 2009; Hogstrand et al., 2009; Sensi et al., 2009; Hershfinkel et al., 2010).

The discovery that Zn is essential to humans has been attributed to Ananda Prasad, who encountered adult men in Iran with features of prepubescent boys presenting growth retardation, testicular atrophy, rough and dry skin, and mental retardation (Prasad et al., 1961). The condition was later found to be caused by geophagia (clay eating), a habit that was common in the area, rendering Zn unavailable for uptake (Prasad, 2009). A recommended daily allowance (RDA) was established in 1974 by the Food and Nutrition Board of the National Research Council of the USA National Academy of Sciences (NAS). Similarly, the Board of Agriculture, NAS, has set nutritional requirements of Zn for fish (NRC, 1993). The nutritional requirement of Zn for channel catfish was set to $20 \, \mathrm{mg \, kg^{-1}}$ feed, while rainbow trout, common carp, and tilapia were considered to require $30 \, \mathrm{mg \, kg^{-1}}$. However, partially because of Zn's perceived additional benefits and concerns about bioavailability and batch variability, fish feeds in the EU are allowed to contain a maximum of $200 \, \mathrm{mg \, kg^{-1}}$ feedstuff (EC, 2003). In fish, Zn deficiency causes anorexia, poor growth, bone deformations, reduced survival, cataracts (Gatlin and Wilson, 1983; Eid and Ghonim, 1994), and exaggerated startling response (own unpublished information).

7.2. Zinc Signaling

7.2.1. METAL-RESPONSIVE TRANSCRIPTION FACTOR-I: THE CELLULAR ZINC SENSOR

Zinc regulates transcription of genes through metal-responsive transcription factor-1 (Mtf1), which is conserved through evolution and is present in most animals including fish (Dalton et al., 2000; Chen et al., 2002). This protein has about 590 amino acids including six Zn-finger domains, two of which bind Zn with lower affinity than the others (Andrews, 2001; Kimura et al., 2009). If there is a rise in the cytosolic Zn^{2+} concentration, the low-affinity Zn binding sites will be filled, enabling Mtf1 to associate with its cognate DNA motif (5'-TGCRCNC-3'), known as a metal response element (MRE) and thereby regulating transcription (Kimura et al., 2009). There may be other mechanisms, such as phosphorylation, that modulate Mtf1 activity, but the Zn fingers are sufficient in regulating DNA binding in response to Zn because a chimeric construct consisting of the DNA binding domain from Mtf1 and transactivation domain from Gal4 mediates Zn-dependent transcription when transfected into cells (Andrews, 2001; Kimura et al., 2009). Several metals, including Cd, Hg, Ag, and Cu, can induce Mtf1-mediated gene expression, but Zn appears to be the only metal activating the protein to any significant extent. Thus, the effects of other metals on Mtf1 activity must be either by alternative mechanisms, such as phosphorylation, or by increasing the cytosolic Zn^{2+} concentration. The former could be achieved through activation of kinases or inhibition of protein phosphatases and the latter by displacement of Zn from proteins. Mtf1 was first identified in mouse as the transcription factor responsible for induction of metallothionein (*Mt*) expression in response to metal exposure (Westin and Schaffner, 1988). Mt is a family of metal binding proteins that can protect cells against metal insult (Hogstrand and Haux, 1991) and function as a redox switch to release Zn^{2+} ions for Zn signaling events (Maret, 1995; Chung et al., 2005b). Functions of Mt in cell biology will be discussed in Sections 10.1 and 13. There have been several searches for Mtf1 target genes and the list is now considerably expanded (Gunes et al., 1998; Lichtlen et al., 2001; Wang et al., 2004; Hogstrand et al., 2008). Using microarray technology with a multifactorial experimental design, including RNAi knockdown of *mtf1* and Zn treatments in zebrafish ZF4 cells, it was shown that regulation of over 1000 genes was Mtf1 dependent (Hogstrand et al., 2008). However, sequence analysis of these genes showed that only 43 of them contained MRE in configurations and locations compatible with Mtf1 responsiveness and it was concluded that remaining genes were likely induced as downstream ripples of the signaling cascade. Of the 43 putative Mtf1 targets, 19 were genes involved in development. There was also a staggering overrepresentation of transcription factors, which made up almost half (48%) of the identified genes. The results strongly suggest that Mtf1 has roles beyond stress responses and that genomic Zn signaling via Mtf1 is of particular importance during embryonic development. These findings are in keeping with earlier findings that *Mtf1* knockout mice die during development from failed organogenesis (Gunes et al., 1998), while excision of *Mtf1* in liver and bone marrow with Cre recombinase after birth is non-lethal but causes leukopenia and sensitivity to metal stress (Wang et al., 2004). Specific functions of Mtf1 during organogenesis open up another hypothetical mechanism of metal toxicity. Exposure of embryos to metals would likely cause inappropriate activation of Mtf1 and it is therefore tempting to speculate that metal toxicity during embryogenesis could involve disruption of Zn signaling through Mtf1.

Zinc activation of Mtf1 is also of importance in the defense against free radical stress in fish as well as in humans. Mt is an important antioxidant and it reduces free radicals while metal thiolate bonds are oxidized and Zn released (Maret, 1994; Krezel et al., 2007). This results in a transient increase in the labile Zn^{2+} concentration of the cell, activation of Mtf1, and consequential expression of several key antioxidant genes, including glutathione peroxidase (*gpx*), glucose-6-phosphate dehydrogenase (*g6pd*), glutathione-*S*-transferase (*gst*), and *mt* itself (Fig. 3.1) (Chung et al., 2005a,b). Using primary rainbow trout gill cells, it was shown that expression of these antioxidant genes was completely abolished if a cell-permeant Zn chelator was introduced, preventing the increase in cytosolic $[Zn^{2+}]$ (Chung et al., 2005b). With the same gill cell culture system it was also shown that the polyphenol caffeic acid, which is the major antioxidant in red wine, protects against oxidative stress by redox cycling, leading to release of protein-bound Zn and subsequent induction of antioxidant proteins (Chung et al., 2006).

7.2.2. NON-GENOMIC ZINC SIGNALING

The existence of non-genomic Zn signaling was probably first proposed by the visionary R. J. P. Williams at the University of Oxford, who correctly predicted that Zn may interact with calcium metabolism and act on cell signaling by inhibition of proteins (Williams, 1984). Because observations of biological actions of Zn were until recently based on experiments with organisms treated with exogenous Zn and often at high concentrations, any such suggestions were generally met with skepticism and the critique that it is difficult to separate a biological function from a pharmacological response. However, with the demonstration of sudden changes in cytosolic Zn^{2+} concentrations in the absence of Zn treatment paradigms, and consequential changes in cell signaling pathways, the notion of Zn participating in various signaling pathways is now rapidly gaining

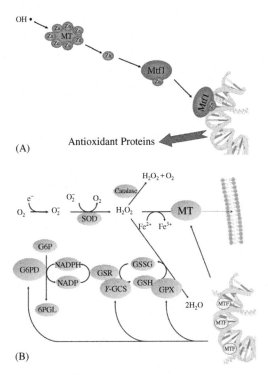

Fig. 3.1. The role of Zn in cell signaling of oxidative stress. (A) Reactive oxygen and nitrogen species oxidize Zn–thiolate bonds in the protein metallothionein (Mt), resulting in Zn release and an increase in kinetically accessible cytosolic Zn^{2+}. This increase in cytosolic Zn^{2+} is detected by the intracellular Zn sensor, metal-responsive transcription factor-1 (Mtf1), which upon binding of Zn to a regulatory Zn finger migrates to the nucleus where it associates with cognate binding motifs, called metal-response elements (MREs), in its target genes. (B) Binding of Zn-activated Mtf1 (shown as MTF) to MREs leads to expression of several genes coding for antioxidant proteins either through direct induction or through downstream events (Chung et al. 2005b). Abbreviations and gene symbols: SOD: superoxide dismutase; MT: metallothionein; G6P: glucose-6-phosphate; 6-phosphogluconolactone; G6PD: glucose-6-phosphate dehydrogenase; NADP: nicotinamide adenine dinucleotide phosphate; NADPH: reduced form of nicotinamide adenine dinucleotide phosphate; GSH: glutathione; GSSG: oxidized form of glutathione; GSR: glutathione reductase; γ-glutamylcysteine synthetase; GPX: glutathione peroxidase.

acceptance (Hershfinkel et al., 2010). One of the most persuasive examples of Zn signaling comes from a study on zebrafish, in which it was shown that the epithelial-to-mesenchyme transition (EMT) in the gastrula organizer was dependent on the presence of a particular Zn importer, called Zip6 (aka Liv-1/ Slc39a6) (Yamashita et al., 2004). During EMT, cells downregulate expression of the cell–cell adhesion protein, E-cadherin (Cdh1), allowing them to migrate into new positions. In the zebrafish embryo this process leads to longitudinal migration of stem cells and extension of the body axis (Fig. 3.2). Silencing of Zip6 stopped downstream signaling for EMT, resulting in a dwarfed embryo (Yamashita et al., 2004). Incidentally, EMT is pathologically activated during progression and metastasis of cancers; human *ZIP6* was originally discovered because it is one of the most upregulated genes in estrogen-dependent breast cancers (Dressman et al., 2001). This biological function appears to be highly evolutionarily conserved

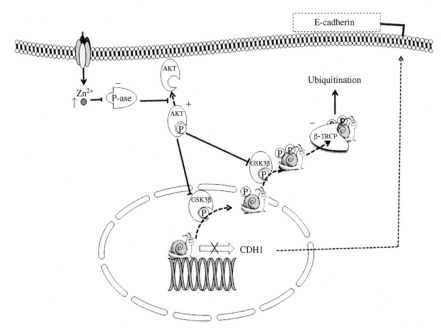

Fig. 3.2. Hypothetical signal transduction in Zn-mediated epithelial–mesenchymal transition (EMT). Zn importer Zip6 is required for Snail nuclearization in the zebrafish gastrula organizer, leading to downregulation of E-cadherin and cell migration (Yamashita et al. 2004). Snail is a transcriptional repressor of *cdh1*, which encodes for E-cadherin. E-cadherin is a membrane protein and its function is to provide cell–cell adhesion. In epithelial cells, phosphorylated AKT phosphorylates GSK-3β, which in turn phosphorylates Snail. Phosphorylated Snail exits the nucleus and is further phosphorylated by AKT in the cytosol where it then binds to βTrcp, is ubiquinated and finally degraded. This results in a constitutive expression of E-cadherin and the cell stays in the epithelium. Cellular influx of Zn^{2+} through Zip6 may inhibit one or several phosphates with Akt as target. This would result in an increased phosphorylation state of Akt, reducing the activity of this serine-threonine kinase and eventually resulting in Snail remaining in the nucleus where it downregulates transcription of *cdh1*. Abbreviations and gene symbols: Zip6: zrt- and Irt-like protein 6; P-ase: phosphatase; cdh1: cadherin 1; AKT: v-akt murine thymoma viral oncogene homologue; GSK3β: glycogen synthase kinase 3 beta; βTrcp: beta-transducin repeat containing protein.

because another Zn transporter of the same family, ZIP10 (aka fear-of-intimacy/Slc39a10), controls cell migration of gonad and glial cell progenitors in *Drosophila* embryos (Pielage et al., 2004; Mathews et al., 2006) and stimulates cell migration and invasiveness in human breast cancer cells (Kagara et al., 2007). In all three species (fruitfly, zebrafish, and human) this process involved downregulation of E-cadherin.

The requirement of Zn for cell division has been well established. It was recently shown that Zn regulates the exit from meiosis in mouse oocytes and that chelation of Zn blocks meiosis past telophase I (Kim et al., 2010). Zinc is also involved in regulation of cell proliferation and in mammalian somatic cells an increase in cytosolic $[Zn^{2+}]$ appears to be required for activation of proliferative protein tyrosine kinases, such as v-src sarcoma (Schmidt-Ruppin A-2) viral oncogene homologue (avian) (SRC), epidermal growth factor receptor (EGFR), insulin receptor substrate 1 and 2 (IRS1/2), and insulin-like growth factor 1 receptor (IGF1R) (Taylor et al., 2008; Hogstrand et al., 2009). The actual regulation by Zn is likely to be through inhibition of specific protein tyrosine phosphatases, resulting in an increased phosphorylation status of protein tyrosine kinases (Haase and Maret, 2003). In some human cancer cell types that rely on protein tyrosine kinase pathways for aggressive growth, the Zn signal for activation seems to come from yet another Zn transporter, ZIP7 (Taylor et al., 2008; Hogstrand et al., 2009). ZIP7 is located in the endoplasmic reticulum (ER) and releases Zn^{2+} into the cytosol by a process that might be gated through facultative phosphorylation of two serine residues. Release of Zn^{2+} from the ER is also a part of cell signaling in mast cells during the response to antigens. Cross-linking of the high-affinity immunoglobulin E receptor (FcjR) triggered release of labile Zn^{2+} into the cytosol from the ER and perinuclear area within minutes of the stimulus (Yamasaki et al., 2007). The wave of Zn^{2+} was dependent on a prior Ca^{2+} release and activation of mitogen-activated protein kinase (MAPK). Again, phosphatases were indicated as the targets of Zn^{2+} signaling, resulting in phosphorylation of ERK1/2 and JNK1/2 (Yamasaki et al., 2007). The importance of Zn for the immune system is well known and Zn signals have been observed also in monocytes and dentritic cells after stimulation with lipopolysaccharide and in T cells after treatment with phorbol esters (Haase and Rink, 2009). Zinc signals have now been shown to be involved in cytokine production in monocytes, maturation in dendritic cells, degranulation of mast cells, and apoptosis in lymphocytes (Haase and Rink, 2009).

There is also strong evidence from the mammalian literature for involvement of Zn signaling in bone and cartilage formation. Zinc deficiency suppresses matrix mineralization and delays osteogenic activity in mouse osteoblastic MC3T3-E1 cells through downregulation of Runx2 (runt-related transcription factor 2) (Kwun et al., 2010). Furthermore, knockout of either of several Zn transporters in mice, including *ZnT1*, *Znt5*, *Zip4*, and *Zip13*, results in skeletal deformities (Inoue et al., 2002; Andrews et al., 2004; Dufner-Beattie et al., 2007; Fukada et al., 2008). *Zip13* −/− mice have a phenotype typified by defects in bone, teeth, and connective tissue and these malformations are linked to changes in the BMP signaling pathway (Dufner-Beattie et al., 2007; Fukada et al., 2008). Taken together, present data suggest that Zn deficiency, caused by reduced Zn uptake or disruption of Zn transporters, leads to dysregulation of mineralizing cells in a process that involves effects on osteoblast differentiation.

The growth-promoting effect of Zn is well established, as illustrated by the retardation in growth during Zn deficiency (Eid and Ghonim, 1994; Prasad, 2009; Davies et al., 2010). The mechanism behind this effect is at least partially known, namely that Zn stimulates cellular glucose uptake. In fact, Zn has an insulin-mimetic effect on animals and can replace insulin in cell culture (Coulston and Dandona, 1980; Tang and Shay, 2001; Haase and Maret, 2004b; Wong et al., 2006). Zinc causes an increase in phosphorylation of the insulin receptor through inhibition of PTP1B, leading to activation of downstream signaling pathways and recruitment of insulin-responsive glucose transporter 4 to the plasma membrane (Tang and Shay, 2001; Haase and Maret, 2004b).

Thus, there is mounting evidence involving diverse pathways that Zn^{2+} operates as a second messenger, similarly to Ca^{2+}. The major difference between Zn^{2+} and Ca^{2+} signaling is that whereas Ca^{2+} signals are in the micromolar range, Zn^{2+} signals operate at low nanomolar concentrations.

7.2.3. ZINC SIGNALING IN THE NERVOUS SYSTEM

Finally, it is worth mentioning that Zn is accumulated in specific areas of the mammalian brain, such as the mossy fibers of the CA3 area of the hippocampus (Palmiter et al., 1996a; Sensi et al., 2009). Zinc is specifically found in synaptic vesicles of certain glutaminergic neurons and is released together with glutamate (Sensi et al., 2009). Synaptically released Zn^{2+} may modulate the activity of N-methyl-D-aspartate (NMDA) and γ-aminobutyric acid type A ($GABA_A$) receptors. Thus, Zn may influence both excitatory and inhibitory synapses in the brain. There is even evidence that Zn^{2+} has its own receptor, which activates a metabotropic response (Besser et al., 2009). It could therefore be argued that Zn^{2+} is a neurotransmitter in its own right. In mammals, Zn is loaded into synaptic vesicles of Zn-enriched neurons by a Zn transporter, called Znt3 (Slc30a3) (Palmiter et al., 1996b). This is one of the few Zn transporters that do not appear to have orthologues in fish (Feeney et al., 2005). Like mammals, there are Zn-enriched neurons in the fish brain (Pinuela et al., 1992) and it would therefore be exciting to find out if another Zn transporter is doing the job of Znt3 in mammals. If so, comparative sequence analysis could reveal what properties of the protein result in targeting of and ability to transport Zn into synaptic vesicles.

8. POTENTIAL FOR BIOCONCENTRATION OF ZINC

8.1. Bioconcentration from the Environment

As previously established, Zn is an essential element. It follows that when Zn availability is low, fish will take up as much of it as needed from their environment and in cases where environmental concentrations are in excess of the requirement, they will attempt to avoid further accumulation. Thus, bioaccumulation of Zn and its flow through the food web cannot be viewed in the same context as non-essential elements, or organic pollutants for that matter. Bioconcentration factors (BCFs) for Zn are therefore meaningless for purposes of regulating Zn concentrations in aquatic environments (McGeer et al., 2003). There is, however, no doubt that Zn accumulation increases as a direct function of environmental levels and this has been documented in numerous laboratory and field studies (Bradley and Sprague, 1985; Hogstrand and Haux, 1990; Hogstrand et al., 1991; Farag et al., 2003; Giguere et al., 2006; Besser et al., 2007; Zheng et al., 2008). The same factors that govern acute Zn toxicity (e.g. water concentrations of DOM, calcium, and H^+) control uptake of Zn from the water (Bradley and Sprague, 1985; Santore et al., 2002; De Schamphelaere et al., 2005). However, it would be expected that under most environmental conditions dietary uptake is the dominating contributor to whole-body Zn status (Renfro et al., 1975; Milner, 1982; Willis and Sunda, 1984; Spry et al., 1988; Niyogi et al., 2007). Thus, Zn bioaccumulation in aquatic life tends to be higher in waters where Zn levels are elevated, but there is no evidence for biomagnification in the food chain because Zn concentrations in higher trophic levels are not higher than those in lower levels (Eisler, 1993; Besser et al., 2007).

8.2. Relationship Between Body Size and Zinc Accumulation in Tissues

In the majority of studies where intraspecies relationships between Zn accumulation and fish size (or mass) have been investigated, either a negative correlation or no relationship at all has been found (Bradley and Sprague, 1985; Newman and Mitz, 1988; Hogstrand et al., 1991; Canli and Atli, 2003; Farkas et al., 2003). Negative correlations between Zn concentrations and fish size within a species have been attributed to a slower metabolism of older fish (Newman and Mitz, 1988). This would relate to the roles of Zn in metabolism and in particular to the fact that highly proliferating cells need more Zn (Beyersmann and Haase, 2001). In fact, a negative correlation between Zn status and age exists also in humans, where elderly people are considered at risk of Zn deficiency (Prasad, 2009). It is still debated whether this is due to a slower metabolism, changes in eating habits, or loss of homeostatic control. In fish, there are a few isolated cases where positive intraspecific correlations between Zn concentrations in tissues and body mass have been observed. A survey was carried out in the Western Indian Ocean, east and west off the coast of Mozambique, in which trace element levels were analyzed in tissues of four large predatory fish species (Kojadinovic et al., 2007). For two of these species, yellowfin tuna (*Thunnus albacores*) and swordfish (*Xiphias gladius*), the concentrations of Zn in both liver and kidney showed positive correlations to body length. In swordfish, the correlations were weak ($r^2 = 0.09$–0.19, $p < 0.03$, $N = 56$), but in yellowfin tuna they were stronger ($r^2 = 0.33$–0.45, $p < 0.001$, $N = 45$). Whether or not these results are reproducible or incidental, there is strong evidence that Zn concentrations do not increase with age or size in most fish species.

8.3. A Case Study of Zinc Accumulation in Perch

In early September of 1987 and 1988, perch (*Perca fluviatilis*) were collected from two areas in southern Sweden. One of these areas consists of a system of lakes and rivers surrounding the small town of Gusum, which was the site of a foundry specializing in brass production from the middle of the seventeenth century until 1968 (Hogstrand et al., 1991). Because the now abandoned foundry reportedly had short stacks, the Zn and Cu emitted to the air precipitated in the immediate surroundings. Consequently, the metal gradient emanating from the site is very steep and in 1987 was declining from 59 μg L^{-1} to 0.54 μg L^{-1} in a distance of only a few kilometers, making the area an excellent natural laboratory for studies of long-term effects of moderate Cu and Zn pollution. The other area is the north-western part of the lake Vänern, at the outlet of the river Asfjorden, which had high concentrations of Zn in the sediment (1000 mg kg^{-1} dry mass), but only modestly elevated levels of total Zn in the water (12–19 μg L^{-1}). The water was soft, which is typical for Sweden. Water Ca^{2+} concentrations ranged from 0.1 to 0.26 mM (hardness approximately 15 to 40 mg $CaCO_3$), with the lowest Ca^{2+} concentration measured in water samples from Vänern and the highest at the two most Zn-contaminated sites in the Gusum area. Colored dissolved organic matter (CDOM) was also measured and this was uniform except at the two most Zn-polluted sites at Gusum, where CDOM was lower than elsewhere. The perch collected at Gusum were part of a study to quantitatively investigate the role of Mt for the sequestration of Cu and Zn in areas of legacy metal pollution (Hogstrand et al., 1991). Zinc concentrations were also analyzed in perch from Vänern, and these results have not been previously published. Fortunately, original data from these studies have been archived and were available for reanalysis of trends regarding Zn accumulation and bioconcentration in perch. In total, there were complete records of fish weights and concentrations of Zn in liver and water for 94 fish and of these 64 were collected at Gusum.

Zinc levels in liver of perch along the Zn gradient increased as a function of total Zn concentrations in the water with little apparent influence of geographical area (Gusum or Vänern) (Fig. 3.3A). However, a 100-fold increase in total water Zn concentration resulted in only a modest, but statistically significant 1.2-fold increase in the liver Zn concentration (Fig. 3.3A) (Hogstrand et al., 1991). This shows that perch are quite capable of regulating Zn accumulation over a wide range of ambient Zn levels. In the fish collected in Vänern the Zn levels in the liver appear to correlate to the slightly elevated Zn concentration in the water rather than the very high levels in the sediment. However, Vänern is a large lake and there were no migration barriers where the perch were collected, so it is possible that they were not resident in the relatively local area of Zn-polluted sediments.

The ability of perch to regulate Zn uptake is further evidenced by the very steep negative exponential model, which described the BCF for the hepatic Zn concentration at each of the sampling stations as related to the total waterborne Zn concentration (Fig. 3.3B). At the site where the total waterborne Zn concentration was 0.56 $\mu g\ L^{-1}$ the average BCF was 44,900, compared to 2260 where the water contained 12 $\mu g\ L^{-1}$. Above this water Zn concentration, the BCF showed a shallow decline with increasing water Zn concentrations, reaching a value of 550 at the most Zn-polluted site, which contained 59 $\mu g\ L^{-1}$ in the water. This negative exponential relationship between BCF and water Zn concentration again suggests that perch regulate Zn well over a range of ambient Zn concentrations. The 20-fold higher BCF at 0.54 $\mu g\ L^{-1}$ compared with 12 $\mu g\ L^{-1}$ is probably indicative of Zn uptake from water being negligible compared to dietary sources at low Zn concentrations. A comparison between BCFs of males and females further shows that the BCFs were not influenced by gender. However, the perch were sampled in September, which is outside the breeding season, and it has been shown that female fish accumulate Zn in liver to a higher extent than males during vitellogenesis (Overnell et al., 1987; Thompson et al., 2003). There was also no influence of fish mass on liver Zn accumulation as analyzed for perch from each sampling site separately and combined for fish from all locations (Fig. 3.3C).

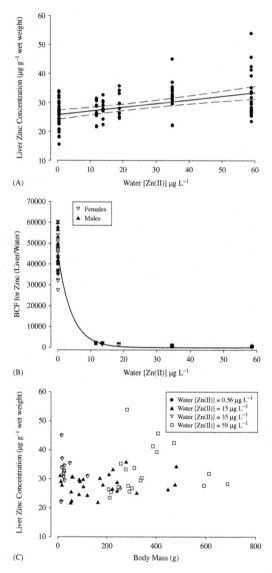

Fig. 3.3. Zinc accumulation in liver of perch (*Perca fluviatilis*) collected from lakes and streams in southern Sweden with different Zn concentrations. (A) Accumulation of Zn in liver increased marginally over a range of total waterborne Zn concentrations from 0.54 $\mu g\,L^{-1}$ to 59 $\mu g\,L^{-1}$. (B) The bioconcentration factor (BCF) decreased steeply at waterborne Zn concentrations from 0.54 $\mu g\,L^{-1}$ to 12 $\mu g\,L^{-1}$, indicating a transition in physiological regulation from promoting to limiting Zn uptake. The fish were collected outside the spawning season and there were no differences in BCF patterns between females and males as denoted. (C) There was no obvious relationship between liver Zn concentration and body mass either when analyzed sitewise or by including fish from different sites with different water Zn concentrations.

The overall conclusion from the literature and this particular case study of perch in Zn contaminated waters is that Zn accumulation in tissues of fish are well regulated over a wide range of total water Zn concentrations (see also Section 8.1 and McGeer et al., 2003). It is now known that this control involves a complex machinery of proteins and peptides, and the nature of these and how they are working together to confer Zn homeostasis will be discussed below.

9. CHARACTERIZATION OF UPTAKE ROUTES

9.1. Sites of Zinc Uptake

Zinc can be absorbed by intestine as well as the gill of fish. The relative importance of gill and gut depends on physiological status, previous Zn exposure, water chemistry, and Zn availability. In plaice (*Pleuronectes platessa*) treated with a series of environmentally realistic waterborne and dietary Zn concentrations, seawater was the source of less than 10% of the total Zn uptake (Milner, 1982). Other studies have come to similar conclusions with regard to Zn uptake in marine fish (Pentreath, 1973, 1976; Renfro et al., 1975; Willis and Sunda, 1984; Zhang and Wang, 2007b). Furthermore, as marine fish drink water some of the waterborne Zn in seawater is taken up by the gut (Zhang and Wang, 2007b). At much higher and environmentally unrealistic concentrations of Zn in seawater, the contribution by waterborne Zn to total Zn uptake increases and may amount to half of the total uptake (Milner, 1982; Willis and Sunda, 1984). In a comprehensive study, freshwater rainbow trout were treated with combinations of Zn in feed (1–590 mg kg^{-1}) and water (7–148 µg L^{-1}) for 16 weeks (Spry et al., 1988). Changing the dietary Zn concentration from 1 mg kg^{-1} to 590 mg kg^{-1} resulted in a 1.7–2.3-fold difference in whole-body Zn content at low and high waterborne Zn concentrations, respectively. Conversely, a change in the Zn concentration of the water from 7 to 148 µg L^{-1} was reflected in a 1.7–2.9-fold increase in whole-body Zn of fish in a pattern that was not obviously related to the dietary Zn level. The highest fold change in whole-body Zn content occurred in the group fed a Zn-adequate diet. In fish treated with adequate dietary Zn (90 mg kg^{-1}) and high waterborne Zn (148 µg L^{-1}), as much as 57% of the Zn uptake was calculated to have originated from the water. Since freshwater fish drink very little, uptake from waterborne Zn probably occurred across the gills. Furthermore, waterborne Zn could restore Zn deficiency symptoms in fish fed Zn-depleted diets. It can be concluded that during most conditions uptake of Zn occurs primarily over the gut and the gill is an auxiliary organ for Zn uptake. However, especially in freshwater fish, uptake across the gill can contribute significantly (>50%) to total Zn absorption if the Zn concentration in the water is high or that in the diet low. Certainly in terms of toxicology of Zn, the gill is very important (see Section 6). It was found that wild-caught yellow perch (*Perca flavescens*) from Zn-impacted waters have a decreased rate of Zn uptake across the gills, compared with those from a cleaner site, but no such difference was observed for Zn uptake from the gut (Niyogi et al., 2007). These findings suggest that it may be more critical to limit Zn accumulation in the gill than to attenuate uptake of Zn across the intestine, and they also indicate that Zn absorption across gill and gut can be regulated independently.

9.2. Zinc Transporters

There are two protein families dedicated to Zn transport in animals, Slc30 (Zn transporter, Znt, family) and Slc39 (Zrt irt-related protein, Zip, family). The expression of Zn transporters in juvenile zebrafish tissues, as indicated by abundances of their respective mRNAs, is shown in Table 3.2. Protein structure and cellular localization of Zn transporting proteins are depicted in Figs. 3.4–3.6. In most instances the experimental evidence from

Table 3.2

Expression of zinc transporters of the Znt (Slc30) and Zip (Slc39) families in tissues of zebrafish as indicated by abundances of their respective transcripts (Feeney et al., 2005)

	Gill	Intestine	Kidney	Ovary	Brain	Eye	Muscle	Liver
znt1	+	+	+	++	−	−	+	+
znt2	−	−	−	−	+	+	−	−
znt4	+	+	+	++	+	+	−	+
znt5	++	++	+	++	−	+	−	+
znt7	+	+	+	++	−	−	−	−
znt8	+	−	−	−	−	+	−	−
znt9	+	−	−	+	−	+	−	−
zip1	+	+	++	++	+	++	+	++
zip3	++	++	+	+	+	+	+	+
zip4	−	++	−	+	−	+	−	−
zip6	+	−	+	++	+	+	−	−
zip7	++	++	++	++	+	++	+	++
zip8	+	+	−	−	−	−	−	−
zip10	++	++	++	++	++	+	−	++
zip13	+	+	+	++	+	+	+	+

Relative abundance of mRNA for each transporter is shown with −, +, and ++, denoting absent or low, present, and abundant, respectively.

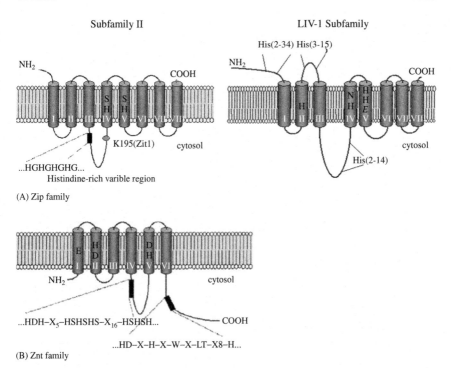

Fig. 3.4. Predicted structures of the Zip (Slc39) and Znt (slc30) families of Zn transporters. (A) The Zip (Slc39) family of transporters has two subfamilies in animals, named Subfamilies II and LIV-1. All members have eight transmembrane domains (TM) and both N- and C-termini are exiting the membrane away from the cytosol. There is also a large cytosolic loop between TM III and IV, which is believed to be of importance for Zn binding because it contains several histidine residues. The LIV-1 subfamily is characterized by having a conserved amino acid motif, HEXPHEXGD, located in TM V. In contrast to subfamily II, the proteins in the LIV-1 subfamily have a long N-terminal stretch of amino acid before the first TM and it contains a putative proteolytic cleavage site just before TM I. Other conserved amino acid residues are indicated for proteins of either subfamily. (B) Proteins in the Znt (Slc30) family of Zn transporters typically have six TM, except for Znt5, which has two isoforms with 12 and 15 TM, respectively. Both termini are located in the cytosol. There is a histidine-rich cytoplasmic loop and a C-terminal tail of varying length between members. Adapted from PhD thesis by Qiu (2004).

these localizations comes from experiments with mammalian cells (Hogstrand et al., 2009), with the exceptions of Znt1 (Slc30a1), Zip1 (Slc39a1), Zip3 (Slc39a3), and Zip7 (Slc39a7; Qiu and Hogstrand, unpublished), which have been localized in cultured fish cells (Qiu et al., 2005; Qiu and Hogstrand, 2005; Balesaria and Hogstrand, 2006). Furthermore, zebrafish *znt5* shows transcriptional responses consistent with a role of Zn uptake at apical membranes of epithelial cells (Fig. 3.6) (Zheng et al., 2008).

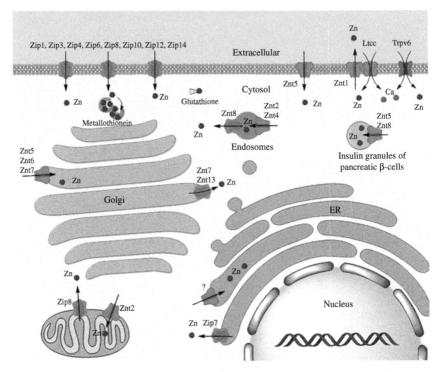

Fig. 3.5. Cellular regulatory proteins for Zn. Zinc may enter the cell through either of a number of Zn transporters, which include Znt5 and several members of the Zip family of proteins. Zinc may also enter through L-type or epithelial calcium channels (Ltcc, Trpv6), of which the latter is believed to be of importance for gill Zn uptake at lethally toxic water Zn concentrations. In the cytosol, Zn is buffered by molecules, such as metallothionein and glutathione, and moved into deep storage sites, including the endoplasmic reticulum, the Golgi apparatus, endosomes, and mitochondria. In pancreatic β-cells, the Zn transporters Znt5 and 8 transport Zn into insulin granules, where it is used to coordinate insulin. Zinc transporters show varying degrees of tissue specificity. The cellular location of Zn transporters shown is mostly based on studies on their mammalian orthologues. See the text for details. **SEE COLOR PLATE SECTION.**

ZNT1 (SLC30a1) was the first Zn transporter to be discovered. It was identified as a protein in rat that conferred resistance to Zn in cell culture (Palmiter and Findley, 1995). It was shown that ZNT1 is a Zn efflux transporter that is ubiquitously expressed in the body. Znt1 was also the first Zn transporter to be functionally characterized in fish, and the orthologues of Japanese pufferfish (*Takifugu rubripes*), common carp (*Cyprinus carpio*), and zebrafish share all important features with rat ZNT1 (Feeney et al., 2005; Balesaria and Hogstrand, 2006; Muylle et al., 2006a). There are now 10 SLC30 paralogues described in mammalian genomes and eight in the

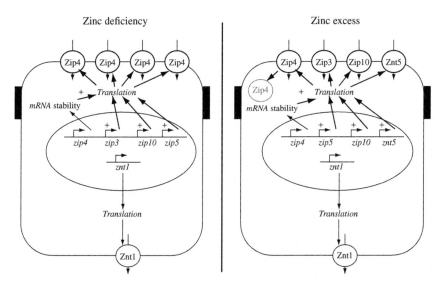

Fig. 3.6. Homeostatic regulation of Zn transporters in transporting epithelia. The transporters shown are expressed in zebrafish intestine and, with exception for Zip4, in zebrafish gills. During Zn deficiency (left) expression of apical Zn importers (top of figure) is upregulated to promote Zn uptake. When there is excess of Zn (right) at the apical side of the epithelium, expression of apical Zn importers is downregulated to limit Zn influx. At the basolateral side of the cells (bottom of figure), the Zn exporter Znt1 is upregulated to remove Zn from the epithelial cells and thereby protect them from Zn cytotoxicity. Regulation of Zn transporters includes transcriptional, translational, and post-translational processes. Zip3, Zip10, Znt1, and Znt5 are regulated at least partially through changes in the rate of transcription. The gene for Zip4 is constitutively expressed, but its translation is positively regulated by Zn deficiency and its internalization and degradation are stimulated by Zn excess.

genomes of Japanese pufferfish, spotted green pufferfish (*Tetraodon nigroviridis*), and zebrafish (Feeney et al., 2005). Of all Znt proteins only Znt1 has been shown to operate as a Zn extrusion system and ZNT2–4 and 6–8 are located to the membranes of intracellular organelles (Palmiter and Huang, 2004; Hogstrand et al., 2009; Lichten and Cousins, 2009; Hershfinkel et al., 2010). ZNT5 occurs as two splice variants, one of which is located at the plasma membrane and is probably a Zn importer, and one that appears to transport Zn into the Golgi apparatus (Cragg et al., 2002; Kambe et al., 2002; Jackson et al., 2007). All of the Slc30 paralogues, except Znt5, have six transmembrane domains, with N- and C-termini being located on the cytoplasmic side (Palmiter and Huang, 2004). Znt5 is much larger and has up to 12 transmembrane domains (Cragg et al., 2002). ZNT5 is also apparently unique among SLC30 proteins in that the shorter of the splice variants can transport Zn into the cytosol. The longer ZNT5 splice variant along with all other SCL30 isoforms examined transport Zn out of the cytosol and either out of the cell (Znt1) or into different organelles.

The Zip family, Slc39, of Zn transporters has 14 members in mammalian genomes and 13 in fish (Feeney et al., 2005). In contrast to the Znt family, which typically transport Zn away from the cytosol, all Zip paralogues appear to transport Zn into the cytosol, at least during normal physiological conditions (Hogstrand et al., 2009; Lichten and Cousins, 2009; Hershfinkel et al., 2010). Also in contrast to the Znt family, most Zip proteins, except Zip7, Zip9, and Zip13, are functional in the plasma membrane and mediate tissue-specific Zn uptake. Zip7 releases Zn from the ER, and Zip9 and Zip13 from the trans-Golgi network. Zip8 may transport Zn into some cells and out of lysosomes in other tissues. Zip proteins have eight transmembrane-spanning domains with both N- and C-termini located away from the cytosol. They all have a long cytoplasmic loop (between transmembrane domains III and IV), which typically has several histidine residues located in clusters. These are believed to act as temporary binding sites for Zn as it transverses the protein. There is a subfamily of these proteins, called the LIV-1 subfamily (Taylor and Nicholson, 2003; Taylor et al., 2007). The members of this subfamily have a long extracellular N-terminal stretch which is also rich in histidine residues and in some members contains a putative proteolytic cleavage site, which may be of importance for post-translational processing (Zhao et al., 2007). Three Zip Zn transporters, Zip1, Zip3, and Zip10, have been functionally characterized in fish and all three were shown to mediate Zn import when ectopically expressed in cells (Qiu et al., 2005; Qiu and Hogstrand, 2005; Zheng et al., 2008). *Zip3* is one of the most abundantly expressed Zn importers in gills and intestine of zebrafish (Feeney et al., 2005; Zheng et al., 2008). Through manipulation of water chemistry and modeling of Zn speciation it was determined that the Zn^{2+} ion is the likely transported species for Zip3 (Qiu and Hogstrand, 2005). Furthermore, Zn transport was stimulated by a slightly acidic medium (pH 5.5–6.5) and inhibited by HCO_3^-.

Although there is considerable redundancy in terms of Zn transporter function, many of the described Zn transporters have distinct biological functions relating to their unique subcellular or tissue distribution (Feeney et al., 2005; Hogstrand et al., 2009; Lichten and Cousins, 2009; Hershfinkel et al., 2010). For example, ZNT3 in mammals transports Zn into synaptic vesicles and ZNT8 into insulin granules of pancreatic β-cells; ZIP7 is the only Zn release transporter in the ER, and ZIP12 is expressed only in fenestrated epithelia. In fish, *znt2* seems to be exclusive to neural tissue, *znt8* is found in the eye, *zip8* is exclusive to the gill, and all Zn transporters show some degree of tissue specificity (see Table 3.2) (Feeney et al., 2005). Thus, the distribution of Zn within the fish and between cellular compartments is regulated by differential expression and activation of a complement of 21 Zn transporters.

In addition to the Znt and Zip protein families, it is quite well established that calcium transporters are generally permeable to Zn. Already in 1977, it was shown that voltage-gated calcium channels in insect muscle are permeable to Zn (Fukuda and Kawa, 1977). It was later shown that blockers of voltage-gated Ca^{2+} channels (VGCC) can inhibit Zn uptake into the gill of the mussel (*Mytilus edulis*) (Vercauteren and Blust, 1999). Similarly, Zn can enter mammalian neurons through VGCC and α-amino-3-hydroxy-5-methyl-4-isoxazole proprionic acid receptors (AMPAR) (Sensi et al., 2009). For fish, the competition of Zn^{2+} with Ca^{2+} for the epithelial calcium channel (Ecac/Trpv6), located on the apical membrane of the gill, is of particular importance because this contributes to the protective effect of hardness against Zn^{2+} toxicity (Hogstrand et al., 1994, 1996b; Qiu and Hogstrand, 2004).

9.3. Branchial Uptake

Ecac (Trp6) belongs to the transient receptor potential (TRP) family of proteins (Hoenderop and Bindels, 2008). It is primarily expressed in the gill where it is responsible for apical calcium entry (Qiu and Hogstrand, 2004; Shahsavarani et al., 2006; Liao et al., 2007). While Ecac (Trpv6) is probably a Zn uptake pathway when Zn in the water is in excess, it is not known whether or not Ecac contributes to nutritional Zn uptake. Feeding rainbow trout with a high-calcium diet reduced both Zn and calcium uptake across the gill (Niyogi and Wood, 2006). Similarly, intraperitoneal injection of calcium decreased influx of both Ca^{2+} and Zn^{2+} across the gill (Hogstrand et al., 1996b). Treatment of rainbow trout with the calciotropic hormone, 1α,25-$(OH)_2D_3$, increased Zn^{2+} uptake across the gill with a concomitant increase in expression of Ecac (Qiu et al., 2007). However, it is possible that at least in this latter case the effect on Zn uptake was mediated by Zn transporters because expression of the Zn importer *zip1* was also increased by 1α,25-$(OH)_2D_3$ treatment (Qiu et al., 2007).

The unidirectional influx of Zn across the gill of rainbow trout follows Michaelis–Menten type kinetics and is highly influenced by the water calcium concentration (Spry and Wood, 1989). Increasing the calcium concentration from 0.5 to 5 mM resulted in more than 10-fold increase in K_M (decreased affinity) from 1.8 to 23 μM (120 to 1500 μg L^{-1}) with a concomitant 1.6-fold increase in J_{max} (maximal transport capacity) for Zn. At an acclimation ambient water calcium concentration of 1 mM, the apparent affinity for branchial Zn influx showed a cyclic variation from 3 to 8 μM (195–520 μg L^{-1}) with about 2 week periodicity, which coincided with a similar fluctuation in calcium uptake (Hogstrand et al., 1998). The affinity of the gill for Zn must be a composite measure for all Zn transporters present in the tissue. The affinities for two fish Zn importers, Zip1 and Zip3, have been determined and they are $0.5\,\mu M$ $(32\,\mu g\,L^{-1})$ and $14\,\mu M$ $(915\,\mu g\,L^{-1})$, respectively, bracketing the range of observed normal Zn affinities of the gill (Qiu and Hogstrand, 2005; Qiu et al., 2005).

When freshwater rainbow trout were transferred to water with a higher Zn concentration ($2.3\,\mu M = 150\,\mu g\,L^{-1}$), there was a rapid decrease in the affinity for Zn (increased K_M) evident already during the first 24 h of Zn exposure, resulting in a reduced Zn uptake (Hogstrand et al., 1995, 1998; Alsop and Wood, 2000). This would suggest that there is a rapid and selective decrease in the activities of high-affinity Zn transporters in response to Zn exposure. In support of this idea, the apical Zn importer, ZIP4, in the mouse intestine is quickly ubiquinated and degraded in response to high dietary Zn supplementation (Mao et al., 2007; Weaver et al., 2007). The dynamics in modulating Zn kinetics across the gill in response to changes in Zn levels was further shown by the demonstration of large differences in Zn binding kinetics to the gill and a large increase in the rapidly exchangeable pool of Zn in gills of Zn-acclimated rainbow trout (Alsop and Wood, 2000). Waterborne exposure of rainbow trout to $2.3\,\mu M$ of Zn also results in a competitive inhibition of calcium influx as described above (Hogstrand et al., 1994, 1995, 1998). However, in moderately hard water (calcium = 1 mM) an increase in K_M of the gill for calcium from 0.04 to 0.15 mM had only a small negative impact on calcium uptake and this could be compensated for by increasing the number of calcium uptake sites (J_{max}). So, overall, physiological acclimation to increased waterborne Zn involves a reduced affinity for both Zn and calcium uptake and an increase in J_{max} for calcium influx.

Following the identification of the molecular entities that mediate Zn and calcium uptake it was of interest to investigate how transcription of Zn transporters in the fish gill may change during acclimation to conditions of high or low Zn availability. A diagram summarizing what is known about regulation of Zn transporters in transporting epithelia in response to Zn excess and depletion is provided in Fig. 3.5. As some of the mechanisms outlined in Fig. 3.5 are based on findings in mammalian systems, there is an underlying assumption that the Zn transporters in fish will respond in the same way. Using the tractability of zebrafish for molecular research, the zebrafish complement of Zn transporters was identified and expression levels were first examined in eight different tissues, including the gill (Feeney et al., 2005). It was found that at the mRNA level at least 13 Zn transporters were expressed in the zebrafish gill and of these *znt5*, *zip3*, *zip7*, and *zip10* were especially abundant (Table 3.2). With the addition of *zip4*, which is known to be essential for homeostatic Zn uptake in mammals, exactly the same set of transporters predominated in the intestine.

This study was followed by an experiment where Zn was either depleted from or supplemented to water and feed for 2 weeks (Zheng et al., 2008). It was found that *znt1*, *znt5*, *zip3*, and *zip10* were all regulated at the mRNA level in response to Zn depletion or supplementation. Expression of *znt1* increased in Zn-supplemented fish and decreased in Zn depletion (Zheng et al., 2008), in keeping with the known transcriptional regulation of this gene by Mtf1 (Muylle et al., 2006a; Hogstrand et al., 2008) and the basolateral location of ZNT1 in the mammalian intestine (McMahon and Cousins, 1998). Expression of *znt5*, *zip3*, and *zip10* was consistent with these transporters operating as Zn importers at the apical membrane of the gill because Zn depletion resulted in upregulation of all three, and Zn supplementation resulted in decreased expression of *znt5* and *zip10*. Of the Zn importers, the strongest transcriptional responses to the changes in Zn availability were observed for *zip10*.

The negative Zn regulation of expression of *zip10* was investigated in detail and it was shown to be mediated by Mtf1, which was demonstrated to function as a repressor for the gill transcript of *zip10* (Hogstrand et al., 2008; Zheng et al., 2008). A cluster of three MREs was identified as being required for Mtf1 transcriptional repression in *zip10* and these were straddling the transcription initiation site. The sequence context making the MREs inhibitory rather than promoting transcription of *zip10* is not clear, but the MREs were found to overlap with SP1 sites and binding of Mtf1 to these MREs may therefore block assembly of the transcription initiation complex. The *zip10* transcript expressed in kidney is regulated by an alternative promoter, which is positively regulated by Zn and Mtf1 (Zheng et al., 2008). Because of known interactions with Ecac in rainbow trout, expression of this gene in response to Zn depletion and excess was also investigated in zebrafish but it was unaffected by Zn treatment. Thus, it can be concluded that, as predicted through physiological studies, expression of Zn importers likely mediating entry of Zn from the water into the gill is decreased during waterborne Zn exposure. The expression of the basolateral Zn efflux protein Znt1 is increased, presumably to protect the gill cells from Zn overload. During Zn depletion, apical Zn transporters are instead upregulated to improve uptake efficiency. Zinc depletion results in a downregulation in mRNA for the basolateral Zn effluxer, Znt1. This is counter-intuitive from an organismal homeostatic point of view because it would be predicted that the fish would benefit from improvement of the transfer of Zn into the bloodstream when Zn availability is low. However, it should be noted that *znt1* mRNA rather than the actual protein was analyzed. It is entirely possible that Znt is also regulated post-translationally to enhance basolateral transfer. Translational and post-translational regulation of the other Zn transporters may also have occurred without being noticed and it should be stressed that only the transcripts for the transporters were measured.

9.4. Gastrointestinal Uptake

Although the intestinal Zn uptake may be quantitatively more important than that across the gills, the latter has received more attention in fish, partially because of its significance for Zn toxicity, but also because of the ease by which gill uptake can be manipulated experimentally. Nevertheless, there are several studies on physiological principles of gastrointestinal Zn uptake in both marine and freshwater fish that allow some conclusions to be drawn. Absorption of Zn in the marine fish winter flounder (*Pseudopleuronectes americanus*), plaice (*Pleuronectes platessa*) and black sea bream (*Acanthopagrus schlegeli*) was found to be highest in the upper small intestine and follow first order kinetics (Pentreath, 1976; Shears and Fletcher, 1983; Zhang and Wang, 2007a). The movement of Zn across the brush-border membrane in winter flounder was inhibited by Cu, Cd, Co, Cr, Ni, Mg, and Hg, but not by calcium. The lack of competition with calcium for uptake is consistent with the absence of *Ecac* expression in the gut of fish (Qiu and Hogstrand, 2004; Shahsavarani et al., 2006), but not with the finding that Zn transport by Zip3 and perhaps other Zn importers is inhibited by calcium and most notably Cu (Qiu and Hogstrand, 2005). Similarly to the marine fish studied, Zn uptake in isolated gut sacs from freshwater rainbow trout intestine was highest in sacs prepared from the anterior and mid intestine (Ojo and Wood, 2007, 2008; Ojo et al., 2009). A 10-fold molar excess of Cu over Zn reduced Zn uptake in mid and posterior sections of the intestine by 50 and 78%, respectively, but not in the anterior part (Ojo et al., 2009). Conversely, a 10-fold Zn over Cu excess inhibited Cu uptake in the same sections but not in the anterior section or in the stomach where Cu uptake was found to be high. A high calcium concentration (100 mM) stimulated Zn binding to mucus and the mucosal epithelium in the stomach, but not in any part of the intestine (Ojo and Wood, 2008). Using an *in vivo* intestinal perfusion technique on freshwater rainbow trout it was found that, compared with the gill, the intestine represents a low-affinity ($K_M = 309$ μM or 20,000 μg L^{-1}) and high-capacity ($J_{max} = 933$ nmol kg^{-1} h^{-1}) uptake pathway for Zn (Glover and Hogstrand, 2002b). The same conclusion was made for gill and gut uptake pathways in yellow perch (Niyogi et al., 2007). In rainbow trout, mucus secretion was greatly stimulated by the presence of unbound Zn in the intestinal lumen (Glover and Hogstrand, 2002b). This mucus moderated Zn uptake by stimulating the uptake at low concentrations of Zn in the intestinal lumen, and inhibiting Zn uptake when luminal Zn concentrations were high. With the same perfusion system it was found that Cu and calcium inhibited entry of Zn into the intestinal epithelium and consequently also the transfer of Zn into the circulation (Glover and Hogstrand, 2003). Calcium, Cu, and Mg reduced Zn-stimulated mucus secretion while mucus secretion was further stimulated by Cd. This is interesting because Cd is a Zn mimic and presumably activates the pathway of mucus secretion otherwise intended to be regulated by Zn.

Amino acid chelates of Zn are often used as feed additives for farmed animals, including fish, in an attempt to improve the efficiency of Zn absorption. However, feeding trials in fish using amino acid and other organic chelates of Zn have not resulted in convincing improvements of absorption and performance indicators above those obtained with inorganic forms of Zn (Kjoss et al., 2006; Davies et al., 2010). The effects of amino acids upon intestinal Zn uptake in freshwater rainbow trout and seawater adapted black sea bream were studied using *in vivo* or *in vitro* perfusion techniques (Glover and Hogstrand, 2002a; Zhang and Wang, 2007a). It was found that Zn bound to L-histidine was taken up by trout intestine at least as efficiently as unbound Zn, but through histidine-facilitated pathways (Glover and Hogstrand, 2002a; Glover et al., 2003). Chelation of Zn by L-cysteine increased the Zn uptake rate in trout and sea bream by 100% and 60%, respectively, compared with that for unbound Zn. The presence of histidine and cysteine strongly influenced the distribution of the newly accumulated Zn in the body. Histidine promoted accumulation of Zn in the intestinal tissue, whereas cysteine caused Zn to specifically accumulate in the blood (Glover and Hogstrand, 2002a).

Thus, uptake of Zn in the gut starts with the diffusion into the unstirred layer followed by binding to the mucus of the intestinal epithelium. The metal is then transported into the epithelial cells either by Zn transporters as the Zn^{2+} ion or bound to amino acids, such as histidine and cysteine. Measurement of mRNA for Zn transporters in the zebrafish intestine by real-time polymerase chain reaction (PCR) indicates that the Zn transporters expressed at the highest levels are *znt5*, *zip3*, *zip4*, *zip7*, and *zip10* (Table 3.2) (Feeney et al., 2005). Based on knowledge of their mammalian orthologues, structural similarities to the mammalian proteins, and homeostatic responses to Zn depletion or supplementation, it is postulated that Zip7 is located in the endoplasmic reticulum (Fig. 3.5) and the others in the brush-border membrane, mediating uptake of unbound Zn (Figs. 3.5 and 3.6) (Feeney et al., 2005; Zheng et al., 2008; Hogstrand et al., 2009). One would imagine that Zn in the lumen of the anterior intestine (where Zn uptake is highest) would be bound to amino acids, such as cysteine and histidine, and it has been shown that Zn bound to these amino acids is taken up by the enterocytes (Glover and Hogstrand, 2002a). So, the question might be asked whether or not Zn uptake by specific Zn transporters is of physiological importance for transfer of Zn across the Zn intestine. The answer is that for fish we do not really know, but in mammals uptake through Zn transporters is essential. In humans there is an autosomal recessive disease called acrodermatitis enteropathica, in which individuals die from Zn deficiency at young age unless their diet is supplemented with high levels of Zn (Ackland and Michalczyk, 2006). This is a monogenetic disease caused by one of several mutations in the brush-border membrane Zn transporter, ZIP4 (Kury et al., 2002; Nakano et al., 2009). Similarly, the homozygote mouse *Zip4* knockout dies during embryo development and the heterozygote is very sensitive to Zn deficiency (Dufner-Beattie et al., 2007). As ZIP4 transports inorganic Zn, these results tell us that inorganic Zn uptake is essential and, thus, that at some point during the uptake phase, Zn has to leave the amino acids in the intestinal lumen. One possibility is that there is first a ligand exchange with mucin, which then donates the Zn to the transporters in the brush-border membrane.

10. CHARACTERIZATION OF INTERNAL HANDLING

10.1. Cellular Zinc Regulation and Homeostatic Responses

Uptake of Zn in fish is inversely proportional to Zn availability and stimulated by cortisol and vitamin D (1,25-dihydroxycholecalciferol) (Hogstrand et al., 1995; Qiu et al., 2007; Bury et al., 2008; Zheng et al., 2008). Cellular Zn influx and efflux are regulated by changing expression and activities of Zn transporters. Regulation of Zn transporters occurs at transcriptional, translational, and post-translational levels (see Fig. 3.6). Many Zn importers are negatively regulated by Zn and the Zn exporter, Znt1, is positively regulated by Zn. While Zip3, Zip10, Znt1, and Znt5 have shown transcriptional responses to changes in cellular Zn levels (Jackson et al., 2007; Zheng et al., 2008), Zn-dependent regulation of Zip4 appears to be entirely post-translational, although this has only been investigated in mammalian systems. Mouse *Zip4* is constitutively expressed, but stability of *Zip4* mRNA is increased during Zn deficiency, leading to an increased rate of translation (Weaver et al., 2007). During conditions of Zn excess, ZIP4 is ubiquitinated, internalized, and degraded, as shown in human cells (Mao et al., 2007).

Because of its ability to influence the activities of a great number of molecules in the cell, the cytosolic Zn^{2+} concentration has to be kept very low. How low has been a matter of a long debate that really has been limited to extrapolations following a series of assumptions. With the advent of specific fluorescent probes for Zn it is now possible to measure the $[Zn^{2+}]$ within a cell and such analyses have shown that (1) the Zn^{2+} concentration fluctuates during different phases of the cell cycle, (2) the resting Zn^{2+} concentration in cells is probably in the picomolar range, and (3) during instances of Zn signaling the Zn^{2+} concentration may reach up to mid-nanomolar levels (Li and Maret, 2009; Colvin et al., 2010). The results from these measurements fit well with the setpoint for activation of the cell's own Zn sensor, Mtf1, which has an activation setpoint in the low nanomolar range (Laity and Andrews, 2007).

Metallothionein (Mt) and glutathione are major Zn-binding molecules in the cytosol (Fig. 3.5) (Jiang et al., 1998; Colvin et al., 2008) and the importance of these for Zn binding in fish cells increases in cells with high Zn load (Hogstrand and Haux, 1990, 1996; Hogstrand et al., 1991; Lange et al., 2002; Muylle et al., 2006b). Mt is a cysteine-rich protein that can bind up to seven Zn atoms and when Mt is isolated from tissues it is saturated with Zn, or Zn in combination with other metals. However, recent detailed analysis has shown that the different Zn binding sites differ by four orders of magnitude in their affinities for Zn and that unsaturated Mt with up to three available Zn binding sites exists in the cell (Krezel and Maret, 2007).

The classic general view of Zn handling by the cell has been that Zn first binds to glutathione and that an overload of the glutathione pool will activate Mtf1, which will then stimulate *de novo* synthesis of Znt1, gamma-glutamylcysteine synthetase (Gcl), and apo-metallothionein (thionein), which will then increase the capacity for Zn extrusion and sequester the Zn excess (Hogstrand and Wood, 1996; Andrews, 2001). It is now known that because Mt in the cell is not normally saturated with metal, it can contribute to the initial buffering of Zn entering the cell (Colvin et al., 2010). However, when experimental data of labile Zn concentrations in the cytosol of cells exposed to Zn were fitted to quantitative mathematical models, it was clear that the buffering capacity of Mt and glutathione could not account for what was observed (Colvin et al., 2008). Instead, the model that best fitted the data was the rapid translocation of the entering Zn to a "deep store" by a vehicle that also had buffering capacity before the Zn reappeared in the cytosol at a later stage. This vehicle was termed a "muffler" to distinguish from a pure buffer, and the deep store was predicted to perhaps be one or several organelles (Fig. 3.5) (Colvin et al., 2008, 2010). Such a model fits very well with observations on movement in rainbow trout hepatocytes (Muylle et al., 2006b). Similar to the studies by Colvin and co-workers, time-dependent intracellular fluxes of Zn were followed after exposure to exogenous Zn and other manipulations. It was found that the

Zn entering the cells was not visible before it appeared in vesicles. The model also explains observations made in tamoxifen-resistant breast cancer cells (TamR) (Taylor et al., 2008; Hogstrand et al., 2009). Adding exogenous Zn to these cells causes a massive activation of growth factor receptors, such as ErB, IGF-1R, EGFR, and c-SRC through Zn-stimulated tyrosine phosphorylation, followed by increased growth and invasive behavior. In Zn-exposed cells where *ZIP7* was silenced by siRNA, there was no labile Zn^{2+} in the cytosol, no growth factor receptor activation, and no effect on growth or invasiveness. The conclusion must be that before Zn has effects on proteins in the cytosol it has to emerge from the ER and it must get there within seconds or minutes following Zn treatment of the cells (Taylor et al., 2008; Hogstrand et al., 2009). Thus, according to the latest understanding, Zn that enters the cell is muffled, perhaps by Mt in combination with glutathione, and then immediately moved into organelles, such as the ER, before it might re-enter the cytosol. What controls this re-entry, if it is at all controlled, is currently unknown. With continuous Zn exposure Mtf1-mediated gene expression is activated, resulting in increased Zn muffling by Mt and export by Znt1, but as opposed to effects on cell signaling, which happen within minutes, gene expression responses take hours to manifest themselves.

10.2. Transport through the Bloodstream

Average concentrations of Zn in plasma of rainbow trout, lake trout (*Salvelinus namaycush*), walleye (*Stizotedion vitreum*), squirrelfish (*Holocentrus marianus*), and whitefish have been determined to range from 6.3 to 15.1 mg L^{-1} (96 to 231 μM) (Bettger et al., 1987). The plasma Zn level in turbot (*Scophtalamus maximus*) is slightly higher at 23 mg L^{-1} (Overnell et al., 1988). The concentration of Zn in red blood cells of rainbow trout was measured to be 45 μg g^{-1} protein but data were not provided to allow a comparison to plasma (Bettger et al., 1987). In squirrelfish, there was a dichotomy in plasma Zn concentration between genders, with females showing significantly higher concentrations (7.5 mg L^{-1}) than males (5.2 mg L^{-1}) (Hogstrand et al., 1996a). The concentration of Zn in the red blood cells of squirrelfish was determined to be 10.6 μg g^{-1} tissue with no difference between males and females (Hogstrand et al., 1996a). Thus, in this species the concentration of Zn in red blood cells was slightly higher than that of plasma. In tilapia (*Oreochromis aureus*) red blood cells and plasma were about equal in Zn concentration, assuming a hematocrit of 30%, but in grass carp (*Ctenopharyngodon idellus*) and silver carp (*Aristichthys nobilis*) the Zn concentrations were calculated to be three times higher in red blood cells than in plasma (Jeng et al., 2007). In rainbow trout, the membrane of the erythrocyte contained about twice the concentration of Zn as compared with the whole blood (Bettger et al., 1987), which may be unexpected because of the high concentration of the Zn-enzyme carbonic anhydrase in red blood cells.

There is no known specific plasma protein, like transferrin for iron or ceruloplasmin for Cu, which distributes Zn among tissues. Instead, as mentioned earlier, distribution of Zn in the body is managed by a complex set of Zn transporters and their expression patterns and activities dictate Zn uptake in different tissues. In rainbow trout plasma, 0.2% has been estimated to be unbound and this corresponds to about 22 $\mu g\ L^{-1}$ (340 nM) (Bettger et al., 1987), but this estimate should be revisited using Zn probes and other more recent technologies. Most of the plasma Zn is bound to albumin, which at least in mammals contains a major high-affinity Zn-binding site (Blindauer et al., 2009), and the rest to α_2-macroglobulin (Falchuk, 1977; Inagaki et al., 2000).

Although it may not be a specific plasma protein for Zn, the egg yolk protein, vitellogenin (Vtg), does bind Zn and transports Zn to the developing oocytes (Montorzi et al., 1994; Falchuk et al., 1995). Presence of Vtg in the plasma of female squirrelfish is probably one reason why plasma Zn concentrations are higher in the females than in the males (Thompson et al., 2002). There is a positive correlation between plasma Zn concentration and sexual maturation in females (Thompson et al., 2002). Indeed, treatment of squirrelfish with 17β-estradiol resulted in a seven-fold increase in plasma Zn and a 76-fold increase in circulating Vtg concentration, resulting in a molar ratio of Zn to Vtg of about 11:1 (Thompson et al., 2002).

10.3. Tissue Distribution

Concentrations of Zn in different tissues of fish have been reviewed in detail before (Hogstrand and Haux, 1991; Eisler, 1993; Hogstrand and Wood, 1996). From the points of uptake, the intestine and the gill, Zn is distributed in the body and taken up by different tissues. The total Zn content of fish is generally 10–40 mg kg^{-1} wet mass. The largest amounts of Zn in the body are found in muscle, bone, and skin, which combined make up 60% of the body's Zn content (Pentreath, 1973, 1976; Wicklund Glynn, 1991). The highest concentrations are present in the eye, where up to 30 mg g^{-1} dry weight has been recorded (Bowness and Morton, 1952; Eckhert, 1983). Most parts of the eye are actually very high in Zn and in some cell types this may be related to the coordination of Zn in melanin. The highest levels are found in the choroid and the lowest in the lens. This is not a peculiarity for fish, but a common theme among vertebrates, although the exact function(s) of Zn in the eye remains to be clarified.

There does not seem to exist any specialized storage organ for Zn, with the exception of the liver, which in female fish stores up Zn ahead of redistribution to the ovaries for incorporation into the developing oocytes (Montorzi et al., 1994; Thompson et al., 2002, 2003). This takes extreme proportions in females of the squirrelfish family (Holecentridae), which accumulate copious amounts of Zn in the liver and later deliver this to the eggs (Hogstrand et al., 1996a; Hogstrand and Haux, 1996; Thompson et al., 2001, 2002, 2003). The highest Zn concentration measured in a squirrelfish liver was 4.6 mg g^{-1} wet weight, which corresponds to about 1.8% of the dry mass. To cope with these enormous amounts of Zn without toxicity, female squirrelfish have extremely high levels of Mt in the liver (Hogstrand et al., 1996a; Hogstrand and Haux, 1996). Males, in contrast, have liver Zn concentrations similar to those of other teleost fish species (27 μg g^{-1} wet mass), as do immature females (94 μg g^{-1} wet mass). The reason for this extraordinary behavior has not been proven, but it is speculated that it has to do with the fact that squirrelfish have extremely large eyes, which take up half of the size of the entire head and as mentioned above, eyes are rich in Zn. Furthermore, nocturnal animals, including fish, often have a light-reflecting layer in the retina, called the tapetum lucidum, and this layer contains exceptionally high levels of Zn (Weitzel et al., 1954). Thus, it is possible that the female squirrelfish supplies large amounts of Zn for eye development in the larvae.

Typical Zn concentrations in gill, intestine, and ovary among different species (exposed and unexposed) are 14–130 μg g^{-1} wet mass (Eisler, 1993; Hogstrand and Wood, 1996; Kojadinovic et al., 2007). In most species and most conditions, the kidney contains 20–100 μg Zn g^{-1} wet weight (Hogstrand and Wood, 1996) but in yellowfin tuna (*Thunnus albacares*) the average dry mass Zn content of the kidney from 86 fish caught in the Mozambique Channel was an amazing 23,500 μg g^{-1} (Kojadinovic et al., 2007). Zinc concentrations in edible muscle of different fish are 4–40 μg g^{-1} wet mass (Eisler, 1993; Hogstrand and Wood, 1996; Kojadinovic et al., 2007).

11. CHARACTERIZATION OF EXCRETION ROUTES

Elimination of Zn from gills, liver, and kidney is fast, but whole-body excretion is slow with a biological half-life in excess of 200 days (Newman and Mitz, 1988; Wicklund Glynn, 1991). Relatively little is known about excretion routes of Zn in fish. A single study has systematically examined Zn excretion in fish and implicated the gill as a possible major excretory route in rainbow trout (Hardy et al., 1987). One-third of the Zn that was eliminated from the fish following a single gavage-fed meal containing [65]Zn left the fish from the head region. No regurgitation was observed, leading to the conclusion that the [65]Zn was excreted across the gills. In mammals, the major route of Zn excretion is through the intestine by secretion of digestive juices and shedding of intestinal cells (King et al., 2000). In rainbow trout administered [65]Zn through diet or injection, less than 1% of the recovered dose was found in the bile (Hardy et al., 1987; Chowdhury et al., 2003). This indicates that Zn is excreted into the intestine, but that the relative contribution of the biliary route might be small. Likewise, excretion of Zn with the urine may represent less than 1% of total Zn losses in fish (Spry and Wood, 1985; Hardy et al., 1987). In mammals, urinary losses are only 15–30% of those lost in feces (King et al., 2000). As noted above, female fish deposit significant amounts of Zn in oocytes, but at least in the squirrelfish this transfer was preceded by a phase of accelerated Zn uptake and accumulation in liver, so it is not known whether there are conditions during which there may be a net elimination of Zn during ovulation.

12. BEHAVIORAL EFFECTS OF ZINC

Metals are known to cause behavioral responses in fish (Atchison et al., 1987). Avoidance is perhaps the most sensitive behavioral response to waterborne Zn exposure (Table 3.3). That is, if given the choice between clean and Zn-supplemented water in an experimental system, the fish will prefer the clean water (Atchison et al., 1987). There are at least three studies reporting that rainbow trout show avoidance reaction to waterborne Zn concentrations orders of magnitude below those causing lethality (Sprague, 1968; Black and Birge, 1980; Svecevicius, 1999). For example, Sprague found that the threshold for Zn avoidance in very soft water (14 mg L^{-1} as $CaCO_3$) was as low as 5.6 µg L^{-1}, which was only 1% of the threshold for incipient mortality in the same water (Sprague, 1968). It is unclear whether the avoidance response to Zn is modified by hardness, because although the two other studies on Zn avoidance were carried out in harder water and report higher effect concentrations, there is no obvious overall correlation to hardness (Table 3.3). Other fish species also show avoidance to relatively modest concentrations of Zn, with the LOEC for lake whitefish (*Coregonus clupeaformis*), Atlantic salmon (*Salmo salar*), and vimba bream (*Vimba vimba*) being 10, 53, and 220 µg L^{-1}, respectively. Other effects on behavior,

Table 3.3
Behavioral responses to zinc in fish species and associated lowest observed
effect concentrations (LOECs)

Species	LOEC (μg L^{-1})	Response	Hardness (mg L^{-1} CaCO$_3$)	pH	Reference
Rainbow trout	5.6	Avoidance	14	7.2	Sprague (1968)
Lake whitefish	10	Avoidance	90	7.6	Scherer and McNicol (1998)
Rainbow trout	10	Avoidance	248	8.0	Svecevicius (1999)
Rainbow trout	47	Avoidance	112	7.6	Black and Birge (1980)
Atlantic salmon	53	Avoidance	18	7.5	Sprague (1964)
Rainbow trout	144	Ventilation rate	25	7.0	Cairns et al. (1982)
Vimba bream	220	Avoidance[a]	120	7.3	Svecevicius (1999)
Brook charr	1390	Cough rate	45	7.5	Drummond and Carlson (1977)
Bluegill	3640	Movement pattern	51	7.8	Waller and Cairns (1972)

[a]Experiment carried out in a stream.

such as ventilation rate, cough rate, and swimming patterns, have been shown to be affected by Zn, albeit at higher concentrations (Table 3.3). Sloman and co-workers tested the effects of different metals on social behavior in juvenile rainbow trout exposed to waterborne metals at 15% of the respective 96 h LC50 and showed that of the several metals tested (including Zn) only Cd had an effect, causing exposed fish to have fewer aggressive encounters and to be subordinate to unexposed individuals (Sloman et al., 2003). However, feeding rainbow trout an experimental diet containing, 1900 mg Zn kg^{-1} feed at 2.5% of their body weight per day for 21 days decreased aggression, assessed as the frequency of strikes against the other fish. This treatment also reduced growth rate by 5–8%.

Although behavioral responses to metals can clearly be sensitive endpoints of effects, they are not commonly used in calculation of NOECs for risk assessments. Changes in behavior could be potentially damaging to fish populations. For example, it might be predicted that fish stock density would be affected if fish are avoiding waters with elevated metal concentrations. Avoidance responses in the laboratory occur at Zn concentrations that quite commonly occur in areas with anthropogenic influence (see Sections 3 and 4). There is therefore a need to investigate the effect of Zn avoidance–preference behavior on fish distribution patterns in the real world. In one such study, distribution of salmonids was actually more influenced by habitat quality and water temperature than by water metal concentrations, even though Zn concentrations ranged from 2 to 243 μg L^{-1} (hardness: 29–94 mg L^{-1} as CaCO$_3$) and, thus, were well within the range of those eliciting an avoidance response in laboratory studies (Table 3.3) (Harper et al., 2009).

13. MOLECULAR CHARACTERIZATION OF ZINC TRANSPORTERS, STORAGE PROTEINS, AND CHAPERONES

Transport of Zn across the plasma membrane and between cellular compartments is mediated by two families of proteins called Znt (Slc30) and Zip (Slc39). A more detailed discussion of these can be found in Section 9.2. There are at least eight Znt paralogues and 13 Zip paralogues in fish and in almost all cases these map directly onto their mammalian orthologues (Feeney et al., 2005). Thus, additional paralogues of the genes coding for these proteins created through genome duplications in fish must have been suppressed through evolution. Members of the Zip family of Zn transporters move Zn into the cytosol, either from the exterior or from organelles. Znt proteins transport Zn away from the cytosol and either into organelles or out of the cell. However, at least one of the Znt proteins, Znt5, can function as a cellular Zn importer.

The principal Zn binding protein in cells is Mt, which is discussed in Section 10.1. The fraction of total tissue Zn bound to Mt varies enormously among tissues and with Zn content, and may be as little as 6% in the gill of an unexposed rainbow trout (Hogstrand et al., 1995) or as much as 74% in the liver of female squirrelfish and soldierfish (Hogstrand and Haux, 1996). Small molecules, such as glutathione, cysteine, and histidine, are probably important Zn ligands in cells but the quantitative roles of these in Zn binding in fish tissues have not been well researched.

14. GENOMIC AND PROTEOMIC STUDIES

Only a handful of articles has been published in which transcriptomic or proteomic technologies have been deployed in the study of Zn biology or toxicology in fish. In a proof-of-principle experiment, ^{32}P-labeled cDNA from gills of Zn exposed or control rainbow trout was hybridized to spotted arrays constructed from a Japanese pufferfish (*Takifugu rubripes*) cDNA library (Hogstrand et al., 2002). Archived clones corresponding to cDNA spots that showed differences in hybridization between treatments were obtained and sequenced to reveal the identities of differentially regulated genes. With this technique, 12 genes were identified that were regulated by Zn treatment. The same samples were also subjected to proteomics analysis using surface-enhanced laser desorption/ionization (SELDI) (Hogstrand et al., 2002). Although proteins could only be tentatively identified, based on their molecular mass and biochemical properties (i.e. Zn-, cationic- and anionic-binding), the method allowed semi-quantitative analysis of protein abundances, which were clearly very different in gills of Zn exposed and control rainbow trout.

The introduction of oligonucleotide-based microarrays vastly improved transcriptomic analysis. An application of this technology was to identify Mtf1 target genes in zebrafish (Hogstrand et al., 2008). A description of the results from this study is also provided in Section 7.2.1. Zebrafish ZF4 cells were transfected with Mtf1 siRNA to investigate the effect of *mtf1* knockdown on global gene expression. The cells were then grown either with or without 10 μM (654 μg L^{-1}) of Zn in the culture medium to reveal the role of Mtf1 in Zn-induced gene expression. It was found that as many as 1012 genes were regulated by Zn only when Mtf1 was present in cells, but only a few of these were determined to be likely Mtf1 targets. Almost half of the Mtf1 targets were developmental genes, which is in keeping with the essentiality of Zn and Mtf1 for embryogenesis.

In two recent studies, zebrafish were treated with either supplementation or depletion of Zn from feed and water, and gene expression profiles of the gills analyzed as the fish were acclimating to the new conditions (Zheng et al., 2010a,b). Changing Zn availability resulted in transcriptional cascades of genes being expressed. These cascades culminated 7 days into the treatment, when the maximum numbers of regulated genes were observed during either treatment, and by day 14 very few genes remained regulated. Over this time-course there was a succession of functional gene categories being regulated, of which "regulation of gene expression" and "development" were prominent groupings. Network analysis and reverse engineering of the transcriptional cascades revealed that early regulation of genes was likely coordinated by relatively few transcription factors, followed by activation of developmental signaling pathways, including hedgehog and bone-morphogenic protein (*bmp*) signaling (Fig. 3.7). This provides clues to the process of acclimation to a change in Zn availability, because Bmp7 is known to be involved in differentiation of ionocytes (Hsiao et al., 2007), the likely sites for Zn uptake at the gill. Zinc supplementation was also shown to have a pronounced influence on lipid metabolism, an effect that seems to have originated in the activation of the nuclear receptor Ppara (Zheng et al., 2010b). Both Zn supplementation and depletion had effects on a number of genes involved in male development, including *wt1*, *nr5a1a*, *cyp11a*, *hsd3b*, and *gata4*, but the functional significance of this is uncertain (Zheng et al., 2010a,b).

15. INTERACTIONS WITH OTHER METALS

Zinc in animals interacts with several other elements and in particular with Cu, Fe, Cd, and Ca. None of these interactions is completely

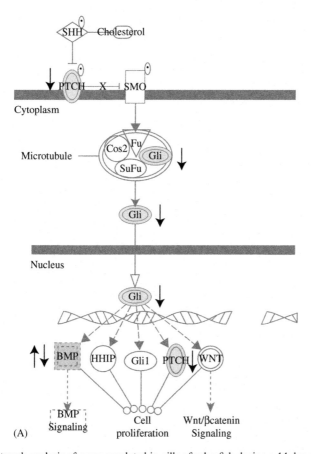

(A)

Fig. 3.7. Network analysis of genes regulated in gills of zebrafish during a 14 day period of Zn supplementation (Zheng et al. 2010b). Zebrafish were treated with Zn-supplemented water (330 μg L^{-1}) and feed (7.8 μg g^{-1} body weight day^{-1}), gills were sampled after 8 h, and 1, 4, 7, and 14 days of treatment, and gills were subjected to transcriptomic analysis by microarray. Expression values at first instance of significant regulation were entered into the Ingenuity Pathway Analysis (IPA) software, which assembles networks of the genes based on canonical pathways as well as on prior knowledge mined from the literature. (A) Regulation of the Sonic Hedgehog (SHH) signaling pathway. PTCH and Gli were downregulated, which was a plausible reason for the differential expression of several Gli target genes of the BMP family. Upregulation and downregulation of genes are shown as upward and downward arrows, respectively. Gene symbols: SHH: sonic hedgehog; PTCH: patched; SMO: smoothed; Fu: fused; Cos2: costal 2; Gli: gli; SuFu: suppressor of fused homologue (*Drosophila*); BMP: bone morphogenic protein; HHIP: hedgehog interacting protein; WNT: wingless-type MMTV integration site family. (B) Significantly enriched network including Bmp family proteins and their interacting partners. The network was generated by uploading the closest human homologue of each regulated zebrafish gene to the IPA software. Icons for proteins encoded by regulated genes are shaded with gene expression values indicated as fold-changes compared with

182 CHRISTER HOGSTRAND

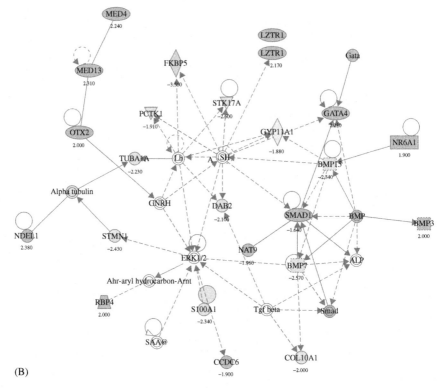

(B)

Fig. 3.7. (*Continued*)

the control with positive and negative values shown for upregulation and downregulation, respectively. Full names for proteins and protein families appearing in the network are provided below, along with fold change values within parentheses: Ahr: aryl hydrocarbon receptor; ALP: ALP family; Arnt: aryl hydrocarbon receptor nuclear translocator; BMP: bone morphogenic protein family; BMP15: bone morphogenetic protein 15 (−2.3); BMP3: bone morphogenetic protein 3 (+2.0); BMP7: bone morphogenetic protein 7 (−2.6); CCDC6: coiled-coil domain containing 6 (+1.9); COL10A1: collagen, type X, alpha 1 (−2.0); CYP11A1: cytochrome P450, family 11, subfamily A, polypeptide 1 (−1.9); DAB2, disabled homologue 2, mitogen-responsive phosphoprotein (*Drosophila*) (−2.1); ERK1/2: extracellular signal-regulated kinase 1/2; FKBP5: FK506 binding protein 5 (−3.4); FSH: follicle-stimulated hormone; Gata: GATA binding protein family; GATA4: GATA binding protein 4 (+2.2); GNRH: gonadotropic hormone-releasing hormone; Lh: luteinizing hormone; LZTR1: leucine-zipper-like transcription regulator 1 (−2.2); MED13: mediator complex subunit 13 (+2.3); MED4: mediator complex subunit 4 (+2.2); NAT9: N-acetyltransferase 9 (GCN5-related) (+2.0); NDEL1: nudE nuclear distribution gene E homologue (*A. nidulans*)-like 1 (+2.4); NR6A1: nuclear receptor subfamily 6, group A, member 1 (+1.9); OTX2: orthodenticle homeobox 2 (+2.0); PCTK1: serine/threonine-protein kinase 1 (−1.9).; RBP4: retinol binding protein 4, plasma (+2.0); S100A1: S100 calcium binding protein A1 (−2.3); SAA@: serum amyloid A1 cluster; Smad: SMAD family; SMAD1: SMAD family member 1 (+1.8); STK17A: serine/threonine kinase 17a (−2.6); STMN1: stathmin 1 (−2.4); Tgf beta: transforming growth factor beta; TUBA1A: tubulin, alpha 1a (−2.2).

understood, but they are likely to occur at several levels, some of which have already been discussed. The interaction with the non-essential element Cd is the most easily explained because Cd is chemically similar to Zn; both have similar size, bind strongly to sulfhydryl groups, and can assume tetrahedrical binding geometries. Because Zn has preference for sulfhydryl groups and Cd binds more avidly to these than does Zn, a good Zn binding site is often an excellent Cd binding site. This can result in inappropriate substitution of Cd for Zn in proteins and any other context in which Zn mimicry may occur.

Interactions between Zn and calcium may happen because Zn is able to traverse biological membranes through a variety of different calcium channels. It is likely that Zn enters the gills through Ecac (Trpv6) when fish are exposed to elevated Zn levels in the water (Fig. 3.5) (Hogstrand et al., 1996b; Qiu and Hogstrand, 2004). Likewise, evidence suggests that during ischemia, Zn^{2+} enters postsynaptic neurons through VGCC and AMPAR, but whether or not this occurs during normal physiological conditions is unclear (Sensi et al., 2009). However, Zn does not only interact with calcium in toxicological contexts, there is cross-talk between these elements that is likely physiologically grounded. For example, the release of Zn^{2+} from the ER in mast cells following antigen activation of FcjR requires a transient in cytosolic Ca^{2+} (Yamasaki et al., 2007). The finding that the high-affinity Ca^{2+}-ATPase, Pmca, is inhibited by free $[Zn^{2+}]$ concentrations well within those present in the cytosol of cells raises the question of whether this calcium efflux protein is regulated by Zn^{2+} (Hogstrand et al., 1996b, 1999). Furthermore, the Zn efflux transporter, ZNT1, has been shown to inhibit the L-type Ca^{2+} channel (LTCA) (Fig. 3.5) (Segal et al., 2004).

Inhibition of Fe and Cu uptake is among the most sensitive adverse effects of excessive dietary Zn intake in mammals (Maret and Sandstead, 2006; Stefanidou et al., 2006). Nutritional trials have shown that Zn in the diet clearly has a negative effect on Fe and Cu uptake in fish (Knox et al., 1984; Eid and Ghonim, 1994), but the mechanism involved remains unsolved. Uptake of non-heme-bound Fe at the brush-border membrane is mediated by the divalent metal transporter, DMT1 (Gunshin et al., 1997). It has been suggested that Zn^{2+} may compete with Fe^{2+} for transport by DMT1, but this is unlikely because it has been conclusively shown that DMT1 is not permeable to Zn^{2+} (Garrick et al., 2003; Mackenzie et al., 2007). This is in line with the observation using gut sacs from rainbow trout that among several elements, Zn had least effect on competition with Fe for apical uptake (Kwong and Niyogi, 2009). Investigating the interactions between Zn and Fe in Caco-2 cells, a cellular model for intestinal transport functions in human, it was found that Zn surprisingly upregulates expression of DMT1, leading to increased apical Fe transfer (Yamaji et al., 2001). This in contrast to the inhibitory effect of Zn on systemic Fe absorption. Thus, in spite of its importance, the nature of the interference of Zn with Fe uptake remains enigmatic.

There are reciprocal inhibitory effects of Zn on Cu uptake, and vice versa, in mid and post intestine of rainbow trout (Glover and Hogstrand, 2003; Nadella et al., 2007; Ojo et al., 2009). There are few experimental data shedding light on the interactions between Zn and Cu uptake. One possibility is that Zip Zn transporters, located in the apical membrane of transporting epithelia, are also involved in Cu transport and that there is a competition between Zn and Cu for uptake. This speculation is supported by the observation that of several cations Cu^{2+} was the strongest inhibitor of Zn^{2+} transport through pufferfish Zip3 (aka FrZip2) (Qiu and Hogstrand, 2005). In fact, a 20-fold excess of Cu^{2+} over $^{65}Zn^{2+}$ completely stopped Zip3-mediated uptake of $^{65}Zn^{2+}$ and this was a stronger inhibitory effect than that elicited by the same fold excess of cold Zn^{2+}.

16. KNOWLEDGE GAPS AND FUTURE DIRECTIONS

There have been some remarkable advances in our understanding of the physiology, biochemistry, and toxicity of Zn in recent years. It is now known in some detail how water chemistry influences Zn toxicity to aquatic life and computer models have been devised that can determine what concentrations are lethal to fish (Santore et al., 2002; Van Sprang et al., 2009). It is also known that Zn is a cofactor of thousands of proteins and a signaling substance involved in a number of biological processes (Hershfinkel et al., 2010). From this it follows that disruption of Zn homeostasis on a whole organism or cellular level can be detrimental to a wide spectrum of systems. Disruption of Zn homeostasis may occur through Zn exposure and depletion, but also through interference with other metals, such as Cd, Hg, and Cu, and even persistent organic pollutants. Indeed, expression of the archetype Zn-responsive protein, Mt, is known to change in response to many noxious stimuli and this is generally regarded as a relatively unspecific stress response. However, the evidence that induction of *mt* transcription in fish by stress factors, such as glucocorticoids, is of major importance is unconvincing. While *mt* gene transcription in rainbow trout is increased by glucocorticoids, the direct cause of the effect is actually an increase in cellular Zn^{2+} influx (Bury et al., 2008). The concept of Ca^{2+} disruption is well established as a mechanism of toxicity for a variety of toxicants, such as Cd, Zn, and halogenated organic pollutants. Similarly, there is ample circumstantial evidence that Zn^{2+} disruption may be a general mechanism of toxicity and this hypothesis deserves serious attention.

From the viewpoint of aquatic toxicology and legislative regulation of Zn concentrations in natural waters, it is evident that exposure to high Zn levels can decrease body stores of Cu and Fe. The interactions between Zn, Cu, Fe, and Cd, are very complex and knowledge of the mechanisms behind these interactions will be important for the understanding of effects from complex metal mixtures.

In terms of toxicology of Zn to fish, the largest gaps in our knowledge seem to be in the mechanisms of toxicity on the chronic side, and especially in seawater, where it would be difficult to explain Zn toxicity as hypocalcemia. Given the mounting evidence that Zn signaling is critically important during embryogenesis, the influence of altered Zn regulation on developmental processes seems like a particularly fertile area of research. Such work would also be of wider significance and contribute to our understanding of developmental biology.

The nutritional requirement for different fish species is about 20–30 mg kg^{-1} feed (NAS, 1979), but analysis of commercial fish feeds used in Norway indicates that European fish feeds contain up to 10 times this amount, with an average of 144 mg kg^{-1} in 2007 (Måge et al., 2007). This discrepancy is because of the perceived beneficial effects of dietary Zn on performance parameters, to safeguard against variability between batches, and to compensate for low availability of Zn from feeds based on some fishmeals that are rich in hydroxyapatite (Davies et al., 2010; Rider et al., 2010). Accordingly, the maximum permissible level of Zn in fish feeds in Europe is 200 mg kg^{-1} (EC, 2003). A consequence of this high-level supplementation is that Zn concentrations are often substantially elevated in the deposition zone underneath fish farm sea cages. Whether or not this has any relevance to sediment ecotoxicity by comparison to the overall deposit of organic material can be debated, but there is clearly a need to find ways to increase the efficiency of Zn uptake from fish feeds. This would best be carried out with a sound understanding of intestinal Zn absorptive physiology and biochemistry. Although a few physiological studies have been carried out on Zn uptake from the gut of fish, our current understanding of the molecular nature of intestinal Zn uptake and its regulation is almost entirely based on extrapolation from mammalian systems and even there, knowledge is patchy.

Yet another area where there is a gap in our knowledge about Zn handling in fish is that of excretion. By extrapolation from human physiology, it could be expected that Zn in fish is excreted principally with the feces, through sloughing of the intestinal epithelium, in the urine, and by secretion of digestive juices. It appears that in rainbow trout the gill may be an important excretory pathway for Zn (Hardy et al., 1987), and this possibility needs to be studied in greater detail and addressed in other species.

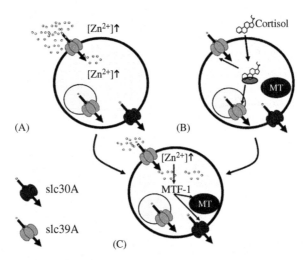

Fig. 3.8. Mechanism of glucocorticoid-stimulated metallothionein (Mt) expression in cultured rainbow trout gill cells. (A) Exposure to Zn leads to increased Zn uptake, which (C) activates Mtf1, leading to expression of Mt and Znt1 and consequential increase in capacity for Zn sequestration and efflux. (B) Cortisol stimulates cytosolic Zn influx from the extracellular compartment and possibly from intracellular stores by a non-genomic mechanism, resulting in (C) Mtf1 activation and expression of Mt and Znt1. There appears to be no direct induction of rainbow trout Mt by corticosteroid receptors (Bury et al., 2008).

Finally, it is evident that tissue levels of Zn are well regulated and that control systems exist to maintain Zn homeostasis. However, such control systems for Zn have been demonstrated only at the cellular level. The calcium regulatory hormones, stanniocalcin and calcitriol, may influence Zn uptake, but this does not mean they are involved in Zn homeostasis (Hogstrand et al., 1996b; Qiu et al., 2007). Similarly, it has been shown that glucocorticoids stimulate Zn influx into cultured gill cells, but this may be more of a response to stress than a control of the whole-body Zn status (Fig. 3.8) (Bury et al., 2008). If there is an integrated endocrine control of Zn status in animals, it is yet to be found.

REFERENCES

Ackland, M. L., and Michalczyk, A. (2006). Zinc deficiency and its inherited disorders – a review. *Genes Nutr.* **1**, 41–49.

Ahrland, S., Chatt, J., and Davies, N. R. (1958). The relative affinities of ligand atoms for acceptor molecules and ions. *Q. Rev. Chem. Soc.* **12**, 265–276.

Alsop, D. H., and Wood, C. M. (2000). Kinetic analysis of zinc accumulation in the gills of juvenile rainbow trout: effects of zinc acclimation and implications for biotic ligand modeling. *Environ. Toxicol. Chem.* **19**, 1911–1918.

Alsop, D. H., McGeer, J. C., McDonald, D. G., and Wood, C. M. (1999). Costs of chronic waterborne zinc exposure and the consequences of zinc acclimation on the gill/zinc interactions of rainbow trout in hard and soft water. *Environ. Toxicol. Chem.* **18**, 1014–1025.

Andreini, C., Banci, L., Bertini, I., and Rosato, A. (2005). Counting the zinc-proteins encoded in the human genome. *J. Proteome Res.* **5**, 196–201.

Andrews, G. K. (2001). Cellular zinc sensors: MTF-1 regulation of gene expression. *Biometals* **14**, 223–237.

Andrews, G. K., Wang, H. B., Dey, S. K., and Palmiter, R. D. (2004). Mouse zinc transporter 1 gene provides an essential function during early embryonic development. *Genesis* **40**, 74–81.

ANZECC (1992). *Australian Water Quality Guidelines.* Australian and New Zealand Environment & Conservation Council, Melbourne.

Atchison, G. J., Henry, M. G., and Sandheinrich, M. B. (1987). Effects of metals on fish behaviour – a review. *Environ. Biol. Fish.* **18**, 11–25.

Avvakumov, G. V., Muller, Y. A., and Hammond, G. L. (2000). Steroid-binding specificity of human sex hormone-binding globulin is influenced by occupancy of a zinc-binding site. *J. Biol. Chem.* **275**, 25920–25925.

Balesaria, S., and Hogstrand, C. (2006). Identification, cloning and characterization of a plasma membrane zinc efflux transporter, TrZnT-1, from fugu pufferfish (*Takifugu rubripes*). *Biochem. J.* **394**, 485–493.

Bengtsson, B. E. (1974). Effect of zinc on growth of the minnow *Phoxinus phoxinus. Oikos* **25**, 370–373.

Benoit, D. A., and Holcombe, G. W. (1978). Toxic effects of zinc on fathead minnows *Pimephales promelas* in soft water. *J. Fish Biol.* **13**, 701–708.

Bervoets, L., and Blust, R. (2000). Effects of pH on cadmium and zinc uptake by the midge larvae *Chironomus riparius. Aquat. Toxicol.* **49**, 145–157.

Besser, J. M., Brumbaugh, W. G., May, T. W., and Schmitt, C. J. (2007). Biomonitoring of lead, zinc, and cadmium in streams draining lead-mining and non-mining areas, Southeast Missouri, USA. *Environ. Monit. Assess.* **129**, 227–241.

Besser, L., Chorin, E., Sekler, I., Silverman, W. F., Atkin, S., Russell, J. T., and Hershfinkel, M. (2009). Synaptically released zinc triggers metabotropic signaling via a zinc-sensing receptor in the hippocampus. *J. Neurosci.* **29**, 2890–2901.

Bettger, W. J., Spryf, D. J., Cockell, K. A., Cho, C. Y., and Hilton, J. W. (1987). The distribution of zinc and copper in plasma, erythrocytes and erythrocyte membranes of rainbow trout (*Salmo gairdneri*). *Comp. Biochem. Physiol. C* **87**, 445–451.

Beyersmann, D. (2002). Homeostasis and cellular functions of zinc. *Materialwiss. Werksttech* **33**, 764–769.

Beyersmann, D., and Haase, H. (2001). Functions of zinc in signaling, proliferation and differentiation of mammalian cells. *Biometals* **14**, 331–341.

Black, J. A., and Birge, W. J. (1980). *An Avoidance Response Bioassay for Aquatic Pollutants.* University of Kentucky, Water Resources Research Institute, Lexington, KY.

Blindauer, C. A., Harvey, I., Bunyan, K. E., Stewart, A. J., Sleep, D., Harrison, D. J., Berezenko, S., and Sadler, P. J. (2009). Structure, properties, and engineering of the major zinc binding site on human albumin. *J. Biol. Chem.* **284**, 23116–23124.

Bodar, C. W., Pronk, M. E., and Sijm, D. T. (2005). The European Union risk assessment on zinc and zinc compounds: the process and the facts. *Integr. Environ. Assess. Manag.* **1**, 301–319.

Bowness, J. M., and Morton, R. A. (1952). Distribution of copper and zinc in the eyes of freshwater fishes and frogs. Occurrence of metals in melanin fractions from eye tissues. *Biochem. J.* **51**, 521–530.

Bradley, R. W., and Sprague, J. B. (1985). Accumulation of zinc by rainbow-trout as influenced by pH, water hardness and fish size. *Environ. Toxicol. Chem.* **4**, 685–694.

Bradley, R. W., DuQuesnay, C., and Sprague, J. B. (1985). Acclimation of rainbow trout, *Salmo gairdneri* Richardson, to zinc: kinetics and mechanism of enhanced tolerance induction. *J. Fish Biol.* **27**, 367–379.

Brinkman, S., and Woodling, J. (2005). Zinc toxicity to the mottled sculpin (*Cottus bairdi*) in high-hardness water. *Environ. Toxicol. Chem.* **24**, 1515–1517.

Brown, B. E. (1977). Effects of mine drainage on the River Hayle, Cornwall a) Factors affecting concentrations of copper, zinc and iron in water, sediments and dominant invertebrate fauna. *Hydrobiology* **52**, 221–233.

Brungs, W. A. (1969). Chronic toxicity of zinc to the fathead minnow, *Pimephales promelas* Rafinesque. *Trans. Am. Fish. Soc.* **98**, 272–279.

Bu-Olayan, A. H., and Thomas, B. V. (2008). Trace metals toxicity and bioaccumulation in mudskipper *Periophthalmus waltoni* Koumans 1941 (Gobiidae: Perciformes). *Turk. J. Fish. Aquat. Sci.* **8**, 215–218.

Bury, N. R., Chung, M. J., Sturm, A., Walker, P. A., and Hogstrand, C. (2008). Cortisol stimulates the zinc signaling pathway and expression of metallothioneins and ZnT1 in rainbow trout gill epithelial cells. *Am. J. Physiol.* **294**, R623–R629.

Cairns, M. A., Garton, R. R., and Tubb, R. A. (1982). Use of fish ventilation frequency to estimate chronically safe toxicant concentrations. *Trans. Am. Fish. Soc.* **111**, 70–77.

Canli, M., and Atli, G. (2003). The relationships between heavy metal (Cd, Cr, Cu, Fe, Pb, Zn) levels and the size of six Mediterranean fish species. *Environ. Pollut.* **121**, 129–136.

CCME (2007). Canadian Water Quality Guidelines for the Protection of Aquatic Life: Summary Table. Updated December, 2007. In *Canadian Environmental Quality Guidelines, 1999*. Winnipeg: Canadian Council of Ministers of the Environment.

Chen, W. Y., John, J. A. C., Lin, C. H., and Chang, C. Y. (2002). Molecular cloning and developmental expression of zinc finger transcription factor MTF-1 gene in zebrafish, *Danio rerio. Biochem. Biophys. Res. Commun.* **291**, 798–805.

Cheng, T., De Schamphelaere, K., Lofts, S., Janssen, C., and Allen, H. E. (2005). Measurement and computation of zinc binding to natural dissolved organic matter in European surface waters. *Anal. Chim. Acta* **542**, 230–239.

Chowdhury, M. J., Grosell, M., McDonald, D. G., and Wood, C. M. (2003). Plasma clearance of cadmium and zinc in non-acclimated and metal-acclimated trout. *Aquat. Toxicol.* **64**, 259–275.

Chung, M. J., Hogstrand, C., and Lee, S. J. (2005a). Cytotoxicity of nitric oxide is alleviated by zinc-mediated expression of antioxidant genes. *Exp. Biol. Med.* **231**, 1555–1563.

Chung, N. J., Walker, P. A., Brown, R. W., and Hogstrand, C. (2005b). Zinc-mediated gene expression offers protection against H_2O_2-induced cytotoxicity. *Toxicol. Appl. Pharmacol.* **205**, 225–236.

Chung, M. J., Walker, P. A., and Hogstrand, C. (2006). Dietary phenolic antioxidants, caffeic acid and Trolox, protect rainbow trout gill cells from nitric oxide-induced apoptosis. *Aquat. Toxicol.* **80**, 321–328.

Clearwater, S. J., Farag, A. M., and Meyer, J. S. (2002). Bioavailability and toxicity of dietborne copper and zinc to fish. *Comp. Biochem. Physiol. C* **132**, 269–313.

Cleven, R. F. M. J., Janus, J. A., Annema, J. A. and Slooff, W. (1993). Integrated criteria document zinc. In *RIVM Report 710401028*. Bilthoven: National Institute for Public Health and the Environment.

Coleman, J. E. (1992). Zinc proteins: enzymes, storage proteins, transcription factors, and replication proteins. *Annu. Rev. Biochem.* **61**, 897–946.

Colvin, R. A., Bush, A. I., Volitakis, I., Fontaine, C. P., Thomas, D., Kikuchi, K., and Holmes, W. R. (2008). Insights into Zn^{2+} homeostasis in neurons from experimental and modeling studies. *Am. J. Physiol.* **294**, C726–C742.

Colvin, R. A., Holmes, W. R., Fontaine, C. P., and Maret, W. (2010). Cytosolic zinc buffering and muffling: their role in intracellular zinc homeostasis. *Metallomics* **2**, 306–317.

Coulston, L., and Dandona, P. (1980). Insulin-like effect of zinc on adipocytes. *Diabetes* **29**, 665–667.

Cragg, R. A., Christie, G. R., Phillips, S. R., Russi, R. M., Kury, S., Mathers, J. C., Taylor, P. M., and Ford, D. (2002). A novel zinc-regulated human zinc transporter, hZTL1, is localized to the enterocyte apical membrane. *J. Biol. Chem.* **277**, 22789–22797.

Cusimano, R. F., Brakke, D. F., and Chapman, G. A. (1986). Effects of pH on the toxicity of cadmium, copper and zinc to steelhead trout (*Salmo gairdneri*). *Can. J. Fish. Aquat. Sci.* **43**, 1497–1503.

Dalton, T. P., Solis, W. A., Nebert, D. W., and Carvan, M. J. (2000). Characterization of the MTF-1 transcription factor from zebrafish and trout cells. *Comp. Biochem. Physiol. B* **126**, 325–335.

Davies, S. J., Rider, S., and Lundebye, A. K. (2010). Selenium and zinc nutrition of farmed fish: new perspectives in feed formulation to optimise health and production. In *Surface Chemistry, Bioavailability and Metal Homeostasis in Aquatic Organisms: An Integrated Approach* (N.R. Bury and R.D. Handy, eds). Vol. 2. Society for Experimental Biology, London.

De Schamphelaere, K. A. C., and Janssen, C. R. (2004). Bioavailability and chronic toxicity of zinc to juvenile rainbow trout (*Oncorhynchus mykiss*): comparison with other fish species and development of a biotic ligand model. *Environ. Sci. Technol.* **38**, 6201–6209.

De Schamphelaere, K. A. C., Lofts, S., and Janssen, C. R. (2005). Bioavailability models for predicting acute and chronic toxicity of zinc to algae, daphnids, and fish in natural surface waters. *Environ. Toxicol. Chem.* **24**, 1190–1197.

Dean, R. J., Shimmield, T. M., and Black, K. D. (2007). Copper, zinc and cadmium in marine cage fish farm sediments: an extensive survey. *Environ. Poll.* **145**, 84–95.

Denton, G. R. W., and Burdon-Jones, C. (1986). Environmental effects on toxicity of heavy metals to two species of tropical marine fish from Northern Australia. *Chem. Ecol.* **2**, 233–249.

Dinnel, P. A., Stober, Q., Link, J. M., Letourneau, M. W., Roberts, W. E., Felton, S. P., and Nakatani, R. E. (1983). *Methodology and Validation of a Sperm Cell Toxicity Test for Testing Toxic Substances in Marine waters.* Fisheries Research Institute, School of Fisheries, University of Washington, Seattle, WA.

Dressman, M. A., Walz, T. M., Lavedan, C., Barnes, L., Buchholtz, S., Kwon, I., Ellis, M. J., and Polymeropoulos, M. H. (2001). Genes that co-cluster with estrogen receptor alpha in microarray analysis of breast biopsies. *Pharmacogenom. J.* **1**, 135–141.

Drummond, R. A. and Carlson, R. W. (1977). *Procedures for Measuring Cough (Gill Purge) Rates of Fish.* Final Report. Washington, DC: US Environmental Protection Agency.

Dufner-Beattie, J., Weaver, B. P., Geiser, J., Bilgen, M., Larson, M., Xu, W., and Andrews, G. K. (2007). The mouse acrodermatitis enteropathica gene Slc39a4 (Zip4) is essential for early development and heterozygosity causes hypersensitivity to zinc deficiency. *Hum. Mol. Genet.* **16**, 1391–1399.

DWAF (1995). *South African Water Quality Guidelines for Coastal Marine Waters.* Vol. 1: *Natural Environment.* Earth, Marine and Atmospheric Science and Technology. CSIR, Stellenbosch.

DWAF (1996). *South African Water Quality Guidelines.* Vol. 7: *Aquatic Ecosystems* (ed. S. Holmes). Pretoria: CSIR Environmental Services.

EC (2003). *Opinion of the Scientific Committee for Animal Nutrition on the Use of Zinc in Feedingstuffs,* Vol. 2010. European Commission, Brussels.

Eckhert, C. D. (1983). Elemental concentrations in ocular tissues of various species. *Exp. Eye Res.* **37**, 639–647.

Eid, A. E., and Ghonim, S. I. (1994). Dietary zinc requirement of fingerling *Oreochromis niloticus. Aquaculture* **119**, 259–264.

Eisler, R. (1993). Zinc hazards to fish, wildlife and invertebrates: a synoptic review. In *Contaminant Hazard Reviews Report (USA), Biological Report.* Washington, DC: Fish and Wildlife Service.

Ellwood, M. J. (2004). Zinc and cadmium speciation in subantarctic waters east of New Zealand. *Mar. Chem.* **87**, 37–58.

EU (2000). Directive 2000/60/EC of the European Parliament and of the Council of 23 October 2000 establishing a framework for Community action in the field of water policy. *Official Journal of the European Union L* **327**, 1–73.

Evans, L. J. (2000). Fractionation and aqueous speciation of zinc in a lake polluted by mining activities, Flin Flon, Canada. *Water Air Soil Pollut.* **122**, 299–316.

Falchuk, K. H. (1977). Effect of acute disease and ACTH on serum zinc proteins. *N. Engl. J. Med.* **296**, 1129–1134.

Falchuk, K. H., Montorzi, M., and Vallee, B. L. (1995). Zinc uptake and distribution in *Xenopus laevis* oocytes and embryos. *Biochemistry* **34**, 16524–16531.

Farag, A. M., Skaar, D., Nimick, D. A., MacConnell, E., and Hogstrand, C. (2003). Characterizing aquatic health using salmonid mortality, physiology, and biomass estimates in streams with elevated concentrations of arsenic, cadmium, copper, lead, and zinc in the Boulder River watershed, Montana. *Trans. Am. Fish. Soc.* **132**, 450–467.

Farkas, A., Salánki, J., and Specziár, A. (2003). Age- and size-specific patterns of heavy metals in the organs of freshwater fish *Abramis brama* L. populating a low-contaminated site. *Water Res.* **37**, 959–964.

Feeney, G. P., Zheng, D., Kille, P., and Hogstrand, C. (2005). *The phylogeny of teleost ZIP and ZnT zinc transporters and their tissue specific expression and response to zinc in zebrafish* **1732**, 88–95. *Biochim. Biophys. Acta* **1732**, 88–95.

Fukada, T., Civic, N., Furuichi, T., Shimoda, S., Mishima, K., Higashiyama, H., Idaira, Y., Asada, Y., Kitamura, H., Yamasaki, S., Hojyo, S., Nakayama, M., Ohara, O., Koseki, H., Dos Santos, H. G., Bonafe, L., Ha-Vinh, R., Zankl, A., Unger, S., Kraenzlin, M. E., Beckmann, J. S., Saito, I., Rivolta, C., Ikegawa, S., Superti-Furga, A., and Hirano, T. (2008). The zinc transporter SLC39A13/ZIP13 is required for connective tissue development; its involvement in BMP/TGF-beta signaling pathways. *PLoS ONE* **3**, e3642.

Fukuda, J., and Kawa, K. (1977). Permeation of manganese, cadmium, zinc and beryllium through calcium channels of an insect muscle membrane. *Science* **196**, 309–311.

Garrick, M. D., Dolan, K. G., Horbinski, C., Ghio, A. J., Higgins, D., Porubcin, M., Moore, E. G., Hainsworth, L. N., Umbreit, J. N., Conrad, M. E., Feng, L., Lis, A., Roth, J. A., Singleton, S., and Garrick, L. M. (2003). DMT1: a mammalian transporter for multiple metals. *Biometals* **16**, 41–54.

Gatlin, D. M., III, and Wilson, R. P. (1983). Dietary zinc requirement of fingerling channel catfish. *J. Nutr.* **113**, 630–635.

Giguere, A., Campbell, P. G. C., Hare, L., and Couture, P. (2006). Sub-cellular partitioning of cadmium, copper, nickel and zinc in indigenous yellow perch (*Perca flavescens*) sampled along a polymetallic gradient. *Aquat. Toxicol.* **77**, 178–189.

Glover, C. N., Bury, N. R., and Hogstrand, C. (2003). Zinc uptake across the apical membrane of freshwater rainbow trout intestine is mediated by high affinity, low affinity, and histidine-facilitated pathways. *Biochim. Biophys. Acta Biomemb.* **1614**, 211–219.

Glover, C. N., and Hogstrand, C. (2002a). Amino acid modulation of *in vivo* intestinal zinc absorption in freshwater rainbow trout. *J. Exp. Biol.* **205**, 151–158.

Glover, C. N., and Hogstrand, C. (2002b). *In vivo* characterisation of intestinal zinc uptake in freshwater rainbow trout. *J. Exp. Biol.* **205**, 141–150.

Glover, C. N., and Hogstrand, C. (2003). Effects of dissolved metals and other hydrominerals on *in vivo* intestinal zinc uptake in freshwater rainbow trout. *Aquat. Toxicol.* **62**, 281–293.

Gunes, C., Heuchel, R., Georgiev, O., Muller, K. H., Lichtlen, P., Bluthmann, H., Marino, S., Aguzzi, A., and Schaffner, W. (1998). Embryonic lethality and liver degeneration in mice lacking the metal-responsive transcriptional activator MTF-1. *EMBO J.* **17**, 2846–2854.

Gunshin, H., Mackenzie, B., Berger, U. V., Gunshin, Y., Romero, M. F., Boron, W. F., Nussberger, S., Gollan, J. L., and Hediger, M. A. (1997). Cloning and characterization of a mammalian proton-coupled metal-ion transporter. *Nature* **388**, 482–488.

Haase, H., and Maret, W. (2003). Intracellular zinc fluctuations modulate protein tyrosine phosphatase activity in insulin/insulin-like growth factor-1 signaling. *Exp. Cell Res.* **291**, 289–298.

Haase, H., and Maret, W. (2004a). Fluctuations of cellular, available zinc modulate insulin signaling via inhibition of protein tyrosine phosphatases. *J. Trace Elem. Med. Biol.* **19**, 37–42.

Haase, H., and Maret, W. (2004b). Protein tyrosine phosphatases as targets of the combined insulinomimetic effects of zinc and oxidants. *Biometals* **18**, 333–338.

Haase, H., and Rink, L. (2009). Functional significance of zinc-related signaling pathways in immune cells. *Annu. Rev. Nutr.* **29**, 133–152.

Hardy, R. W., Sullivan, C. V., and Koziol, A. M. (1987). Absorption, body distribution, and excretion of dietary zinc by rainbow trout (*Salmo gairdneri*). *Fish Physiol. Biochem.* **3**, 133–143.

Harper, D. D., Farag, A. M., Hogstrand, C., and MacConnell, E. (2009). Trout density and health in a stream with variable water temperatures and trace element concentrations: does a cold-water source attract trout to increased metal exposure? *Environ. Toxicol. Chem.* **28**, 800–808.

Hershfinkel, M., Aizenman, E., Andrews, G. and Sekler, I. (2010). Zinc bells rang in Jerusalem! *Sci. Signal.* **3**, mr2.

Hirano, T., Murakami, M., Fukada, T., Nishida, K., Yamasaki, S., and Suzuki, T. (2008). Roles of zinc and zinc signaling in immunity: zinc as an intracellular signaling molecule. *Adv. Immunol.* **97**, 149–176.

Hoenderop, J. G. J., and Bindels, R. J. M. (2008). Calciotropic and magnesiotropic TRP channels. *Physiology* **23**, 32–40.

Hogstrand, C., and Haux, C. (1990). Metallothionein as an indicator of heavy-metal exposure in 2 subtropical fish species. *J. Exp. Mar. Biol. Ecol.* **138**, 69–84.

Hogstrand, C., and Haux, C. (1991). Binding and detoxification of heavy metals in lower vertebrates with reference to metallothionein. *Comp. Biochem. Physiol. C.* **100**, 137–141.

Hogstrand, C., and Haux, C. (1996). Naturally high levels of zinc and metallothionein in liver of several species of the squirrelfish family from Queensland, Australia. *Mar. Biol.* **125**, 23–31.

Hogstrand, C., and Wood, C. M. (1996). The physiology and toxicology of zinc in fish. In: *Aquatic Toxicology* (E.W. Taylor, ed.), pp. 61–84. Cambridge University Press, Cambridge.

Hogstrand, C., Lithner, G., and Haux, C. (1991). The importance of metallothionein for the accumulation of copper, zinc and cadmium in environmentally exposed perch, *Perca-fluviatilis*. *Pharmacol. Toxicol.* **68**, 492–501.

Hogstrand, C., Wilson, R. W., Polgar, D., and Wood, C. M. (1994). Effects of zinc on the kinetics of branchial calcium-uptake in fresh-water rainbow-trout during adaptation to waterborne zinc. *J. Exp. Biol.* **186**, 55–73.

192   CHRISTER HOGSTRAND

Hogstrand, C., Reid, S. D., and Wood, C. M. (1995). Ca^{2+} versus Zn^{2+} transport in the gills of fresh-water rainbow-trout and the cost of adaptation to waterborne Zn^{2+}. *J. Exp. Biol.* **198**, 337–348.

Hogstrand, C., Gassman, N. J., Popova, B., Wood, C. M., and Walsh, P. J. (1996a). The physiology of massive zinc accumulation in the liver of female squirrelfish and its relationship to reproduction. *J. Exp. Biol.* **199**, 2543–2554.

Hogstrand, C., Verbost, P. M., Bonga, S. E., and Wood, C. M. (1996b). Mechanisms of zinc uptake in gills of freshwater rainbow trout: interplay with calcium transport. *Am. J. Physiol.* **270**, R1141–R1147.

Hogstrand, C., Webb, N., and Wood, C. M. (1998). Covariation in regulation of affinity for branchial zinc and calcium uptake in freshwater rainbow trout. *J. Exp. Biol.* **201**, 1809–1815.

Hogstrand, C., Verbost, P. M., and Bonga, S. E. W. (1999). Inhibition of human erythrocyte Ca^{2+}-ATPase by Zn^2. *Toxicology* **133**, 139–145.

Hogstrand, C., Balesaria, S., and Glover, C. N. (2002). Application of genomics and proteomics for study of the integrated response to zinc exposure in a non-model fish species, the rainbow trout. *Comp. Biochem. Physiol. B.* **133**, 523–535.

Hogstrand, C., Zheng, D., Feeney, G., Cunningham, P., and Kille, P. (2008). Zinc-controlled gene expression by metal-regulatory transcription factor 1 (MTF1) in a model vertebrate, the zebrafish. *Biochem. Soc. Trans.* **36**, 1252–1257.

Hogstrand, C., Kille, P., Nicholson, R. I., and Taylor, K. M. (2009). Zinc transporters and cancer: a potential role for ZIP7 as a hub for tyrosine kinase activation. *Trends Mol. Med.* **15**, 101–111.

Holst, B., Egerod, K. L., Schild, E., Vickers, S. P., Cheetham, S., Gerlach, L. O., Storjohann, L., Stidsen, C. E., Jones, R., Beck-Sickinger, A. G., and Schwartz, T. W. (2007). GPR39 signaling is stimulated by zinc ions but not by obestatin. *Endocrinology* **148**, 13–20.

Hsiao, C.-D., You, M.-S., Guh, Y.-J., Ma, M., Jiang, Y.-J., and Hwang, P.-P. (2007). A positive regulatory loop between *foxi3a* and *foxi3b* is essential for specification and differentiation of zebrafish epidermal ionocytes. *PLoS ONE* **2**, e302.

Inagaki, K., Mikuriya, N., Morita, S., Haraguchi, H., Nakahara, Y., Hattori, M., Kinosita, T., and Saito, H. (2000). Speciation of protein-binding zinc and copper in human blood serum by chelating resin pre-treatment and inductively coupled plasma mass spectrometry. *Analyst* **125**, 197–203.

Inoue, K., Matsuda, K., Itoh, M., Kawaguchi, H., Tomoike, H., Aoyagi, T., Nagai, R., Hori, M., Nakamura, Y., and Tanaka, T. (2002). Osteopenia and male-specific sudden cardiac death in mice lacking a zinc transporter gene, Znt5. *Hum. Mol. Genet.* **11**, 1775–1784.

Jackson, K. A., Helston, R. M., McKay, J. A., O'Neill, E. D., Mathers, J. C., and Ford, D. (2007). Splice variants of the human zinc transporter ZnT5 (SLC30A5) are differentially localized and regulated by zinc through transcription and mRNA stability. *J. Biol. Chem.* **282**, 10423–10431.

Jansen, R. A. G., van Leeuwen, H. P., Cleven, R. F. M. J., and van den Hoop, M. A. G. T. (1998). Speciation and lability of zinc(II) in river waters. *Environ. Sci. Technol.* **32**, 3882–3886.

Janssen, C. R., De Schamphelaere, K., Heijerick, D., Muyssen, B., Lock, K., Bossuyt, B., Vangheluwe, M., and Van Sprang, P. (2000). Uncertainties in the environmental risk assessment of metals. *Hum. Ecol. Risk Assess.* **6**, 1003–1018.

Jeng, S. S., Yau, J. Y., Chen, Y. H., Lin, T. Y., and Chung, Y. Y. (2007). High zinc in the erythrocyte plasma membranes of common carp *Cyprinus carpio*. *Fish. Sci.* **73**, 421–428.

Jiang, L. J., Maret, W., and Vallee, B. L. (1998). The glutathione redox couple modulates zinc transfer from metallothionein to zinc-depleted sorbitol dehydrogenase. *Proc. Natl. Acad. Sci. U.S.A.* **95**, 3483–3488.

Kagara, N., Tanaka, N., Noguchi, S., and Hirano, T. (2007). Zinc and its transporter ZIP10 are involved in invasive behavior of breast cancer cells. *Cancer Sci.* **98**, 692–697.

Kambe, T., Narita, H., Yamaguchi-Iwai, Y., Hirose, J., Amano, T., Sugiura, N., Sasaki, R., Mori, K., Iwanaga, T., and Nagao, M. (2002). Cloning and characterization of a novel mammalian zinc transporter, zinc transporter 5, abundantly expressed in pancreatic beta cells. *J. Biol. Chem.* **277**, 19049–19055.

Khan, A. T., Graham, T. C., Ogden, L., Ali, S., Thompson, S. S. J., Shireen, K. F., and Mahboob, M. (2007). A two-generational reproductive toxicity study of zinc in rats. *J. Environ. Sci. Health B* **42**, 403–415.

Khan, F. R., Irving, J. R., Bury, N. R., and Hogstrand, C. (2011). Differential tolerance of two *Gammarus pulex* populations transplanted from different metallogenic regions to a polymetal gradient. *Aquat. Toxicol.* **102**, 95–103.

Kim, A. M., Vogt, S., O'Halloran, T. V., and Woodruff, T. K. (2010). Zinc availability regulates exit from meiosis in maturing mammalian oocytes. *Nat. Chem. Biol.* **6**, 674–681.

Kimura, T., Itoh, N., and Andrews, G. K. (2009). Mechanisms of heavy metal sensing by metal response element-binding transcription factor-1. *J. Health Sci.* **55**, 484–494.

King, J. C., Shames, D. M., and Woodhouse, L. R. (2000). Zinc homeostasis in humans. *J. Nutr.* **130**, 1360S–1366S.

Kjoss, V. A., Wood, C. M., and McDonald, D. G. (2006). Effects of different ligands on the bioaccumulation and subsequent depuration of dietary Cu and Zn in juvenile rainbow trout (*Oncorhynchus mykiss*). *Can. J. Fish. Aquat. Sci.* **63**, 412–422.

Knox, D., Cowey, C. B., and Adron, J. W. (1984). Effects of dietary zinc intake upon copper metabolism in rainbow trout (*Salmo gairdneri*). *Aquaculture* **40**, 199–207.

Kojadinovic, J., Potier, M., Le Corre, M., Cosson, R. P., and Bustamante, P. (2007). Bioaccumulation of trace elements in pelagic fish from the Western Indian Ocean. *Environ. Pollut.* **146**, 548–566.

Krezel, A., and Maret, W. (2007). Dual nanomolar and picomolar Zn(II) binding properties of metallothionein. *J. Am. Chem. Soc.* **129**, 10911–10921.

Krezel, A., Hao, Q., and Maret, W. (2007). Zinc/thiolate redox biochemistry of metallothionein and the control of zinc ion fluctuations in cell signaling. *Arch. Biochem. Biophys.* **463**, 188–200.

Kury, S., Dreno, B., Bezieau, S., Giraudet, S., Kharfi, M., Kamoun, R., and Moisan, J. P. (2002). Identification of SLC39A4, a gene involved in acrodermatitis enteropathica. *Nat. Genet.* **31**, 239–240.

Kwong, R. W. M., and Niyogi, S. (2009). The interactions of iron with other divalent metals in the intestinal tract of a freshwater teleost, rainbow trout (*Oncorhynchus mykiss*). *Comp. Biochem. Physiol. C.* **150**, 442–449.

Kwun, I. S., Cho, Y. E., Lomeda, R. A., Shin, H. I., Choi, J. Y., Kang, Y. H., and Beattie, J. H. (2010). Zinc deficiency suppresses matrix mineralization and retards osteogenesis transiently with catch-up possibly through Runx 2 modulation. *Bone* **46**, 732–741.

Laity, J. H., and Andrews, G. K. (2007). Understanding the mechanisms of zinc-sensing by metal-response element binding transcription factor-1 (MTF-1). *Arch. Biochem. Biophys.* **463**, 201–210.

Lange, A., Ausseil, O., and Segner, H. (2002). Alterations of tissue glutathione levels and metallothionein mRNA in rainbow trout during single and combined exposure to cadmium and zinc. *Comp. Biochem. Physiol. C.* **131**, 231–243.

Legret, M., and Pagotto, C. (2006). Heavy metal deposition and soil pollution along two major rural highways. *Environ. Technol.* **27**, 247–254.

Li, Y., and Maret, W. (2009). Transient fluctuations of intracellular zinc ions in cell proliferation. *Exp. Cell Res.* **315**, 2463–2470.

Liao, B. K., Deng, A. N., Chen, S. C., Chou, M. Y. and Hwang, P. P. (2007). Expression and water calcium dependence of calcium transporter isoforms in zebrafish gill mitochondrion-rich cells. *BMC Genom.* **8**.

Lichten, L. A., and Cousins, R. J. (2009). Mammalian zinc transporters: nutritional and physiologic regulation. *Annu. Rev. Nutr.* **29**, 153–176.

Lichtlen, P., Wang, Y., Belser, T., Georgiev, O., Certa, U., Sack, R., and Schaffner, W. (2001). Target gene search for the metal-responsive transcription factor MTF-1. *Nucleic Acid Res.* **29**, 1514–1523.

Luoma, S. N., and Rainbow, P. S. (2008). *Metal Contamination in Aquatic Environments: Science and Lateral Management.* Cambridge University Press, New York.

Mackenzie, B., Takanaga, H., Hubert, N., Rolfs, A., and Hediger, M. A. (2007). Functional properties of multiple isoforms of human divalent metal-ion transporter 1 (DMT1). *Biochem. J.* **403**, 59–69.

Måge, A., Julshamn, K., Hemre, G. I. and Lunestad, B. T. (2007). Årsrapport 2007. In *Overvakningsprogram for førvarer til fisk og andre akvatiske dyr.* Bergen: National Institute of Nutrition and Seafood Research, Norway.

Mallatt, J. (1985). Fish gill structural changes induced by toxicants and other irritants: a statistical review. *Can. J. Fish. Aquat. Sci.* **42**, 630–648.

Mao, X. Q., Kim, B. E., Wang, F. D., Eide, D. J., and Petris, M. J. (2007). A histidine-rich cluster mediates the ubiquitination and degradation of the human zinc transporter, hZIP4, and protects against zinc cytotoxicity. *J. Biol. Chem.* **282**, 6992–7000.

Maret, W. (1994). Oxidative metal release from metallothionein via zinc-thiol/disulfide interchange. *Proc. Natl. Acad. Sci. U.S.A.* **91**, 237–241.

Maret, W. (1995). Metallothionein/disulfide interactions, oxidative stress, and the mobilization of cellular zinc. *Neurochem. Int.* **27**, 111–117.

Maret, W. (2001). Crosstalk of the group IIa and IIb metals calcium and zinc in cellular signaling. *Proc. Natl. Acad. Sci. U.S.A.* **98**, 12325–12327.

Maret, W., and Li, Y. (2009). Coordination dynamics of zinc in proteins. *Chem. Rev.* **109**, 4682–4707.

Maret, W., and Sandstead, H. H. (2006). Zinc requirements and the risks and benefits of zinc supplementation. *J. Trace Elem. Med. Biol.* **20**, 3–18.

Mathews, W. R., Ong, D., Milutinovich, A. B., and Van Doren, M. (2006). Zinc transport activity of Fear of Intimacy is essential for proper gonad morphogenesis and DE-cadherin expression. *Development* **133**, 1143–1153.

McDonald, D. G., and Wood, C. M. (1993). Branchial mechanisms of acclimation to metals in freshwater fish. In: *Fish Ecophysiology* (J.C. Rankin and F.B. Jensen, eds), pp. 197–321. Chapman & Hall, London.

McGeer, J. C., Brix, K. V., Skeaff, J. M., DeForest, D. K., Brigham, S. I., Adams, W. J., and Green, A. (2003). Inverse relationship between bioconcentration factor and exposure concentration for metals: implications for hazard assessment of metals in the aquatic environment. *Environ. Toxicol. Chem.* **22**, 1017–1037.

McMahon, R. J., and Cousins, R. J. (1998). Regulation of the zinc transporter ZnT-1 by dietary zinc. *Proc. Natl. Acad. Sci. U.S.A.* **95**, 4841–4846.

Milner, N. J. (1982). The accumulation of zinc by O-group plaice, *Pleuronectes platessa* (L.), from high concentrations in sea water and food. *J. Fish Biol.* **21**, 325–336.

Miranda, E. R., and Dey, C. S. (2004). Effect of chromium and zinc on insulin signaling in skeletal muscle cells. *Biol. Trace Elem. Res.* **101**, 19–36.

Mohanty, M., Adhikari, S., Mohanty, P., and Sarangi, N. (2009). Effect of waterborne zinc on survival, growth, and feed intake of Indian major carp, *Cirrhinus mrigala* (Hamilton). *Water Air Soil Pollut.* **201**, 3–7.

Montorzi, M., Falchuk, K. H., and Vallee, B. L. (1994). *Xenopus laevis* vitellogenin is a zinc protein. *Biochem. Biophys. Res. Commun.* **200**, 1407–1413.

Mulkidjanian, A. Y. (2009). On the origin of life in the Zinc world: I. Photosynthesizing, porous edifices built of hydrothermally precipitated zinc sulfide as cradles of life on Earth. *Biol. Direct* **4**.

Mulkidjanian, A. Y., and Galperin, M. Y. (2009). On the origin of life in the zinc world. 2. Validation of the hypothesis on the photosynthesizing zinc sulfide edifices as cradles of life on Earth. *Biol. Direct* **4**.

Murakami, M., and Hirano, T. (2008). Intracellular zinc homeostasis and zinc signaling. *Cancer Sci.* **99**, 1515–1522.

Muylle, F., Robbens, J., De, C. W., Timmermans, J. P., and Blust, R. (2006a). Cadmium and zinc induction of ZnT-1 mRNA in an established carp cell line. *Comp. Biochem. Physiol. C.* **143**, 242–251.

Muylle, F. A. R., Adriaensen, D., De Coen, W., Timmermans, J. P., and Blust, R. (2006b). Tracing of labile zinc in live fish hepatocytes using FluoZin-3. *Biometals* **19**, 437–450.

Muyssen, B. T. A., and Janssen, C. R. (2005). Importance of acclimation to environmentally relevant zinc concentrations on the sensitivity of *Daphnia magna* toward zinc. *Environ. Toxicol. Chem.* **24**, 895–901.

Nadella, S. R., Grosell, M., and Wood, C. M. (2007). Mechanisms of dietary Cu uptake in freshwater rainbow trout: evidence for Na-assisted Cu transport and a specific metal carrier in the intestine. *J. Comp. Physiol. B.* **177**, 433–446.

Naito, W., Kamo, M., Tsushima, K., and Iwasaki, Y. (2010). Exposure and risk assessment of zinc in Japanese surface waters. *Sci. Total Environ.* **408**, 4271–4284.

Nakano, H., Nakamura, Y., Kawamura, T., Shibagaki, N., Matsue, H., Aizu, T., Rokunohe, D., Akasaka, E., Kimura, K., Nishizawa, A., Umegaki, N., Mitsuhashi, Y., Shimada, S., and Sawamura, D. (2009). Novel and recurrent nonsense mutation of the SLC39A4 gene in Japanese patients with acrodermatitis enteropathica. *Br. J. Dermatol.* **161**, 184–186.

NAS (1979). *Zinc.* United States National Academy of Sciences, National Research Council, Subcommittee on Zinc. Baltimore, MD: University Park Press.

NEPA (1989). *Water Quality Standard for Fisheries.* GB11607-89. Beijing: China National Environmental Protection Agency.

Newman, M. C., and Mitz, S. V. (1988). Size dependence of zinc elimination and uptake from water by mosquitofish *Gambusia affinis* (Baird and Girard). *Aquat. Toxicol.* **12**, 17–32.

Niyogi, S., and Wood, C. M. (2006). Interaction between dietary calcium supplementation and chronic waterborne zinc exposure in juvenile rainbow trout, *Oncorhynchus mykiss. Comp. Biochem. Physiol. C.* **143**, 94–102.

Niyogi, S., Pyle, G. G., and Wood, C. M. (2007). Branchial versus intestinal zinc uptake in wild yellow perch (*Perca flavescens*) from reference and metal-contaminated aquatic ecosystems. *Can. J. Fish. Aquat. Sci.* **64**, 1605–1613.

Niyogi, S., Kent, R., and Wood, C. M. (2008). Effects of water chemistry variables on gill binding and acute toxicity of cadmium in rainbow trout (*Oncorhynchus mykiss*): a biotic ligand model (BLM) approach. *Comp. Biochem. Physiol. C.* **148**, 305–314.

NRC (1993). *Nutrient Requirements of Fish.* National Academy Press, Washington, DC.

Nriagu, J. O., and Pacyna, J. M. (1988). Quantitative assessment of worldwide contamination of air, water and soils by trace metals. *Nature* **333**, 134–139.

Ojo, A. A., and Wood, C. M. (2007). *In vitro* analysis of the bioavailability of six metals via the gastro-intestinal tract of the rainbow trout (*Oncorhynchus mykiss*). *Aquat. Toxicol.* **83**, 10–23.

Ojo, A. A., and Wood, C. M. (2008). *In vitro* characterization of cadmium and zinc uptake via the gastro-intestinal tract of the rainbow trout (*Oncorhynchus mykiss*): interactive effects and the influence of calcium. *Aquat. Toxicol.* **89**, 55–64.

Ojo, A. A., Nadella, S. R., and Wood, C. M. (2009). *In vitro* examination of interactions between copper and zinc uptake via the gastrointestinal tract of the rainbow trout (*Oncorhynchus mykiss*). *Arch. Environ. Contam. Toxicol.* **56**, 244–252.

Overnell, J., McIntosh, R., and Fletcher, T. C. (1987). The levels of liver metallothionein and zinc in plaice, *Pleuronectes platessa* L., during the breeding season, and the effect of estradiol injection. *J. Fish Biol.* **30**, 539–546.

Overnell, J., Fletcher, T. C., and McIntosh, R. (1988). The apparent lack of effect of supplementary dietary zinc on zinc metabolism and metallothionein concentrations in the turbot, *Scophthalmus maximus* (Linnaeus). *J. Fish Biol.* **33**, 563–570.

Palmiter, R. D., and Findley, S. D. (1995). Cloning and functional characterization of a mammalian zinc transporter that confers resistance to zinc. *EMBO J.* **14**, 639–649.

Palmiter, R. D., and Huang, L. P. (2004). Efflux and compartmentalization of zinc by members of the SLC30 family of solute carriers. *Pflug. Arch. Eur. J. Physiol.* **447**, 744–751.

Palmiter, R. D., Cole, T. B., and Findley, S. D. (1996a). ZnT-2, a mammalian protein that confers resistance to zinc by facilitating vesicular sequestration. *EMBO J.* **15**, 1784–1791.

Palmiter, R. D., Cole, T. B., Quaife, C. J., and Findley, S. D. (1996b). ZnT-3, a putative transporter of zinc into synaptic vesicles. *Proc. Natl. Acad. Sci. U.S.A.* **93**, 14934–14939.

Paquin, P. R., Gorsuch, J. W., Apte, S., Batley, G. E., Bowles, K. C., Campbell, P. G. C., Delos, C. G., Di Toro, D. M., Dwyer, R. L., Galvez, F., Gensemer, R. W., Goss, G. G., Hogstrand, C., Janssen, C. R., McGeer, J. M., Naddy, R. B., Playle, R. C., Santore, R. C., Schneider, U., Stubblefield, W. A., Wood, C. M., and Wu, K. B. (2002). The biotic ligand model: a historical overview. *Comp. Biochem. Physiol. C* **133**, 3–35.

Passerini, A., Andreini, C., Menchetti, S., Rosato, A., and Frasconi, P. (2007). Predicting zinc binding at the proteome level. *BMC Bioinform.* **8**, 39.

Pentreath, R. J. (1973). The accumulation and retention of ^{65}Zn and ^{54}Mn by the plaice, *Pleuronectes platessa* L.. *J. Exp. Mar. Biol. Ecol.* **12**, 1–18.

Pentreath, R. J. (1976). Some further studies on accumulation and retention of Zn-65 and Mn-54 by plaice, *Pleuronectes platessa* L.. *J. Exp. Mar. Biol. Ecol.* **21**, 179–189.

Pielage, J., Kippert, A., Zhu, M., and Klämbt, C. (2004). The *Drosophila* transmembrane protein Fear-of-intimacy controls glial cell migration. *Dev. Biol.* **275**, 245–257.

Pinuela, C., Baatrup, E., and Geneser, F. A. (1992). Histochemical distribution of zinc in the brain of the rainbow-trout, *Oncorhynchus-mykiss*. *Anat. Embryol.* **186**, 275–284.

Porter, F. C. (1991). *Zinc Handbook: Properties, Processing, and Use in Design.* Marcel Dekker, New York.

Prasad, A. S. (2009). Impact of the discovery of human zinc deficiency on health. *J. Am. Coll. Nutr.* **28**, 257–265.

Prasad, A. S., Halsted, J. A., and Nadimi, M. (1961). Syndrome of iron deficiency anemia hepatosplenomegaly, hypogonadism, dwarfism and geophagia. *Am. J. Med.* **31**, 532–546.

Preciado, H. F., and Li, L. Y. (2006). Evaluation of metal loadings and bioavailability in air, water and soil along two highways of British Columbia, Canada. *Water Air. Soil Pollut.* **172**, 81–108.

Qiu, A. (2004). *Function and regulation of zinc transporters in model fish species.* PhD thesis, King's College, London.

Qiu, A. D., and Hogstrand, C. (2004). Functional characterisation and genomic analysis of an epithelial calcium channel (ECaC) from pufferfish, *Fugu rubripes*. *Gene* **342**, 113–123.

Qiu, A. D., and Hogstrand, C. (2005). Functional expression of a low-affinity zinc uptake transporter (FrZIP2) from pufferfish (*Takifugu rubripes*) in MDCK cells. *Biochem. J.* **390**, 777–786.

Qiu, A., Shayeghi, M., and Hogstrand, C. (2005). Molecular cloning and functional characterization of a high-affinity zinc importer (DrZIP1) from zebrafish (*Danio rerio*). *Biochem. J.* **388**, 745–754.

Qiu, A., Glover, C. N., and Hogstrand, C. (2007). Regulation of branchial zinc uptake by 1 alpha,25-(OH)(2)D-3 in rainbow trout and associated changes in expression of ZIP1 and ECaC. *Aquat. Toxicol.* **84**, 142–152.

Rainbow, P. S., Malik, I., and O'Brien, P. (1993). Physicochemical and physiological effects on the uptake of dissolved zinc and cadmium by the amphipod crustacean *Orchestia gammarellus*. *Aquat. Toxicol.* **25**, 15–30.

Rauch, J. N. (2009). Global mapping of Al, Cu, Fe, and Zn in-use stocks and in-ground resources. *Proc. Natl. Acad. Sci. U.S.A.* **106**, 18920–18925.

Renfro, W. C., Fowler, S. W., Heyraud, M., and La Rosa, J. (1975). Relative importance of food and water in long-term zinc-65 accumulation by marine biota. *J. Fish. Res. Bd Can.* **32**, 1339–1345.

Rider, S. A., Davies, S. J., Jha, A. N., Clough, R., and Sweetman, J. W. (2010). Bioavailability of co-supplemented organic and inorganic zinc and selenium sources in a white fishmeal-based rainbow trout (*Oncorhynchus mykiss*) diet. *J. Anim. Physiol. Anim. Nutr.* **94**, 99–110.

Rozan, T. F., Lassman, M. E., Ridge, D. P., and Luther, G. W. (2000). Evidence for iron, copper and zinc complexation as multinuclear sulphide clusters in oxic rivers. *Nature* **406**, 879–882.

Santore, R. C., Mathew, R., Paquin, P. R., and DiToro, D. (2002). Application of the biotic ligand model to predicting zinc toxicity to rainbow trout, fathead minnow, and *Daphnia magna*. *Comp. Biochem. Physiol. C* **133**, 271–285.

Sappal, R., Burka, J., Dawson, S., and Kamunde, C. (2009). Bioaccumulation and subcellular partitioning of zinc in rainbow trout (*Oncorhynchus mykiss*): cross-talk between waterborne and dietary uptake. *Aquat. Toxicol.* **91**, 281–290.

SCHER (2007). *Scientific Opinion on the Risk Assessment Report on Zinc, Environmental Part.* 29 November 2007. Brussels: Scientific Committee on Health and Environmental Risks.

Scherer, E., and McNicol, R. E. (1998). Preference–avoidance responses of lake whitefish (*Coregonus clupeaformis*) to competing gradients of light and copper, lead, and zinc. *Water Res.* **32**, 924–929.

Segal, D., Ohana, E., Besser, L., Hershfinkel, M., Moran, A., and Sekler, I. (2004). A role for ZnT-1 in regulating cellular cation influx. *Biochem. Biophys. Res. Comm.* **323**, 1145–1150.

Sensi, S. L., Paoletti, P., Bush, A. I., and Sekler, I. (2009). Zinc in the physiology and pathology of the CNS. *Nat. Rev. Neurosci.* **10**, 780–791.

Shahsavarani, A., and Perry, S. F. (2006). Hormonal and environmental regulation of epithelial calcium channel in gill of rainbow trout (*Oncorhynchus mykiss*). *Am. J. Physiol.* **291**, R1490–R1498.

Shahsavarani, A., McNeill, B., Galvez, F., Wood, C. M., Goss, G. G., Hwang, P. P., and Perry, S. F. (2006). Characterization of a branchial epithelial calcium channel (ECaC) in freshwater rainbow trout (*Oncorhynchus mykiss*). *J. Exp. Biol.* **209**, 1928–1943.

Sharir, H., Zinger, A., Nevo, A., Sekler, I., and Hershfinkel, M. (2010). Zinc released from injured cells is acting via the Zn^{2+}-sensing receptor, ZnR, to trigger signaling leading to epithelial repair. *J. Biol. Chem.* **285**, 26097–26106.

Shears, M. A., and Fletcher, G. L. (1983). Regulation of Zn^{2+} uptake from the gastrointestinal tract of a marine teleost, the winter flounder (*Pseudoplueronectes americanus*). *Can. J. Fish. Aquat. Sci.* **40**, 197–205.

Skidmore, J. F., and Tovell, P. W. A. (1972). Toxic effects of zinc sulphate on the gills of rainbow trout. *Water Res.* **6**, 217–230.

Sloman, K. A., Baker, D. W., Ho, C. G., McDonald, D. G., and Wood, C. M. (2003). The effects of trace metal exposure on agonistic encounters in juvenile rainbow trout, *Oncorhynchus mykiss*. *Aquat. Toxicol.* **63**, 187–196.

South West Water (1983). *Chemical Survey of R. Hayle.* Rectorate of Engineering and Environmental Services, South West Water.

198 CHRISTER HOGSTRAND

Sprague, J. B. (1964). Avoidance of copper–zinc solutions by young salmon in the laboratory. *J. Water Pollut. Control Fed.* **36**, 990–1004.

Sprague, J. B. (1968). Avoidance reactions of rainbow trout to zinc sulphate solutions. *Water Res.* **2**, 367–372.

Spry, D. J., and Wood, C. M. (1985). Ion flux rates, acid–base status, and blood gases in rainbow trout, *Salmo gairdneri*, exposed to toxic zinc in natural soft water. *Can. J. Fish. Aquat. Sci.* **45**, 2206–2215.

Spry, D. J., and Wood, C. M. (1989). A kinetic method for the measurement of zinc influx *in vivo* in the rainbow trout, and the effects of waterborne calcium on flux rates. *J. Exp. Biol.* **142**, 425–446.

Spry, D. J., Hodson, P. V., and Wood, C. M. (1988). Relative contributions of dietary and waterborne zinc in the rainbow trout, *Salmo gairdneri*. *Can. J. Fish. Aquat. Sci.* **45**, 32–41.

Stefanidou, M., Maravelias, C., Dona, A., and Spiliopoulou, C. (2006). Zinc: a multipurpose trace element. *Arch. Toxicol.* **80**, 1–9.

Storjohann, L., Holst, B., and Schwartz, T. W. (2008). Molecular mechanism of Zn^{2+} agonism in the extracellular domain of GPR39. *FEBS Lett.* **582**, 2583–2588.

Svecevicius, G. (1999). Fish avoidance response to heavy metals and their mixtures. *Acta Zool. Lituan. Hydrobiol.* **9**, 103–113.

Tang, X.-h., and Shay, N. F. (2001). Zinc has an insulin-like effect on glucose transport mediated by phosphoinositol-3-kinase and Akt in 3T3-L1 fibroblasts and adipocytes. *J. Nutr.* **131**, 1414–1420.

Taylor, K. M., and Nicholson, R. I. (2003). The LZT proteins; the LIV-1 subfamily of zinc transporters. *Biochim. Biophys. Acta Biomembr.* **1611**, 16–30.

Taylor, K. M., Morgan, H. E., Smart, K., Zahari, N. M., Pumford, S., Ellis, I. O., Robertson, J. F., and Nicholson, R. I. (2007). The emerging role of the LIV-1 subfamily of zinc transporters in breast cancer. *Mol. Med.* **13**, 396–406.

Taylor, K. M., Vichova, P., Jordan, N., Hiscox, S., Hendley, R., and Nicholson, R. I. (2008). ZIP7-mediated intracellular zinc transport contributes to aberrant growth factor signaling in antihormone-resistant breast cancer cells. *Endocrinology* **149**, 4912–4920.

Thompson, E. D., Olsson, P. E., Mayer, G. D., Haux, C., Walsh, P. J., Burge, E., and Hogstrand, C. (2001). Effects of 17 beta-estradiol on levels and distribution of metallothionein and zinc in squirrelfish. *Am. J. Physiol.* **280**, R527–R535.

Thompson, E. D., Mayer, G. D., Walsh, P. J., and Hogstrand, C. (2002). Sexual maturation and reproductive zinc physiology in the female squirrelfish. *J. Exp. Biol.* **205**, 3367–3376.

Thompson, E. D., Mayer, G. D., Balesaria, S., Glover, C. N., Walsh, P. J., and Hogstrand, C. (2003). Physiology and endocrinology of zinc accumulation during the female squirrelfish reproductive cycle. *Comp. Biochem. Physiol. A* **134**, 819–828.

Todd, A. S., Brinkman, S., Wolf, R. E., Lamothe, P. J., Smith, K. S., and Ranville, J. F. (2009). An enriched stable-isotope approach to determine the gill–zinc binding properties of juvenile rainbow trout (*Oncorhynchus mykiss*) during acute zinc exposures in hard and soft waters. *Environ. Toxicol. Chem.* **28**, 1233–1243.

Tsushima, K., Naito, W., and Kamo, M. (2010). Assessing ecological risk of zinc in Japan using organism- and population-level species sensitivity distributions. *Chemosphere* **80**, 563–569.

USDoI (1998). Guidelines for the interpretation of the biological effects of selected constituents in biota, water, and sediment. In *National Irrigation Water Quality Program Information Report No. 3*, pp. 184–198. US Department of the Interior. http://minerals.usgs.gov/minerals/pubs/commodity/zinc/index.html

USEPA (1980). *Exposure and Risk Assessment for Zinc.* EPA/440/4-81/016. Washington, DC: Monitoring and Data Support Division, Office of Water Regulations and Standards, US Environmental Protection Agency.

USEPA (1987). *Ambient Water Quality Criteria for Zinc – 1987.* EPA-440/5-87-003. Washington, DC: Criteria and Standards Division, Office of Water Regulations and Standards, US Environmental Protection Agency.

USEPA (2002). *National Recommended Water Quality Criteria: 2002.* EPA-822-R-02-047. Washington, DC: Office of Science and Technology, US Environmental Protection Agency.

USGS (2010a). *Mineral Information – Zinc.* Statistics and Information, US Geological Survey. Washington, DC: United States Government Printing Office.

USGS (2010b). *Zinc.* Mineral Commodity Summaries, US Geological Survey.

Vallee, B. L. (1986). A synopsis of zinc biology and pathology. In *Zinc Enzymes* (H. Gray, ed.), Vol. 1, pp. 1–15. Birkhäuser, Basel.

Van Sprang, P. A., Verdonck, F. A. M., Van Assche, F., Regoli, L., and De Schamphelaere, K. A. C. (2009). Environmental risk assessment of zinc in European freshwaters: a critical appraisal. *Sci. Total Environ.* **407**, 5373–5391.

Vega, M., Pardo, R., Herguedas, M. M., Barrado, E., and Castrillejo, Y. (1995). Pseudopolarographic determination of stability-constants of labile zinc-complexes in fresh-water. *Anal. Chim. Acta* **310**, 131–138.

Vercauteren, K., and Blust, R. (1999). Uptake of cadmium and zinc by the mussel *Mytilus edulis* and inhibition by calcium channel and metabolic blockers. *Mar. Biol.* **135**, 615–626.

Waller, W. T., and Cairns, J. (1972). Use of fish movement patterns to monitor zinc in water. *Water Res* **6**, 257–269.

Wang, Y., Wimmer, U., Lichtlen, P., Inderbitzin, D., Stieger, B., Meier, P. J., Hunziker, L., Stallmach, T., Forrer, R., Rulicke, T., Georgiev, O., and Schaffner, W. (2004). Metal-responsive transcription factor-1 (MTF-1) is essential for embryonic liver development and heavy metal detoxification in the adult liver. *FASEB J.* **18**, 1071–1079.

Wang, Y., Chen, P., Cui, R., Si, W., Zhang, Y., and Ji, W. (2010). Heavy metal concentrations in water, sediment, and tissues of two fish species (*Triplohysa pappenheimi, Gobio hwanghensis*) from the Lanzhou section of the Yellow River, China. *Environ. Monit. Assess.* **165**, 97–102.

Weaver, B. P., Beattie, J. D., Kambe, T., and Andrews, G. K. (2007). Novel zinc-responsive post-transcriptional mechanisms reciprocally regulate expression of the mouse Slc39a4 and Slc39a5 zinc transporters (Zip4 and Zp5). *Biol. Chem.* **388**, 1301–1312.

Weitzel, G., Strecker, F., Roester, U., Buddecke, E., and Fretzdorff, A. (1954). Zink im Tapetum lucidum. *Hoppe Seylers Z. Physiol. Chem.* **296**, 19–30.

Westin, G., and Schaffner, W. (1988). A zinc-responsive factor interacts with a metal-regulated enhancer element (MRE) of the mouse metallothionein-I gene. *EMBO J.* **7**, 3763–3770.

Wicklund Glynn, A. (1991). Cadmium and zinc kinetics in fish: studies on water-borne [109]Cd and [65]Zn turnover and intracellular distribution in the minnow, *Phoxinus phoxinus Pharmacol. Toxicol.* **68**, 485–491.

Williams, R. J. (1984). Zinc: what is its role in biology? *Endeavour* **8**, 65–70.

Willis, J. N., and Sunda, W. G. (1984). Relative contributions of food and water in the accumulation of zinc by 2 species of marine fish. *Mar. Biol.* **80**, 273–279.

Wong, V. V. T., Nissom, P. M., Sim, S.-L., Yeo, J. H. M., Chuah, S.-H., and Yap, M. G. S. (2006). Zinc as an insulin replacement in hybridoma cultures. *Biotechnol. Bioeng.* **93**, 553–563.

Woodling, J., Brinkman, S., and Albeke, S. (2002). Acute and chronic toxicity of zinc to the mottled sculpin *Cottus bairdi. Environ. Toxicol. Chem.* **21**, 1922–1926.

Xue, H. B., and Sigg, L. (1994). Zinc speciation in lake waters and its determination by ligand-exchange with EDTA and differential-pulse anodic-stripping voltammetry. *Anal. Chim. Acta* **284**, 505–515.

Yamaji, S., Tennant, J., Tandy, S., Williams, M., Srai, S. K. S., and Sharp, P. (2001). Zinc regulates the function and expression of the iron transporters DMT1 and IREG1 in human intestinal Caco-2 cells. *FEBS Lett.* **507**, 137–141.

Yamasaki, S., Sakata-Sogawa, K., Hasegawa, A., Suzuki, T., Kabu, K., Sato, E., Kurosaki, T., Yamashita, S., Tokunaga, M., Nishida, K., and Hirano, T. (2007). Zinc is a novel intracellular second messenger. *J. Cell Biol.* **177**, 637–645.

Yamashita, S., Miyagi, C., Fukada, T., Kagara, N., Che, Y. S., and Hirano, T. (2004). Zinc transporter LIVI controls epithelial–mesenchymal transition in zebrafish gastrula organizer. *Nature* **429**, 298–302.

Zhang, L., and Wang, W. X. (2007a). Gastrointestinal uptake of cadmium and zinc by a marine teleost *Acanthopagrus schlegeli*. *Aquat. Toxicol.* **85**, 143–153.

Zhang, L., and Wang, W. X. (2007b). Waterborne cadmium and zinc uptake in a euryhaline teleost *Acanthopagrus schlegeli* acclimated to different salinities. *Aquat. Toxicol.* **84**, 173–181.

Zhang, Y., Liu, Y., Huang, X., Liu, X., Jiao, B., Meng, Z., Zhu, P., Li, S., Lin, H., and Cheng, C. H. K. (2008). Two alternatively spliced GPR39 transcripts in seabream: molecular cloning, genomic organization, and regulation of gene expression by metabolic signals. *J. Endocrinol.* **199**, 457–470.

Zhao, L., Chen, W., Taylor, K. M., Cai, B., and Li, X. (2007). LIV-1 suppression inhibits HeLa cell invasion by targeting ERK1/2-Snail/Slug pathway. *Biochem. Biophys. Res. Commun.* **363**, 82–88.

Zheng, D., Feeney, G. P., Kille, P., and Hogstrand, C. (2008). Regulation of ZIP and ZnT zinc transporters in zebrafish gill: zinc repression of ZIP10 transcription by an intronic MRE cluster. *Physiol. Genom.* **34**, 205–214.

Zheng, D., Kille, P., Feeney, G., Cunningham, P., Handy, R., and Hogstrand, C. (2010a). Dynamic transcriptomic profiles of zebrafish gills in response to zinc depletion. *BMC Genom.* **11**, 548.

Zheng, D., Kille, P., Feeney, G., Cunningham, P., Handy, R., and Hogstrand, C. (2010b). Dynamic transcriptomic profiles of zebrafish gills in response to zinc supplementation. *BMC Genom.* **11**, 553.

4

IRON

NICOLAS R. BURY

DAVID BOYLE

CHRISTOPHER A. COOPER

1. Chemical Speciation in Freshwater and Seawater
2. Sources of Iron and Economic Importance
3. Environmental Situations of Concern
4. A Survey of Acute and Chronic Ambient Water Quality Criteria in Various Jurisdictions in Freshwater and Seawater
    4.1. United Kingdom
    4.2. European Union
    4.3. United States
    4.4. Canada
5. Mechanisms of Toxicity
    5.1. Waterborne Toxicity
    5.2. Free Radical Production
6. Essentiality or Non-Essentiality of Iron: Evidence For and Against
7. Potential for Bioconcentration and/or Biomagnification of Iron
8. Characterization of Uptake Routes
    8.1. Iron Uptake in Marine Fish Intestine
    8.2. Iron Uptake Across the Gill
9. Characterization of Internal Handling
    9.1. Transferrin and Transferrin Receptor
    9.2. Mitochondrial Heme and Fe/S Cluster Protein Synthesis
    9.3. Ferritin
    9.4. Immune Response
10. Characterization of Excretion Routes
11. Behavioral Effects of Iron
12. Molecular Characterization of Epithelial Iron Transporters and Hepcidin
    12.1. Molecular Characteristics of Fish Epithelial Iron Transporters
    12.2. Hepcidin
13. Genomic and Proteomic Studies
14. Interactions with Other Metals
15. Knowledge Gaps and Future Directions

Homeostasis and Toxicology of Essential Metals: Volume 31A
FISH PHYSIOLOGY

Iron (Fe) is essential for life, being involved in oxygen transfer, respiratory chain reactions, DNA synthesis, and immune function. The essentiality of this element for all vertebrate life means that uptake pathways and internal handling strategies are conserved between mammals and fish, and the zebrafish has been used extensively to understand vertebrate Fe metabolism at a mechanistic level. This review provides an overview of aquatic Fe speciation and the way this governs bioavailability and influences toxicity. It focuses on the mechanisms of Fe uptake across the gills and intestine, highlighting the geochemical constraints that fish have to overcome to acquire Fe from the water and diet. Molecular characteristics of fish iron importer [solute carrier (slc) 11/natural resistant-associated macrophage protein (Nramp)/DMT] and exporter (slc40a1/ferroportin/ IREG) proteins that facilitate this iron uptake process are explained in detail. Also included is a description of the internal Fe handling proteins that store iron and transport it around the body, the role of hepcidin in controlling iron metabolism, and how Fe plays a key role in defense against pathogens in fish.

1. CHEMICAL SPECIATION IN FRESHWATER AND SEAWATER

Iron (Fe) is a transition metal found at concentrations ranging from $ng\,L^{-1}$ in marine environments (e.g. Barbeau et al., 2001) to $mg\,L^{-1}$ in freshwater that receive acid mine drainage (e.g. Winterbourn et al., 2000). It exists in waters in different oxidation states depending on certain extrinsic factors (e.g. Fig. 4.1A). Ferrous iron [Fe(II)] is more soluble than ferric iron [Fe(III)], forms weaker bonds with complexing agents, and is generally more bioavailable to eukaryotes (Stumm and Morgan, 1996). However, at neutral pH and in well-oxygenated waters, Fe(III) is more thermodynamically stable and Fe(II) may account for approximately 0.2–0.7% of unfiltered Fe concentration in the top 5 m of circumneutral lakes during the daytime (Emmenegger et al., 2001). The half-life of Fe(II) can be in the region of seconds in well-oxygenated alkaline conditions (Emmenegger et al., 1998), but may be far greater (minutes to hours) in more favorable redox conditions, such as acidic streams (e.g. Sections 3 and 5; McKnight et al., 2001; Sherman, 2005; Gammons et al., 2005; Teien et al., 2008; Duckworth et al., 2009a).

The vast majority of the Fe(III) forms insoluble non-bioavailable Fe oxides, a term used to include Fe(III) oxides, Fe(III) oxohydroxides, and Fe(III) hydroxides, that precipitate out of solution (Fig. 4.1A) (Stumm and Morgan, 1996; Tipping et al., 2002; Cooper and Bury, 2007). Iron oxides

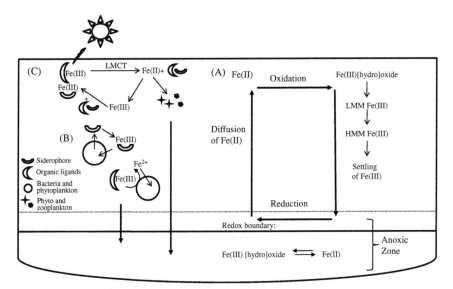

Fig. 4.1. Iron speciation in water. (A) In oxygenated circumneutral water Fe(II) is oxidized to Fe(III) which forms low molecular mass (LMM) Fe(III) [hydro]oxides that over time form high molecular mass (HMM) Fe(III) [hydro]oxides precipitates. The precipitates settle out and the redox conditions in the benthos favor the conversion of Fe(III) to Fe(II), which may then diffuse into the water column (adapted from Stumm and Morgan, 1996). (B) In response to low Fe concentrations some phytoplankton and bacteria secrete compounds called siderophores that have a strong affinity for Fe(III) that increases Fe(III) solubility. These organisms can access this Fe source via specialized Fe(III)-siderophore uptake mechanisms. Other phytoplankton and bacteria possess membrane-bound ferric chelate reductase that effectively binds to Fe(III)–ligand complex reducing Fe(III) to Fe(II) that is available for uptake. (C) Sunlight can induce ligand to metal charge transfer (LMCT) that enables electron transfer from Fe(III) to the ligand resulting in Fe(II) release. This Fe(II) may be oxidized and rebind to the ligand, but it is also taken up by microorganisms.

accumulate in the sediments of lakes and oceans where the redox condition in the subsurface anoxic zone favors Fe(III) to Fe(II) formation. Bioturbation, physical resuspension, and upwelling return Fe(II) to the more productive upper reaches of these water bodies (Burdige, 1993).

This aquatic Fe geochemical cycling is an oversimplification of Fe speciation in terms of the bioavailable fraction (e.g. Fig. 4.1B) (Barbeau et al., 2001; Barbeau, 2006). To circumvent a reduction in Fe availability due to the formation of Fe oxides a number of organisms produce biotic compounds that bind to Fe(III) generating a pool of soluble Fe complexes (Duckworth et al., 2009a, b). For example, in the marine environment where concentrations of dissolved Fe in the open ocean can be as low as 5.6 ng L^{-1} (100 pM) (Barbeau, 2006), some species of bacteria (Barbeau et al., 2001) and phytoplankton (Hutchins et al., 1999) secrete a group of compounds known as siderophores, which are low molecular weight chelators with a high affinity for Fe(III) (Armstrong and Van Baalen, 1979; Barbeau et al., 2001; Amin et al., 2009). Organisms that secrete siderophores are able to take up the siderophore–Fe complexes via specialized import mechanisms (Macrellis et al., 2001). A number of other eukaryotic phytoplankton and yeast possess a membrane-bound ferric chelate reductase that effectively binds to the Fe(III)–ligand complex reducing Fe(III) to Fe(II), in so doing increasing the concentration of Fe(II) at the site of uptake (Jones et al., 1987; Yehuda et al., 1996; Robinson et al., 1999). Although relatively common in marine waters, Duckworth et al. (2009b) first reported the existence of siderophores in freshwater, suggesting that they are a ubiquitous biogeochemical agent in oxygen-rich environments. The significance for marine ecosystems of the evolution of mechanisms to capture and sequester Fe for primary productivity within the aquatic environment in Fe-poor regions cannot be underestimated (Boyd et al., 2000).

In the majority of freshwater bodies, dissolved organic matter (DOM), e.g. humic and fulvic acids, is important with respect to chemical speciation, mobility, and bioavailability of trace metals (Hamilton-Taylor et al., 2002). Iron(III)–DOM complexes are present in freshwater rivers and are important in maintaining Fe solubility (Tipping et al., 2002; Lofts et al., 2008). The proportion found in this form is dependent on pH and temperature (Lofts et al., 2008).

Whenever Fe(III) forms an organic complex, Fe may have the potential to undergo ligand to metal charge transfer (LMCT) (Barbeau et al., 2001; Barbeau, 2006). In the LMCT process the electrons of the ligand and transition metal become shared, which creates an energy difference between them. The ensuing elevation of negative energy subsequently generates a cleavage between the metal and the ligand, resulting in the metal being released in a reduced state, e.g. Fe(III) to Fe(II) (Bellelli et al., 2001; Russo et al., 2003). Extrinsic factors, such as sunlight, and more specifically the ultraviolet (UV) light spectra, can initiate this process, in both marine and freshwaters. Barbeau et al. (2001) have shown that when Fe(III) is bound to a marine siderophore, known as aquachelin, and exposed to natural sunlight, the complex is photolyzed and the Fe(III) is reduced to Fe(II). The resulting Fe(II) either is converted back to Fe(III) and rebinds to the ligand or is available for biological uptake (Fig. 4.1C) (Barbeau et al., 2001). Similarly, Fe(III) can also be reduced via LMCT when it is bound by freshwater-derived humic acid and exposed to UV (Fukushima and Tatsumi, 1999). The significance of photochemical cycling of Fe(III) in both marine and freshwater environments for bioavailability to fish has yet to be fully understood (Cooper and Bury, 2007).

2. SOURCES OF IRON AND ECONOMIC IMPORTANCE

Five percent of the Earth's crust is Fe, where it is found as predominantly magnetite (Fe_3O_4), and hematite (Fe_2O_3), as well as goethite [FeO(OH)], kinonite [FeO(OH).n(H_2)], or siderite ($FeCO_3$). Iron is arguably the most important metal to be mined in the world, with 98% of that extracted being used in the production of steel, a key component for the majority of manufacturing, transport, and building industries. The ubiquitous nature of Fe means there is a wide mining base in over 50 countries producing approximately 2300 million tonnes per year. The production and export market is dominated by China, Brazil, Australia, and India (Table 4.1).

In more recent times Fe has been found to have further economic importance in remediating contaminated water (Zhang, 2003). Zero-valent iron (Fe0) is increasingly being used to clean up groundwater contaminated with a large array of organic chemicals (e.g. chlorinated methanes, chlorinated benzenes, pesticides, polychlorinated hydrocarbons), other metals, as well as nitrates and phosphate pollution from agricultural runoff (e.g. Zhang, 2003; Theron et al., 2008; Geng et al., 2009). The Fe0 has been

Table 4.1
Iron ore worldwide production (million tonnes) for 2008 and 2009

	Mine production	
	2008	2009
China	824	900
Brazil	355	380
Australia	342	370
India	220	260
Russia	100	85
Ukraine	73	56
South Africa	49	53
Iran	32	33
Canada	31	27
USA	54	26
Kazakhstan	23	21
Sweden	24	18
Venezuela	21	16
Mexico	12	12
Mauritania	11	11
Other countries	47	47
Total world production	2220	2300

Source: US Geological Survey, Mineral Commodities Summaries, January 2010. http://minerals.usgs.gov/minerals/pubs/commodity/iron_ore/mcs-2010-feore.pdf

shown to be a reductant and catalyst, and halogenated hydrocarbons in the presence of Fe0 are reduced to benign hydrocarbons (Zhang, 2003). In the USA there are 1500 Superfund sites and in 2003 it was estimated that the cost for cleaning up each site is $25 million (Zhang, 2003); the abundance of Fe0 potentially makes an economically viable option for remediation.

3. ENVIRONMENTAL SITUATIONS OF CONCERN

Concerns over elevated Fe in the aquatic environment are largely focused on discharges from iron ore mining activities and acid mine drainage that contain a plethora of metals in addition to Fe. Even in regions where iron ore or coal mining has ceased, the legacy of discharge from abandoned mines continues for many years. Iron, and other metals, from these mines accumulate in estuarine sediments (Pirrie et al., 1997), and a general acidification of the water due to the acid mine discharge affects the solubility and bioavailability of metals. In contrast, in the marine environment, Fe is the limiting factor in primary productivity in open oceans (Boyd et al., 2000), reducing food availability for pelagic fishes.

Rust-colored Fe oxide precipitates characterize oxygen-rich circumneutral pH streams receiving both mine drainage and natural inputs high in Fe. The major environmental concern in this situation is the smothering of the river benthos (Fig. 4.2). Iron precipitates may deposit on the respiratory gill surfaces, affecting many macroinvertebrate populations (Gray, 1997; Gray and Delaney, 2010) and fish (e.g. Section 5.1), leading to reduced biodiversity. Indeed, it is recognized that the low species abundance in wetlands immediately receiving mine drainage high in Fe is due to Fe oxide

(A)   (B)   (C)

Fig. 4.2. Images of (A) the mixing zone between the Red River, Camborne, Cornwall, UK, and the discharge from Dolcoath mine; (B) the extent of Fe oxide downstream of this point; and (C) the Fe oxide precipitation downstream of the mixing zone over a 3 month period following the explant of containers containing substrate mimicking that of the river (V. Fowler, T. Geatches and N. Bury, unpublished images). **SEE COLOR PLATE SECTION.**

precipitation (Batty et al., 2005). In streams with lower buffering capacity the acidic mine drainage will keep water pH levels low, resulting in the maintenance of high concentrations of ionic metal species in solution (Liang and Thomson, 2009). In these situations it is the combined potency of the acid and metal mixture that is toxic, rather than a single element. Since Fe (II) is more readily available it is believed to be more toxic (Vuori, 1995), probably owing to a greater accumulation in tissues increasing the likelihood of free radical production and internal oxidative damage (see Section 5.2).

There are concerns about the water quality in lakes receiving iron ore mining discharge, an effluent that has been considered relatively benign (Payne et al., 1998). Field studies by Payne et al. (1998, 2001) have studied lake trout (*Salvelinus namaycush*) residing in Wabush Lake, Newfoundland, Canada, which received mine tailings rich in magnetite and hematite. These trout showed signs of skin bleaching ("bleached fish syndrome"), as well as increased liver inflammation and perturbed hematology. A mechanistic understanding of the bleached fish syndrome phenomenon has not been established, but it is presumably due to increased oxidative stress (e.g. Section 5.2). The authors suggest that regulations on iron ore effluent should be reconsidered, but currently the lake trout bleaching syndrome appears to be unique and further study to assess the environmental risk and hazards of iron ore mining effluent is required.

4. A SURVEY OF ACUTE AND CHRONIC AMBIENT WATER QUALITY CRITERIA IN VARIOUS JURISDICTIONS IN FRESHWATER AND SEAWATER

4.1. United Kingdom

In 1989 the Department of the Environment (DoE) in the UK required Fe to be treated in the same manner as other list II substances under the Dangerous Substances Directive (76/464/EEC). Environmental quality standards (EQS) of 1 mg Fe L^{-1} for freshwater and saltwater were proposed (WRc, 1988) and entered into legislation in England and Wales in 1989 (DoE, 1989). Statutory EQS for Fe are long-term annual average concentrations of "dissolved"; Fe, defined as that which passes through a 0.45 μm filter. Subsequent re-review of these EQS recommended retention of the standard but the addition of a proviso that the integrity of biological communities be assessed in waters where annual average concentrations routinely exceed 0.3 mg Fe L^{-1} (WRc, 1998). However, short-term EQS for

Fe are considered redundant owing to the expected long-term or near continuous release of Fe into the environment (WRc, 1998).

4.2. European Union

The Dangerous Substances Directive (76/464/EEC and daughter directives) will be repealed in 2013 and replaced with the European Water Framework Directive (WFD; 2000/60/EC). Under the WFD, Fe is designated an Annex VIII substance and candidate for specific pollutant status. As such, derivation of predicted no effect concentrations (PNEC) for Fe and implementation of these thresholds as EQS are the responsibility of individual member states where Fe is identified as being discharged to water in significant quantities.

4.3. United States

In the USA, Fe is designated a non-priority pollutant with a national recommended freshwater quality criterion of 1 mg L^{-1} for total Fe (USEPA, 1976, 1986). This criterion continuous concentration (CCC) is an estimate of the highest concentration in surface water to which an aquatic community can be exposed indefinitely without resulting in an unacceptable effect. Data are considered insufficient to derive a criterion for saltwater.

4.4. Canada

The national water quality guideline for Fe for Canada is 0.3 mg L^{-1} based on measured total Fe concentration (CCREM, 1987). This guideline is a continuous exposure criterion based on toxicity data for the most sensitive species of plants and animals found in Canadian waters. Provincial and territorial jurisdictions within Canada may, however, implement their own water quality guidelines specific to the requirements of their respective jurisdictions.

The principal differences among water quality standards (Table 4.2) are the fraction of Fe, dissolved or total, specified in the legislation and the implementation of a saltwater standard in some UK waters. Use of dissolved Fe concentration for EQS in UK waters is based on the premise that the dissolved metal ion is intrinsic to toxicity. This is despite evidence to the contrary in the Fe literature (e.g. Section 3 and Vuori, 1995), regardless of the efficacy of 0.45 μm filtration in assessing the dissolved Fe concentration. Furthermore, in alkaline marine waters, Fe is predicted to rapidly flocculate and become removed from the water column (Fox and

Table 4.2
Water quality criteria for iron (for protection against chronic, lifetime toxicity)
for selected countries

Country	Freshwater	Saltwater
UK	1 mg L^{-1a}	1 mg L^{-1a}
USA	1 mg L^{-1b}	
Canada	0.3 mg L^{-1b}	

[a]Dissolved.
[b]total.

Wofsy, 1983). The protective value of a saltwater standard for dissolved Fe is unclear. In Canada and the USA, toxicity of colloidal Fe is implicitly considered in freshwater standards for total Fe. However, the breadth of data used to derive a freshwater criterion in the USA has attracted recent criticism owing to its assessment of a single field study and laboratory-based toxicity tests with few fish species (Linton et al., 2007). A paucity of data has also been cited for the absence of saltwater standards in the USA and Canada (EPA 440/9-76-023, July, 1976; CCREM, 1987) and water quality standards for Fe in Australia and New Zealand (ANZECC, 2000).

5. MECHANISMS OF TOXICITY

Toxicity of elevated Fe concentrations can occur via external exposure from the water or diet and internally by a breakdown in Fe homeostatic regulation leading to tissue Fe overload. Excessive Fe in artificial diets (> 86 mg Fe kg^{-1} added as $FeSO_4.7H_2O$) has been shown to induce liver and kidney histopathology, reduce growth rate, and increase mortalities in rainbow trout (Desjardins et al., 1987). Iron deficiency will cause respiratory problems associated with anemia, increased likelihood of bacterial infection, and potential susceptibility to other divalent metal toxicity (see Section 13).

5.1. Waterborne Toxicity

Toxicity of Fe in water is closely related to speciation and the interaction of those Fe species with the body and gill surface. In acidic environments and at low oxygen concentrations, Fe is predominantly found in the ferrous state. It is thought that Fe(II) is more readily available and potentially toxic to aquatic organisms owing to Fe overloading (Vuori, 1995; Bury and Grosell, 2003a), but the condition where Fe(II) is prevalent (i.e. acidic/low oxygen water) is seldom encountered by fish.

The toxicity of Fe in the mixing zone between acid mine drainage and river water is dependent on the formation of Fe(III) oxides, which is related to the oxygen content and pH of the water. For example, in waters of pH 5, Fe may be found as low molecular mass (LMM) Fe(II) OH^- species, and as pH and oxygen rises, LMM Fe(III) formation increases exponentially (Teien et al., 2008). However, the rate of oxidation is also dependent on the chemistry of the donor and receiving waters, and is lowered in the presence of SO_4^{2-}, Cl^- (Tamura et al., 1976) and organic material (Tipping et al., 2002). In the mixing zone, LMM Fe(II) concentration may prevail for a number of hours (Duckworth et al., 2009a). Teien et al. (2008) showed that 0.5 mg LMM Fe(II) L^{-1} at pH 6.3 was non-toxic to Atlantic salmon (*Salmo salar*), but as pH rose to 6.7 mortality increased, which coincided with an increase in the formation of LMM Fe(III) species (Teien et al., 2008). The mechanism of toxicity was not fully elucidated. However, despite the lack of visible rust-colored Fe precipitates that characterize sites of high Fe loading and high molecular mass Fe(III) formation, Fe was shown to accumulate on the gills of the fish in the experiments where death occurred. This would support the hypothesis that the toxic mode of action is via smothering of the epithelium, thereby interfering with respiratory gas exchange (see Section 3).

A number of other studies have also demonstrated that the respiratory surfaces are the site of Fe toxicity in fish: Grobler et al. (1989) found that tilapia (*Tilapia sparrmanii*) on exposure to Fe show signs of coughing and spluttering. Dalzell and MacFarlane (1999) compared the toxicity of commercial Fe slurry (which is used to aggregate algal blooms in drinking water treatment plants) to that of pure $Fe_2(SO_4)_3$ and found that both treatments resulted in heavy deposits of Fe oxide on the gills of brown trout (*Salmo trutta*), which subsequently caused respiratory failure. Peuranen et al. (1994) showed that exposure of brown trout to pH 5 and 6 and Fe at 2 mg L^{-1} caused severe branchial damage with a fusion of the lamellae and hypertrophy of the epithelium increasing the diffusion distance from the water to the blood. Although these types of alterations to branchial architecture are symptomatic of a general response to an aquatic irritant (Mallett, 1985), the accumulation of Fe precipitates on the gill surface will further increase the diffusion distance for gases, thereby enhancing the respiratory stress. Indeed, a decrease in oxygen consumption was measured in these brown trout (Peuranen et al., 1994).

There is little direct evidence that Fe acts as a specific ionoregulatory toxicant in fish. Peuranen et al. (1994) found that plasma Na^+ and Ca^{2+} content decreased in brown trout exposed to 2 mg L^{-1} Fe, and similar ionoregulatory disturbances were seen by Lappivaara et al. (1999) in whitefish (*Coregonus lavaretus*) exposed to 8 mg Fe L^{-1} in natural water rich in humic acid. In contrast, Gonzalez et al. (1990) found no effect on plasma

Na^+ concentrations in brook trout (*Salvelinus fontinalis*) exposed to 0.3 mg L^{-1} Fe for 2 days, but did observe a reduction in whole-body sodium content. The evidence would point to the effect on ion regulation being due to the general stress response evoked by respiratory distress upon Fe exposure. Notably, in the study of Lapivaara et al. (1999) the ionoregulatory disturbance was accompanied by an increase in plasma cortisol and lactate, both indicators of stress.

Mizuno et al. (2004) observed that elevated Fe concentration during fertilization causes hardening of the eggs of shishamo (*Spirinchus lanceolatus*) smelt. The resulting increase in egg pressure caused by chorion hardening reduced the hatching rate, and egg pressure was further increased in the presence of tannins (Mizuno et al., 2004). These findings may also have a significant bearing on salmonid spawning, which occurs primarily in headstreams that typically receive water rich in tannin and Fe from upland catchments, as well as benthic spawners (e.g. bullhead, *Cottus gobio*) in the lower catchment that receive considerable input of Fe from groundwater (Lautz and Fanelli, 2008; Duckworth et al., 2009a). Concerns over the effects of Fe precipitation on salmonid egg hatching success were raised in the 1970s when Smith and Sykora (1976) observed the mortality of brook trout and coho salmon eggs coated with Fe particulates.

5.2. Free Radical Production

Intracellular Fe toxicity is associated with the ability of Fe to alter redox states. Iron is involved in Fenton chemistry (Eq. 1) acting as a catalyst for hydroxyl radicals (OH^-) and hydrogen peroxide (H_2O_2) formation:

$$Fe^{2+} + H_2O_2 = Fe^{3+} + OH^- + OH \tag{1}$$

Hydroxyl radicals are highly reactive molecules and can cause peroxidation of lipid membranes, damage nucleic acids and affect antioxidant enzyme activity (Li et al., 2009). This disruption to macromolecule structure and activity can be so severe as to result in cell death and tissue injuries (Brewer, 2010). Excessive Fe overload resulting in DNA damage has been linked to a greater risk of cancer in humans (Huang, 2003). The potential toxicity of free radicals, which are also natural products of aerobic respiration (Brewer, 2010), is reduced by free radical scavengers such as glutathione, catalases, and glutathione-*S*-transferase. The disruption to normal cellular processes is rare and Fe toxicity in humans is often associated with genetic disease (e.g. during disease associated with aging) (Brewer, 2010). Evidence for similar disease in wild fish populations is extremely rare, presumably because such a disease would reduce the fitness of the organisms, resulting in susceptibility to predators.

Evidence for oxidative damage due to Fe in fish comes from the field studies of Payne et al. (1998, 2001), where lake trout kept in cages in a lake receiving iron ore mining effluent showed increased DNA oxidative damage, while the natural fish population showed signs of bleached fish syndrome, liver oxidative damage, and inflammation. More recently, evidence for branchial and internal oxidative damage caused by Fe has been observed in a study assessing the effects of waterborne non-valent Fe0 (Li et al., 2009). Li and colleagues (2009) observed dose-dependent alterations in superoxide dismutase and malondialdehyde in embryonic medaka (*Oryzias latipes*) exposed to non-valent Fe0. In adults the gills and intestine showed histopathology while the brain and liver changed antioxidant status, indicating that waterborne non-valent Fe0 entered and damaged fish (Li et al., 2009).

6. ESSENTIALITY OR NON-ESSENTIALITY OF IRON: EVIDENCE FOR AND AGAINST

Iron is essential for life in an oxygen-rich environment. It is integral in the oxygen binding metalloprotein hemoglobin and it forms part of cytochrome *c* oxidases that make up the respiratory chain, acting as an electron donor or acceptor. It also plays a role in DNA synthesis and the host's response to pathogens. The daily requirement of mammals is 28 mg Fe $kg^{-1} day^{-1}$ (Conrad et al., 1999; Andrews, 2005), and 14 mg Fe $kg^{-1} day^{-1}$ for zebrafish (Bury and Grosell, 2003b). However, Fe requirements may vary considerably depending on the species of fish (Gaitlin and Wilson, 1986; Zibdeh et al., 2001; Shiau and Su, 2002), and within aquaculture the addition of Fe salts to the diet varies between 30 and 170 mg kg^{-1} dry weight (Watanabe et al., 1997). A potential negative impact of the addition of large quantities of metal salts to aquaculture feed to meet dietary requirements is the accumulation of metals in the vicinity of fish farms owing to the large quantities of feces produced in a small area or from uneaten food. Iron, along with other metals, has been shown to be elevated in the water column and sediments immediately around fish farms (Sutherland et al., 2007; Basaran et al., 2010).

7. POTENTIAL FOR BIOCONCENTRATION AND/OR BIOMAGNIFICATION OF IRON

Bioconcentration of Fe poses little risk to the health of fish populations. Branchial Fe uptake in zebrafish exposed to 0.92 μg Fe L^{-1} (16.5 nmol L^{-1}

as $^{59}FeCl_3$) in ion-poor water was characterized by rapid accumulation (within 2 h) at the gill and subsequent decrease with concomitant partitioning to the body (Bury and Grosell, 2003b). Excretion in zebrafish was also rapid: $>50\%$ of ^{59}Fe in 24 h. Addition of dithiothreitol, a reducing agent, increased Fe uptake, indicating reduction of Fe^{3+} to Fe^{2+} as the rate-limiting step of branchial Fe uptake (Bury and Grosell, 2003b). Gregorović et al. (2008) also demonstrate deposition of Fe in liver of carp during long-term exposure to 1 mg Fe L^{-1}, although the Fe salt used, Fe-dextran, is often used to treat Fe deficiency in humans because of its elevated bioavailability.

Tissue Fe concentrations vary considerably between species (Table 4.3) and can be greatly influenced by seasons (Dural et al., 2007; Ersoy and Çelik, 2010). However, owing to Fe's essentiality and requirement for hemoglobin function, tissue Fe values are typically in the mg kg^{-1} range, and it would thus be expected that bioconcentration factors (BCFs) will be high in pristine conditions. Indeed, several studies which have measured gill tissue Fe burdens and water Fe concentrations have reported high Fe BCFs

Table 4.3
Examples of tissue iron concentrations measured in freshwater and marine fish species

Species	Tissue			
	Muscle	Liver	Gill	Whole body
Marine				
Sparus aurata	0.51−0.81[a]	148−213[a]	152±62[e]	
	19.6±7.8[e]	256±109[e]	61−136[h]	
	7.2−16.5[h]	48−230[h]		
Chelidonichthys lucernusd	0.8−2.2[a]	49−91[a]		
Upeneus molluccensis	1.6−2.6[a]	27−453[a]		
Solea solea	0.4−1.1[a]	93−158[a]		
Merluccius merluccius	0.2−1.52[a]	12−33[a]		
Saurida undosquamis	0.4−0.8[a]	128−233[a]		
Melangrammus aeglefinus	15.6±1.9[c]			
Sardinella aurita	28.4±4.6[d]	161±46[d]		
	40.6±30.69[d]	146±31[d]		
	53.6±22.5[d]	331±51[d]		
Mullus surmuleteus	7.0±0.87[d]	97±26[d]		
	15.3±1.5[d]	165±12[d]		
		144±91[d]		
Lithognathus mormyrus	7.7±0.7[d]	111±21[d]	275±67[e]	
	22.9±4.3[d]	210±138[d]		
	21.0±5.4[d]	146±23[d]		
	38.7±18.3[e]	370±252[e]		

(*Continued*)

Table 4.3 (Continued)

Species	Muscle	Liver	Gill	Whole body
		Tissue		
Mugil cephalus	90.1 ± 15.6^d	243 ± 81^d	$132-345^h$	
	11.0 ± 2.2^d	302 ± 85^d		
	25.6 ± 6.1^d	171 ± 56^d		
	$7.2-11^h$	$88-383^h$		
Pagellus erythrinus	34.4 ± 27.1^d	110 ± 40.8^d		
	47.6 ± 19.5^d	150 ± 17^d		
	14.1 ± 0.9^d	101 ± 33^d		
	42.9 ± 17.3^d	204 ± 42.3^d		
Synodus suras	5.9 ± 0.8^d	111 ± 10.8^d		
	10.3 ± 2.0^d	47.3 ± 3.7^d		
Aspitrigla cuculus	37.1 ± 3.0^d	516 ± 363^d		
Epinephelus alexandrinus	16.8 ± 3.7^d	61.5 ± 16.5^d		
Atherina hepsetus	78 ± 37^e	393 ± 171^e	793 ± 411^e	
Trigla cuculus	30.7 ± 10.2^e	582 ± 208^e	499 ± 339^e	
Sardina pilchardus	39.6 ± 8.6^e	225.4 ± 51.5^e	227 ± 32^e	
Scomberesox saurus	29.8 ± 16.2^e	407 ± 145^e	885 ± 514^e	
Dicentrarchus labrax	$7.8-11.1^h$	$62.3-188.9^h$	$103-341^h$	
Antarctic fish				
Pagothenia borchgrevinki	$1.4-6.0^b$	$10-114^b$		$4.0-9.3^b$
Notothenia corriceps	$9.5 \pm 8.5 \, ♀^j$	$116 \pm 115 ♀^j$		
	$17.9 \pm 18.7 ♂^j$	$242 \pm 186 ♂^j$		
Freshwater				
Cyprinus carpio	$1.5-9.1^f$	$45-220^f$	$36-43^f$	
Barbus xanthopterus	5.3 ± 2.0^g	95 ± 39^g	68 ± 51^g	
Barbus rajanorum mystaceus	4.0 ± 1.0^g	95 ± 36^g	62 ± 35^g	
Salmo trutta		$188-540^i$		
		$374-990^i$		
		$394-1040^i$		
		$384-1020^i$		

Values represent mg/kg dry weight (except for those derived from a, b, d, and j where they represent mg/kg wet weight) as either a range (−) or mean±standard error. Multiple measurements for the same species from the same study represent either different sample sites or season.

[a]Erosy and Çelik (2010).
[b]Honda et al. (1983).
[c]Roy and Lall (2006).
[d]Tepe (2009).
[e]Canli and Atli (2003).
[f]Tekin-Ozan and Kir (2008).
[g]Alhas et al. (2009).
[h]Dural et al. (2007).
[i]Lamas et al. (2007).
[j]Marquez et al. (1998).

in fish, some of which have been used to inform water quality criteria. A BCF of 50 has been reported for muscle of *Oreochromis mossambicus* in River Cauvery, southern India, although low aqueous Fe concentrations, mean 0.126 mg Fe L^{-1}, suggest Fe burdens are within homeostatic limits or dietary in origin (Ayyadurai et al., 1994). Similarly, Fe contamination in polluted Manchar Lake, Pakistan, was associated with high metal concentrations in sediment (up to 17 g kg^{-1} dry weight) and relatively low concentrations in ambient water (3 mg L^{-1} total Fe), typical of lentic ecosystems (Arain et al., 2008).

Biomagnification also appears unlikely under most environmental conditions. In fact, the opposite may be true. Concentrations of Fe in invertebrates were unrelated to feeding strategy and were lower than in algae and bryophytes in acid mine drainage in impacted New Zealand streams receiving up to 32.6 mg Fe L^{-1} (Winterbourn et al., 2000). Furthermore, no effect of low pH on Fe burdens in invertebrates and plants was observed despite elevated dissolved Fe in the water column. In an artificial food chain examining transfer of metals from sewage sludge to algae, shrimp and carp, total body burdens of Fe diminished with increasing trophic level (Wong and Tam, 1984). Omission of shrimp from the experimental food chain led to elevated Fe in carp indicative of trophic dilution. Similar data have been presented for marine fish. Using ^{15}N as a marker of trophic level, Nfon et al. (2009) measured the magnification of Fe in the metal-impacted Baltic Sea. Successive trophic enrichments of ^{15}N in zooplankton, mysids, and herring were associated with a decrease in total Fe, indicative of biodilution in a marine food chain. Together, these data suggest that a combination of homeostatic regulation and abiotic factors limits aqueous and dietary Fe uptake in fish.

8. CHARACTERIZATION OF UPTAKE ROUTES

In the absence of a regulated mechanism for Fe excretion, uptake is tightly controlled, so as to maintain Fe homeostasis (Shi and Camus, 2006). In fish there are two entry routes, via the diet and water (summarized in Fig. 4.3A and B). Within the diet, Fe can be found primarily as either heme-bound or non-heme bound Fe, with the majority of dietary Fe uptake studies in fish and mammals focusing on non-heme bound uptake. Currently, there is debate over the presence of an intestinal heme transporter, with the candidate protein slc46a1 (heme carrier protein 1) (Shayeghi et al., 2005) being a potential folate transporter (Qiu et al., 2006; Laftah et al., 2009). This section will describe the whole-animal Fe uptake

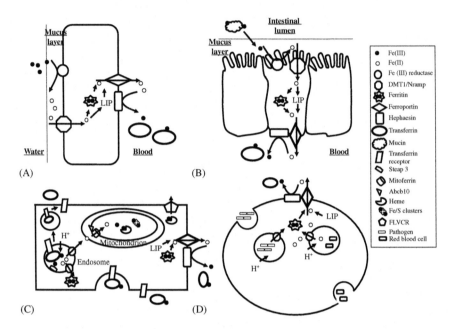

Fig. 4.3. Iron (Fe) uptake pathways and internal Fe handling in fish. (A) Aquatic Fe compounds traverse the mucous layer covering the gill surface where either a membrane ferric reductase or extrinsic Fe reducing agents (e.g. ascorbate) converts Fe(III) to Fe(II) and Fe(II) is the substrate for the Fe^{2+}/H^+ symporter divalent metal transporter 1 (DMT1). Once inside the cell Fe is stored as ferritin or enters the labile iron pool (LIP). Export of Fe(II) occurs via ferroportin, which is linked to a membrane-bound ferric oxidase called hephaestin that converts Fe(II) back to Fe(III). Hephaestin homologues are still to be identified in fish. Fe(III) binds to transferrin and circulates in the body. (B) In the intestine Fe bound to mucins traverses the mucosal layer; thereafter the uptake process is similar to the gill. Caption (C) represents a cell from an internal tissue. Fe(III)-transferrin binds to the membrane-bound transferrin receptor that is internalized by endocytosis. An endosomal proton pump increases the internal acidity causing the release of Fe-transferrin from its receptor and Fe(III) from transferrin. An internal ferric reductase, termed Steap 3, converts Fe(III) to Fe(II) and this is pumped out of the endosome by DMT1. The transferrin receptor is recycled and the apo-transferrin re-enters the circulation. Fe either enters the LIP, ferritin, is exported by ferroportin, or travels to the mitochondria. In all cells the mitochondria is the site of heme and Fe/S cluster formation. It is unclear how Fe crosses the outer mitochondrion membrane, but it crosses the inner membrane via mitoferrin, facilitated in some way by an ATP-binding cassette transporter Abcb10. Fe(II) is delivered to the enzyme ferrochelatase where it is oxidized and enters heme or Fe/S cluster synthesis. Heme leaves the cell via a feline leukemia virus subgroup c receptor (FLVCR). (D) Macrophages play two key roles, the recycling of Fe from senescent red blood cells and defense against pathogens. Senescent RBCs are engulfed by macrophages. A proton pump decreases the acidity of the endosyme lyzing the RBC and Fe is released by heme oxygenase 1. Currently, piscine Fe(II)/H$^+$ transporter characteristics suggest they are symporters and it is hypothesized that a proton pump on the membrane of the endosome decreases internal pH enabling Fe(II)/H+ symport activity depriving the pathogen of Fe. Fe either enters the LIP or is stored in ferritin before export via ferroportin.

processes, and a more detailed molecular characterization of the proteins involved in this uptake will be described in Section 12. To aid the understanding of the particular challenges associated with piscine Fe uptake, a brief description of uptake process and transport proteins involved in mammalian Fe epithelial uptake is provided. However, a large number of homologues for the mammalian transport protein exist in fish (e.g. Table 4.4 for zebrafish) and it is likely that the majority of the uptake mechanisms are conserved in vertebrates.

In mammals, the acidic environs of the stomach cause Fe to dissociate from non-heme Fe complexes. The stomach secretes mucins that bind the free Fe(III), maintaining solubility throughout the intestinal tract (Powell et al., 1999a, b). In the duodenum, the mucin–Fe(III) complex traverses the mucosal covering of the epithelium. Before entry into the enterocytes, Fe (III) is reduced to Fe(II) either via an external reducing agent such as ascorbate or via an apical bound membrane ferric reductase, known as duodenal cytochrome *b* (Dcytb) (McKie et al., 2001; Latunde-Dada et al., 2002). Currently, no fish ferric reductase homologues to mammalian Dcytb have been cloned; however, its activity has been recorded (Carriquiriborde et al., 2004). Iron(II) is a substrate for slc11 a2, more commonly known as natural associated macrophage protein 2 (Nramp2) or divalent metal transporter 1 (DMT1), a metal/proton symporter (Gunshin et al., 1997). The characteristics of the uptake of Fe by freshwater fish intestine suggest that it is also taken up in the ferrous state, presumably by a similar DMT1 pathway (Carriquiriborde et al., 2004; Kwong and Niyogi, 2008; Kwong et al., 2010). On entry to the cell Fe may enter a labile intracellular Fe pool, or more

Table 4.4
Zebrafish mutants used to decipher the role of genes in human iron-related diseases

Name of mutation	Gene target	Disease	Reference
Chianti	Transferrin receptor 1a	Transferrin transporter	Wingert et al. (2004)
Frascati	Mitoferrin	Mitochondria iron importer	Shaw et al. (2006)
Chardonnay	Divalent metal transporter 1	Iron importer	Donovan et al. (2002)
Weissherbst	Ferroportin	Enterocyte Fe export	Donovan et al. (2000)
Sauternes	Aminolevulinate synthetase-2	Heme biosyntheiss	Brownlie et al. (1998)
Zinfandel	Globin locus	Disrupts globin functioning	Brownlie et al. (2003)

probably it is incorporated into the cell's iron storage molecule ferritin (e.g. Section 9.3). It is unclear precisely how Fe travels to the site of export from epithelial cells. It has been postulated that Fe-DMT1 endocytosis and translocation of the vesicle to the basolateral membrane may play a role (Ma et al., 2002). However, Fe leaves the cell via the ferrous iron exporter slc40 a1, more commonly known as ferroportin or IREG1 (Abboud and Haile, 2000; Donovan et al., 2000; McKie et al., 2000), which is coupled to the membrane bound multi-copper ferroxidase hephaestin (Vulpe et al., 1999) that facilitates the oxidation of Fe(II) to Fe(III) and transfer to transferrin. Characterization and localization of a piscine hephaestin molecule are required to confirm this export process, but homologues to another copper-containing enzyme involved in Fe metabolism, the ferroreductase ceruloplasmin (Sharp, 2004), have been identified in fish (Korzh et al., 2001; Yada et al., 2004). Coordination of Fe transfer from ferroportin to apotransferrin is an intricate process that remains to be demonstrated. The process, however, is inhibited by drugs that disrupt intracellular vesicular trafficking (Moriya and Linder, 2006), supporting the hypothesis that vesicular trafficking plays an important role in shuttling Fe from the apical to basolateral membranes (Ma et al., 2002). Systemic Fe is found bound to transferrin and it is in this form that Fe is presented to the internal organs.

The challenge is to relate the molecular functionality of the known Fe transport proteins defined by *in vitro* studies to the *in vivo* uptake process. There are two situations where it is predicted that conditions may severely hamper Fe uptake in fish, but contrary to these predictions empirical evidence shows that fish have evolved mechanisms to overcome these challenges (Roeder and Roeder, 1966; Andersen, 1997; Bury and Grosell, 2003a, b; Bury et al., 2003; Cooper et al., 2006b, 2007; Cooper and Bury, 2007). These two conditions are the alkaline intestine of most marine fish that may reduce the efficacy of the enterocyte Fe^{2+}/H^+ symporter (Kwong et al., 2010; Section 12.1) and reduced Fe^{2+} bioavailability in both marine and freshwaters limiting uptake across the gill.

8.1. Iron Uptake in Marine Fish Intestine

Marine teleost fish have an intestine that is highly alkaline (Walsh et al., 1991; Wilson and Grosell, 2003; Wilson et al., 2009). This is because to avoid dehydration in a hyperosmotic environment marine teleosts drink high volumes of seawater (Wilson and Grosell, 2003). Intestinal water uptake causes divalent ions (Ca^{2+} and Mg^{2+}) to concentrate in the lumen, and to avoid divalent metal toxicity marine fish secrete vast quantities of bicarbonate to remove Ca^{2+} and Mg^{2+} via the production of Ca^{2+} and

Mg^{2+} bicarbonate precipitates (Wilson et al., 2009; Whittamore et al., 2010). These luminal conditions pose two potential problems for Fe uptake. Firstly, the high bicarbonate may also cause the formation of Fe bicarbonates that precipitate out of solution; the formation of other metal (Cd, Zn, and Cu) carbonate precipitates has been observed in the intestine of the seawater-adapted eel (*Anguilla anguilla*), and a member of the scorpionfish family *Scorpaena* sp. (Noël-Lambot, 1981). Secondly, the increase in alkalinity due to the bicarbonate secretion would be predicted to reduce the efficacy of an $Fe(II)/H^+$ symporter.

The first point of contact between ingested Fe and the intestinal epithelium is the overlying mucous layer. A hypothesis is that the mucous layer may aid in retaining Fe solubility in the presence of elevated bicarbonate concentrations. In mammals, the Fe binding glycoprotein gastroferrin is secreted along the intestine and binds Fe(III) to maintain Fe solubility (Powell et al., 1999a, b). However, the metal binding properties of fish intestinal mucus have yet to be intensively studied, and only a few studies have considered its role in metal capture and presentation to the epithelial transport. Glover and Hogstrand (2002, 2003) showed that mucus enhances zinc (Zn) uptake at low luminal Zn concentrations in the freshwater rainbow trout. An *in vitro* study on the intestine of a marine teleost, the Gulf toadfish (*Opsanus beta*), found that the mucous layer had an equal affinity for both Fe(II) and Fe(III). However, when Fe was presented as Fe(II), uptake rates were significantly higher than when the Fe was presented as Fe(III) (Cooper et al., 2006a). Concurrent with this, another *in vitro* intestinal study on a euryhaline fish species, the European flounder (*Platichthys flesus*), also found that intestinal Fe uptakes rates were much higher when the Fe was first reduced to Fe(II) (Bury et al., 2001). In both of these studies 1 mM ascorbate was used to reduce the Fe from Fe(III) to Fe(II). This reducing milieu may have prevented Fe carbonate formation, thereby increasing Fe accumulation in the intestine. Most teleost fish are unable to synthesize ascorbic acid because they lack the enzyme that is required for the last step of ascorbic acid biosynthesis, gluconolactone oxidase (GLO) (Fracalossi et al., 1998; Moreau and Dabrowski, 1998a). This enzyme is present, however, in lower vertebrates such as lamprey (Moreau and Dabrowski, 1998b) and sturgeon (*Acipenser transmontanus*) (Moreau and Dabrowski, 2003). Therefore, the only source of ascorbate for teleosts is via the diet and the daily requirement varies among species (El Nagger and Lovell, 1991; Fracalossi et al., 1998).

The alkalinity of the marine fish intestine will inhibit proton symporter activity. To overcome this, marine fish would have to evolve either a transport process not reliant on this proton gradient and/or physiological processes that ensure a proton gradient is maintained in the boundary layer close to the transporter. The only piscine homologues to mammalian DMT1 to be fully characterized in a *Xenopus* oocyte expression system are from the freshwater rainbow trout; these were found to import Fe at similar rates at pH 5.0 and 7.5, with uptake rates only dropping off at pH 8 and 9 (e.g. Section 12) (Cooper et al., 2007). The intracellular pH of oocytes is around pH 7.4 (Kim et al., 2005), which indicates that only a slight pH gradient was required for Fe import. This observation is supported by studies of Fe uptake in isolated enterocytes from freshwater rainbow trout that show similar maximum uptake rates at pH 6.0 and 7.4, which are greater than that measured at pH 8.2 (Kwong et al., 2010). This contrasts with the mammalian DMT1, where a rise in pH from 5.1 to 7.0 reduced Fe uptake by a factor of 3 (Mackenzie et al., 2006). It is difficult to extrapolate the transport properties of a freshwater-adapted rainbow trout DMT1 to the marine fish intestine, and it will require further characterization work of a marine fish DMT1 to establish its favored pH conditions for Fe transport. However, a pH change from pH 5.5 to pH 7.0 does not alter the rate of Fe accumulation in the intestine of the marine Gulf toadfish (Cooper et al., 2006a). Evidence supports the following hypothesis – that as a result of chemical constraints, teleost fish Fe importers have evolved unique characteristics that enable them to function over a relatively wide range of pH values.

8.2. Iron Uptake Across the Gill

Fish can sequester Fe from the water column, and similarly to the situation in the gut, Fe(II) is more bioavailable than Fe(III) at the gills (Roeder and Roeder, 1966; Bury and Grosell, 2003a, b; Bury et al., 2003; Cooper et al., 2006b, 2007; Cooper and Bury, 2007). Section 1 describes aquatic Fe speciation where Fe typically exists in the aqueous environment predominantly as either Fe oxide or organic ligand complex (Fig. 4.1). But there is also evidence that photochemical cycling of Fe produces a small pool of readily bioavailable Fe(II) concentrations (Fukushima and Tatsumi, 1999; Barbeau et al., 2001). Currently, there is no known mechanism by which fish can access the large pool of ferric oxide, and it must be assumed that they are able to acquire Fe from the other two pools [Fe(II) or organic–Fe complex].

In freshwater, humic acids complex metals, thereby reducing availability and protecting fish against elevated metal concentrations (e.g. Richards and Playle, 1998). The opposite appears to hold for Fe and in experiments where Fe is complexed to humic acids, Fe was shown to be readily available for uptake by the freshwater fish gill (Cooper and Bury, 2007). In contrast, other chelating agents (e.g. citric acid, nitrilotriacetic acid, and the synthetic

Fe chelator desferrioxiamne) with higher binding affinities for Fe prevented Fe accumulation into the body (Cooper and Bury, 2007). The uptake rate of Fe bound to HA did not differ between fish kept in the dark or light, suggesting that photochemical cycling of Fe(III) was not responsible for the uptake and that the fish gill is able to access Fe bound to organic compounds with a relatively low affinity for Fe (Cooper and Bury, 2007). However, there is a caveat, as currently this is the only study to assess the possibility of LMCT-derived Fe(II) uptake in fish.

Iron uptake across the gill is considerably less than the uptake across the intestine, which suggests that the latter is the predominant route of Fe uptake (Cooper et al., 2006b). In zebrafish (*Danio rerio*), higher rates of Fe uptake across the intestine, when compared to the gill, also correlated with higher transcript (mRNA) concentrations of DMT1 and ferroportin (Cooper et al., 2006b). Although a change in mRNA concentration does not necessarily mean a change in the respective protein expression, it appears that the gill and intestine may have different roles in Fe homeostasis. During periods of low dietary Fe intake the role of the gill to maintain Fe homeostasis becomes more significant. Indeed, Cooper and Bury (2007), inferred that rainbow trout could potentially acquire approximately 85% of their daily recommended intake of Fe across the gill. Therefore, even if for the majority of time the gills play a supplementary role in whole-body Fe homeostasis, they are capable of obtaining almost all of the required Fe intake from the water if necessary.

It is becoming evident that other epithelia are capable of importing metals. For example; cadmium (Cd) uptake in fish can occur via the olfactory rosette (Gottofrey and Tjälve, 1991; Sloman et al., 2003), and mercury (Hg), manganese (Mn) and nickel (Ni) have also been shown to accumulate in brain tissue via the olfactory system of mammals (reviewed by Tjälve and Henrikkson, 1999). The stomach has also been shown to be a site of metal uptake (e.g. Ojo and Wood, 2007). It seems reasonable that olfactory epithelia and stomach may be capable of importing Fe, but to date there are no examples in the literature and therefore these possibilities warrant future research.

9. CHARACTERIZATION OF INTERNAL HANDLING

Iron circulates in blood bound to transferrin. There is a substantial body of literature available on mammalian Fe internal handling (see review by Garrick and Garrick, 2009). Because the majority of Fe's biological uses are conserved between and within phyla, the mechanism of internal Fe handling in mammals is more than likely to be replicated in fish. The following section describing the molecular characterizations of the proteins involved in internal handling and storage demonstrates the close link between the piscine proteins and mammalian counterparts. Indeed, in recent years the study of the genes involved in Fe handling by zebrafish (Table 4.4) has provided the medical field with insights into Fe-related human genetic diseases (Lumsden et al., 2007), the mechanisms of hepcidin regulation of vertebrate Fe homeostasis (Fraenkel et al., 2009), transferrin receptor functioning (Wingert et al., 2004), the mechanisms of Fe uptake and export (Donovan et al., 2000, 2002), and heme biosynthesis (Nilsson et al., 2009).

9.1. Transferrin and Transferrin Receptor

Transferrin (tf) is a glycoprotein of approximately 690 amino acids that binds Fe(III) very efficiently and transports Fe via the blood to internal organs (Fig. 4.4). It is primarily synthesized in the liver, but is also produced

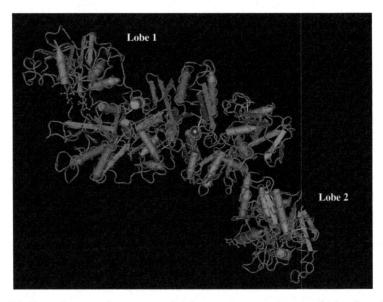

Fig. 4.4. Three-dimensional structure of iron transport protein, transferrin. Transferrin reversibly binds iron in each of two lobes (B1 and B2) and delivers it to cells by a receptor-mediated, pH-dependent process. The binding and release of iron result in a large conformational change in which two subdomains in each lobe close or open with a rigid twisting motion around a hinge (Wally et al., 2006). Cylinders and arrows represent strands and helices, respectively. Note that the arrows on the strands and helix cylinders always point in the N-to-C direction. Protein structures MMDB ID 40108 (Wally et al., 2006).

in the brain, testes, ovary, kidney, and spleen (Bowman et al., 1988; Neves et al., 2009; Liu et al., 2010). Transferrin is comprised of two globular domains, each of which has an Fe(III) binding site (Lambert et al., 2005). Homologues to the human transferrin have been found in a large number of fish groups (Lee et al., 1998; Yang and Gui, 2004; Scudiero et al., 2007). Most circulating Fe is bound to transferrin, and only under severe disruption to Fe homeostasis is it detected in the serum as non-transferrin bound iron (NTBI) (Gosriwatana et al., 1999). However, to the authors' knowledge, the presence of NTBI has not been observed in fish. Uptake of the Fe–transferrin complex (Fe-tf) from the serum to the organs occurs via the transferrin receptor (tfr) (Fig. 4.3C). In zebrafish there are two tfr isoforms, tfr1, that has two paralogues, tfr1a and tfr1b – as a consequence of the whole genome duplication event that has occurred in the teleost lineage (Jaillon et al., 2004), and tfr2 (Wingert et al., 2004). tfr1a is highly expressed in differentiating erthryocytes facilitating the delivery of Fe for hemoglobin synthesis, whereas tfr1b is more widely distributed, probably enabling the delivery of Fe for heme synthesis necessary in all cells (Wingert et al., 2004). Expression of tfr2 is located to hepatocytes and precursor erythroid cells (Wingert et al., 2004) and has a lower affinity for Fe-Tf (Kawabata et al., 1999).

Binding of the Fe-tf to the tfr induces clathrin-mediated endocytosis (Fig. 4.3C) (Andrews and Schmidt, 2007). A V-type ATPase on the endosome membrane inwardly pumps protons to acidify the vesicle, thereby ensuring the release of Fe(III) from transferrin. In mouse models Steap 3 acts as a ferrireductase in the endosome (Sendamarai et al., 2008) converting Fe(III) to Fe(II), and the ferrous Fe is exported from the endosome by DMT1, where it either enters the labile Fe pool, is bound up to ferritin or is transported into the mitochondria for heme synthesis. Genbank searches for Steap3 homologues in fish species show proteins with high sequence similarity in *Tetraodon negrividis* (N. Bury, personal observation). The endosome is recycled, thus restoring the apo-tf/tfr complex back to the membrane, and the apo-tf is released owing to changes in pH (Wingert et al., 2004).

9.2. Mitochondrial Heme and Fe/S Cluster Protein Synthesis

In all cells Fe is transferred to the mitochondria where it is required for heme and Fe/S cluster protein synthesis, and in erythroid tissues this leads to the synthesis of hemoglobin (Fig. 4.3C). In teleosts, hemoglobin synthesis is generally located in the reticuloendothelial cells in the stroma of the spleen and within the kidney (Agius and Roberts, 2003), but synthesis has also been reported in other tissues (Macchi et al., 1992). The transport of the Fe(II) across the mitochondria inner membrane occurs via mitoferrin, identified in the zebrafish mutant *frascati*, whose phenotype is impaired heme synthesis (Shaw et al., 2006). Further mammalian studies show that a mitochondrial membrane ATP-binding cassette transporter, Abcb10, physically interacts with mitoferrin to facilitate Fe(II) import into the mitochondria of developing erythrocytes (Chen et al., 2009). Iron(II) is delivered to the enzyme ferrochelatase, where it catalyzes the incorporation of Fe into protoporphyrin XI, leading to the synthesis of heme (Andrews, 2009). The export of heme from the cytoplasm occurs via the feline leukemia virus subgroup c receptor (FLVCR) (Quigley et al., 2004; Keel et al., 2008).

9.3. Ferritin

The major intracellular Fe storage protein is the 450 kDa ferritin molecule, which keeps Fe in a soluble bioavailable non-toxic form in the cytoplasm (Fig. 4.5). It consists of 24 subunits that fold as four helix bundles to form a spherical shell within which up to 4500 atoms of Fe can be stored (Granier et al., 2003; Arosio and Levi, 2010). In mammals two proteins

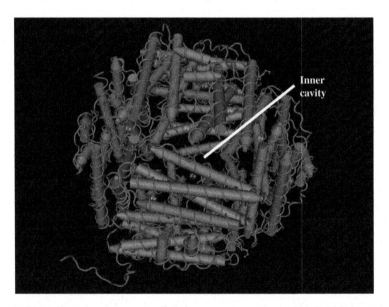

Fig. 4.5. Three-dimensional structures of the iron storage protein, ferritin. Ferritin is a polymer made of 24 subunits; each of them (15–20 kDa) consists of a four α-helix bundle to which a fifth small α-helix is attached at the C-terminal end. The 24 subunits assemble into a spherical shape, with an inner cavity. The cavity can host up to 4500 iron ions stored as hydrous ferric oxide–phosphate (Granier et al. 2003). Protein structures taken from MMDB ID: 79758 (Weeratunga et al., 2010).

make up the ferritin molecule, heavy (H-chain) and light (L-chain) chain. Iron enters the ferritin molecule as Fe^{2+} via small hydrophobic and hydrophilic channels. A key feature that distinguishes the H-chain is a di-Fe binding site in the fourth helix that also interacts with oxygen and is responsible for the ferrioxidase activity of the protein (Guo et al., 1998). In contrast, the L-chain is ferrioxidase inactive, but owing to a set of negatively charged amino acids it promotes nucleation of Fe micelles formation within the ferritin molecule (Juan and Aust, 1998). The release of Fe from the molecule occurs through the same channels as it entered following reduction to Fe^{2+} within the sphere (Huang et al., 2004); the physiological reducing agent has yet to be unequivocally identified.

The amino acid sequences of the H- and L-chains are highly conserved in mammals and homologues to H-chain have been identified in a number of fish species (e.g. Scudiero et al. 2008). However, there is little sequence information for the presence of L-chain subunits in fish and the only data available are from rainbow trout (Miguel et al., 1991). In lower vertebrates there is a third subunit, termed middle chain (M-chain) based on its electrophoretic mobility (Dickey et al., 1987). The M-chain possesses residues that have both ferrioxidase activity and the nucleation micelle formation sites (Giorgi et al., 2008). M-chain ferritin subunits have been identified in lamprey (*Lampetra fluviatilius*) (Andersen et al., 1998) and Atlantic salmon (*Salmo salar*) (Andersen et al., 1995), and Scudiero et al. (2007) identified an H-chain and two M-chain subunits in the zebrafish genome. In the Antarctic fish species *Trematomus bernacchii* (Mignogna et al., 2002) and *T. newnesi* (Giorgi et al., 2008) the ferritin of the spleen exists as a homopolymer of M-chains, while the liver ferritin consists of both H- and M-chains.

9.4. Immune Response

Iron plays an important role in the defense against bacterial infection: Vidal et al. (1993) demonstrated that transcripts of a member of the slc11 a1 family termed natural resistance-associated macrophage protein (Nramp) were detected only in the reticuloendothelial organs (spleen and liver) of mice. Nramp was also highly expressed in purified macrophages and macrophage cell lines, and was upregulated following infection with intracellular parasites (Vidal et al., 1993; Forbes and Gros, 2001). These findings have subsequently been confirmed in fish species. For example, the expression of Nramp occurs in late endosomes of pufferfish (*Takifugu rubripes*) (Sibthorpe et al., 2004) and channel catfish (*Ictalurus punctatus*), Nramp transcript levels increase after the fish and catfish monocyte cells (42TA) are exposed to lipopolysaccharides (LPS) (Chen et al., 2002), and channel catfish infected with the bacterium *Edwardsiella ictaluri* show increased Nramp mRNA expression 48 h postinfection (Elibol-Flemming et al., 2009). Similar studies on red sea bream (*Pagrus major*) (Chen et al., 2004) and striped bass (*Morone saxatilus*) (Burge et al., 2004), among others, show increased transcript levels of the respective Nramp genes following a bacterial challenge, demonstrating the significance of these proteins in fish for defense against pathogens.

Exactly how Nramp assists in pathogen mortality is still debated (Techau et al., 2007). However, the appearance of Nramp in late endosome membranes following endocytosis of the pathogen (Sibthorpe et al., 2004) suggests that it is likely to be involved in regulating the supply of Fe to the pathogen. Mammalian Nramps involved in pathogen defense are $Fe(II)/H^+$ antiporters (e.g. Goswami et al., 2001; Techau et al., 2007); in contrast, the only fish Nramps to have been characterized suggest they are $Fe(II)/H^+$ symporters (Cooper et al., 2007; Techau et al., 2007; see Section 12). The antibacterial properties of an Nramp transporter will depend on the direction of the proton gradient between the endosome and the cytoplasm. In the case of fish, if the gradient favors Fe import (e.g. cytoplasm pH < pH of the endosome), then the accumulation of Fe in the phagosome will stimulate Fe-catalyzed Haber–Weiss or Fenton reactions and hydroxyl radical-mediated inhibition of bacterial growth (Zwilling et al., 1999). However, it is more likely that proton pump activity on the endosome membrane will decrease internal pH and stimulate Fe(II) export, depleting the endosomal compartment of this essential metal. In this case, antimicrobial activity is due to Fe starvation (Fig. 4.3D) (Gomes and Appelberg, 1998; Techau et al., 2007). Future studies are required to elucidate which hypothesis is correct in fish.

Macrophages also play a significant role in recycling Fe from senescent red blood cells (RBCs) (Fig. 4.3D). The RBCs are endocytosed and protons are pumped into the endosome, decreasing pH and causing lysis of the RBC. Iron is released from heme due to the activities of heme-oxygenase 1 (Poss and Tonegawa, 1997). In fish $Fe(II)/H^+$ symporter activity suggests that acidification of the endosome is important for Fe release (Fig. 4.3D). In mammalian models both Nramp1 and DMT1 are implicated in endosomal Fe export (Gruenheid et al., 1997; Soe-Lin et al., 2010). The metal then either enters the Fe labile pool or is stored in ferritin and finally released via ferroportin to the systemic Fe pool bound to transferrin (Fig. 4.3D) (Knutson et al., 2003). The majority of Fe required for erythropoiesis is predominantly met by RBC destruction and Fe recycling (Cavill, 2002). It is within the regions of erythropoiesis in fish that structures termed melano-macrophages appear; these are the sites of macrophage congregations and recycling of Fe (Agius and Roberts, 2003; Leknes, 2007).

10. CHARACTERIZATION OF EXCRETION ROUTES

Although Fe can exit the epithelial cell via ferroportin 1, there appears to be no regulated mechanism for Fe excretion from the body (Latunde-Dada et al., 2002). It is a widely held view that the control of whole-body Fe status is mainly dependent on tight regulation of Fe uptake from the diet and water. Iron recycling inside the body is efficient and only small amounts of Fe are eliminated via the liver (i.e. biliary routes) (LeSage et al., 1986) and to some extent the kidney (Ferguson et al., 2003). For example, rainbow trout injected with ^{59}Fe do not lose Fe via their feces or urine over a 30 day period, and all Fe is either stored in the liver or used for RBC synthesis (Walker and Fromm, 1976). However, in mammals 12–14 μg Fe kg^{-1} day^{-1} is lost through the sloughing of gastrointestinal epithelial cells (Cook, 1990). A similar sloughing process may also occur in the intestine of fish. Accumulation of Fe in the intestinal epithelia of Gulf toadfish is biphasic; after an initial phase of accumulation of Fe on to the epithelia over approximately 120 min, accumulation of Fe slows down and remains constant for a further 60 min. During this period lumen Fe concentrations increase, after which Fe continues to accumulate (Cooper et al., 2006). The period in which Fe accumulation slows corresponds to a phase of Fe sloughing.

11. BEHAVIORAL EFFECTS OF IRON

Evidence for the effects of Fe on behavior of fish is sparse. In one of a few studies, Updegraff and Sykora (1976) assessed avoidance of lime-neutralized ferrous sulfate by Coho salmon (*Oncorhynchus kisutch*) fry in a simulation of waters receiving treated acid-mine drainage water. Using an experimental trough equipped with inflows at both ends and a centralized outflow, the authors were able to generate discrete zones receiving either clean water or Fe(III) hydroxide precipitates. With increasing Fe concentration (0.75–6.45 mg Fe L^{-1}), actively swimming salmon clearly avoided Fe-enriched water. Further trials with fry acclimated to ferric hydroxide precipitates for several months elicited similar behavior, suggesting that chemoresponsiveness in coho salmon fry is unaffected at these concentrations, and that physiological acclimation to environments with high concentrations of Fe precipitates appears unlikely.

While no studies are available for fish, the wider vertebrate literature have demonstrated links between Fe status, brain functioning, and behavior. Much of this focus has been on human Fe deficiency (e.g. Shafir et al., 2008);

these effects may be relevant to fish biology given the conserved biochemical pathways in brains of vertebrates. In the brain, Fe has essential roles in myelination (Badaracco et al., 2010) and as a cofactor for enzymes involved in monoaminergic neurotransmitter activity (Goodwill et al., 1998). Iron deficiency, especially during early development, has been shown to alter monoamine profiles in the brain with corresponding effects on cognitive function, locomotion and associated behaviors in mammals (Lubach and Coe, 2008; Coe et al., 2009). Since changes in brain monoamine concentrations underpin similar complex behaviors in fish, e.g. social status (Winberg et al., 1997) and avoidance (Hoglund et al., 2005), Fe deficiency may also elicit similar responses in fish as in mammals and this could be an important area of further research, especially for aquaculture.

12. MOLECULAR CHARACTERIZATION OF EPITHELIAL IRON TRANSPORTERS AND HEPCIDIN

12.1. Molecular Characteristics of Fish Epithelial Iron Transporters

Divalent metal transport 1 was discovered via *Xenopus* oocyte expressional cloning from rat DNA (Gunshin et al., 1997) and positional cloning to identify genes responsible for microcytic anemic mice (Fleming et al., 1997). It is predicted to have 12 transmembrane domains (TM) with a consensus Fe transport motif between TM8 and TM9 (Gunshin et al., 1997) (Fig. 4.6). *Xenopus* expression studies demonstrated that DMT acts as a Fe^{2+}/H^+ symporter, and that inward current could also be generated in the presence of Zn, Cd, Mn, Cu, and Co, suggesting a much wider divalent metal substrate range (see Section 13).

To date, four variants of DMT1 have been discovered in mammals, the difference between them being the presence or the absence of an iron response element (IRE) in the 3′ or 5′ untranslated region (UTR) (Tchernitchko et al., 2002; Tabuchi et al., 2002; Mims and Prchal, 2005). Similarly, the two isoforms of the DMT1 homologue identified in the pufferfish (*Takifugu rubripes*) also differ in one possessing a 3′ UTR IRE and the other not (Sibthorpe et al., 2004). IREs are *cis*-regulatory mRNA motifs whose consensus sequence is CNNNNNCAGTG (Saeij et al., 1999). The interaction of the cytosolic iron regulatory protein (IRP) with the IRE is dependent on the Fe status of the cell. If the labile Fe concentrations are high, then 4Fe–4S cluster formations assemble in the IRP, preventing IRP binding to the IRE (Wingert et al., 2005). In Fe-deficient conditions, the IRP/IRE complex is formed (Muckenthaler et al., 2008). IRP/IRE formation in the gene's 5′ region inhibits translation, whereas binding to

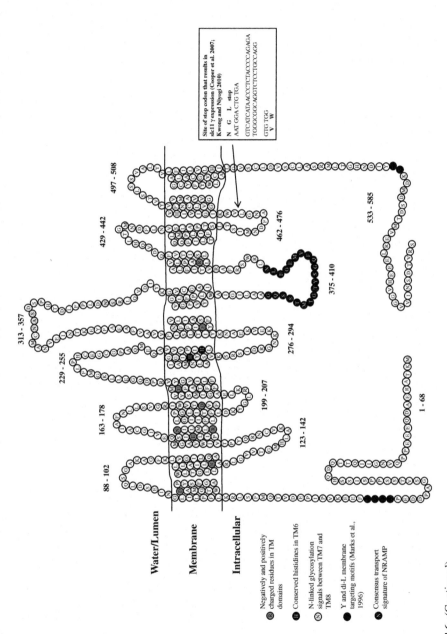

Fig. 4.6. (Continued)

the 3′ UTR stabilizes transcript formation (Kato et al., 2007). It is the IRP/ IRE complexes along with hepcidin (see Section 12.2) that control cellular Fe metabolism and this mechanism of Fe homeostatic control appears conserved in vertebrates, with IREs and IRP proteins being present in the lamprey (*Lampetra fluviatilis*) (Andersen et al., 1998).

Molecular evidence suggests that the slc11 family of proteins (e.g. Nramp1 and Nramp2/DMT1) is highly conserved and has been found in a number of teleost species, including rainbow trout (Dorschner and Phillips, 1999; Cooper et al., 2007), common carp (*Cyprinus carpio*) (Saeij et al., 1999), channel catfish (Chen et al., 2002), zebrafish (Donovan et al., 2002; Cooper and Bury, 2007) and pufferfish (Sibthorpe et al., 2004). Phylogenetic analysis shows that these teleost slc11 share a greater homology to mammalian Nramp2/DMT1 than Nramp1 (Techau et al., 2007). To date, however, only a few studies have linked the function of teleost slc11 with Fe uptake. Cooper et al. (2007) cloned two cDNA isoforms of scl11 from rainbow trout gills, known as slc11β and slc11γ, and both isoforms have recently been shown to be expressed along the entire intestinal tract of freshwater rainbow trout (Kwong et al., 2010). The difference between the two clones was that slc11γ isoform contained a 52 bp insert in the exon between TM10 and TM11 that contained a stop codon resulting in the transcribed protein lacking the last two TM domains (Fig. 4.6). Expression of the two isoforms in *Xenopus* oocytes showed that they act as ferrous Fe importers (Cooper et al., 2007). However, the rainbow trout slc11 differed from its mammalian counterparts. Mammalian DMT1 may function at pH 7.4, but the Fe transport capacity is greatly reduced (Mackenzie et al., 2006). In contrast, the pH range for maximum Fe transport in the rainbow trout was from 5 to 7.5. Techau et al. (2007) took a different approach and developed a yeast complementation assay to determine slc11 family

---

Fig. 4.6 (Continued)

Schematic representation of the iron import protein solute carrier 11, also known as natural resistant-associated macrophage protein (Nramp) or divalent metal transporter 1 (DMT). The amino acid sequence represents that which encodes for rainbow trout Nramp α (Dorschner and Phillips, 1999), and the figure is based on one produced by Lam-Yuk-Tseng et al. (2003). The 12 transmembrane domains are predicted from hydropathy profiling, (www.ch.embnet.org/ software/TMPRED) and the 13 predicted intracellular and extracellular regions are identified via their amino acid numbering. The amino acids that define the negatively and positively charged residues within predicted TM domains, the conserved histidine (H) residues in TM6, asparagine (N) linked glycosylation signals in the TM7–TM8 extracytoplasmic loop, the predicted membrane targeting motifs and consensus transport signature common to slc11 (Nramp/DMT) orthologues are represented on the figure. The arrow indicates the site of the 52 base pair insertion (inset box) in the TM10–TM11 intracytoplasmic loop that possesses a stop codon that results in the expression of a truncated slc11 protein termed slc11 γ (Cooper et al., 2007; Kwong et al., 2010).

transporting activity. The results corroborated those for rainbow trout scl11 isoforms and also showed that the pufferfish slc11 are Fe^{2+}/H^+ symporters.

The cellular Fe exporter ferroportin was simultaneously identified in the mouse and via positional cloning in anemic zebrafish (Abboud and Haile, 2000; Donovan et al., 2000; McKie et al., 2000). It has been localized to the basolateral membrane of duodenal enterocytes (Abboud and Haile, 2000; McKie et al., 2000), and was initially predicted to possess 10 TM (Donovan et al., 2000). However, this prediction is controversial because subsequent topological analysis and tagging of putative intermembrane regions suggest a 12 TM structure (Liu et al., 2005) (Fig. 4.7). The identification of the evolutionarily conserved hepcidin binding region also supports the 12 TM structure (De Dominico et al., 2008), because hepcidin is known to bind ferroportin extracellularly and in the original predicted zebrafish ferroportin topology this region was located intracellularly. The Fe transporting site of ferroportin has yet to be identified, but there are several mutations in the intracellular loop linking TM4 and TM5 that cause Fe transport activity to be severely affected (Lee and Beutler, 2009) (Fig. 4.7).

12.2. Hepcidin

The main regulator of Fe absorption and distribution throughout the organs of the body is the peptide hepcidin (Nemeth and Ganz, 2009), originally known as liver-expressed antimicrobial peptide (LEAP-1) because of its role in controlling pathogenicity (Shi and Camus, 2006). The hepcidin gene encodes an 84–amino acid prepropeptide, which is cleaved to form a 20–26 amino acid active peptide that is characterized by eight cysteines that form disulfide bridges, creating a hairpin structure (Nemeth and Ganz, 2009). The main site of synthesis is the major Fe regulatory organ, the liver, but localized expression in other tissues is also observed, especially in fish (Martin-Antonio et al., 2009). The N-terminal region possesses a metal binding sequence ATCUN, which has been shown to interact with Cu(II) and Ni(II) and is important for the Fe regulatory properties of the molecule (Melino et al., 2005).

The transcription of hepcidin in hepatocytes is regulated by a variety of stimuli, including cytokines (tumor necrosis factor-α, interleukin-6), erythropoietic activity, Fe stores, and hypoxia (De Domenico et al., 2007). Hepcidin controls Fe levels by regulating the absorption of dietary Fe from the intestine, the release of recycled Fe from macrophages, and the movement of stored Fe from hepatocytes (Nicolas et al., 2001; Nemeth et al., 2004). Hepcidin can directly interact post-translationally with ferroportin (Fig. 4.7) and transferrin receptor 2 (TfR2). Nemeth et al. (2004) reported hepcidin binding to ferroportin 1 causing internalization

232

NICOLAS R. BURY *ET AL.*

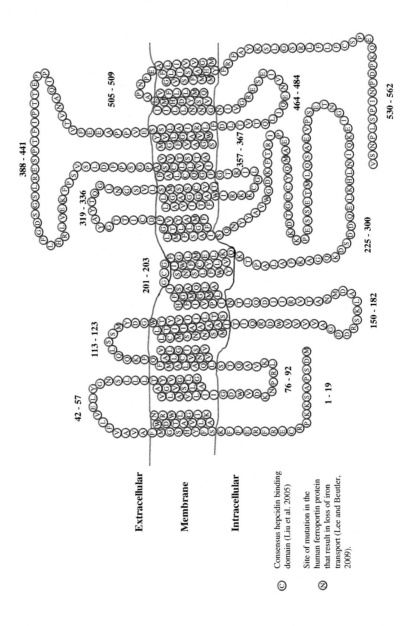

Fig. 4.7. Schematic representation of solute carrier slc40a1 also known as ferroportin or IREG1. The amino acid sequence represents that which encodes for the zebrafish ferroportin (Donovan et al., 2000). The 12 transmembrane domains and predicted intracellular and extracellular regions are identified via their amino acid numbering based on the membrane topology derived by Liu et al. (2005). The extracellular conserved hepcidin binding site is shown, as well as those amino acids that are conserved between zebrafish and human ferroportin protein and if mutated reduce iron transport

and degradation and resulting in a decrease in Fe export from the cell. The hepcidin binding site is predicted to be located between TM6 and TM7 of ferroportin in zebrafish (Fig. 4.7) and is highly conserved within the vertebrates (De Domenico et al., 2008).

Wallace et al. (2005), in mice, and Fraenkl et al. (2009), in zebrafish, showed that expression of hepcidin requires correct functioning of tfr2. This suggests a feedback loop whereby increased circulatory Fe is taken up by hepatocytes via tfr2, which in turn stimulates hepcidin synthesis and release, resulting in a reduction in ferroportin activity. This causes an increase in intracellular Fe within the enterocytes, hepatocytes, macrophages, and presumably gill cells, although this has not been ascertained. The antimicrobial properties of ferroportin are related to the decrease in Fe export from cells as a result of decreased transporter activity, which limits the circulatory concentrations of Fe available to invading pathogens.

The dual functionality of hepcidin as a regulator of Fe uptake and antimicrobial activity was demonstrated in sea bass (*Dicentrarchus labrax*) (Rodrigues et al., 2006). However, the control of these processes may be more complicated than the mammalian models suggest owing to a multitude of hepcidin isoforms that have been identified in a number of fish species: five in the winter flounder (*Pseudopleuronectes americanus*) (Douglas et al., 2003), seven in the black porgy (*Acanthopagrus schlegelii*) (Yang et al., 2007), and four in the redbanded seabream (*Pagrus auriga*) (Martin-Antonio et al., 2009), all showing sequence diversity (Padhi and Verghese, 2007). In the redbanded seabream differential expression occurs and hepcidin 1 (termed HAMP1 in their study) is ubiquitously expressed; HAMP2 in the kidney, spleen, and intestine, and HAMP3 and 4 in the liver (Martin-Antonio et al., 2009). LPS treatment induced gene expression of all four HAMP isoforms, but there were significant differences in the temporal pattern of expression for each HAMP isoform during the LPS treatment (Martin-Antonio et al., 2009).

In the Antarctic notothenoid fish there are also four hepcidin variants; within this group there are two distinct types. Type I is similar to those of mammals, whereas type II possesses only four cysteine residues (Xu et al., 2008) (Fig. 4.8). The type II hepcidin was not found in extant members of the closest teleost group to the Notothenoid fishes, the Antarctic eelpouts *Lycodichthys dearborni*, but a structurally distinct four-cysteine hepcidin was isolated, suggesting that this type of hepcidin is positively selected for in fish occupying freezing waters (Xu et al., 2008). Xu et al. (2008) further postulate that the diversity of the hepcidin found in the teleosts is in response to the diverse challenges to Fe metabolism and infection posed by the aquatic environment.

Human	DTHFPI	C	IF	CC	G	CC	HRSK	C	GM	CC	KT
Mouse	DTHFPI	C	IF	CC	G	CC	NNSQ	C	GM	CC	KT
Zebrafish	QSHLSL	C	RF	CC	K	CC	RNKG	C	GY	CC	KF
Catfish	QSHLSL	C	RY	CC	N	CC	KNKG	C	GF	CC	RF
Redbanded Seabream	QSHISM	C	YW	CC	N	CC	RANKG	C	GY	CC	KF
Winter Flounder	HISHISL	C	RW	CC	N	CC	KANKG	C	GF	CC	KF
Salmon	QSHLSL	C	RW	CC	N	CC	HNKG	C	GF	CC	KF
Dissostichus	RRRK	C	KF	CC	N	CC	SNI	C	QT	CC	TRRF (Type I)
mawsoni	GIK	C	RFR	C	RRGV			C	GLY	C	KKRFG (Type II)

Fig. 4.8. Alignment of representatives of fish, human and mouse hepcidin peptides. All sequences possess eight conserved cysteine (bold) and glycine (underlined) residues. The exception are the Antarctic notothenioids, represented by *Dissostichus mawsoni*, which possess two isoforms, one with eight cysteines and a glutamine replacing the glycine, and the other containing four cysteines (Xu et al., 2008). In addition, fish species may possess multiple hepcidin isoforms (e.g. Douglas et al., 2003; Yang et al., 2007; Xu et al., 2008; Martin-Antonio et al., 2009).

13. GENOMIC AND PROTEOMIC STUDIES

To the authors' knowledge there are no genomic and proteomic studies that specifically explore fish tissue responses of fish exposed to or deprived of Fe. However, several omic studies have identified alterations in the expression of Fe homeostatic genes and proteins in the liver of zebrafish exposed to arsenic (Lam et al., 2006) and brominated flame retardants (Kling et al., 2008), and Atlantic cod (*Gadus morhua*) immune tissues exposed to formalin-killed, atypical *Aeromonas salmonicida* (Feng et al., 2009).

14. INTERACTIONS WITH OTHER METALS

Promiscuous metal uptake via DMT1 could have detrimental effects on both essential metal homeostasis and metal toxicity (Cooper et al., 2007). Gunshin et al. (1997) were the first to conclude that DMT1 had a broad substrate range that included Fe, Zn, Mn, Co, Cd, Cu, Ni, and Pb, as these metals evoked an inward current, although actual uptake of these metals was not measured. Conflicting data exist (post Gunshin et al., 1997) regarding which other divalent metals, apart from Fe, can use DMT1 for cellular import. Some studies have shown that only Ni (Tallkvist et al., 2003), Cd (Bannon et al., 2003), or Cu (Arredondo et al., 2003) competed with Fe for uptake. One study showed that Cd, not Pb, competed with Fe for uptake via DMT1 (Elisma and Jumarie, 2001), whereas another study demonstrated that Pb did interfere with Fe uptake (Smith et al., 2002). These contrasting results reflect the expression system used, or the species from which the DMT1 clone was isolated. What is evident is that potentially non-essential metals can access the cell via DMT1. For example, DMT1 has been linked with Cd uptake in many studies (Bressler et al., 2004). High rates of Cd were imported by *Xenopus* oocytes expressing human DMT1 (Okubo et al., 2003) and scallop (*Mizuhopecten yessoensis*) DMT1 (Toyohara et al., 2005). Furthermore, a dietary study on rats found that those fed an Fe-deficient diet upregulated DMT1 in the duodenum, which coincided with an increase in radiolabelled Cd uptake (Ryu et al., 2004).

The piscine literature presents a similar story. Kwong and Niyogi (2009) found that all of the divalent metals except Co inhibited intestinal Fe absorption in rainbow trout, and the magnitude of inhibition followed the order of: Ni(II)~Pb(II)>Cd(II)~Cu(II)>Zn(II). More recently, this has been confirmed in isolated rainbow trout enterocytes, where Fe^{2+} uptake is severely inhibited by Pb and to a lesser extent by Cd (Kwong et al., 2010). Furthermore, Nadella et al. (2007) observed the inhibition of Cu uptake in rainbow trout gut sacs by Fe suggesting that a part of dietary Cu uptake is via DMT 1. Molecular characterization of rainbow trout slc11βand γ also shows that Cd, Mn, Zn, Pb, and Co all significantly inhibited Fe uptake in *Xenopus* oocytes expressing these transport proteins (Cooper et al., 2007). However, inhibition of Fe uptake does not necessarily mean that the other metal is transported by the protein expressed, and Cooper et al. (2007) found that only *Xenopus* oocytes expressing slc11β accumulated radiolabeled Cd. It will be important to assess whether there are other divalent metal uptake differences between the two isoforms. Both isoforms are expressed throughout the intestinal tract of rainbow trout (Kwong et al., 2010), but it will also be of interest to determine whether these isoforms are differentially expressed in other tissues.

The promiscuity of DMT1 suggests that there may be an increase in risks to non-essential divalent metals in circumstances where fish are depleted of Fe. In a study by Cooper et al. (2006a), zebrafish fed a diet low in Fe transported approximately 15% more Cd across the intestine into the body and liver than those fed the normal Fe diet, which correlated with an increase in DMT1 expression in both the gut and gill (Cooper et al., 2006a). These data imply that in an *in vivo* system, non-essential metals can use DMT1 to enter the cell and if DMT1 is upregulated, in a response to maintain Fe homeostasis, these metals could accumulate and cause acute or chronic metal toxicity. More recently, a study has also shown that Cd can increase ferroportin 1 gene expression in macrophage cells, indicating that there is a far more complex interaction between Fe and non-essential metal homeostasis (Park and Chung, 2009).

The interaction between Cu and Fe uptake is exemplified by the Cu-rich ferroxidase hephaestin and ceruloplasmin essential for the oxidation of Fe (II), following transport from the cell by ferroportin (Vulpe et al., 1999) and mobilization of Fe from storage tissues (Sharp, 2004), respectively. A reduction in copper concentrations leads to anemia in mammals (Sharp, 2004). In fish the interaction between these two essential metals and gene expression of Cu transporters has further been studied by Craig et al. (2009). A high Fe diet increases the expression of the copper importer, copper transporter 1 (ctr1), as well as the copper exporter ATP7a in the gills and gut. These responses would contribute to increase Cu loading in these tissues and presumably ensure that sufficient Cu is present for Cu ferroxidase synthesis to assist in the regulation of Fe homeostasis.

15. KNOWLEDGE GAPS AND FUTURE DIRECTIONS

Iron is essential for life and, therefore, a vast amount of research has already been conducted to determine aquatic speciation, uptake mechanisms, homeostasis and cellular function. However knowledge gaps exist. From an Fe-uptake perspective there have been relatively few *in vivo* mechanistic studies on fish Fe acquisition from water, but we know it occurs and can make a significant contribution to whole-body Fe acquisition (Roeder and Roeder, 1966; Bury and Grosell, 2003a; Cooper and Bury, 2007; Cooper et al., 2007). Iron speciation in water is complex and a key challenge is to understand what forms of aquatic Fe fish can acquire. Are fish capable of secreting siderophore-like compounds to capture Fe, or does the mucus covering the gill possess properties that facilitate the capture of Fe–ligand complexes? What is the significance of Fe(II) derived from ligand metal charge transfer for fish larvae that are a component of the mesoplankton?

Given that Fe(III) dominates in oxygenated circumneutral waters, knowledge is also required on the mechanism for ferric reductase at the gill. Homologues to the mammalian DcytB have yet to be identified in fish, and other extrinsic reducing factors may be present; preliminary studies suggest the presence of ascorbate in the mucus of fish (Cooper and Bury, unpublished data). Similar questions are also pertinent to Fe uptake at the intestine, including whether the stomach may also play a role in Fe acquisition, similar to that observed for other metals in fish (Ojo and Wood, 2007). More specifically to the intestine, identifying the compound within the mucus of marine fish that enables Fe and other metals to remain in solution amidst the elevated bicarbonate, and the methods by which enterocytes access this source of metal need further study. The identification of other transport routes such as heme–Fe import is required. In this regard, the presence of a ferric uptake pathway facilitated by a protein termed mobilferrin (Simovich et al., 2003) may have gone out of favor in the model of mammalian Fe acquisition, but has never been completely refuted (Garrick and Garrick, 2009).

Study of a number of Fe-related diseases in zebrafish has rapidly increased our knowledge of internal Fe metabolism and homeostatic control (Table 4.4). Therefore, further medical breakthroughs are possible using the zebrafish as a vertebrate model. Despite conserved homeostatic mechanisms among vertebrates, some fish species (at present an estimated 24,000 or more species) may have evolved unique mechanisms to meet unusual environmental pressures. Understanding the reason why a number of fish species have retained multiple hepcidin isoforms will help us to understand the original role of this peptide (Douglas et al., 2003; Padhi and Verghese, 2007; Xu et al., 2008; Martin-Antonio et al., 2009).

It is often thought that the risk from Fe toxicity to wildlife is minimal, but Payne et al. (1998, 2001) have questioned the need to re-evaluate the guidelines for allowable Fe concentrations after observing bleached fish syndrome in lakes receiving iron ore effluent. Aquatic toxicity of Fe is associated with smothering of the respiratory surfaces with Fe precipitates, and the identification of situations in which Fe precipitates form is necessary to improve on environmental hazard assessment. Teien et al. (2008) showed that only a small change in pH from 6.3 to 6.7 results in a significant increase in Fe deposition on the gills of fish causing increased mortality. Understanding the biogeochemical processes of Fe speciation in the hyporheic region will determine the situations in which Fe precipitates may form in the benthic region of streams and affect those species that lay eggs and whose larvae develop in the gravels. Linton et al. (2007) highlight the paucity in the number of studies assessing aquatic Fe toxicity and question the validity of setting dissolved Fe criteria for regulations. Thus, it may be time to re-evaluate the environment risk management of Fe and the ecological hazard it poses (Linton et al., 2007).

REFERENCES

Abboud, S., and Haile, D. J. J. (2000). A novel mammalian iron-regulated protein involved in intracellular iron metabolism. *Biol. Chem.* **275**, 19906–19912.
Agius, C., and Roberts, R. J. (2003). Melano-macrophage centres and their role in fish pathology. *J. Fish Dis.* **26**, 499–509.

Alhas, E., Oymak, S. A., and Akin, H. K. (2009). Heavy metal concentrations in two barb, *Barbus xanthopterus* and *Barbus rajanorum mystaceus* from Atatürk Dam Lake, Turkey. *Environ. Monit. Assess.* **148**, 11–18.

Amin, S. A., Green, D. H., Frithjof, C. K., and Carrano, C. J. (2009). Vibroferrin, an unusual marine siderophore: iron binding, photochemistry and biological implications. *Inorg. Chem.* **48**, 11451–11458.

Andersen, O. (1997). Accumulation of waterborne iron and expression of ferritin and transferrin in early developmental stages of brown trout (*Salmo trutta*). *Fish Physiol. Biochem.* **16**, 223–231.

Andersen, O., Delhi, A., Standal, H., Giskegjerde, T. A., Karstensen, R., and Rorvik, K. A. (1995). 2 ferritin subunits of Atlantic salmon (*Salmo salar*) – cloning of the liver cDNAs and antibody preparation. *Mol. Mar. Biol. Biotechnol.* **4**, 164–170.

Andersen, O., Pantopoulos, K., Kao, H. T., Muckenthaler, M., Youson, J. H., and Pieribone, V. (1998). Regulation of iron metabolism in the sanguivore lamprey *Lampetra fluviatilis* – molecular cloning of two ferritin subunits and two iron-regulatory proteins (IRP) reveals evolutionary conservation of the iron-regulatory element (IRE)/IRP regulatory system. *Eur. J. Biochem.* **254**, 223–229.

Andrews, N. C. (2005). Molecular control of iron metabolism. *Best Pract. Res. Clin. Haematol.* **8**, 159–169.

Andrews, N. C. (2009). ABCs of erythroid mitochondrial iron uptake. *Proc. Natl. Acad. Sci. U.S.A.* **106**, 16012–16013.

Andrews, N. C., and Schmidt, P. J. (2007). Iron homeostasis. *Annu. Rev. Physiol.* **69**, 69–85.

ANZECC. (2007). *Australian and New Zealand Guidelines for Fresh and Marine Water Quality.* Australian and New Zealand Environment and Conservation Council, Canberra.

Arain, M. B., Kazi, T. G., Jamali, M. K., Jalbani, N., Afridi, H. I., and Shah, A. (2008). Total dissolved and bioavailable elements in water and sediment samples and their accumulation in *Oreochromis mossambicus* of polluted Manchar Lake. *Chemosphere* **70**, 1845–1856.

Armstrong, J. E., and Van Baalen, C. (1979). Iron transport in microalgae: the isolation and biological activity of a hydroxamate siderophore from blue–green alga. *J. Gen. Microbiol.* **111**, 253–262.

Arosio, P., and Levi, S. (2010). Cytosolic and mitochondrial ferritins in the regulation of cellular iron homeostasis and oxidative damage. *Biochim. Biophys. Acta* **1800**, 783–792.

Arredondo, M., Muñoz, P., Mura, C. V., and Nùñez, M. T. (2003). DMT1, a physiologically relevant apical Cu1+ transporter of intestinal cells. *Am. J. Physiol.* **284**, C1525–C1530.

Ayyadurai, K., Swaminathan, C. S., and Krishnasamy, V. (1994). Studies on heavy metal pollution in the finfish, *Oreochromis mossambicus* from River Cauvery. *Indian J. Environ. Health* **36**, 99–103.

Badaracco, M. E., Siri, M. V. R., and Pasquini, J. M. (2010). Oligodendrogenesis: the role of iron. *Biofactors* **36**, 98–102.

Bannon, D. I., Abounader, R., Lees, P. S. J., and Bressler, J. P. (2003). Effect of DMT1 knockdown on iron, cadmium and lead uptake in Caco-2 cells. *Am. J. Physiol.* **284**, C44–C50.

Barbeau, K. (2006). Photochemistry of organic iron(III) complexing ligands in oceanic systems. *Photochem. Photobiol.* **82**, 1505–1516.

Barbeau, K., Rus, E. L., Bruland, K. W., and Butler, A. (2001). Photochemical cycling of iron in the surface ocean mediated by microbial iron (III)-binding ligands. *Nature* **413**, 409–413.

Basaran, A. K., Aksu, M., and Ozdemire, E. (2010). Impacts of the fish farms on the water column nutrient concentrations and accumulation of heavy metals in the sediments in the eastern Aegean Sea (Turkey). *Environ. Monit. Assess* **162**, 439–451.

Batty, L. C., Atkin, L., and Manning, D. A. C. (2005). Assessment of the ecological potential of mine-water treatment wetlands using a baseline survey of macroinvertebrate communities. *Environ. Pollut.* **138**, 412–419.

Bellelli, A., Brunori, M., Brzezinski, P., and Wilson, M. T. (2001). Photochemically induced electron transfer. *Methods* **24**, 139–152.

Bowman, B. H., Yang, F., and Adrian, G. S. (1988). Transferrin: evolution and genetic regulation and expression. *Adv. Genet.* **25**, 1–38.

Boyd, P. W., Watson, A. J., Law, C. S., Abraham, E. R., Trull, T., Murdoch, R., Bakker, D. C., Bowie, A. R., Buesseler, K. O., Chang, H., Charette, M., Croot, P., Downing, K., Frew, R., Gall, M., Hadfield, M., Hall, J., Harvey, M., Jameson, G., LaRoche, J., Liddicoat, M., Ling, R., Maldonado, M. T., McKay, R. M., Nodder, S., Pickmere, S., Pridmore, R., Rintoul, S., Safi, K., Sutton, P., Strzepek, R., Tanneberger, K., Turner, S., Waite, A., and Zeldis, J. (2000). A mesoscale phytoplankton bloom in the polar Southern Ocean stimulated by iron fertilization. *Nature* **407**, 695–702.

Bressler, J. P., Olivi, L., Cheong, J. H., Kim, Y., and Bannon, D. (2004). Divalent metal transporter 1 in lead and cadmium transport. *Ann. N. Y. Acad. Sci.* **1012**, 142–152.

Brewer, G. J. (2010). Risks of copper and iron toxicity during aging in humans. *Chem. Res. Toxicol.* **23**, 319–326.

Brownlie, A., Donovan, A., Pratt, S. J., Paw, B. H., Oates, A. C., Brugnara, C., Witkowska, H. E., Sassa, S., and Zon, L. I. (1998). Positional cloning of the zebrafish sauternes gene: a model for congenital sideroblastic anaemia. *Nat. Genet.* **20**, 244–250.

Brownlie, A., Hersey, C., Oates, A. C., Paw, B. H., Falick, A. M., Witkowska, H. E., Flint, J., Higgs, D., Jessen, J., Bahary, N., Zhu, H., Lin, S., and Zon, L. (2003). Characterization of embryonic globin genes of the zebrafish. *Dev. Biol.* **1**, 48–61.

Burdige, D. J. (1993). The biogeochemistry of manganese and iron reduction in marine-sediments. *Earth Sci. Rev.* **35**, 249–284.

Burge, E. J., Gauthier, D. T., Ottinger, C. A., and Van Veld, P. A. (2004). Mycobacterium – inducible NRAMP in striped bass (*Morone saxatilis*). *Infect. Immun.* **72**, 1626–1636.

Bury, N., and Grosell, M. (2003a). Iron acquisition by teleost fish. *Comp. Biochem. Physiol. C* **135**, 97–105.

Bury, N., and Grosell, M. (2003b). Waterborne iron acquisition by a freshwater teleost fish, zebrafish *Danio rerio*. *J. Exp. Biol.* **206**, 3529–3535.

Bury, N. R., Grosell, M., Wood, C. M., Hogstrand, C., Wilson, R. W., Rankin, J. C., Busk, M., Lecklin, T., and Jensen, F. B. (2001). Intestinal iron uptake in the European flounder (*Platichthys flesus*). *J. Exp. Biol.* **204**, 3779–3787.

Bury, N. R., Walker, P. A., and Glover, C. N. (2003). Review: Nutritive metal uptake in teleost fish. *J. Exp. Biol.* **206**, 11–23.

Canli, M., and Atli, G. (2003). The relationships between heavy metal (Cd, Cr, Cu, Fe, Pb, Zn) levels and the size of six Mediterranean fish species. *Environ. Pollut.* **121**, 129–136.

Carriquiriborde, P., Handy, R. D., and Davies, S. J. (2004). Physiological modulation of iron metabolism in rainbow trout (*Oncorhynchus mykiss*) fed low and high iron diets. *J. Exp. Biol.* **207**, 75–86.

Cavill, I. (2002). Erythropoiesis and iron. *Best Pract. Res. Clin. Haematol.* **15**, 399–409.

CCREM (Canadian Council of Resource and Environment Ministers) (1987). Canadian Water Quality Guidelines. Prepared by the task force on water quality guidelines. Government of Canada, Winnipeg.

Chen, H., Waldbieser, G. C., Rice, C. D., Elibol, B., Wolters, W. R., and Hanson, L. A. (2002). Isolation and characterization of channel catfish natural resistance associated macrophage protein gene. *Dev. Comp. Immunol.* **26**, 517–531.

Chen, S. L., Xu, M. Y., Ji, X. S., and Yu, G. C. (2004). Cloning and characterization of natural resistance associated macrophage protein (Nramp) cDNA from red sea bream (*Pagrus major*). *Fish Shellfish Immunol.* **17**, 305–313.

Chen, W., Paradkar, P. N., Li, L., Pierce, E. L., Langer, N. B., Takahashi-Makise, N., Hyde, B. B., Shirihai, O. S., Ward, D. M., Kaplan, J., and Paw, B. H. (2009). Abcb10 physically interacts with mitoferrin-1 (Slc25a37) to enhance its stability and function in the erythroid mitochondria. *Proc. Natl. Acad. Sci. U.S.A.* **106**, 16263–16268.

Coe, C. L., Lubach, G. R., Bianco, L., and Beard, J. L. (2009). A history of iron deficiency anemia during infancy alters brain monoamine activity later in juvenile monkeys. *Dev. Psychobiol.* **51**, 301–309.

Conrad, M. E., Umbreit, J. N., and Moore, E. G. (1999). Iron absorption and transport. *Am. J. Med. Sci.* **318**, 213–229.

Cook, J. D. (1990). Adaptation in iron metabolism. *Am. J. Clin. Nutr.* **51**, 301–308.

Cooper, C. A., and Bury, N. R. (2007). The gills as an important uptake route for the essential nutrient iron in freshwater rainbow trout *Oncorhynchus mykiss. J. Fish Biol.* **71**, 115–128.

Cooper, C. A., Grosell, M., and Bury, N. R. (2006a). The effects of pH and the iron redox state on iron uptake in the intestine of a marine teleost fish, gulf toadfish (*Opsanus beta*). *Comp. Biochem. Physiol. A* **143**, 292–298.

Cooper, C. A., Handy, R. D., and Bury, N. R. (2006b). The effects of dietary iron concentration on gastrointestinal and branchial assimilation of both iron and cadmium in zebrafish (*Danio rerio*). *Aquat. Toxicol.* **79**, 167–175.

Cooper, C. A., Shayeghi, M., Techau, M. E., Capdevila, D. M., MacKenzie, S., Durrant, C., and Bury, N. R. (2007). Analysis of the rainbow trout solute carrier 11 family reveals iron import $\leq$ pH 7.4 and a functional isoform lacking transmembrane domains 11 and 12. *FEBS Lett.* **581**, 2599–2604.

Council of European Communities. (2007). Directive concerning the discharge of certain dangerous substances to the aquatic environment, 4 May 1976. *Official Journal* **L129**,18 May 1976, 76/464/EEC.

Craig, P. M., Galus, M., Wood, C. M., and McClelland, G. B. (2009). Dietary iron alters waterborne copper-induced gene expression in soft water acclimated zebrafish (*Danio rerio*). *Am. J. Physiol.* **296**, R362–R373.

Dalzell, D. J. B., and MacFarlane, N. A. A. (1999). The toxicity of iron to brown trout and effects on the gills: a comparison of two grades of iron sulphate. *J. Fish Biol.* **55**, 301–315.

De Domenico, I., Ward, D. M., and Kaplan, J. (2007). Hepcidin regulation: ironing out the details. *J. Clin. Invest.* **117**, 1755–1758.

De Domenico, I., Nemeth, E., Nelson, J. M., Phillips, J. D., Ajioka, R. S., Kay, M. S., Kuschner, J. P., Ganz, T., Ward, D. M., and Kaplan, J. (2008). The hepcidin-binding site on ferroportin is evolutionarily conserved. *Cell Metab.* **8**, 146–156.

Department of the Environment (1989). *Water and the Environment.* DoE Circular 7/89. HMSO, London.

Desjardins, L. M., Hicks, B. D., and Hilton, J. W. (1987). Iron catalyzed oxidation of trout diets and its effect on the growth and physiological response of rainbow trout. *Fish Physiol. Biochem.* **3**, 173–182.

Dickey, L. F., Sreedharan, S., Theil, E. C., Didsbury, J. R., Wang, Y. H., and Kaufman, R. E. (1987). Differences in the regulation of messenger RNA for housekeeping and specialized-cell ferritin – a comparison of 3 distinct ferritin complementary DNAs, the corresponding subunits, and identification of the 1st processed pseudogene in amphibia. *J. Biol. Chem.* **262**, 7901–7907.

Donovan, A., Brownlie, A., Zhou, Y., Shepard, J., Pratt, S. J., Moynihan, J., Paw, B. H., Drejer, A., Barut, B., Zapata, A., Law, T. C., Brugnara, C., Lux, S. E., Pinkus, G. S., Pinkus, J. L., Kingsley, P. D., Palis, J., Fleming, M. D., Andrews, N. C., and Zon, L. L. (2000). Positional cloning of zebrafish ferroportin 1 identifies a conserved vertebrate iron exporter. *Nature* **403**, 776–781.

Donovan, A., Brownlie, A., Dorschner, M. O., Zhou, Y., Pratt, S. J., Paw, B. H., Phillips, R. B., Thisse, C., Thisse, B., and Zon, L. I. (2002). The zebrafish mutant gene chardonnay (cdy) encodes divalent metal transporter 1 (DMT1). *Blood* **100**, 4655–4659.

Dorschner, M. O., and Phillips, R. B. (1999). Comparative analysis of two Nramp loci from rainbow trout. *DNA Cell Biol.* **18**, 573–583.

Douglas, S. E., Gallant, J. W., Liebscher, R. S., Dacanay, A., and Tsoi, S. C. M. (2003). Identification and expression analysis of hepcidin-like antimicrobial peptides in bony fish. *Dev. Comp. Immunol.* **27**, 589–601.

Duckworth, O. W., Holmström, S. J. M., Pena, J., and Spositio, G. (2009a). Biogeochemistry of iron oxidation in a circumneutral freshwater habitat. *Chem. Geol.* **260**, 149–158.

Duckworth, O. W., Bargar, J. R., and Spositio, G. (2009b). Coupled biogeochemical cycling of iron and manganese as mediated by microbial siderophores. *Biometals* **22**, 605–613.

Dural, M., Ziya Lugal Göksu, M., and Akif Özak, A. (2007). Investigation of heavy metal levels in economically important fish species captured from the Tuzla lagoon. *Food Chem.* **102**, 415–421.

EC. (2007). Directive 2000/60/EC of the European Parliament and of the Council of 23 October 2000 establishing a framework for Community action in the field of water policy. *Official Journal of the European Communities* **L327**, 1–72. (22/12/2000).

El Nagger, G., and Lovell, R. T. (1991). L-Ascorbyl-2-monophosphate has equal antiscorbutic activity as L-ascorbic acid but L-ascorbyl-2-sulfate is inferior to L-ascorbic acid for channel catfish. *J. Nutr.* **21**, 1622–1626.

Elibol-Flemming, B., Waldbieser, G. C., Wolters, W. R., Boyle, C. R., and Hanson, L. A. (2009). Expression analysis of selected immune-relevant genes in channel catfish during *Edwardsiella ictaluri* infection. *J. Aquat. Anim. Health* **21**, 23–35.

Elisma, F., and Jumarie, C. (2001). Evidence for cadmium uptake through Nramp2: metal speciation studies with Caco-2 cells. *Biochem. Biophys. Res. Commun.* **285**, 662–668.

Emmenegger, L., King, D. W., Sigg, L., and Sulzberger, B. (1998). Oxidation kinetics of Fe(II) in a eutrophic Swiss lake. *Environ. Sci. Technol.* **32**, 2990–2996.

Emmenegger, L., Schonenberger, R. R., Sigg, L., and Sulzberger, B. (2001). Light-induced redox cycling of iron in circumneutral lakes. *Limnol. Oceanogr.* **46**, 49–61.

Ersoy, B., and Çelik, M. (2010). The essential and toxic elements in tissues of six commercial demersal fish from Eastern Mediterranean Sea. *Food Chem. Toxicol.* **48**, 1377–1382.

Feng, C. Y., Johnson, S. C., Hori, T. S., Rise, M., Hall, J. R., Gamperl, A. K., Hubert, S., Kimball, J., Bowman, S., and Rise, M. L. (2009). Identification and analysis of differentially expressed genes in immune tissues of Atlantic cod stimulated with formalin-killed, atypical *Aeromonas salmonicida*. Physiol. Genom. **37**, 149–163.

Ferguson, C. J., Wareing, M., Delannoy, M., Fenton, R., McLarnon, S. J., Ashton, N., Cox, A. G., McMahon, R. F., Garrick, L. M., Green, R., Smith, C. P., and Riccardi, D. (2003). Iron handling and gene expression of the divalent metal transporter, DMT1, in the kidney of the anemic Belgrade (b) rat. *Kidney Int.* **64**, 1755–1764.

Fleming, M. D., Trenor, C. C., Su, M. A., Foernzler, D., Beier, D. R., Dietrich, W. F., and Andrews, N. C. (1997). Microcytic anaemia mice have a mutation in Nramp2, a candidate iron transporter gene. *Nat. Genet.* **16**, 383–386.

Forbes, J. R., and Gros, P. (2001). Divalent-metal transport by NRAMP proteins at the interface of host–pathogen interactions. *Trends Microbiol.* **9**, 397–403.

Fox, L. E., and Wofsy, S. C. (1983). Kinetics of removal of iron colloids from estuarine waters. *Geochim. Cosmochim. Acta* **47**, 211–216.

Fracalossi, D. M., Allen, M. E., Nichols, D. K., and Oftedal, O. T. (1998). Oscars, *Astronotus ocellatus*, have a dietary requirement for vitamin C. *J. Nutr.* **128**, 1745–1751.

Fraenkel, P. G., Gibert, Y., Holzheimer, J. L., Lattanzi, V. J., Burnett, S. F., Dooley, K. A., Wingert, R. A., and Zon, L. I. (2009). Transferrin-A modulates hepcidin expression in zebrafish embryos. *Blood* **113**, 2843–2850.

Fukushima, M., and Tatsumi, K. (1999). Light acceleration of iron (III) reduction by humic acid in the aqueous solution. *Colloids Surf. A* **155**, 249–258.

Gaitlin, D. M., and Wilson, R. P. (1986). Characterization of iron deficiency and the dietary iron requirements of fingerling channel catfish. *Aquaculture* **52**, 191–198.

Gammons, C. H., Nimick, D. A., Parker, S. R., Cleasby, T. E., and McCleskey, R. B. (2005). Diel behavior of iron and other heavy metals in a mountain stream with acidic to neutral pH: Fisher Creek, Montana, USA. *Geochim. Cosmochim. Acta* **69**, 2505–2516.

Garrick, M. D., and Garrick, L. M. (2009). Cellular iron transport. *Biochim. Biophys. Acta* **1790**, 309–325.

Geng, B., Jin, Z. H., Li, T. L., and Qi, X. H. (2009). Kinetics of hexavalent chromium removal from water by chitosan-Fe-0 nanoparticles. *Chemosphere* **75**, 825–830.

Giorgi, A., Mignogna, G., Bellapadrona, G., Gattoni, M., Chiaraluce, R., Consalvi, V., Chiancone, E., and Stefanini, S. (2008). The unusual co-assembly of H- and M-chains in the ferritin molecule from the Antarctic teleosts *Trematomus bernacchii* and *Trematomus newneisi*. *Arch. Biochem. Biophys.* **478**, 69–74.

Glover, C. N., and Hogstrand, C. (2002). In vivo characterization of intestinal zinc uptake in freshwater rainbow trout. *J. Exp. Biol.* **205**, 141–150.

Glover, C. N., and Hogstrand, C. (2003). Effects of dissolved metals and other hydrominerals on in vivo intestinal zinc uptake in freshwater rainbow trout. *Aquat. Toxicol.* **62**, 281–293.

Gomes, M. S., and Appelberg, R. (1998). Evidence for a link between iron metabolism and Nramp1 gene function in innate resistance against *Mycobacterium avium*. *Immunology* **95**, 165–168.

Gonzalez, R. J., Grippo, R. S., and Dunson, W. A. (1990). The disruption of sodium balance in brook trout *Salvelinus fontinalis* (Mitchell), by manganese and iron. *J. Fish Biol.* **37**, 765–774.

Goodwill, K. E., Sabatier, C., and Stevens, R. C. (1998). Crystal structure of tyrosine hydroxylase with bound cofactor analogue and iron at 2.3 Å resolution: self-hydroxylation of Phe300 and the Pterin-binding site. *Biochemistry* **37**, 13437–13445.

Gosriwatana, I., Loreal, O., Lu, S., Brissot, P., Porter, J., and Hider, R. C. (1999). Quantification of non-transferrin-bound iron in the presence of unsaturated transferrin. *Anal. Biochem.* **273**, 212–220.

Goswami, T., Bhattacharjee, A., Babal, P., Searle, S., Moore, E., Li, M., and Blackwell, J. M. (2001). Natural-resistance-associated macrophage protein 1 is an H^+/bivalent cation antiporter. *Biochem. J.* **354**, 511–519.

Gottofrey, J., and Tjälve, H. (1991). Axonal transport of cadmium in the olfactory nerve of the pike. *Pharmacol. Toxicol.* **69**, 242–252.

Granier, T., Langlois d'Estaintot, B., Gallois, B., Chevalier, J.-M., Precigoux, G., Santambrogio, P., and Arosio, P. (2003). Structural description of the active sites of mouse L-chain ferritin at 1.2 Å resolution. *J. Biol. Inorg. Chem.* **8**, 105–111.

Gray, N. F. (1997). Environmental impact and remediation of acid mine drainage: a management problem. *Environ. Geol.* **30**, 62–71.

Gray, N. F., and Delaney, E. (2010). Measuring community response of benthic macroinvertebrates in an erosional river impacted by acid mine drainage by use of a simple model. *Ecol. Indic.* **10**, 668–675.

Gregorović, G., Kralj-Klobučar, N., and Kopiar, N. (2008). Histological and morphometric study on the tissue and cellular distribution of iron in carp *Cyprinus carpio* L. during chronic waterborne exposure. *J. Fish Biol.* **72**, 1841–1846.

Grobler, E., Du Preez, H. H., and Van Vuren, J. H. J. (1989). Toxic effects of zinc and iron on the routine oxygen-consumption of *Tilapia sparrmanii* (Cichlidae). *Comp. Biochem. Physiol. C* **94**, 207–214.

Gruenheid, S., Canonne-Hergaux, F., Gauthier, S., Hackam, D. J., Grinstein, S., and Gros., P. (1999). The iron transport protein NRAMP2 is an integral membrane glycoprotein that colocalizes with transferrin in recycling endosomes. *J. Exp. Med.* **189**, 831–841.

Gunshin, H., Mackenzie, B., Berger, U. V., Gunshin, Y., Romero, M. F., Boron, W. F., Nussberger, S., Gollan, J. L., and Hediger, M. A. (1997). Cloning and characterisation of a mammalian proton coupled metal ion transporter. *Nature* **388**, 482–488.

Guo, J. H., Juan, S. H., and Aust, S. D. (1998). Mutational analysis of the four alpha-helix bundle iron-loading channel of rat liver ferritin. *Arch. Biochem. Biophys.* **352**, 71–77.

Hamilton-Taylor, J., Posthill, A. S., Tipping, E., and Harper, P. M. (2002). Laboratory measurements and modeling of metal–humic interactions under estuarine conditions. *Geochim. Cosmochim. Acta* **66**, 403–415.

Hoglund, E., Weltzien, F. A., Schjolden, J., Winberg, S., Ursin, H., and Doving, K. B. (2005). Avoidance behaviour and brain monoamines in fish. *Brain Res.* **1032**, 104–110.

Honda, K., Sahrul, M., Hidika, H., and Tatsukawa, R. (1983). Organ and tissue distribution of heavy metals, and their growth-related changes in Antarctic fish *Pagothenia borchgrevinki. Agric. Biol. Chem.* **47**, 2521–2532.

Huang, X. (2003). Iron overload and its association with cancer risk in humans: evidence for iron as a carcinogenic metal. *Mutat. Res.* **533**, 153–171.

Huang, H. Q., Xiao, Z. Q., Chen, X., Lin, Q. M., Cai, Z. W., and Chen, P. (2004). Characteristics of structure, composition, mass spectra, and iron release from the ferritin of shark liver (*Sphyrna zygaena*). *Biophys. Chem.* **111**, 213–222.

Hutchins, D., Witter, A., Butler, A., and Luther, G. (1999). Competition among marine phytoplankton for different chelated iron species. *Nature* **400**, 858–861.

Jaillon, O., Aury, J. M., Brunet, F., Petit, J. L., Stange-Thomann, N., Mauceli, E., Bouneau, L., Fischer, C., Ozouf-Costaz, C., Bernot, A., Nicaud, S., Jaffe, D., Fisher, S., Lutfalla, G., Dossat, C., Segurens, B., Dasilva, C., Salanoubat, M., Levy, M., Boudet, N., Castellano, S., Anthouard, V., Jubin, C., Castelli, V., Katinka, M., Vacherie, B., Biémont, C., Skalli, Z., Cattolico, L., Poulain, J., De Berardinis, V., Cruaud, C., Duprat, S., Brottier, P., Coutanceau, J. P., Gouzy, J., Parra, G., Lardier, G., Chapple, C., McKernan, K. J., McEwan, P., Bosak, S., Kellis, M., Volff, J. N., Guigó, R., Zody, M. C., Mesirov, J., Lindblad-Toh, K., Birren, B., Nusbaum, C., Kahn, D., Robinson-Rechavi, M., Laudet, V., Schachter, V., Quétier, F., Saurin, W., Scarpelli, C., Wincker, P., Lander, E. S., Weissenbach, J., and Roest Crollius, H. (2004). Genome duplication in the teleost fish *Tetraodon nigroviridis* reveals the early vertebrate proto-karyotype. *Nature* **431**, 946–957.

Jones, G. J., Pakenik, B. P., and Morell, F. M. (1987). Trace metal reduction by phytoplankton: the role of the plasmalemma redox enzymes. *J. Phycol.* **23**, 237–244.

Juan, S. H., and Aust, S. D. (1998). The effect of putative nucleation sites on the loading and stability of iron in ferritin. *Arch. Biochem. Biophys.* **350**, 259–265.

Kato, J., Kobune, M., Ohkubo, S., Fujikawa, K., Tanaka, M., Takimoto, R., Takada, K., Takahari, D., Kawano, Y., Kohgo, Y., and Niitsu, Y. (2007). Iron/IRP-1-dependent regulation of mRNA expression for transferrin receptor, DMT1 and ferritin during human erythroid differentiation. *Exp. Hematol.* **35**, 879–887.

Kawabata, H., Yang, R., Hirama, T., Vuong, P. T., Kawano, S., Gombart, A. F., and Koeffler, H. P. (1999). Molecular cloning of transferrin receptor 2. A new member of the transferrin receptor-like family. *J. Biol. Chem.* **274**, 20826–20832.

Keel, S. B., Doty, R. T., Yang, Z., Quigley, J. G., Chen, J., Knoblaugh, S., Kingsley, P. D., De Domenico, I., Vaughn, M. B., Kaplan, J., Palis, J., and Abkowitz, J. L. (2008). A heme export protein is required for red blood cell differentiation and iron homeostasis. *Science* **319**, 825–828.

Kim, K. H., Shcheynikov, N., Wang, Y., and Muallem, S. (2005). SLC26A7 is a Cl⁻ channel regulated by intracellular pH. *J. Biol. Chem.* **280**, 6463–6470.

Kling, P., Norman, A., Andersson, P. L., Norrgren, L., and Förlin, L. (2008). Gender-specific proteomic responses in zebrafish liver following exposure to a selected mixture of brominated flame retardants. *Ecotoxicol. Environ. Saf.* **71**, 319–327.

Knutson, M. D., Vafa, M. R., Haile, D. J., and Wessling-Resnick, M. (2003). Iron loading and erythrophagocytosis increase ferroportin 1 (FPN1) expression in J774 macrophages. *Blood* **102**, 4191–4197.

Korzh, S., Emelyanov, A., and Korzh, V. (2001). Developmental analysis of ceruloplasmin gene and liver formation in zebrafish. *Mech. Dev.* **103**, 137–139.

Kwong, R. W., and Niyogi, S. (2008). An *in vitro* examination of intestinal iron absorption in a freshwater teleost, rainbow trout (*Oncorhynchus mykiss*). *J. Comp. Physiol. B* **178**, 963–975.

Kwong, R. W. M., and Niyogi, S. (2009). The interactions of iron with other divalent metals in the intestinal tract of a freshwater teleost, rainbow trout (*Oncorhynchus mykiss*). *Comp. Biochem. Physiol. C* **150**, 442–449.

Kwong, R. W. M., Andrés, J. A., and Niyogi, S. (2010). Molecular evidence and physiological characterization of iron absorption in isolated enterocytes of rainbow trout (*Oncorhynchus mykiss*): implications for dietary cadmium and lead absorption. *Aquat. Toxicol.* **99**, 343–350.

Laftah, A. H., Latunde-Dada, G. O., Fakih, S., Hider, R. C., Simpson, R. J., and McKie, A. T. (2009). Haem and folate transport by proton-coupled folate transporter/haem carrier protein 1 (SLC46A1). *Br. J. Nutr.* **101**, 1150–1156.

Lam, S. H., Winata, C. L., Tong, Y., Korzh, S., Lim, W. S., Korzh, V., Spitsbergen, J., Mathavan, S., Miller, L. D., Liu, E. T., and Gong, Z. (2006). Transcriptome kinetics of arsenic-induced adaptive response in zebrafish liver. *Physiol. Genom.* **27**, 351–361.

Lamas, S., Fernández, J. A., Aboal, J. R., and Carballeria, A. (2007). Testing the use of juvenile *Salmo trutta* L. as a biomonitor of heavy metal pollution in freshwater. *Chemosphere*, **67**, 221–228.

Lambert, L. A., Perri, H., Halbrooks, P. J., and Mason, A. B. (2005). Evolution of the transferrin family: conservation of residues associated with iron and anion binding. *Comp. Biochem. Physiol. B* **142**, 129–141.

Lam-Yuk-Tseng, S., Govoni, G., Forbes, J., and Gros, P. (2003). Iron transport by Nramp2/DMT: pH regulation of transport by histidines in transmembrane domain 6. *Blood* **101**, 3699–3707.

Lappivaara, J., Kiviniemi, A., and Oikari, A. (1999). Bioaccumulation and subchronic physiological effects of waterborne iron overload on whitefish exposed in humic and nonhumic water. *Arch. Environ. Contam. Toxicol.* **37**, 196–204.

Latunde-Dada, G. O., Westhuizen, J. V. D., Vulpe, C. D., Anderson, G. J., Simpson, R. J., and McKie, A. T. (2002). Molecular and functional roles of duodenal cytochrome *b* (Dcytb) in iron metabolism. *Blood Cell Mol. Dis.* **29**, 356–360.

Lautz, L. K., and Fanelli, R. M. (2008). Seasonal biogeochemical hotspots in the streambed around restoration structures. *Biogeochemistry* **91**, 85–104.

Lee, J. Y., Tada, T., Hirono, I., and Aoki, T. (1998). Molecular cloning and evolution of transferrin cDNAs in salmonids. *Mol. Mar. Biol. Biotechnol.* **7**, 287–293.

Lee, P. L., and Beutler, E. (2009). Regulation of hepcidin and iron-overload disease. *Annu. Rev. Pathol. Mech. Dis.* **4**, 489–515.

Leknes, I. L. (2007). Melano-macrophage centres and endocytic cells in kidney and spleen of pearl gouramy and platyfish (Anabantidae, Poeciliidae: Teleostei). *Acta Histochem.* **109**, 164–168.

LeSage, G. D., Kost, L. J., Barham, S. S., and LaRusso, N. F. (1986). Biliary excretion of iron from hepatocyte lysosomes in the rat. A major excretory pathway in experimental iron overload. *J. Clin. Invest.* **77**, 90–97.

Li, H., Zhou, Q., Wu, Y., Fu, J., Wang, T., and Jiang, G. (2009). Effects of waterborne nano-iron on medaka (*Oryzias latipes*): antioxidant enzymatic activity, lipid peroxidation and histopathology. *Ecotoxicol. Environ. Saf.* **72**, 684–692.

Liang, H. C., and Thomson, B. M. (2009). Minerals and mine drainage. *Water Environ. Res.* **81**, 1615–1663.

Linton, T. K., Pacheco, M. A. W., McIntyre, D. O., Clement, W. H., and Goodrich-Mahoney, J. (2007). Development of bioassessment-based benchmarks for iron. *Environ. Toxicol. Chem.* **26**, 1291–1298.

Liu, H., Takano, T., Abernathy, J., Wang, S. L., Sha, Z. X., Jiang, Y. L., Terhune, J., Kucuktas, H., Peatman, E., and Liu, Z. J. (2010). Structure and expression of transferrin gene of channel catfish, *Ictalurus punctatus*. *Fish Shellfish Immunol.* **28**, 159–166.

Liu, X-b., Yang, F., and Haile, D. J. (2005). Functional consequences of ferroportin 1 mutations. *Blood Cells Mol. Dis.* **35**, 33–46.

Lofts, S., Tipping, E., and Hamilton-Taylor, J. (2008). The chemical speciation of Fe(III) in freshwaters. *Aquat. Geochem.* **14**, 337–358.

Lubach, G. R., and Coe, C. L. (2008). Selective impairment of cognitive performance in the young monkey following recovery from iron deficiency. *J. Dev. Behav. Pediatr.* **29**, 11–17.

Lumsden, A. L., Henshall, T. L., Dayan, S., Lardelli, M. T., and Richards, R. I. (2007). Huntingtin-deficient zebrafish exhibit defects in iron utilization and development. *Hum. Mol. Genet.* **16**, 1905–1920.

Ma, Y., Specian, R. D., Yeh, M., Rodriguez-Paris, J., and Glass, J. (2002). The transcytosis of divalent metal transporter and apo-transferrin during iron uptake in intestinal epithelium. *Am. J. Physiol.* **283**, G965–G974.

Macchi, G. J., Romano, L. A., and Christiansen, H. E. (1992). Melanomacrophage centers in the whitemouth croaker, *Micropogonias furnieri*, as biological indicators of environmental changes. *J. Fish Biol.* **40**, 971–973.

Mackenzie, B., Ujwal, M. L., Chang, M. H., Romero, M. F., and Hediger, M. A. (2006). Divalent metal-ion transporter DMT1 mediates both H^+-coupled Fe^{2+} transport and uncoupled fluxes. *Pflugers Arch.* **451**, 544–558.

Macrellis, H. M., Trick, C. G., Rue, E. L., Smith, G., and Bruland, K. W. (2001). Collection and detection of natural iron-binding ligands from seawater. *Mar. Chem.* **76**, 175–187.

Mallett, J. (1985). Fish gill structure induced by toxicants and other irritants: a statistical review. *Can. J. Fish. Aquat. Sci.* **42**, 630–648.

Marks, M. S., Woodruff, L., Ohno, H., and Bonifacino, J. S. (1996). Protein targeting by tyrosine- and di-leucine-based signals: evidence for distinct saturable components. *J. Cell Biol.* **135**, 341–354.

Márquez, M., Vodopivez, C., Casaux, R., and Curtosi, A. (1998). Metal (Fe, Zn, Mn and Cu) levels in Antarctic fish *Notothenia coriiceps*. *Polar Biol.* **20**, 404–408.

Martin-Antonio, B., Jimenez-Cantizano, R. M., Salas-Leiton, E., Infante, C., and Manchado, M. (2009). Genomic characterization and gene expression analysis of four hepcidin genes in the redbanded seabream (*Pagrus auriga*). *Fish Shellfish Immunol.* **26**, 483–491.

McKie, A. T., Marciani, P., Rolfs, A., Brennan, K., Wehr, K., Barrow, D., Miret, S., Bomford, A., Peters, T. J., Farzaneh, F., Hedigre, M. A., Hentze, M. W., and Simpson, R. J. (2000). A novel duodenal iron regulated transporter IREG1, implicated in the basolateral transfer of iron to the circulation. *Mol. Cell* **5**, 299–309.

McKie, A. T., Barrow, D., Latunde-Dada, G. O., Rolfs, A., Sager, G., Mudaly, E., Mudaly, M., Richardson, C., Barlow, D., Bomford, A., Peters, T. J., Raja, K. B., Shirali, S., Hediger, M. A., Farzaneh, F., and Simpson, R. J. (2001). An iron regulated ferric reductase associated with the absorption of dietary iron. *Science* **291**, 1755–1759.

McKnight, D. M., Kimball, B. A., and Runkel, R. L. (2001). pH dependence of iron photoreduction in a rocky mountain stream affected by acid mine drainage. *Hydrol. Process.* **15**, 1979–1992.

Melino, S., Garlando, L., Patamia, M., Paci, M., and Petruzzelli, R. (2005). A metal-binding site is present in the amino terminal region of the bioactive iron regulator hepcidin-25. *J. Pept. Res.* **66**, 65–71.

Mignogna, G., Chiaraluce, R., Consalvi, V., Cavallo, S., Stefanini, S., and Chiancone, E. (2002). Ferritin from the spleen of the Antarctic teleost *Trematomus bernacchii* is an M-type homopolymer. *Eur. J. Biochem.* **269**, 1600–1606.

Miguel, J. L., Pablos, M. I., Agapito, M. T., and Recio, J. M. (1991). Isolation and characterization of ferritin from the liver of the rainbow trout (*Salmo gairdneri* R.). *Biochem. Cell Biol.* **69**, 735–741.

Mims, M. P., and Prchal, J. T. (2005). Divalent metal transporter 1. *Hematology* **10**, 339–345.

Mizuno, S., Sasaki, Y., Omoto, N., and Imada, K. (2004). Elimination of adhesiveness in the eggs of shishamo smelt *Spirinchus lanceolatus* using kaolin treatment to achieve high hatching rate in an environment with a high iron concentration. *Aquaculture* **242**, 713–726.

Moreau, R., and Dabrowski, K. (1998a). Fish acquired ascorbic acid synthesis prior to terrestrial vertebrate emergence. *Free Rad. Biol. Med.* **25**, 989–990.

Moreau, R., and Dabrowski, K. (1998b). Body pool and synthesis of ascorbic acid in adult sea lamprey (*Petromyzon marinus*): an agnathan fish with gulonolactone oxidase activity. *Proc. Natl. Acad. Sci. U.S.A* **95**, 10279–10282.

Moreau, R., and Dabrowski, K. (2003). Alpha-Tocopherol downregulates gulonolactone oxidase activity in sturgeon. *Free Rad. Biol. Med.* **34**, 1326–1332.

Moriya, M., and Linder, M. C. (2006). Vesicular transport and apotransferrin in intestinal iron absorption, as shown in the Caco-2 cell model. *Am. J. Physiol.* **290**, G301–G309.

Muckenthaler, M. U., Galy, B., and Hentze, M. W. (2008). Systemic iron homeostasis and the iron-responsive element/iron-regulatory protein (IRE/IRP) regulatory network. *Annu. Rev. Nutr.* **28**, 197–213.

Nadella, S. R., Grosell, M., and Wood, C. M. (2007). Mechanisms of dietary Cu uptake in freshwater rainbow trout: evidence for Na-assisted Cu transport and a specific metal carrier in the intestine. *J. Comp. Physiol. B* **177**, 433–446.

Nemeth, E., and Ganz, T. (2009). The role of hepcidin in iron metabolism. *Acta Haematol.* **122**, 78–86.

Nemeth, E., Tuttle, M. S., Powelson, J., Vaughn, M. B., Donovan, A., Ward, D. M., Ganz, T., and Kaplan, J. (2004). Hepcidin regulates cellular iron efflux by binding to ferroportin and inducing its internalization. *Science* **306**, 2090–2093.

Neves, J. V., Wilson, J. M., and Rodrigues, P. N. (2009). Transferrin and ferritin response to bacterial infection: the role of the liver and brain in fish. *Dev. Comp. Immunol.* **33**, 848–857.

Nfon, E., Cousins, I. T., Järvinen, O., Mukherjee, A. B., Verta, M., and Broman, D. (2009). Trophodynamics of mercury and other trace elements in a pelagic food chain from the Baltic Sea. *Sci. Total Environ.* **407**, 6267–6274.

Nicolas, G., Bennoun, M., Devaux, I., Beaumont, C., Grandchamp, B., Kahn, A., and Vaulont, S. (2001). Lack of hepcidin gene expression and severe tissue iron overload in upstream stimulatory factor 2 (USF2) knockout mice. *Proc. Natl. Acad. Sci. U.S.A.* **98**, 8780–8785.

Nilsson, R., Schultz, I. J., Pierce, E. L., Soltis, K. A., Naranuntarat, A., Ward, D. M., Baughman, J. M., Paradkar, P. N., Kingsley, P. D., Culotta, V. C., Kaplan, J., Palis, J., Paw, B. H., and Mootha, V. K. (2009). Discovery of genes essential for heme biosynthesis through large-scale gene expression analysis. *Cell Metab.* **10**, 119–130.

Noël-Lambot, F. (1981). Presence in the intestinal lumen of marine fish of corpuscles with high cadmium-, zinc- and copper-binding capacity: a possible mechanism of heavy metal tolerance. *Mar. Ecol. Prog. Ser.* **4**, 175–181.

Ojo, A. A., and Wood, C. M. (2007). In vitro analysis of the bioavailability of six metals via the gastro-intestinal tract of the rainbow trout (*Oncorhynchus mykiss*). *Aquat. Toxicol.* **83**, 10–23.

Okubo, M., Yamada, K., Hosoyamada, M., Shibasaki, T., and Endou, H. (2003). Cadmium transport by human Nramp2 expressed in *Xenopus laevis* oocytes. *Toxicol. Appl. Pharmacol.* **187**, 162–167.

Padhi, A., and Verghese, B. (2007). Evidence for positive Darwinian selection on the hepcidin gene of Perciform and Pleuronectiform fishes. *Mol. Divers.* **11**, 119–130.

Park, B. Y., and Chung, J. (2009). Cadmium increases ferroportin-1 gene expression in J774 macrophage cells via the production of reactive oxygen species. *Nutr. Res. Pract.* **3**, 192–199.

Payne, J. F., Malins, D. C., Gunselman, S., Rahimtula, A., and Yeats, P. A. (1998). DNA oxidative damage and vitamin A reduction in fish from a large lake system in Labrador, Newfoundland, contaminated with iron-ore mine tailings. *Mar. Environ. Res.* **46**, 289–294.

Payne, J. F., French, B., Hamoutene, D., Yeats, P., Rahimtula, A., Scruton, D., and Andrews, C. (2001). Are metal mining effluent regulations adequate: identification of a novel bleached fish syndrome in association with iron-ore mining effluents in Labrador, Newfoundland. *Aquat. Toxicol.* **52**, 311–317.

Peuranen, S., Vuorinen, P. J., Vuorinen, M., and Hollender, A. (1994). The effects of iron, humic acids and low pH on the gills and physiology of brown trout (*Salmo trutta*). *Ann. Zool. Fenn.* **31**, 389–396.

Pirrie, D., Camm, G. S., Sear, L. G., and Hughes, S. H. (1997). Mineralogical and geochemical signature of mine waste contamination, Tresillian river, Fal estuary, Cornwall, UK. *Environ. Geol.* **29**, 58–65.

Poss, K. D., and Tonegawa, S. (1997). Heme oxygenase 1 is required for mammalian iron reutilization. *Proc. Natl. Acad. Sci. U.S.A.* **94**, 10919–10924.

Powell, J. J., Jugdaohsingh, R., and Thompson, R. P. H. (1999a). The regulation of mineral absorption in the gastrointestinal tract. *Proc. Nutr. Soc.* **58**, 147–153.

Powell, J. J., Whitehead, M. W., Ainley, C. C., Kendall, M. D., Nicholson, J. K., and Thompson, R. P. H. (1999b). Dietary minerals in the gastrointestinal tract: hydroxypolymerisation of aluminium is regulated by luminal mucins. *J. Inorg. Biochem.* **75**, 167–180.

Qiu, A., Jansen, M., Sakaris, A., Min, S. H., Chattopadhyay, S., Tsai, E., Sandoval, C., Zhao, R., Akabas, M. H., and Goldman, I. D. (2006). Identification of an intestinal folate transporter and the molecular basis for hereditary folate malabsorption. *Cell* **127**, 917–928.

Quigley, J. G., Yang, Z., Worthington, M. T., Phillips, J. D., Sabo, K. M., Sabath, D. E., Berg, C. L., Sassa, S., Wood, B. L., and Abkowitz, J. L. (2004). Identification of a human heme exporter that is essential for erythropoiesis. *Cell* **118**, 757–766.

Richards, J. G., and Playle, R. C. (1998). Cobalt binding to gills of rainbow trout (*Oncorhynchus mykiss*): an equilibrium model. *Comp. Biochem. Physiol. C* **119**, 185–197.

Robinson, N. J., Procter, C. M., Connolly, E. L., and Guerinot, M. L. (1999). A ferric chelate reductase for iron uptake from soils. *Nature* **397**, 694–697.

Rodrigues, P. N., Vázquez-Dorado, S., Neves, J. V., and Wilson, J. M. (2006). Dual function of fish hepcidin: response to experimental iron overload and bacterial infection in sea bass (*Dicentrarchus labrax*). *Dev. Comp. Immunol.* **30**, 1156–1167.

Roeder, M., and Roeder, R. H. (1966). Effect of iron on the growth rates of fishes. *J. Nutr.* **90**, 86–90.

Roy, P. K., and Lall, S. P. (2006). Mineral nutrition of haddock *Melanogrammus aeglefins* (L.): a comparison of wild and cultured stock. *J. Fish Biol.* **68**, 1460–1472.

Russo, N., Salahub, D. R., and Witko, M. (2003). *Metal–Ligand Interactions*. Springer, New York.

Ryu, D. Y., Lee, S. J., Park, D. W., Choi, B. S., Klaassen, C. D., and Park, J. D. (2004). Dietary iron regulates intestinal cadmium absorption through iron transporters in rats. *Toxicol. Lett.* **15**, 19–25.

Saeij, J. P. J., Wiegertjes, G. F., and Stet, R. J. M. (1999). Identification and characterisation of a fish natural resistance associate macrophage protein (Nramp) cDNA. *Immunogenetics* **50**, 50–60.

Scudiero, R., Trinchella, F., Riggio, M., and Parisi, E. (2007). Structure and expression of genes involved in transport and storage of iron in red-blooded and hemoglobin-less Antarctic notothenioids. *Gene* **397**, 1–11.

Scudiero, R., Trinchella, F., and Parisi, E. (2008). Iron metabolism genes in Antarctic notothenioids: a review. *Mar. Genom.* **1**, 79–85.

Sendamarai, A. K., Ohgami, R. S., Fleming, M. D., and Lawrence, C. M. (2008). Structure of the membrane proximal oxidoreductase domain of human Steap3, the dominant ferrireductase of the erythroid transferrin cycle. *Proc. Natl. Acad. Sci. U.S.A.* **105**, 7410–7415.

Shafir, T., Angulo-Barroso, R., Jing, Y. Z., Angelilli, M. L., Jacobson, S. W., and Lozoff, B. (2008). Iron deficiency and infant motor development. *Early Hum. Dev.* **84**, 479–485.

Sharp, P. (2004). The molecular basis of copper and iron interactions. *Proc. Nutr. Soc.* **63**, 563–569.

Shaw, G. C., Cope, J. J., Li, L., Corson, K., Hersey, C., Ackermann, G. E., Gwynn, B., Lambert, A. J., Wingert, R. A., Traver, D., Trede, N. S., Barut, B. A., Zhou, Y., Minet, E., Donovan, A., Brownlie, A., Balzan, R., Weiss, M. J., Peters, L. L., Kaplan, J., Zon, L. I., and Paw, B. H. (2006). Mitoferrin is essential for erythroid iron assimilation. *Nature* **440**, 96–100.

Shayeghi, M., Latunde-Dada, G. O., Oakhill, J. S., Laftah, A. H., Takeuchi, K., Halliday, N., Khan, Y., Warley, A., McCann, F. E., Hider, R. C., Frazer, D. M., Anderson, G. J., Vulpe, C. D., Simpson, R. J., and McKie, A. T. (2005). Identification of an intestinal heme transporter. *Cell* **122**, 789–801.

Sherman, D. M. (2005). Electronic structures of iron(III) and manganese(IV) (hydr)oxides minerals: thermodynamics of photochemical reductive dissolution in aquatic environments. *Geochim. Cosmochim. Acta* **69**, 3249–3255.

Shi, J. S., and Camus, A. C. (2006). Hepcidins in amphibians and fishes: antimicrobial peptides or iron-regulatory hormones? *Dev. Comp. Immunol.* **30**, 746–755.

Shiau, S. Y., and Su, L. W. (2002). Ferric citrate is half as effective as ferrous sulphate in meeting the iron requirement of juvenile tilapia, *Orechromis niloticus* × *O. aureus*. *J. Nutr* **133**, 483–488.

Sibthorpe, D., Baker, A. M., Gilmartin, B. J., Blackwell, J. M., and White, J. K. (2004). Comparative analysis of two slc11 (NRAMP) loci in *Takifugu rubripes*. *DNA Cell Biol.* **23**, 45–58.

Simovich, M., Hainsworth, L. N., Fields, P. A., Umbreit, J. N., and Conrad, M. E. (2003). Localization of the iron transport proteins Mobilferrin and DMT-1 in the duodenum: the surprising role of mucin. *Am. J. Hematol.* **74**, 32–45.

Sloman, K. A., Scott, G. R., Diao, Z. Y., Rouleau, C., Wood, C. M., and McDonald, D. G. (2003). Cadmium affects the social behaviour of rainbow trout, *Oncorhynchus mykiss*. *Aquat. Toxicol.* **65**, 171–185.

Smith, E. J., and Sykora, J. L. (1976). Early developmental effects of lime-neutralized iron hydroxide suspensions on brook trout and coho salmon. *Trans. Am. Fish. Soc.* **105**, 308–312.

Smith, M. W., Shenoy, B., Debnam, E. S., Dashwood, M. R., Churchill, L. J., and Srai, S. K. (2002). Divalent metal inhibition of non-haem iron uptake across the rat duodenal brush border membrane. *Br. J. Nut.* **88**, 51–56.

Soe-Lin, S., Apte, S. S., Mikhael, M. R., Kayembe, L. K., Nie, G., and Ponka, P. (2010). Both Nramp1 and DMT1 are necessary for efficient macrophage iron recycling. *Exp. Hematol.* **38**, 609–617.

Stumm, W., and Morgan, J. J. (1996). *Aquatic Chemistry* (3rd edn.). John Wiley & Sons, New York.

Sutherland, T. F., Petersen, S. A., and Levings, C. D. (2007). Distinguishing between natural and aquaculture-derived sediment concentrations of heavy metals in the Broughton Archipelago, British Columbia. *Mar. Pollut. Bull.* **54**, 1451–1460.

Tabuchi, M., Tanaka, N., Nishida-Kitayama, J., Ohno, H., and Kishi, F. (2002). Alternative splicing regulates the subcellular localisation of divalent metal transporter 1 isoforms. *Mol. Biol. Cell* **13**, 4371–4387.

Tallkvist, J., Bowlus, C. L., and lonnerdal, B. (2003). Effect of iron treatment on nickel absorption and gene expression of the divalent metal transporter (DMT1) by human intestinal Caco-2 cells. *Pharmacol. Toxicol.* **92**, 121–124.

Tamura, H., Goto, K., and Nagayama, M. (1976). Effect of anions on oxygenation of ferrous iron in neutral solutions. *J. Inorg. Nucl. Chem.* **38**, 113–117.

Tchernitchko, D., Bourgeois, M., Martin, M., and Beaumont, C. (2002). Expression of the two mRNA isoforms of the iron transporter Nramp2/DMT1 in mice and function of the iron responsive element. *Biochem. J.* **363**, 449–455.

Techau, M. E., Valdez-Taubas, J., Popoff, J. F., Francis, R., Seaman, M., and Blackwell, J. M. (2007). Evolution of differences in transport function in Slc11a family members. *J. Biol. Chem.* **242**, 35646–35656.

Teien, H. C., Garmo, O. A., Atland, A., and Salbu, B. (2008). Transformation of iron species in mixing zones and accumulation on fish gills. *Environ. Sci. Technol.* **42**, 1780–1786.

Tekin-Özan, S., and Kir, I. (2008). Seasonal variations of heavy metals in some organs of carp (*Cyprinus carpio* L., 1758) from Beyşehir Lake (Turkey). *Environ. Monit. Assess* **138**, 201–206.

Tepe, Y. (2009). Metal concentrations in eight fish species from Aegean and Mediterranean Seas. *Environ. Monit. Assess* **159**, 501–509.

Theron, J., Walker, J. A., and Cloete, T. E. (2008). Nanotechnology and water treatment: applications and emerging opportunities. *Crit. Rev. Microbiol.* **34**, 43–69.

Tipping, E., Rey-Castro, C., Bryan, S. E., and Hamiltion-Taylor, J. (2002). Al(III) and Fe(III) binding by humic substances in freshwaters, and implications for trace metal speciation. *Geochim. Cosmochim. Acta* **66**, 3211–3224.

Tjälve, H., and Henrikkson, I. (1999). Uptake of metals in the brain via olfactory pathways. *Neurotoxicology* **20**, 181–195.

Toyohara, H., Yamamoto, S., Hosoi, M., Takagi, M., Hayashi, I., Nakao, K., and Kaneko, S. (2005). Scallop DMT functions as a Ca^{2+} transporter. *FEBS Lett.* **579**, 2727–2730.

Updegraff, K. F., and Sykora, J. L. (1976). Avoidance of lime-neutralized iron hydroxide solutions by coho salmon in laboratory. *Environ. Sci. Technol.* **10**, 51–54.

USEPA. (1976). *Quality Criteria for Water: 1976.* US Environmental Protection Agency, Washington, DC, EPA 440/9-76-023.

USEPA. (1976). *Quality Criteria for Water: 1986.* US Environmental Protection Agency, Washington, DC, EPA 440/5-86-001.

Vidal, S. M., Malo, D., Vogan, K., Skamene, E., and Gros, P (1993). Natural resistance to infection with intracellular parasites: isolation of a candidate for Bcg. *Cell* **73**, 469–485.

Vulpe, C. D., Kuo, Y. M., Murphy, T. L., Cowley, L., Askwith, C., Libina, N., Gitschier, J., and Anderson, G. J. (1999). Hephaestin, a ceruloplasmin homologue implicated in intestinal iron transport, is defective in the sla mouse. *Nat. Genet.* **21**, 195–199.

Vuori, K. M. (1995). Direct and indirect effects of iron on river ecosystems. *Ann. Zool. Fenn.* **32**, 317–329.

Walker, R. L., and Fromm, P. O. (1976). Metabolism of iron by normal and iron deficient rainbow trout. *Comp. Biochem. Physiol. A* **55**, 311–318.

Wallace, D. F., Summerville, L., Lusby, P. E., and Subramaniam, V. N. (2005). First phenotypic description of transferrin receptor 2 knockout mouse, and the role of hepcidin. *Gut* **54**, 980–986.

Wally, J., Halbrooks, P. J., Vonrhein, C., Rould, M. A., Everse, S. J., Mason, A. B., and Buchanan, S. K. (2006). The crystal structure of iron-free human serum transferrin provides insight into inter-lobe communication and receptor binding. *J. Biol. Chem.* **281**, 24934–24944.

Walsh, P. J., Blackwelder, P., Gill, K. A., Danult, E., and Mommsen, T. P. (1991). Carbonate deposits in marine fish intestine: a new source of biomineralisation. *Limnol. Oceanogr.* **36**, 1227–1232.

Watanabe, T., Kiron, V., and Satoh, S. (1997). Trace minerals in fish nutrition. *Aquaculture* **151**, 185–207.

Water Research Council (WRc) (1988). *Proposed Environmental Quality Standards for List II Substances in Water: Iron* (eds. G. Mance and J. A. Campbell). Technical Report TR258. WRc, Medmenham, Buckinghamshire.

Water Research Council (WRc) (1998). *An Update to Proposed Environmental Quality Standards for Iron in Water* (eds. P. Whitehouse, E. Dixon, S. Blake and K. Bailey). Final report DETR 4471/1 to the Department of the Environment, Transport and the Regions. WRc, Medmenham, Buckinghamshire.

Weeratunga, S. K., Lovell, S., Yao, H., Battaile, K. P., Fischer, C. J., Gee, C. E., and Rivera, M. (2010). Structural studies of bacterioferritin B from *Pseudomonas aeruginosa* suggest a gating mechanism for iron uptake via the ferroxidase center. *Biochemistry* **49**, 1160–1175.

Whittamore, J. M., Cooper, C. A., and Wilson, R. W. (2010). HCO_3^- secretion and $CaCO_3$ precipitation play major roles in intestinal water absorption in marine teleost fish *in vivo*. *Am. J. Physiol.* **298**, R877–R886.

Wilson, R. W., and Grosell, M. (2003). Intestinal bicarbonate secretion in marine teleost fish – source of bicarbonate, pH sensitivity, and consequences for whole animal acid–base and calcium homeostasis. *Biochim. Biophys. Acta* **1681**, 163–174.

Wilson, R. W., Millero, F. J., Taylor, J. R., Walsh, P. J., Christensen, V., Jennings, S., and Grosell, M. (2009). Contribution of fish to the marine inorganic carbon cycle. *Science* **323**, 359–362.

Winberg, S., Winberg, Y., and Fernald, R. D. (1997). Effect of social rank on brain monoaminergic activity in a cichlid fish. *Brain Behav. Evolut.* **49**, 230–236.

Wingert, R. A., Brownlie, A., Galloway, J. L., Dooley, K., Fraenkel, P., Axe, J. L., Davidson, A. J., Barut, B., Noriega, L., Sheng, X., Zhou, Y., and Zon, L. I. (2004). The chianti zebrafish mutant provides a model for erythroid-specific disruption of transferrin receptor 1. *Development* **131**, 6225–6235.

Wingert, R. A., Galloway, J. L., Barut, B., Foott, H., Fraenkel, P., Axe, J. L., Weber, G. J., Dooley, K., Davidson, A. J., Schmid, B., Paw, B. H., Shaw, G. C., Kingsley, P., Palis, J., Schubert, H., Chen, O., Kaplan, J., and Zon, L. I. (2005). Deficiency of glutaredoxin 5 reveals Fe-S clusters are required for vertebrate haem synthesis. *Nature* **436**, 1035–1039.

Winterbourn, M. J., McDiffett, W. F., and Eppley, S. J. (2000). Aluminum and iron burdens of aquatic biota in New Zealand streams contaminated by acid mine drainage: effects of trophic level. *Sci. Total Environ* **254**, 45–54.

Wong, M. H., and Tam, F. Y. (1984). Sewage sludge for cultivating freshwater algae and the fate of heavy metal at higher trophic organisms. IV. Heavy metal contents in different trophic levels. *Arch. Hydrobiol.* **100**, 423–430.

Xu, Q. H., Cheng, C. H. C., and Hu, P. (2008). Adaptive evolution of hepcidin genes in antarctic notothenioid fishes. *Mol. Biol. Evol.* **25**, 1099–1112.

Yada, T., Muto, K., Azuma, T., and Ikuta, K. (2004). Effects of prolactin and growth hormone on plasma levels of lysozyme and ceruloplasmin in rainbow trout. *Comp. Biochem. Physiol.* C **139**, 57–96.

Yang, L., and Gui, J. F. (2004). Positive selection on multiple antique allelic lineages of transferrin in the polyploid *Carassius auratus*. *Mol. Biol. Evol.* **21**, 1264–1277.

Yang, M., Wang, K. J., Chen, J. H., Qu, H. D., and Li, S. J. (2007). Genomic organization and tissue-specific expression analysis of hepcidin-like genes from black porgy (*Acanthopagrus schlegelii* B.). *Fish Shellfish Immunol.* **23**, 1060–1071.

Yehuda, Z., Shenker, M., Romheld, V., Marschner, H., Hadar, Y., and Chen, Y. (1996). The role of ligand exchange in the uptake of iron from microbial siderophores by gramineous plants. *Plant. Phycol.* **112**, 1273–1280.

Zhang, W. X. (2003). Nanoscale iron particles for environmental remediation: an overview. *J. Nanopart. Res.* **5**, 323–332.

Zibdeh, M., Matsui, S., and Furuichi, M. (2001). Requirements of tiger pufferfish, *Takifugu rubripes*, for dietary iron. *J. Fac. Agr. Kyushu Univ.* **45**, 473–479.

Zwilling, B. S., Kuhn, D. E., Wikoff, L., Brown, D., and Lafuse, W. (1999). Role of iron in Nramp1-mediated inhibition of mycobacterial growth. *Infect. Immun.* **67**, 1386–1392.

5

NICKEL

GREG PYLE
PATRICE COUTURE

Nickel (Ni) is the 22nd most abundant element and is ubiquitous in marine and freshwater ecosystems. Nickel concentrations increase in aquatic systems that receive inputs from urban and industrial effluents. At pH

Homeostasis and Toxicology of Essential Metals: Volume 31A
FISH PHYSIOLOGY

values common to most aquatic systems, Ni speciation is dominated by Ni^{2+} with increasing hydroxide complexation with increasing pH values. Under oxic conditions, most Ni is either bound to dissolved organic matter or adsorbed onto insoluble Fe or Mn oxyhydroxides. Under anoxic conditions, it forms insoluble sulfides. Nickel is well established as an essential nutrient for plants and terrestrial animals, but not in aquatic animals. However, evidence is mounting to suggest that Ni is probably essential in fish. Nickel toxicity to fish has received relatively little research attention compared to other more toxic metals, such as Cu or Cd. Nickel can be taken up by fish through the gills or olfactory epithelium during waterborne exposures or through the gut during dietary exposures, is transported throughout the fish in the blood while bound to albumins and short peptides, and preferentially accumulates in the kidneys. Acute Ni toxicity is associated with branchial lesions that cumulatively increase the diffusive distance across the gill epithelium, leading to impaired respiratory function. In the kidneys, Ni causes lesions in the renal tubules and antagonizes Mg reabsorption, probably because Ni and Mg share uptake transporters. Cellular damage likely results from the cumulative effects of Ni-induced oxidative damage. Although Ni may not be as acutely toxic as other metals, it does have the capacity to be genotoxic and is therefore potentially hazardous to fish. The ecological implications of Ni contamination of natural freshwater ecosystems are poorly understood. Given the increased global demand for Ni and corresponding potential for increased anthropogenic inputs, it is important to develop a deeper understanding of the basic physiology of Ni in fish, both as a putative essential nutrient and as a toxicant.

1. NICKEL SPECIATION IN FRESHWATER AND SALTWATER

Although nickel (Ni) can exist in any one of several oxidation states, including 0, −1, +1, +3, and +4, its +2 oxidation state dominates in natural systems (Fig. 5.1) (Nriagu, 1980). Using the metal classification of Nieboer and Richardson (1980), Ni shows characteristics that are intermediate between class A and class B but with stronger tendencies towards class B than A. In other words, Ni is a borderline metal with class B character, which suggests that it has a stronger affinity for oxygen and nitrogen than it has for sulfur. Like other metals in the first transition series (V, Cr, Mn, Fe, and Co), Ni is octahedrally coordinated as $Ni[(H_2O)_6]^{2+}$ in aqueous systems (Richter and Theis, 1980). Under oxic conditions, Ni exists primarily as the free aquo species and the hydrous oxides of Fe and Mn control speciation. Hydrous Mn oxides (i.e. Mn oxyhydroxides) are far more important with

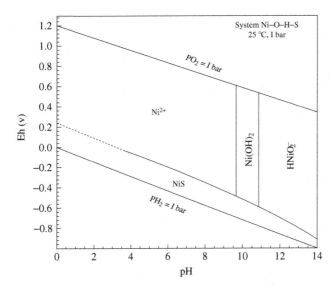

Fig. 5.1. Eh-pH diagram of the partial speciation of Ni in an Ni–O–H–S system assuming all species are dissolved. The graph shows the relationship between electron activity (Eh) and hydrogen ion activity (pH), and provides a simple depiction of the stability of Ni species as a function of pH. Redrawn from Nieminen et al. (2007).

regard to Ni speciation than Fe oxyhydroxides [such as $Fe(OH)_3$] because the former are not affected by pH whereas the latter are (Richter and Theis, 1980; Green-Pedersen et al., 1997). Nickel tends to adsorb to Fe oxyhydroxides more strongly at higher pH values owing to increased electrostatic attraction between the negatively charged oxide surfaces and positively charged Ni cations (Green-Pedersen et al., 1997). At lower pH, competition between Ni^{2+} and H^+ causes Ni dissociation from hydrous oxides. Under hypoxic or anoxic conditions, sulfides control Ni speciation through the formation of insoluble Ni sulfides.

In waters having pH values that are typical for most aquatic systems, ranging between pH 5 and 9, the free divalent Ni cation, Ni^{2+}, is the dominant species in the absence of dissolved organic carbon (DOC). Under oxic conditions, Ni may form inorganic complexes with the following anions in order of decreasing affinity: $OH^- > SO_4^{2-} > Cl^- > NH_3$ (Richter and Theis, 1980). In saltwater systems, Ni complexes involving Cl^- and SO_4^{2-} are more important than in freshwater systems (Turner et al., 1998; Schaumloffel, 2005). At higher pH levels, Ni hydroxide species dominate Ni speciation. Like other metals, Ni forms complexes with carbonate, such as $NiCO_3$; however, Ni carbonate species are relatively unimportant in natural systems, unlike the case for other metals. Formation of carbonate species by other metals, such as Cu or Pb, will free up binding sites on Mn oxyhydroxides to accommodate Ni adsorption (Richter and Theis, 1980). Consequently, although Ni carbonate species play relatively minor roles in Ni speciation, the presence of dissolved carbonate plays an important, albeit indirect, role in Ni speciation (Green-Pedersen et al., 1997).

In natural freshwaters, however, some 99.9% of dissolved Ni is bound to organic complexes (Xue et al., 2001). The proportion of Ni bound to organic complexes in seawater is much less, comprising approximately 35% of dissolved Ni (Nimmo et al., 1989). Organic Ni complexes are either labile or non-labile depending on the nature of the organic material and the chemistry of the bulk solution, such as pH, the presence of other metals, competing inorganic complexing agents, or major cations such as Ca^{2+} or Mg^{2+} (Mandal et al., 2000; Hassan et al., 2008). Nickel becomes increasingly labile by increasing any of these confounding factors. Alternatively, Ni can form very stable (non-labile) organic complexes having stability constants (log K) near 18, especially in seawater (Van den Berg and Nimmo, 1987).

The reaction rates for Ni to form organic complexes are very slow (Xue et al., 2001) because of its highly stable electron configuration: $3d^8 4s^2$ (Kasprzak, 1987). In seawater, and to a lesser extent in freshwater, these organic complexes are very stable (Nimmo et al., 1989). Nickel bound to highly stable organo-complexes may not distribute to all available ligands in the system for an extremely long period, if ever (Xue et al., 2001). Consequently, in natural waters the Ni distribution may never actually achieve equilibrium. Upon first entering an aquatic system, Ni may maintain its original speciation for a long time, thereby disrupting the existing distribution of Ni in the system. Similarly, Ni^{2+} entering an acidic system may require extended periods to form organo-complexes. The net result of this disequilibrium is that the most toxic, free divalent cation, Ni^{2+}, may persist for a much longer period than other metals under the same conditions. In light of the slow reaction rates for Ni to form organo-complexes in natural waters, and because natural systems may never truly achieve equilibrium, some authors have suggested that kinetic models may be preferable to equilibrium models for estimating Ni speciation in natural waters (Celo et al., 2001; Guthrie et al., 2003, 2005; Chakraborty et al., 2006).

2. NICKEL SOURCES AND ECONOMIC IMPORTANCE

Nickel (atomic number 28, atomic mass 58.71) is a common group III transition metal that comprises 0.0099% of the Earth's crust, making it the seventh most abundant transition metal and 22nd most abundant element (Greenwood and Earnshaw, 1984). It is typically found in ultrabasic igneous rocks ranging in content from 0.016% in basalt to 0.20% in periodotite (Birge and Black, 1980). Nickel occurs in two commercially available ores called laterites, commonly found in New Caledonia, Cuba, and Australia, and sulfides, which are commonly found in Canada, Russia, and South Africa. Laterite ores are formed from rock weathering and are typically found near the surface, whereas sulfide ores originate from geological processes deep within the Earth before depositing in the Earth's crust (Reck et al., 2008). Laterite ores are typically associated with oxide and silicate ores (such as garnierite), whereas sulfide ores are associated with other metals including Cu, Co, Au, and Ag. More than half of all the primary Ni mined in the world is from Canada, Russia, and Australia. The world's largest Ni deposit is in Sudbury, Ontario, Canada, which supplies some 25% of world demand for primary Ni.

Nickel's economic importance lies in its unique physicochemical characteristics, the most important of which include its strength, high-temperature stability, corrosion resistance, malleability, ductility, heat and electrical conductive properties, and aesthetic properties. Over 60% of all Ni produced is used in stainless steel production. Other uses of Ni include corrosion-resistant industrial equipment, building materials, medical equipment (including cardiac stents), food and beverage storage containers, electromagnetic shielding, electroplating, battery production, jewelry and coinage, electronic storage media, catalysts, inks and dyes, ceramics, and strong magnets (Eisler, 1998; Reck et al., 2008). Like many other metals, Ni production has increased exponentially over the past century to satisfy consumer demand (Fig. 5.2) (Reck et al., 2008). Consequently, Ni is becoming an increasingly important contaminant of concern in aquatic ecosystems. The estimated world Ni reserve is about 140 Tg, which suggests that at the current rate of extraction there is no impending Ni shortage over the next several decades and primary Ni extraction is likely to continue well into the future (Reck et al., 2008).

Nickel is ubiquitous in aquatic environments owing to natural weathering and geochemical processes (Schaumloffel, 2005). The concentration of Ni in natural waters varies depending on geological factors, and ranges from 0.2 to 0.7 $\mu g\,L^{-1}$ in the open ocean and 0.1 to 10 $\mu g\,L^{-1}$ in unpolluted freshwaters (Chau and Kulikovsky-Cordeiro, 1995). In areas naturally high in Ni, the background concentration can be as high as 10 mg L^{-1}, whereas most industrially contaminated waters have Ni concentrations ranging between 50 and, 2000 $\mu g\,L^{-1}$ (Chau and Kulikovsky-Cordeiro, 1995). Anthropogenic Ni can enter aquatic systems via fallout from airborne particulate matter, surface runoff near industrial and urban areas, industrial effluents released directly into aquatic systems, or wastewater treatment facilities (Schaumloffel, 2005).

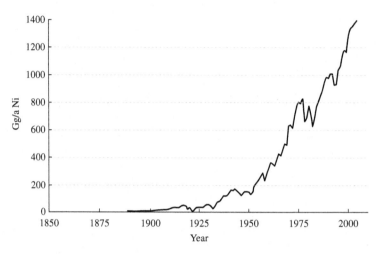

Fig. 5.2. Global Ni production over the past 150 years in gigagrams of Ni per year (Gg a^{-1} Ni). Figure from Reck et al. (2008).

3. ENVIRONMENTAL SITUATIONS OF CONCERN

Environmental contamination by Ni occurs primarily in areas where Ni is mined, including Australia, Canada, Cuba, Russia, and South Africa. The most significant environmental Ni contamination exists in Sudbury, Ontario, Canada, home to the most productive Ni mining operation in the world. Mining in the Sudbury area began in earnest around 1886 and continues to this day (Winterhalder, 1995). The Sudbury industrial zone of impact extends over 17,000 km^2 around major smelting operations and affects approximately 7,000 lakes (Keller and Gunn, 1995). Nickel concentrations in water bodies in and around the Sudbury area range between 7 and 338 μg L^{-1} (Pyle et al., 2005). Although remediation efforts have led to significant improvements to environmental conditions in Sudbury area lakes, metal contamination (mainly by Cu, Ni, and Zn) will continue to be an environmental concern for centuries to come (Arnott et al., 2001).

4. ENVIRONMENTAL CRITERIA

Table 5.1 summarizes the ambient water quality criteria for Ni in fresh and salt waters for the limited number of national or supranational jurisdictions around the world where they exist, i.e. the USA, Canada, Australia/New Zealand, and the European Union (EU). In the first three

Table 5.1
Ambient water quality guidelines for nickel available worldwide

Jurisdiction	Reference	Acute (μg Ni L^{-1})	Chronic (μg Ni L^{-1})	Notes
USA	USEPA (2005)	120	13	Hardness 20 mg L^{-1}
		260	29	Hardness 50 mg L^{-1}
		842	93	Hardness 200 mg L^{-1}
		74	8.2	Saltwater
Canada	CCME (2007)		25	Hardness 20 mg L^{-1}
			65	Hardness 90 mg L^{-1}
			150	Hardness 200 mg L^{-1}
Australia/	ANZGFMWQ		10	Hardness 20 mg L^{-1}
New	(2000)		21	Hardness 50 mg L^{-1}
Zealand[a]			76	Hardness 200 mg L^{-1}
			70	Saltwater
European	European		20	European directives specify
Union	Parliament			that member states may
	(2008)			take hardness and other
				factors affecting
				bioavailability into
				account

Hardness is presented as CaCO$_3$ equivalents.
[a]Values selected are for the protection of 95% of species.

jurisdictions, the freshwater values for chronic toxicity are hardness dependent and increase at higher hardness values. Although the EU, within its REACH program, is developing Ni guidelines that will consider water chemistry, currently the recommended safe value for wildlife, approved in a directive from the EU parliament, is set at 20 μg L^{-1} with a note that member states are encouraged to modulate this value as a function of hardness and other water chemistry parameters and to also take into account natural (background) concentrations. Neither the EU nor Canada currently proposes an Ni criterion for saltwater. The Ni criterion for saltwater proposed by Australia/New Zealand is similar to their value for very hard freshwater. In contrast, the value recommended by the US Environmental Protection Agency (EPA) for saltwater is lower than its recommendation for low hardness freshwater. One reason for this apparent discrepancy is because some marine invertebrates used to derive the EPA guidelines show adverse effects to Ni at very low concentrations (Eisler, 1998). Of all Ni guidelines available for the protection of aquatic life, only the EPA proposes values for acute exposure and they are about nine-fold higher than values for chronic exposures. Excluding the EU guideline, which is too limited to be comparable to guidelines available elsewhere, the Canadian values are the most permissive while the Australia/New Zealand values are the most stringent. Note that the approach taken by the latter countries do not provide criteria for acute or chronic exposures, but trigger values for the protection of a percentage of all biota, ranging between 80 and 99% (95% trigger values are presented in Table 5.1).

Although tissue residue criteria have recently been proposed for amphibian embryos (Perez-Coll et al., 2008), no such criteria are available for fish. In yellow perch, Couture and Pyle (2008) attempted to determine thresholds of tissue Ni concentrations that would efficiently discriminate fish from Ni contaminated or clean environments. However, they found that tissue Ni concentration thresholds could not be established that allowed for a clear discrimination between clean and contaminated fish (unlike the case for Cd and Cu where thresholds could be established). When the Ni concentration threshold was set at the top 10th percentile in liver and kidney, only 8–18% of the fish exceeding the threshold came from Ni-contaminated lakes. The authors proposed that this odd result reflects the variable capacity of yellow perch to regulate tissue Ni concentrations depending on their region of origin.

5. MECHANISMS OF TOXICITY

5.1. Acute Toxicity

Nickel toxicity is influenced by water hardness, pH, total suspended solids, salinity, fish species, and developmental stage (Birge and Black, 1980). Very generally, acute Ni toxicity (96 h LC50; i.e. the median lethal concentration: the concentration of a toxicant required to kill 50% of the test animals after a 96 h exposure) is approximately 4–14 mg L^{-1} in soft water (20–50 mg L^{-1} as $CaCO_3$) and 24–44 mg L^{-1} in hard water (over 50 mg L^{-1} as $CaCO_3$) (Birge and Black, 1980). These acute toxicity concentration ranges are orders of magnitude (mg L^{-1} vs μg L^{-1}) higher than those required to induce acute toxicity for several other metals. For example, in soft water (40 mg L^{-1} as $CaCO_3$) the juvenile rainbow trout 96 h LC50 for Cd is 1.5 μg L^{-1} and for Cu it is 18 μg L^{-1} (Buhl and Hamilton, 1991). Therefore, the acute toxicity of Ni is relatively low compared to most other metals of environmental concern (Table 5.2).

The 96 h LC50 for adult fathead minnows exposed to Ni in soft and hard water is 4 and 24 mg L^{-1}, respectively (Birge and Black, 1980). Goldfish (*Carassius auratus*) showed 100% mortality when exposed for 48 h to 10 mg L^{-1} in soft freshwater or 8.1 mg L^{-1} in hard freshwater. However, to

Table 5.2

Acute nickel toxicity in several species of fish

Reference	Species	Common name	Life stage	Endpoint	Ni conc. (mg L^{-1})	Hardness (mg L^{-1})	pH
Birge and Black (1980)	*Pimephales promelas*	Fathead minnow	Adult	96 h LC50	4.0	Soft	
					24	Hard	
	Carassius auratus	Goldfish	Adult	96 h LC50	10	Soft	
					8.1	Hard	
Brown and Dalton (1970)	*Oncorhynchus mykiss*	Rainbow trout	1 year	96 h LC50	32	240	7.4
Brown (1968)			Unknown	48 h LC50	18	10	
			Unknown	48 h LC50	90	300	
Pickering and Henderson (1966)	*Poecilia reticulata*	Guppy	6 months	96 h TL$_m$	4.45	20	7.5
	Pimephales promelas	Fathead minnow	Unknown		4.88	20	7.5
	Lepomis macrochirus	Bluegill	Unknown		5.27	20	7.5
	Carassius auratus	Goldfish	Unknown		9.82	20	7.5
	Pimephales promelas	Fathead minnow	Unknown		43	360	8.2
	Lepomis macrochirus	Bluegill	Unknown		39.6	360	8.2
Rehwoldt et al. (1971)	*Morone saxatilis*	Striped bass	<20 cm	96 h TL$_m$	6.2	53	7.8
	Lepomis gibbosus	Pumpkinseed	<20 cm		8.1		
	Cyprinus carpio	Common carp	<20 cm		10.6		
	Anguilla rostrata	American eel	<20 cm		13.0		
	Morone americana	White perch	<20 cm		13.6		
	Fundulus diaphanus	Banded killifish	<20 cm		46.2		
Pyle et al. (2002)	*Pimephales promelas*	Fathead minnow	<24 h	96 h LC50	0.45	20	6.9
			<24 h		0.5	40	7.0
			<24 h		2.3	140	7.6
Buhl and Hamilton (1991)	*Thymallus arcticus*	Arctic grayling	Alevin	96 h LC50	8.2	41	7.4
			Juvenile		8.7		
	Oncorhynchus kisutch	Coho salmon	Alevin		16.7		
			Juvenile		18		
	Oncorhynchus mykiss	Rainbow trout	Alevin		25.1		

(Continued)

Table 5.2 (Continued)

Reference	Species	Common name	Life stage	Endpoint	Ni conc. (mg L^{-1})	Hardness (mg L^{-1})	pH
Pyle (2000)	Pimephales promelas	Fathead minnow	Juvenile	96 h LC50	7.8	112	7.9
	Oncorhynchus mykiss	Rainbow trout	<24 h		2.4		
	Esox lucius	Northern pike	Juvenile		51.2		
	Catastomus commersoni	White sucker	<24 h		>3		
			<24 h		17.9		
Birge and Black, 1980	Oncorhynchus mykiss	Rainbow trout	Embryo	96 h LC50	0.05	100	7.5
	Ictalurus punctatus	Channel catfish	Embryo		0.71		
	Micropterus salmoides	Largemouth bass	Embryo		2.06		
	Carassius auratus	Goldfish	Embryo		2.78		

achieve 100% mortality in seawater, 259 mg Ni L^{-1} were required (Birge and Black, 1980). This result demonstrates that Ni is far less toxic in seawater than it is in freshwater, probably because of the protection conferred by Na$^+$, Ca^{2+}, and particularly Mg^{2+} competing for physiologically sensitive binding sites on the fish and the reduced bioavailability of Ni in the presence of these cations (Hall and Anderson, 1995). However, some marine invertebrates have demonstrated adverse effects to Ni at concentrations as low as 30 μg L^{-1} (Eisler, 1998).

Like several other metals of environmental concern, water quality can influence Ni toxicity to fish (Table 5.2). Several studies have demonstrated that Ni is much less toxic to fish in hard water than in soft water (see Section 3) (Pickering and Henderson, 1966; Brown and Dalton, 1970; Birge and Black, 1980; Schubauer-Berigan et al., 1993; Pyle et al., 2002b). It is thought that water hardness (i.e. Ca^{2+} and Mg^{2+}) protects fish against metal toxicity (generally) by competing for physiologically sensitive binding sites, such as those on the gill (Playle et al., 1992; Erickson et al., 1996). As such, waterborne Ni toxicity can be consistently predicted on the basis of the amount of Ni bound to or accumulated in gill tissue (Meyer et al., 1999). Fewer studies have demonstrated the effect of pH on Ni toxicity. The effect of pH on Ni toxicity is associated with the speciation of Ni (see Section 1). As with other metals, it is assumed that Ni^{2+}, which dominates at lower pH conditions, is the most bioavailable, and presumably the most toxic, form. Therefore, Ni toxicity should be highest at lower pH, but studies have demonstrated the opposite. In very hard water (300 mg L^{-1} as CaCO$_3$), Ni was most toxic to larval (<24 h) fathead minnows when pH was highest (96 h LC50 3.1 mg L^{-1}; pH 8–8.5) and least toxic when pH was low (96 h LC50 >4000 mg L^{-1}; pH 6–6.5) (Schubauer-Berigan et al., 1993). This pH effect on Ni toxicity was more pronounced in invertebrates tested in the same study. In another study, Ni was more toxic to larval (<24 h) fathead minnows at pH 7 than it was at pH 5; however, at pH 8.5 Ni toxicity was reduced by about two orders of magnitude relative to pH 5 (Pyle et al., 2002b). It is possible that another metal species, such as NiOH$^+$, may be more bioavailable at pH 7 relative to pH 5 and therefore a more toxic species than the free divalent cation, Ni^{2+}, in much the same way as CuOH$^+$ (Erickson et al., 1996) is bioavailable and contributes to Cu toxicity in fish (Grosell, Chapter 2). Another possible explanation for reduced Ni toxicity in exposure water having lower pH is that the H$^+$ ion may be successfully outcompeting Ni for physiologically sensitive binding sites (McDonald et al., 1991). In either case, more research is necessary to determine why fish may derive some modest protective effect against Ni toxicity at lower pH exposure conditions.

Nickel toxicity varies among fish species. In a study examining the relative toxicity among four species of fish (Table 5.2), Ni sensitivity followed the order: guppy (*Poecilia reticulata*; most sensitive) > fathead minnow > bluegill sunfish (*Lepomis macrochirus*) > goldfish (least sensitive), and median lethal Ni concentration ranged between 4.5 and 9.8 mg L^{-1} (Pickering and Henderson, 1966). In another study examining several different species, relative Ni toxicity followed the order striped bass (*Morone saxatilis*; most sensitive to Ni) > pumpkinseed (*Lepomis gibbosus*) > common carp (*Cyprinus carpio*) > American eel (*Anguilla rostrata*) > white perch (*Morone americana*) > banded killifish (*Fundulus diaphanus*; least sensitive) and ranged between 6.2 and 46.2 mg L^{-1} (Rehwoldt et al., 1971). In this case, Ni toxicity ranged between 6.2 and 13.6 mg L^{-1} for the first five freshwater species and jumped to 46.2 mg L^{-1} for the estuarine species. The relatively large discrepancy in relative toxicity may reflect basic physiological mechanisms appropriate to an estuarine species for regulating Na$^+$ concentrations, which may also play a role in mitigating (or regulating) Ni uptake and subsequent toxicity (Terreros et al., 1988; Hall and Anderson, 1995). In a study examining the relative toxicity of Ni to early life-stage fish of species typically found in northern Canadian lakes, Pyle (2000) found that larval fathead minnows were more sensitive than larval northern pike, white suckers, or rainbow trout. Moreover, juvenile rainbow trout were less sensitive to Ni compared to all other species tested. An earlier study investigating the acute toxicity of Ni to embryonic or juvenile life stages of rainbow trout, channel catfish, largemouth bass, and goldfish demonstrated that the rainbow trout was the most sensitive species (Birge and Black, 1980). Given the amount of research attention given to rainbow trout to establish North American water quality criteria, it may not be a good species choice for establishing water quality criteria for Ni until these toxicity discrepancies can be resolved.

In several studies that compare the relative toxicities of metals, Ni typically ranks quite low. In a comparison among five metals (Cd, Cu, Ni, Pb, and Zn), Ni was generally less toxic than Cd, Cu or Pb and equally toxic as Zn under acute exposure conditions to several species of fish (Pickering and Henderson, 1966). The relatively low toxicity of Ni compared to other metals may be explained by Ni's low gill-binding affinity (log K) relative to other metals (Niyogi and Wood, 2004). Similar results were observed by Rehwoldt et al. (1971), but the relative sensitivities were clearly species dependent. For example, Cu was more toxic than Ni in all test species, but Zn was more toxic than Ni in banded killifish and common carp, but equally toxic in striped bass, white perch, and American eel. Nickel was more toxic than Zn only in pumpkinseed. Buhl and Hamilton (1991) determined Ni to be the fourth most toxic metal of eight tested (in order of most toxic to least toxic, Cd > Ag > Hg > Ni > Au > arsenite > selenite > selenate > Cr(VI) in three fish species; however, the mean 96 h LC50 for Ni was around 10 mg L^{-1} (i.e. the toxicity was not very high).

The developmental stage of the fish is also important in determining the toxicity of Ni. In most cases, earlier life stages are more sensitive to Ni than later life stages (Birge and Black, 1980; Nebeker et al., 1985; Alam and Maughan, 1992); however, this, too, may vary with species (Table 5.2). Alevins of Arctic grayling (*Thymallus arcticus*) and coho salmon (*Oncorhynchus kisutch*) are equally sensitive to waterborne Ni as conspecific juveniles (Hedtke et al., 1982; Buhl and Hamilton, 1991). However, rainbow trout juveniles were about three times more sensitive to Ni than alevins (Buhl and Hamilton, 1991). Although several studies have demonstrated higher juvenile sensitivity to metals relative to alevins among salmonid species, this result highlights another discrepancy between rainbow trout and other salmonid species with respect to evaluating Ni toxicity.

Acute waterborne Ni exposure (12.9 mg L^{-1} for 96 h) interferes with Mg^{2+} reabsorption by the rainbow trout kidney, resulting in decreased plasma Mg^{2+} concentrations owing to increased Mg^{2+} loss through urine (Pane et al., 2005). This effect of Ni specifically antagonizing Mg^{2+} reabsorption appears to be strictly associated with acute Ni exposure. If fish are pre-exposed to a low concentration of waterborne Ni for a prolonged period (441 µg L^{-1} for 36 days) before being subjected to the acute Ni exposure, this antagonism disappears. That is, in contrast to naïve fish, an acute Ni challenge following a chronic Ni exposure causes an increase in the efficiency of Mg reabsorption by the kidney, possibly by Ni-induced upregulation of renal tubule Mg^{2+} transporters or an Ni-induced change in transporter kinetic properties (Pane et al., 2006; Pane et al., 2006a). Nickel may be behaving as an analogue of Mg in fish kidneys, resulting in highly efficient renal Ni reabsorption through increased Mg transporters after an acute Ni challenge (Pane et al., 2005).

Carp exposed to acute Ni concentrations (40 mg L^{-1}) demonstrated significant protein loss coupled to increased protease activity and free amino acid concentrations in gills and kidneys (Sreedevi et al., 1992a). This result suggests that acute Ni exposure results in proteolytic activity. This effect may have resulted from oxidative damage to lysosomal membranes causing the release of protease (a lysosomal enzyme) causing the breakdown and reduction in intracellular proteins.

5.2. Chronic Toxicity

The sublethal effects of Ni are not well documented for fish. In an early study on Ni effects on fathead minnow reproduction, Pickering (1974) determined that in hard water (207 mg L^{-1} as CaCO$_3$; pH 7.8) Ni concentrations below 1.6 mg L^{-1} were not sufficient to affect survival or growth in 6-week-old fish. Fecundity and egg hatchability were significantly reduced at concentrations at or above 730 μg L^{-1} Ni. However, neither egg hatchability nor fry survival was affected at Ni concentrations below 380 μg L^{-1}. Fish embryos and larvae exposed to sublethal Ni concentrations showed no teratogenic effects at Ni concentrations at or below those that impaired egg hatchability; however, above those concentrations rainbow trout, channel catfish, and goldfish demonstrated an increased teratogenic frequency (up to 29%) when egg hatchability was impaired by 50% or more. Largemouth bass were not as sensitive as the other species tested, given that the frequency of teratogenesis was not elevated for largemouth bass at any Ni concentration tested (Birge and Black, 1980). Common carp, however, appear to be susceptible to Ni-induced teratogenesis, given that 3, 4, and 7 mg Ni L^{-1} (hardness 128 mg L^{-1} as CaCO$_3$; pH 7.8) induced 23, 50, and 100% teratogenic frequency (Blaylock and Frank, 1979).

Egg hatchability appears to be a specific target of waterborne Ni exposure. Early studies demonstrated that egg hatchability may actually be a more sensitive endpoint than other chronic endpoints for Ni-exposed fish embryos (e.g. survival, growth). Fathead minnow egg hatchability is reduced at 730 μg L^{-1} (Pickering, 1974), common carp at about 6000 μg L^{-1} (Blaylock and Frank, 1979), and Atlantic salmon (*Salmo salar*) at 100 μg L^{-1} (Grande and Andersen, 1983). However, Dave and Xiu (1991) found that Ni specifically targeted egg hatching in zebrafish (*Danio rerio*) and significantly reduced hatchability at 40 μg L^{-1} and the variability of the response was low. Decreased egg hatchability in Ni-exposed embryos has also been reported by others for zebrafish (Scheil and Köhler, 2009). Dave and Xiu (1991) speculated that Ni may act directly on some aspect related to the hatching process in order to induce such a specific effect on zebrafish egg hatching.

Both waterborne and dietary Ni exposure can induce severe morphological and histopathological damage to fish tissues. In a study examining the effects of sublethal concentrations of waterborne Cd, Cr, and Ni on rainbow trout gills, only Ni caused significant reduction in gill diffusion capacity immediately following the exposure (Hughes et al., 1979). However, diffusion capacity recovered after 21 days in clean water. The banded gourami (*Colisa fasciata*) exposed to approximately 14 mg L^{-1} Ni for 96 h showed severe histopathological damage to the gill structure, including hypertrophy of respiratory and mucus cells, epithelial lifting, lamellar hyperplasia and clubbing, lamellar fusion, cellular necrosis, and an elevated frequency of pyknotic cells (i.e. a marked condensation of chromatin typically associated with necrosis or apoptosis) (Nath and Kumar, 1989).

Silver carp (*Hypophthalmichthys molitrix*) exposed to 5.7 mg L^{-1} of water-borne Ni for up to 30 days showed histopathological damage to gills, liver, intestine, and kidney (Athikesavan et al., 2006). In each tissue, histopathological anomalies developed over time such that the most severe lesions were observed in fish exposed to sublethal waterborne Ni for the longest period. Lesions included epithelial degeneration, increased frequency of pyknosis, necrosis, and general damage to the basic structure of the tissue. Lamellar fusion and epithelial lifting were also observed in gills, hepatocyte rupture and cellular vacuolization were observed in livers, a reduction of surface area was evident in intestines, and glomerular and renal tubule damage occurred in posterior kidneys. In lake whitefish chronically exposed to elevated concentrations of dietary Ni (up to 1 mg g^{-1} of food) for up to 104 days, focal necrosis and damaged bile ducts were observed in livers and damaged glomeruli, renal tubules, and collecting ducts were observed in kidneys (Ptashynski et al., 2002).

Many of the observed morphological effects associated with Ni exposure are reflected in physiological effects. For example, waterborne Ni appears to exert its toxic effect in fish by impairing respiratory function (Pane et al., 2003, 2004a, b). Rainbow trout exposed to 11.7 mg L^{-1} waterborne Ni for 117 h showed no impaired branchial ionoregulatory disturbance of Na^+, Cl^-, Ca^{2+}, or Mg^{2+} (Pane et al., 2003). However, arterial blood showed a marked decrease in oxygen tension, an increase in plasma carbon dioxide (CO_2) tension, and the development of a respiratory-related acidosis. This is an interesting result given that several other metals of environmental concern, such as Cd, Cu, Zn, and others, exert their toxic effects by inducing an ionoregulatory disturbance typically at the gill (Wood, 2001). It appears as though the toxic action of Ni does not involve a disruption of ionoregulation. However, Sayer et al. (1991) reported a small net loss of Ca^{2+} from brown trout (*Salmo trutta*) exposed for 72 h to 10 or 50 μg L^{-1} waterborne Ni in soft (20 μM Ca), acidic (pH 5.6) water, although the Ca^{2+} loss may be more of a reflection of exposure conditions than Ni exposure. An increased diffusive distance across respiratory surfaces, a reduced surface area for branchial gas exchange, and reduced branchial blood flow all contribute to a marked reduction in high-performance gas exchange in strenuously exercised rainbow trout held chronically in Ni-contaminated water at sublethal concentrations, but not in resting trout held under the same conditions (Pane et al., 2004a).

The European eel (*Anguilla anguilla*) is a catadromous fish showing a strong capacity for osmoregulation. Aquaporins are water transporters which are upregulated or downregulated in osmoregulatory tissues depending on osmoregulatory requirements. In a study where all four eel aquaporin orthologues were expressed in *Xenopus* oocytes, MacIver et al. (2009) demonstrated that a very short (2 min) acute (60 mg L^{-1}) exposure to Ni inhibited 88% of the activity of one of the orthologues. Although no other evidence is currently available to support a role of Ni in impairing osmoregulation, the study by MacIver et al. (2009) provides a mechanism by which Ni could disrupt osmoregulation in eels and other fish.

Hematological disturbances measured in Ni-exposed fish may also reflect Ni-induced hypoxia. Pane et al. (2003) observed elevated hematocrit and plasma lactate and a significant reduction in splenic hemoglobin in rainbow trout exposed for up to 60 h to 15.6 mg Ni L^{-1} in moderately hard water (140 mg L^{-1} as $CaCO_3$; pH 7.9). Increased hematocrit through splenic discharge probably reflects a homeostatic response to hypoxia to ensure continued oxygen delivery to peripheral tissues. The hyperlactacidemia probably reflects a reliance on anaerobic respiration to meet physiological energetic requirements under Ni-induced hypoxia. Similar hematological effects were observed for Ni-exposed tilapia (*Tilapia nilotica*), such as increased hematocrit and hemoglobin concentration, coupled with leuko-penia and lymphopenia, which may cause the fish to become secondarily susceptible to disease or infection (Ghazaly, 1992).

Fish chronically exposed to waterborne Ni concentrations demonstrate alterations in carbohydrate metabolism (Chaudhry, 1984; Chaudhry and Nath, 1985; Ghazaly, 1992; Alkahem, 1995; Jha and Jha, 1995; Canli, 1996). Increasing waterborne Ni concentrations yield marked decreases in liver and muscle glycogen reserves with concomitant increases in plasma glucose and lactic acid concentrations. Although the fluctuations in carbohydrate metabolism associated with Ni exposure may reflect a general stress response to the Ni exposure, mediated through the Ni-induced stimulation of glucocorticoids, the hyperlactacidemia may be a further reflection of Ni-induced hypoxia. Fish suffering from Ni-induced hypoxia may have to switch to anaerobic ATP production, which is less efficient and more glucose consumptive than aerobic ATP production. Consequently, Ni may be inducing glucocorticoids to mobilize liver and muscle glycogen, causing a plasma hyperglycemia, as a means of supporting anaerobic ATP production in response to Ni-induced hypoxia.

In mammals, one of the most important mechanisms of Ni toxicity is the induction of superoxide radicals (Kasprzak, 1987; Eisler, 1998; Lloyd and Phillips, 1999; Denkhaus and Salnikow, 2002; Leonard et al., 2004). These reactive oxygen species readily bind to DNA, proteins, and other important biomolecules, causing cellular and molecular damage and physiological dysfunction. Several antioxidant systems are available to mitigate this damage, and induction of these systems is often an indicator of oxidative stress. Nickel-induced oxidative stress has not been thoroughly investigated in fish, in contrast to mammals for which links between Ni exposure, oxidative stress and DNA damage have been reported. In fish, Ni-induced DNA damage in the form of DNA-protein cross-links reflects the potential for genotoxic Ni effects and a biomarker for Ni exposure (Kuykendall et al., 2009).

Fish exposed to sublethal Ni concentrations ($8 \, mg \, L^{-1}$) showed a significant increase in structural and total protein content, indicating that biosynthetic activity may have increased (Sreedevi et al., 1992a). This effect of Ni-induced biosynthetic activity (measured as the activity of nucleoside diphosphate kinase; NDPK) was also observed together with elevated aerobic metabolic activity in larval fathead minnows exposed to low waterborne Ni concentrations (i.e. cytochrome c oxidase activity) (Lapointe and Couture, 2010). Time to hatch was also reported to be affected by Ni. In agreement with two earlier studies (Pyle, 2000; Pyle et al., 2002a), Lapointe and Couture (2010) reported that exposure to $250 \, \mu g \, L^{-1}$ Ni reduced time to hatch in fathead minnow, possibly due to an Ni-induced increase in metabolism as suggested by higher activities of enzyme indicators of aerobic and biosynthetic capacities. Hence, in contrast to other metals, chronic Ni exposure appears to increase aerobic capacities. Indeed, studies of wild yellow perch support this hypothesis. Couture et al. (2008b) reported that the activity of aerobic enzymes was higher in Ni-contaminated yellow perch. They hypothesized that the enhanced activity of aerobic enzymes is a compensation for oxidative damage to mitochondrial membranes. Pierron et al. (2009) also reported a positive correlation between liver Ni concentration and aerobic enzyme activity in wild yellow perch. They suggested that higher aerobic capacities could reflect a compensation of direct effects of Ni on protein function or, alternately, simply match higher metabolic demands or Ni-induced detoxification and repair. Evidence of a direct stimulation of aerobic enzymes by Ni has been reported. In a recent study, Garceau et al. (2010) observed that the activities of both citrate synthase and cytochrome c oxidase were enhanced by simple additions of Ni in goldfish tissue homogenates. The mechanisms for this direct activation are not known. However, unlike Cd or Cu, there is no evidence that high concentrations of Ni disrupt fish mitochondrial respiration *in vitro* (Garceau et al., 2010).

6. NICKEL ESSENTIALITY

Nickel essentiality in plants and microorganisms is well established (Gerendás et al., 1999; Phipps et al., 2002). There are eight known

Ni-containing enzymes, seven of which are involved with the generation or utilization of gases including carbon monoxide (CO), carbon dioxide (CO_2), methane, hydrogen, ammonia (NH_3), and oxygen (Ragsdale, 2009). These Ni-containing enzymes include CO dehydrogenase (interconverts CO and CO_2), acetyl Co-A synthase (utilizes CO), acireductone dioxygenase (generates CO), hydrogenase (both utilizes and consumes H_2), methyl Co-M reductase (generates methane), urease (generates NH_3 and CO_2), and Ni-superoxide dismutase (generates oxygen) (Ragsdale, 2009). Consequently, the biochemistry of Ni plays a critical role in the global carbon, nitrogen, and oxygen cycles (Ragsdale, 1998, 2007, 2009). The other Ni-containing enzyme, glyoxylase I, converts methylglyoxal, which forms covalent adducts with DNA, to lactate. All of the known Ni-containing enzymes have been isolated from plants or microorganisms, and none is known from animal tissues.

Nickel essentiality in animals has been difficult to establish because no Ni-containing biomolecule has yet been isolated from their tissues. Nickel deficiency has been studied in six groups of vertebrate animals: chicken, cow, goat, pig, rat, and sheep (Nielsen, 2000). However, the results of many of these studies are suspect because some of the reported effects of Ni deficiency may reflect pharmacological effects of Ni or effects of Ni related to the experimental animal's nutritional status as opposed to true Ni deficiency (Nielsen, 1993, 2000). In other studies, Ni deficiency has been associated with a reduction in growth, prolonged gestation period, reduced reproductive output, depressed plasma glucose concentrations, anemia, skin lesions, reduced hemoglobin concentrations, reduced hematocrit, and reduced enzyme activity (Anke et al., 1984; Nielsen, 2000; Denkhaus and Salnikow, 2002; Phipps et al., 2002; Muyssen et al., 2004). Nickel has also been implicated in the redistribution of other mineral nutrients, such as Ca, Fe, and Zn (Nielsen, 2000), and may also be involved in the proper function of vitamin B_{12} (Anke et al., 1984; Nielsen, 1991, 1993).

The evidence for Ni's essentiality in aquatic animals is circumstantial and equivocal. Far less research attention has been directed towards establishing the effects of dietary Ni deficiency in aquatic animals than has been directed at the terrestrial species discussed above (Phipps et al., 2002; Muyssen et al., 2004). In fish, the best evidence of possible Ni essentiality rests with the observations that Ni concentrations in fish tissues remain relatively constant despite wide fluctuations in environmental concentrations (Tjälve et al., 1988; Ray et al., 1990), Ni can be reabsorbed by the kidney (Sreedevi et al., 1992b; Ptashynski and Klaverkamp, 2002), and Ni uptake from food and water appears to be regulated (Lapointe and Couture, 2009).

Nickel can be found in all animal tissues at low concentrations even in the absence of obvious contaminant Ni exposure (Nielsen, 1987). This observation may indicate that Ni is actively participating in the normal function of fish cells even in the absence of an elevated environmental Ni load. Fish exposed to experimentally elevated waterborne or dietary Ni concentrations preferentially accumulate Ni in kidneys relative to other tissues (see below) (Ghazaly, 1992; Sreedevi et al., 1992b; Ptashynski and Klaverkamp, 2002; Pane et al., 2004a, b, 2005). This preferential deposition pattern may reflect renal clearance and one mechanism by which Ni is regulated in fish. Gulf toadfish maintained similar kidney Ni concentrations after being exposed for 72 h to 12.6 or 35.2 mg L^{-1} waterborne Ni or having Ni artificially infused (23.5 µg kg^{-1} h^{-1}) directly into their arteries, which suggests homeostatic control of Ni concentrations in fish kidneys (Pane et al., 2006b). Nickel can also be deposited in other structures, such as scales (Ptashynski and Klaverkamp, 2002) or granules (Lapointe and Couture, 2009; Lapointe et al., 2009), as a means of regulating potentially toxic concentrations from accumulating at physiologically sensitive sites and inducing toxicity. Nickel exposure may also induce metallothionein production in Ni-exposed fish (Ptashynski et al., 2002), given its consistent association with a heat-stable subcellular fraction in Ni-exposed fish (Giguère et al., 2006; Campbell et al., 2008; Lapointe and Couture, 2009; Lapointe et al., 2009).

Fish appear to be able to actively regulate dietary Ni uptake. Rainbow trout (*Oncorhynchus mykiss*) that were pre-exposed to low concentrations of waterborne Ni downregulated intestinal Ni uptake from their food (Chowdhury et al., 2008). Despite relatively large variations in dietary Ni content, these rainbow trout maintained relatively constant Ni concentrations (in gill, kidney, liver, bile, stomach, and scale tissues), which suggests that Ni uptake is highly regulated. Lapointe and Couture (2009) showed that fathead minnows (*Pimephales promelas*) fed an Ni-contaminated diet maintained constant whole body Ni concentrations over the duration of an 8 day exposure.

These observations suggest that Ni concentrations in fish are under rather tight physiological control. This level of physiological regulation is not known for non-essential metals and is consistent with similar observations in other known essential metals, such as Zn (Hogstrand, Chapter 3) or Cu (Grosell, Chapter 2). Consequently, despite the circumstantial nature of the evidence for essentiality of Ni in fish, there is some reason to speculate that Ni is indeed essential. If it turns out that Ni is essential for fish, the dietary requirements are likely to be considerably lower than background Ni concentrations in a typical fish diet (Nielsen, 1993). In other words, Ni deficiency is likely to be a rare occurrence in natural environments.

7. POTENTIAL FOR BIOMAGNIFICATION OR BIOCONCENTRATION OF NICKEL

There is no evidence for either biomagnification or bioconcentration of Ni in aquatic ecosystems. Muyssen et al. (2004) reported a negative relationship between exposure concentration and bioconcentration factors in fish after reviewing the available literature. They suggested that this negative relationship can be explained by fish actively regulating Ni uptake and elimination processes.

8. CHARACTERIZATION OF UPTAKE ROUTES

8.1. Gills

The mechanism of Ni uptake is not currently known. However, fish exposed to relatively high concentrations of waterborne Ni tend to show concentration-dependent Ni accumulation in plasma, suggesting that Ni can cross the gill epithelium and enter the bloodstream (Pane et al., 2006b; Chowdhury et al., 2008). At lower exposure concentrations, branchial Ni uptake kinetics is saturable and probably involves a low-capacity, high-affinity transporter (Brix et al., 2004). Branchial Ni uptake can be facilitated by certain lipophilic organo-nickel complexes, which can be an environmental concern around some industrial operations (Tjälve and Borg-Neczak, 1994).

8.2. Gut

Dietary Ni is taken up primarily through the stomach and mid-intestine (Ojo and Wood, 2007; Leonard et al., 2009). In the stomach, Ni is taken up by a high-affinity, low-capacity transporter, whereas in the mid-intestine it is taken up by a lower affinity, but much higher capacity transporter (Leonard et al., 2009). In the anterior intestine, Ni uptake is passive (Leonard et al., 2009). Dietary Ni assimilation efficiency in fathead minnows fed Ni-contaminated invertebrates was calculated to be 10–11%, despite having 50–85% of total foodborne Ni in a bioavailable form (Lapointe et al., 2009).

9. INTERNAL HANDLING OF NICKEL

9.1. Biotransformation

No studies were identified that discussed the biotransformation of Ni in either the mammalian literature or the fish literature.

9.2. Transport through the Bloodstream

Most studies examining the vascular transport of Ni are based on mammals. Very few studies focus on fish. Nickel is transported by blood in any one of four different forms: (1) as a free, divalent cation, Ni^{2+}, (2) as a small, ultrafilterable complex, (3) as a protein complex, or (4) bound to blood cells (Kasprzak, 1987). Nickel is not evenly distributed among these forms and is dependent on the fish species as well as the Ni binding affinity to plasma albumins. Some 90% of all circulating Ni is bound to plasma albumins, with the remaining Ni bound to free amino acids or small plasma peptides. Nickel has a high affinity for cysteine (Cys) and histidine (His), and small peptides with high concentrations of Cys and His residues (Kowalik-Jankowska et al., 2007), including nickeloplasmin, an α-macroglobulin that has a particularly high binding affinity for Ni, but low binding capacity (Kasprzak, 1987). The physiological significance of nickeloplasmin is not yet fully understood.

9.3. Accumulation in Specific Organs

Although Ni preferentially deposits in bone, gills, and kidneys, several studies have documented Ni accumulation patterns in various fish tissues. Elevated Ni concentrations have been observed in several tissues, including gill, kidney, skeleton, white muscle, liver, brain, heart, stomach, intestine, skin, scales, and gonads, following either a waterborne or dietary Ni exposure (Sreedevi et al., 1992b; Canli and Kargin, 1995; Ptashynski et al., 2001, 2002; Ptashynski and Klaverkamp, 2002). However, the blood plasma represents the main sink for Ni once taken up by the fish (Pane et al., 2004a). Consequently, Ni accumulation patterns in several of the tissues listed above can be accounted for by blood-bound Ni via tissue vascularization (Pane et al., 2004a, b). Although most Ni accumulation can be attributed to plasma trapping in various tissues, Ni accumulation in the gill and kidney is unrelated to plasma Ni and is more a reflection of preferential deposition and accumulation via intracellular processes (Pane et al., 2004a, b). Furthermore, bones and scales preferentially accumulate dietary Ni in addition to preferential accumulation in the kidneys (Ptashynski and Klaverkamp, 2002).

9.4. Subcellular Partitioning

Once Ni is removed from the albumin fraction in blood plasma, it can go on to form a complex with L-histidine (Eisler, 1998). The Ni–histidine complex can then pass through cell membranes, allowing Ni to be taken up into cells. Once taken up, Ni can target the nucleus and nucleolus in addition to other subcellular components.

Research into the subcellular partitioning of Ni in fish is scarce, and most of the available research has only recently been published (Giguère et al., 2006; Campbell et al., 2008; Lapointe and Couture, 2009; Lapointe et al., 2009). In general, this research has focused on two main subcellular fractions: the metal-sensitive fraction and the metal-detoxified fraction (Campbell et al., 2008). The metal-sensitive (or heat-denatured) fraction includes physiologically sensitive biomolecules such as glutathione, metalloenzymes, DNA/RNA, and the cellular organelles. The metal-detoxified fraction (or heat-stable fraction) includes the various cellular subsystems in place to protect the metal-sensitive systems against metal intoxication, and includes the metallothioneins, lysosomes, granules, and membrane-bound vesicles. Metals associated with the metal-sensitive fraction may lead to toxicity, whereas metals associated with the metal-detoxified fraction are thought to be sequestered from the metal-sensitive sites in the cell and are considered detoxified.

Nickel was observed in both hepatic subcellular fractions in wild yellow perch (*Perca flavescens*) collected from several lakes along a metal contamination gradient (Giguère et al., 2006). Unlike subcellular distribution patterns observed for Cd, most of the Ni was found in the metal-sensitive fraction, and only a small portion of Ni was observed in the metal-detoxified fraction. Among the constituents that comprise the metal-sensitive fraction, Ni showed a particularly high affinity for the heat-denaturable proteins (e.g. enzymes). The authors speculate that there is no threshold Ni concentration below which Ni is completely sequestered from the metal-sensitive fraction and consequently the fish is never completely protected against potential Ni toxicity (Campbell et al., 2008). Wild fish that are chronically exposed to metals such as Ni may trade off the cost of complete detoxification against the relative cost of partial metal binding to metal-sensitive constituents which may be tolerable under the right environmental conditions.

In fathead minnows exposed for 8 days to waterborne or dietary Ni, or both, most of the internalized Ni was associated with the metal-detoxified fraction, cellular constituents such as heat-stable proteins, granules, and cellular debris (Lapointe and Couture, 2009). Subcellular Ni partitioning may also be influenced by prey type. For example, fathead minnows fed Ni-contaminated prey (*Daphnia magna* or *Tubifex tubifex*) showed differential Ni distribution among subcellular compartments (Lapointe et al., 2009). Fish fed Ni-contaminated *D. magna* had a significantly higher proportion of assimilated Ni associated with the metal-sensitive fraction than fish fed Ni-contaminated *T. tubifex*.

9.5. Detoxification and Storage Mechanisms

Given that Ni may indeed behave like an essential nutrient (see Section 6), it is reasonable to expect active detoxification and storage mechanisms for Ni. However, very little work has been conducted in this area. Metallothioneins are a family of low molecular weight (6–10 kDa), cysteine-rich, heat-stable proteins that function to regulate divalent cations (such as metals) *in vivo* (Roesijadi, 1992, 1994). Metals bound to metallothioneins are thought to be removed from any metal-sensitive physiological processes in the cell. Therefore, metallothioneins are thought to play a role in protecting the fish against metal intoxication.

In mammals, the evidence for Ni-induced metallothionein production and subsequent Ni sequestration by metallothionein is mixed (Denkhaus and Salnikow, 2002). In fish, however, Ni-induced metallothionein production is dependent on the route of exposure. Rainbow trout exposed for 7 days to $6.6 \, mg \, L^{-1}$ of waterborne Ni showed elevated levels of metallothionein in gills relative to control fish (Pyle, 2000). Lake whitefish fed up to $1000 \, \mu g \, Ni \, g^{-1}$ of food for up to 104 days showed elevated metallothionein production in the intestine (Ptashynski et al., 2002). Several studies have also demonstrated that Ni occurs in subcellular fractions associated with metallothioneins (see Section 9.4). These studies suggest Ni-induced metallothionein production in Ni-exposed fish and point to another possible mechanism by which fish regulate Ni uptake at the gills and intestines.

Fish may also divert excess Ni burdens to non-essential, non-metal-sensitive tissues, such as bones and scales. Lake whitefish chronically exposed to $1000 \, \mu g \, L^{-1}$ food of dietary Ni for up to 104 days showed a dose-dependent increase in Ni concentrations of the skeleton and scales (Ptashynski and Klaverkamp, 2002). The mechanism for such storage remains unknown.

9.6. Homeostatic Control

A hallmark of metal essentiality is homeostatic regulation. Although essentiality has not been shown definitively for Ni (see Section 6), Ni concentration and tissue distribution do appear to be under homeostatic control. Nickel is taken up through the gills during a waterborne Ni exposure (Pane et al., 2003), or through the digestive tract during dietary exposure (Ptashynski et al., 2001; Ptashynski and Klaverkamp, 2002). The mechanism of Ni uptake in either case is currently unknown, but may involve phagocytosis for insoluble Ni species (Denkhaus and Salnikow, 2002) or a Mg^{2+} transporter or proton-coupled divalent metal transporter,

DMT1, similar to what has been described in mammals (Gunshin et al., 1997; Chowdhury et al., 2008). Nickel uptake kinetics are saturable only at low exposure concentrations, which suggests the possible involvement of low-capacity branchial binding sites (Brix et al., 2004). These binding sites might be similar to those described for Cu, which include high-affinity, low-capacity sites and low-affinity, high-capacity sites (Taylor et al., 2002). At higher waterborne Ni concentrations, uptake kinetics do not appear to be saturable (Brix et al., 2004).

It is not currently known whether Ni uptake through the intestinal epithelium is saturable. However, it does appear that intestinal Ni uptake is regulated and occurs primarily in the stomach and mid-intestine (Ojo and Wood, 2007; Leonard et al., 2009). Whitefish (*Coregonus clupeaformis*) fed a diet that was artificially contaminated with Ni showed the highest Ni concentrations on day 10 of a 104 day exposure (Ptashynski and Klaverkamp, 2002). After day 10, intestinal Ni concentrations decreased owing to the possible engagement of protective mechanisms. One possible protective mechanism is that Ni is absorbed into the intestinal mucosa by the same mechanism as that described for Fe in mammals (Tallkvist and Tjalve, 1997), and the Ni-laden mucosa is eventually desquamated and excreted (Ptashynski and Klaverkamp, 2002).

The first putative demonstration of active regulation of Ni uptake (i.e. homeostatic control) involves a complex interaction between the gills and gastrointestinal tracts of Ni-exposed rainbow trout (Chowdhury et al., 2008). In this case, rainbow trout that were pre-exposed to relatively low concentrations of waterborne Ni demonstrated a significant decrease in subsequent intestinal Ni uptake from a dietary Ni source. This result suggests that Ni uptake sites in the intestinal epithelium were downregulated as a consequence of the pre-exposure to waterborne Ni. This homeostatic phenomenon has been observed for other essential metals, such as Cu (Kamunde et al., 2002) and Zn (Chowdhury et al., 2003), but not with non-essential metals, such as Cd (Chowdhury et al., 2003).

Once taken up, Ni tends to accumulate preferentially in the kidney (Ptashynski et al., 2001; Ptashynski and Klaverkamp, 2002; Pane et al., 2004a, b; Chowdhury et al., 2008). This preference for the kidney probably reflects another homeostatic control mechanism as a means to eliminate excess Ni. Rainbow trout that have been chronically exposed to waterborne Ni have demonstrated both renal Ni secretion and reabsorption (Pane et al., 2005, 2006a, c). The observation of renal Ni reabsorption is strong evidence of homeostatic control of internal Ni. Moreover, fish chronically exposed to Ni will reduce their Ni reabsorption rate relative to Ni-naïve fish (Pane et al., 2006c). No direct evidence of Ni regulation exists for wild fish. However, yellow perch from Ni-contaminated lakes around Sudbury, Canada

(Ni concentrations ranged between 0.9 and 175 $\mu g\,L^{-1}$), appear better at regulating their internal Ni concentrations than fish from another area where Ni concentrations are lower (Rouyn-Noranda; Ni concentrations ranged between <0.2 and 3.5 $\mu g\,L^{-1}$), because Sudbury fish had lower tissue Ni accumulation than Rouyn-Noranda fish exposed to similar or lower aqueous Ni concentrations (Couture et al., 2008a). Hence, the authors suggested that unlike Rouyn-Noranda fish, Sudbury fish may have evolved a capacity for Ni regulation due to a higher historical (background) environmental Ni concentration in this richly mineralized basin compared to Rouyn-Noranda. Whether this study demonstrates pollution-induced selective pressure or individual acclimation remains to be clearly demonstrated. In support of the hypothesis that these fish may have evolved some capacity for Ni regulation, Bourret et al. (2008) have shown that yellow perch from Sudbury are genetically distinct from those of Rouyn-Noranda and that within-population genetic diversity decreased along with increases in Cd contamination.

10. CHARACTERIZATION OF EXCRETION ROUTES

10.1. Gills

In freshwater fish, gills are unlikely to serve as an important route of Ni excretion. Nickel concentrations in rainbow trout gill tissues increased significantly with increasing waterborne Ni exposure concentrations (Pane et al., 2004b). However, when rainbow trout were infused with Ni, only a very small proportion of plasma-borne Ni was incorporated into gill cells. Most of the Ni that was present in the gills following the infusion could be accounted for via plasma trapping. These results indicate that branchial Ni excretion is unlikely.

10.2. Gut

Although Ni excretion by the gut was not directly measured, a study by Pane et al. (2004b) suggests that it occurs. They demonstrated that Ni accumulation patterns varied in the gut as a function of exposure type. Nickel accumulation was higher in gut tissues following a waterborne Ni exposure than following Ni infusion. Although the authors could not discount that fish exposed to waterborne Ni increased their drinking rates to account for the elevated intestinal Ni concentrations, plasma trapping, if it occurred, was insufficient to prevent Ni from being incorporated into gut tissue cells (Pane et al., 2003, 2004b). Moreover, biliary Ni excretion is not important in freshwater fish and cannot account for the elevated Ni concentrations observed in gut tissues of water-exposed fish. The implication is that the gut (stomach and intestine) may be an important excretory route for Ni.

10.3. Liver and Bile

Biliary Ni excretion may not be an important mechanism in freshwater fish. After a 117 h exposure to 11.6 mg Ni L^{-1}, rainbow trout did not significantly accumulate Ni in either the liver or the bile (Pane et al., 2003). Wild yellow perch inhabiting Ni-contaminated waters demonstrate elevated liver Ni concentrations, but it is not clear whether the Ni in these fish is being cleared (Pyle et al., 2005; Couture et al., 2008a). No studies are currently available to support significant biliary excretion in fish.

10.4. Kidney

The kidneys represent a site of preferential Ni accumulation in freshwater fish. However, Ni is reabsorbed very efficiently by the kidneys. In one study, some 98% of the Ni filtered by the glomeruli was reabsorbed (Pane et al., 2005). Consequently, renal excretion is not an important route of Ni clearance in freshwater fish. In contrast, renal Ni excretion is much more important in marine fish. At 72 h after Ni infusion, gulf toadfish (*Opsanus beta*) excreted 30% of the infused Ni via the kidneys (Pane et al., 2006b). These observations are consistent with the hypothesis that Ni transport in kidneys is facilitated by a Mg^{2+} transporter (Pane et al., 2006a). These Mg^{2+} transporters are important for Mg clearance in marine fish and therefore contribute to the kidney's role in Ni excretion in species such as the gulf toadfish (see Section 5.1).

11. CHEMOSENSORY AND BEHAVIORAL EFFECTS

It has long been known that exposure to elevated concentrations of waterborne metals can interfere with normal fish behavior (Atchison et al., 1987). Behaviors affected by metals include basic locomotory function (i.e. hypoactivity or hyperactivity), avoidance from or attraction to areas having elevated contaminant concentrations, or behaviors associated with the perception of important chemical cues, such as sex pheromones that mediate reproduction, food cues associated with foraging, or predator odors that elicit predator-avoidance behaviors. The ecological relevance of these subtle effects cannot be overstated. Chemosensory and behavioral effects in fish following exposures to metals other than Ni (e.g. Cu) have been shown to occur at concentrations that are far below those required to induce overt toxicity (Pyle and Mirza, 2007; Pyle and Wood, 2007). Chemosensory and behavioral effects resulting from exposure to Ni have received comparatively little research attention.

The Ni avoidance threshold for rainbow trout in soft water was determined to be 24 μg L^{-1} in a gradient assay (Giattina et al., 1982). This behavioral response, unlike Cu, was not dependent on the slope of the gradient; rainbow trout exposed to Ni in either a steep- or shallow-gradient assay yielded the same response to waterborne Ni. Rainbow trout were significantly attracted to low Ni concentrations (6 μg L^{-1}) relative to Ni-free water, but avoided higher Ni concentrations (>19 μg L^{-1}). Attraction appears to be an appropriate response to a potentially essential nutrient at low concentrations and avoidance is appropriate for a potentially toxic contaminant at higher concentrations. Because very little information exists on the mechanism of Ni toxicity to fish olfactory systems, it is difficult to determine whether or not this attraction to low Ni concentrations can be attributed to chemosensory impairment. Moreover, no work has been conducted to establish the potential for Ni-induced chemosensory dysfunction in fish, especially as it pertains to the perception of ecologically relevant chemical cues.

A study that exposed northern pike (*Esox lucius*) to 400 μg L^{-1} of radiolabeled Ni sealed in their olfactory chambers for up to 10 days demonstrated that Ni could be taken up into olfactory sensory neurons and transported towards the olfactory bulb via slow, anterograde axonal transport (Tallkvist et al., 1998). This transport likely involves Ni binding to molecules having a molecular weight of less than 250, such as L-histidine, or other cytosolic constituents. Because of the slow transport rate of Ni (approximately 3 mm day^{-1}, and about 20 times slower than for Cd or Mn) (Gottofrey and Tjälve, 1991; Tjälve et al., 1995) no Ni was observed in pike neural tissues downstream of olfactory sensory neurons owing to the 10 day limit of the exposure period. However, Ni was observed in downstream regions of Ni-exposed rat brains following intranasal instillation, suggesting that Ni can pass from primary to secondary or tertiary olfactory neurons (Henriksson et al., 1997). The mechanism for interneural transmission of Ni, and whether or not it occurs in aquatic animals, is currently unknown.

Brown et al. (1982) observed a slight reduction in rainbow trout neurophysiological [electroencephalogram (EEG)] responses to L-serine after short-term (30 min) exposure to concentrations of Ni at or above 60 mg L^{-1} (1 mM). This inhibited EEG response to L-serine was the weakest among the eight metals tested on rainbow trout (in order of inhibitory efficacy: Ag > Hg > Cu > Cd > Zn > Pb > Co > Ni).

12. GENOMIC, PROTEOMIC, AND GENOTOXIC EFFECTS

The genotoxicity of Ni in mammals is well known. Exposure to Ni triggers lipid peroxidation and decreases glutathione peroxidase (an enzyme involved in the protection against oxidative stress) activity. Since Ni exposure is associated with an increase in tissue Fe levels, Ni-related oxidative stress and damage may be caused by Fe accumulation. Carcinogenicity would follow on from DNA damage induced by oxidative stress (Stohs and Bagchi, 1995; Denkhaus and Salnikow, 2002). No such evidence is available for fish, except for one publication in which De Luca et al. (2007) reported that Ni induced oxidative stress in rainbow trout erythrocytes, and another that demonstrated Ni induced DNA-protein cross-links in erythrocytes (Kuykendall et al., 2009).

Nickel has been reported to modify the transcription level of about 1300 genes in yeast, among which about 700 were downregulated (Takumi et al., 2010). In strong support of the mammalian literature, genes induced included those involved in response to oxidative stress, DNA damage repair, and iron metabolism. There is very little information on modifications of Ni-induced gene transcription level in fish. Pierron et al. (2009) did not report any relationship between tissue Ni concentration and gene transcription level in wild yellow perch chronically exposed to a polymetallic mixture along a contamination gradient, even though the activity of cytochrome c oxidase as well as total tissue protein concentrations were positively correlated with Ni concentration in liver. Acute (300 µg L^{-1}) 48 h aqueous Ni exposure, however, has been shown to induce the transcription of one isoform of metallothionein (*MLMT-IB*) in liver, kidney and spleen, but not intestine, of the mud loach (*Misgurnus mizolepis*) (Cho et al., 2009). Metallothionein isoform *MLMT-IA* was also induced in kidney and spleen, but its transcription level was decreased in liver following acute Ni exposure. In the former two tissues, Ni was a stronger inducer of the *IB* isoform than other metals tested (Cd, Cu, Cu, Fe, Mn, and Zn). Although the tissue specificity of the induction of metallothionein by Ni is not surprising, the isoform-specific changes in transcription following Ni exposure suggest that genomic approaches could not only improve our understanding of the mechanisms of Ni toxicity but also be used as tools to identify Ni-specific signatures at the level of transcriptomics.

13. NICKEL INTERACTION WITH OTHER METALS

The most well-known metal interaction with Ni is the specific antagonism of Mg. Although Ni–Mg antagonism has been observed in several taxonomic groups, including bacteria, fungi, birds, and mammals (Eisler, 1998), it has only recently been described in fish (see Section 6.3; Pane et al., 2005, 2006c; Leonard et al., 2009). It may be that Ni^{2+} serves as an analogue of Mg^{2+} and shares Mg^{2+} transporters. Nickel also interferes with Fe^{2+} and Fe^{3+} absorption in the mid- and posterior intestinal segments of fish, probably by inhibition of DMT1 (Kwong and Niyogi, 2009). Ni was the strongest inhibitor of intestinal Fe uptake in rainbow trout among six metals tested (Ni, Pb, Cd, Cu, Zn, and Co).

Other interactions of Ni with other metals are largely known from studies involving non-piscine vertebrates. Nickel can interfere with metalloenzyme function by competing for metal cofactors, such as Ca, Fe, Mg, Mn, and Zn (Kasprzak, 1987). Nickel can also bind to and activate calmodulin, a protein involved in Ca-mediated signal transduction (Kasprzak, 1987; Eisler, 1998). Consequently, Ni interactions with Ca occur in cells where Ca flux is mediated by calmodulin signaling (e.g. olfactory tissues). Calcium, Cu, Mg, Mn, and Zn inhibit Ni binding to DNA and reduce Ni-induced tumorigenesis in mammalian models (Eisler, 1998).

14. KNOWLEDGE GAPS AND FUTURE DIRECTIONS

Nickel is less studied than other metals with respect to fish. This is probably because other metals, such as Cu or Cd, are potentially more acutely toxic to aquatic animals, thereby warranting the research attention. However, as this review has demonstrated, Ni is ubiquitous in its distribution, is known to occur at elevated concentrations around human industrial and urban activities, and has potential for inducing respiratory and genotoxic effects in fish. More research attention is required. Because Ni is well recognized as a contact allergen and a powerful carcinogen in mammals, most of the research attention has focused on mammalian systems with an eye towards extrapolating effects from model organisms to humans. The literature available for these mammalian studies is relatively vast compared to what is available for aquatic systems, and should serve as a guide for directing future research into Ni effects on fish.

Recent research has demonstrated that fish inhabiting environments high in Ni can maintain relatively low tissue Ni concentrations, probably through homeostatic regulatory processes (Couture and Pyle, 2008). More research should be directed towards establishing whether or not Ni is an essential nutrient for fish. As discussed, mounting evidence suggests that Ni is indeed essential. However, final confirmation requires identifying an Ni-activated biomolecule(s) and/or establishing and characterizing an Ni-deficiency syndrome. Understanding Ni as an environmental toxicant can only be fully appreciated with a better understanding of Ni's role as a micronutrient. Given that Ni is probably both a micronutrient and a toxicant under different conditions, understanding the difference between the two at a molecular or cellular level will be important. Contemporary cellular and molecular techniques can provide mechanistic insights into fundamental processes involving Ni, such as branchial or intestinal uptake routes, detoxification pathways, or basic metabolism. Because Ni is a respiratory toxicant, these same techniques should be used in both rested and exercised animals.

Additional research should also be directed towards understanding the ecological implications to fish populations inhabiting Ni-contaminated environments. In particular, Ni toxicity to other trophic levels (from phytoplankton to benthic invertebrates) may have severe indirect consequences for fish. Moreover, increasing evidence suggests that elevated metals can interfere with chemical communication systems among aquatic animals, which may have very serious ecological implications (Scott and Sloman, 2004; Lurling and Scheffer, 2007). At present, almost nothing is known about Ni effects on chemosensation. Given that Ni is known to be taken up via olfactory pathways (Tallkvist et al., 1998) and that it can interfere with Ca flux by interacting with calmodulin (Eisler, 1998), Ni is likely to affect critical chemical communication systems in fish.

Although Ni has been studied in fish for over 30 years, continued research into the basic physiology and toxicology of Ni to fish is sure to reveal many interesting, if not surprising, insights.

REFERENCES

Alam, M. K., and Maughan, O. E. (1992). The effect of malathion, diazinon, and various concentrations of zinc, copper, nickel, lead, iron, and mercury on fish. *Biol. Trace Elem. Res.* **34**, 225–236.
Alkahem, H. F. (1995). Effects of nickel on carbohydrate metabolism of *Oreochromis niloticus*. *DIRASAT (Pure and Applied Sciences)* **22B**, 83–88.
Anke, M., Groppel, B., and Kronemann, H. (1984). Nickel – an essential element. *IARC Scientific Publications* **53**, 339–365.
ANZGFMQ (2000). Aquatic ecosystems. In: *Australian and New Zealand Guidelines for Fresh and Marine Water Quality*, Vol. 1. *The Guidelines*. Australian and New Zealand Guidelines for Fresh and Marine Water Quality. http://www.mincos.gov.au/__data/assets/pdf_file/0019/316126/wqg-ch3.pdf
Arnott, S. E., Yan, N. D., Keller, B., and Nichols, K. (2001). The influence of drought-induced acidification on the recovery of plankton in Swan Lake (Canada). *Ecol. Appl.* **11**, 747–763.
Atchison, G. J., Henry, M. G., and Sandheinrich, M. B. (1987). Effects of metals on fish behavior: a review. *Environ. Biol. Fish.* **18**, 11–25.

Athikesavan, S., Vincent, S., Ambrose, T., and Velmurugan, B. (2006). Nickel induced histopathological changes in the different tissues of freshwater fish, *Hypophthalmichthys molitrix* (Valenciennes). *J. Environ. Biol.* **37**, 391–395.

Birge, W. J., and Black, J. A. (1980). Aquatic toxicology of nickel. In: *Nickel in the Environment* (J.O. Nriagu, ed.), pp. 349–366. John Wiley & Sons, New York.

Blaylock, B. G., and Frank, M. L. (1979). A comparison of the toxicity of nickel to the developing eggs and larvae of carp (*Cyprinus carpio*). *Bull. Environ. Contam. Toxicol.* **21**, 604–611.

Bourret, V., Couture, P., Campbell, P. G. C., and Bernatchez, L. (2008). Evolutionary ecotoxicology of wild yellow perch (*Perca flavescens*) populations chronically exposed to a polymetallic gradient. *Aquat. Toxicol.* **86**, 76–90.

Brix, K. V., Keithly, J., DeForest, D. K., and Laughlin, J. (2004). Acute and chronic toxicity of nickel to rainbow trout (*Oncorhynchus mykiss*). *Environ. Toxicol. Chem.* **23**, 2221–2228.

Brown, R. E. E., Thompson, B. E., and Hara, T. J. (1982). Chemoreception and aquatic pollutants. In: *Chemoreception in Fishes* (T.J. Hara, ed.), pp. 363–393. Elsevier, Amsterdam.

Brown, V. M. (1968). The calculation of the acute toxicity of mixtures of poisons to rainbow trout. *Wat. Res.* **2**, 723–733.

Brown, V. M., and Dalton, R. A. (1970). The acute lethal toxicity to rainbow trout of mixtures of copper, phenol, zinc and nickel. *J. Fish Biol.* **2**, 211–216.

Buhl, K. J., and Hamilton, S. J. (1991). Relative sensitivity of early life stages of Arctic grayling, coho salmon, and rainbow trout to nine inorganics. *Ecotoxicol. Environ. Saf.* **22**, 184–197.

Campbell, P. G. C., Kraemer, L. D., Giguère, A., Hare, L., and Hontela, A. (2008). Subcellular distribution of cadmium and nickel in chronically exposed wild fish: inferences regarding metal detoxification strategies and implications for setting water quality guidelines for dissolved metals. *Hum. Ecol. Risk Assess.* **14**, 290–316.

Canadian Council of Ministers of the Environment (CCME) (2007). Canadian Water Quality Guidelines for the Protection of Aquatic Life. In: *Canadian Environmental Quality Guidelines*. Winnipeg, MN: Canadian Council of Ministers of the Environment. http://www.ccme.ca/assets/pdf/aql_summary_7.1_en.pdf

Canli, M. (1996). Effects of mercury, chromium and nickel on glycogen reserves and protein levels in tissues of *Cyprinus carpio*. *Turk. J. Zool* **20**, 161–168.

Canli, M., and Kargin, F. (1995). A comparative study on heavy metal (Cd, Cr, Pb and Ni) accumulation in the tissue of the carp *Cyprinus carpio* and the Nile fish *Tilapia nilotica*. *Turk. J. Zool.* **19**, 165–171.

Celo, V., Murimboh, J., Salam, M. S., and Chakrabarti, C. L. (2001). A kinetic study of nickel complexation in model systems by adsorptive cathodic stripping voltammetry. *Environ. Sci. Technol.* **35**, 1084.

Chakraborty, P., Gopalapillai, Y., Murimboh, J., Fasfous, I. I., and Chakrabarti, C. L. (2006). Kinetic speciation of nickel in mining and municipal effluents. *Anal. Bioanal. Chem.* **386**, 1803–1813.

Chau, Y. K., and Kulikovsky-Cordeiro, O. T. R. (1995). Occurrence of nickel in the Canadian environment. *Environ. Rev.* **3**, 95–120.

Chaudhry, H. S. (1984). Nickel toxicity on carbohydrate metabolism of a freshwater fish. *Colisa fasciatus. Toxicol. Lett.* **20**, 115–121.

Chaudhry, H. S., and Nath, K. (1985). Nickel induced hyperglycemia in the freshwater fish, *Colisa fasciatus. Wat. Air Soil Pollut.* **24**, 173–176.

Cho, Y. S., Lee, S. Y., Kim, K. Y., and Nam, Y. K. (2009). Two metallothionein genes from mud loach *Misgurnus mizolepis* (Teleostei; Cypriniformes): gene structure, genomic organization, and mRNA expression analysis. *Comp. Biochem. Physiol. B. Biochem. Mol. Biol.* **153**, 317–326.

Chowdhury, M. J., Grosell, M., McDonald, D. G., and Wood, C. M. (2003). Plasma clearance of cadmium and zinc in non-acclimated and metal-acclimated trout. *Aquat. Toxicol.* **64**, 259–275.

Chowdhury, M. J., Bucking, C., and Wood, C. M. (2008). Pre-exposure to waterborne nickel downregulates gastrointestinal nickel uptake in rainbow trout: indirect evidence for nickel essentiality. *Environ. Sci. Technol.* **42**, 1359–1364.

Couture, P., and Pyle, G. G. (2008). Live fast and die young: metal effects on condition and physiology of wild yellow perch from two metal-contamination gradients. *Hum. Ecol. Risk Assess.* **14**, 73–96.

Couture, P., Busby, P., Rajotte, J., Gauthier, C., and Pyle, G. (2008a). Seasonal and regional variations of metal contamination and condition indicators in yellow perch (*Perca flavescens*) along two polymetallic gradients. I. Factors influencing tissue metal concentrations. *Hum. Ecol. Risk Assess.* **14**, 97–125.

Couture, P., Rajotte, J., and Pyle, G. (2008b). Seasonal and regional variations of metal contamination and condition indicators in yellow perch (*Perca flavescens*) along two polymetallic gradients. III. Energetic and physiological indicators. *Hum. Ecol. Risk Assess.* **14**, 146–165.

Dave, G., and Xiu, R. (1991). Toxicity of mercury, copper, nickel, lead, and cobalt to embryos and larvae of zebrafish, *Brachydanio rerio*. *Arch. Environ. Contam. Toxicol.* **21**, 126–134.

De Luca, G., Gugliotta, T., Parisi, G., Romano, P., Geraci, A., Romano, O., Scuteri, A., and Romano, L. (2007). Effects of nickel on human and fish red blood cells. *Biosci. Rep.* **27**, 265–273.

Denkhaus, E., and Salnikow, K. (2002). Nickel essentiality, toxicity, and carcinogenicity. *Crit. Rev. Oncol. Hematol.* **42**, 35–56.

Eisler, R. (1998). *Nickel Hazards to Fish, Wildlife, and Invertebrates: A Synoptic Review.* Biological Science Report USGS/BRD/BSR-1998-0001. US Geological Survey, Patuxent Wildlife Research Center, 95.

Erickson, R. J., Benoit, D. A., Mattson, V. R., Nelson, J. H. P., and Leonard, E. N. (1996). The effects of water chemistry on the toxicity of copper to fathead minnows. *Environ. Toxicol. Chem.* **15**, 181–193.

European Parliament (2008). *Directive 2008/105/EC of the European Parliament and of the Council of 16 December 2008 on environmental quality standards in the field of water policy, amending and subsequently repealing Directives 82/176/EEC, 83/513/EEC, 84/156/EEC, 84/491/EEC, 86/280/EEC and amending Directive 2000/60/EC.* http://register.consilium.europa.eu/pdf/en/08/st03/st03644.en08.pdf

Garceau, N., Pichaud, N., and Couture, P. (2010). Inhibition of goldfish mitochondrial metabolism by *in vitro* exposure to Cd, Cu and Ni. *Aquat. Toxicol.* **107**, 107–112.

Gerendás, J., Polacco, J. C., Freyermuth, S. K., and Sattelmacher, B. (1999). Significance of nickel for plant growth and metabolism. *J. Plant Nutr. Soil Sci.* **162**, 241–256.

Ghazaly, K. S. (1992). Sublethal effects of nickel on carbohydrate metabolism, blood and mineral contents of *Tilapia nilotica*. *Wat. Air Soil Pollut.* **64**, 525–532.

Giattina, J. D., Garton, R. R., and Stevens, D. G. (1982). Avoidance of copper and nickel by rainbow trout as monitored by a computer-based data acquisition system. *Trans. Am. Fish. Soc.* **111**, 491–504.

Giguère, A., Campbell, P. G. C., Hare, L., and Couture, P. (2006). Sub-cellular partitioning of cadmium, copper, nickel and zinc in indigenous yellow perch (*Perca flavescens*) sampled along a polymetallic gradient. *Aquat. Toxicol.* **77**, 178–189.

Gottofrey, J., and Tjälve, H. (1991). Axonal transport of cadmium in the olfactory nerve of the pike. *Pharmacol. Toxicol.* **69**, 242–252.

Grande, M., and Andersen, S. (1983). Lethal effects of hexavalent chromium, lead and nickel on young stages of Atlantic salmon (*Salmo salar* L.) in soft water. *Vatten (Sweden)* **39**, 405–416.

Green-Pedersen, H., Jensen, B. T., and Pind, N. (1997). Nickel adsorption on MnO_2, $Fe(OH)_3$, montmorillonite, humic acid and calcite: a comparative study. *Environ. Technol.* **18**, 807–815.

Greenwood, N. N., and Earnshaw, A. (1984). *Chemistry of the Elements*. Butterworth-Heinemann, Cambridge.

Gunshin, H., Mackenzie, B., Berger, U. V., Gunshin, Y., Romero, M. F., Boron, W. F., Nussberger, S., Gollan, J. L., and Hediger, M. A. (1997). Cloning and characterization of a mammalian proton-coupled metal-ion transporter. *Nature* **388**, 482–488.

Guthrie, J. W., Mandal, R., Salam, M. S. A., Hassan, N. M., Murimboh, J., Chakrabarti, C. L., Back, M. H., and Grégoire, D. C. (2003). Kinetic studies of nickel speciation in model solutions of a well-characterized humic acid using the competing ligand exchange method. *Anal. Chim. Acta* **480**, 157–169.

Guthrie, J. W., Hassan, N. M., Salam, M. S. A., Fasfous, I. I., Murimboh, C. A., Murimboh, J., Chakrabarti, C. L., and Grégoire, D. C. (2005). Complexation of Ni, Cu, Zn, and Cd by DOC in some metal-impacted freshwater lakes: a comparison of approaches using electrochemical determination of free-metal-ion and labile complexes and a computer speciation model, WHAM V and VI. *Anal. Chim. Acta* **528**, 205–218.

Hall, L. W., Jr., and Anderson, R. D. (1995). The influence of salinity on the toxicity of various classes of chemicals to aquatic biota. *Crit. Rev. Toxicol.* **25**, 281–346.

Hassan, N., Murimboh, J. D., and Chakrabarti, C. L. (2008). Kinetic Speciation of Ni (II) in model solutions and freshwaters: competition of Al (III) and Fe (III). *Wat. Air Soil Pollut.* **193**, 131–146.

Hedtke, J. L., Robinson Wilson, E., and Weber, L. J. (1982). Influence of body size and developmental stage of coho salmon (*Oncorhynchus kisutch*) on lethality of several toxicants. *Fund. Appl. Toxicol.* **2**, 67–72.

Henriksson, J., Tallkvist, J., and Tjälve, H. (1997). Uptake of nickel into the brain via olfactory neurons in rats. *Toxicol. Lett.* **91**, 153–162.

Hughes, G. M., Perry, S. F., and Brown, V. M. (1979). A morphometric study of effects of nickel, chromium and cadmium on the secondary lamellae of rainbow trout gills. *Wat. Res.* **13**, 665–679.

Jha, B. S., and Jha, M. M. (1995). Biochemical effects of nickel chloride on the liver and gonads of the freshwater climbing perch, *Anabas testudineus* (Bloch.). *Proc. Natl. Acad. Sci. India* **65**, 39–46.

Kamunde, C., Grosell, M., Higgs, D., and Wood, C. M. (2002). Copper metabolism in actively growing rainbow trout (*Oncorhynchus mykiss*): interactions between dietary and waterborne copper uptake. *J. Exp. Biol.* **205**, 279–290.

Kasprzak, K. S. (1987). Nickel. *Adv. Mod. Environ. Toxicol.* **11**, 145–183.

Keller, W., and Gunn, J. M. (1995). Lake water quality improvements and recovering aquatic communities. In: *Restoration and Recovery of an Industrial Region* (J.M. Gunn, ed.), pp. 67–80. Springer, New York.

Kowalik-Jankowska, T., Kozlowski, H., Farkas, E., and Sovago, I. (2007). Nickel ion complexes of amino acids and peptides. In: *Nickel and Its Surprising Impact in Nature* (A. Sigel, H. Sigel and R.K.O. Sigel, eds), pp. 63–108. John Wiley & Sons, Chichester.

Kuykendall, J. R., Miller, K. L., Mellinger, K. M., Cain, A. J., Perry, M. W., Bradley, M., Jarvi, E. J., and Paustenbach, D. J. (2009). DNA-protein cross-links in erythrocytes of freshwater fish exposed to hexavalent chromium or divalent nickel. *Arch. Environ. Contam. Toxicol.* **56**, 260–267.

Kwong, R. W., and Niyogi, S. (2009). The interactions of iron with other divalent metals in the intestinal tract of a freshwater teleost, rainbow trout (*Oncorhyncus mykiss*). *Comp. Biochem. Physiol. C.* **150**, 442–449.

Lapointe, D., and Couture, P. (2009). Influence of the route of exposure on the accumulation and subcellular distribution of nickel and thallium in juvenile fathead minnows (*Pimephales promelas*). *Arch. Environ. Contam. Toxicol.* **57**, 571–580.

Lapointe, D., and Couture, P. (2010). Accumulation and effects of nickel and thallium in early-life stages of fathead minnows (*Pimephales promelas*). *Ecotoxicol. Environ. Saf.* **73**, 572–578.

Lapointe, D., Gentes, S., Ponton, D. E., Hare, L., and Couture, P. (2009). Influence of prey type on nickel and thallium assimilation, subcellular distribution and effects in juvenile fathead minnows (*Pimephales promelas*). *Environ. Sci. Technol.* **43**, 8665–8670.

Leonard, E. M., Nadella, S. R., Bucking, C., and Wood, C. M. (2009). Characterization of dietary Ni uptake in the rainbow trout. *Oncorhynchus mykiss. Aquat. Toxicol.* **93**, 205–216.

Leonard, S. S., Harris, G. K., and Shi, X. (2004). Metal-induced oxidative stress and signal transduction. *Free Radic. Biol. Med.* **37**, 1921–1942.

Lloyd, D. R., and Phillips, D. H. (1999). Oxidative DNA damage mediated by copper (II), iron (II) and nickel (II) Fenton reactions: evidence for site-specific mechanisms in the formation of double-strand breaks, 8-hydroxydeoxyguanosine and putative intrastrand cross-links. *Mutat. Res. Fundam. Mol. Mech. Mutag.* **424**, 23–36.

Lurling, M., and Scheffer, M. (2007). Info-disruption: pollution and the transfer of chemical information between organisms. *Trends Ecol. Evol.* **22**, 374–379.

MacIver, B., Cutler, C. P., Yin, J., Hill, M. G., Zeidel, M. L., and Hill, W. G. (2009). Expression and functional characterization of four aquaporin water channels from the European eel (*Anguilla anguilla*). *J. Exp. Biol.* **212**, 2263–2856.

Mandal, R., Salam, M. S. A., Murimboh, J., Hassan, N. M., Chakrabarti, C. L., Back, M. H., and Gregoire, D. C. (2000). Competition of Ca (II) and Mg (II) with Ni (II) for binding by a well-characterized fulvic acid in model solutions. *Environ. Sci. Technol.* **34**, 2201–2208.

McDonald, D. G., Cavdek, V., and Ellis, R. (1991). Gill design in freshwater fishes: interrelationships among gas exchange, ion regulation, and acid–base regulation. *Physiol. Zool.* **64**, 103–123.

Meyer, J. S., Santore, R. C., Bobbitt, J. P., Debrey, L. D., Boese, C. J., Paquin, P. R., Allen, H. E., Bergman, H. L., and Ditoro, D. M. (1999). Binding of nickel and copper to fish gills predicts toxicity when water hardness varies, but free-ion activity does not. *Environ. Sci. Technol.* **33**, 913–916.

Muyssen, B. T. A., Brix, K. V., DeForest, D. K., and Janssen, C. R. (2004). Nickel essentiality and homeostasis in aquatic organisms. *Environ. Rev.* **12**, 113–131.

Nath, K., and Kumar, N. (1989). Nickel-induced histopathological alterations in the gill architecture of a tropical freshwater perch, *Colisa fasciatus* (Bloch & Schn.). *Sci. Tot. Environ.* **80**, 293–296.

Nebeker, A. V., Savonen, C., and Stevens, D. G. (1985). Sensitivity of rainbow trout early life stages to nickel chloride. *Environ. Toxicol. Chem.* **4**, 233–239.

Nieboer, E., and Richardson, D. H. S. (1980). The replacement of the nondescript term "heavy metals" by a biologically and chemically significant classification of metal ions. *Environ. Pollut.* **1**, 3–26.

Nielsen, F. H. (1987). Nickel. In: *Trace Elements in Human and Animal Nutrition* (W. Mertz and E.J. Underwood, eds), pp. 245–273. Academic Press, New York.

Nielsen, F. H. (1991). Nutritional requirements for boron, silicon, vanadium, nickel, and arsenic: current knowledge and speculation. *FASEB J.* **5**, 2661.

Nielsen, F. H. (1993). Is nickel nutritionally important? *Nutr. Today* **28**, 14.

Nielsen, F. H. (2000). Importance of making dietary recommendations for elements designated as nutritionally beneficial, pharmacologically beneficial, or conditionally essential. *J. Trace Elem. Exp. Med.* **13**, 113–129.

Nieminen, T. M., Ukonmaanaho, L., Rausch, N., and Shotyk, W. (2007). Biogeochemistry of nickel and its release into the environment. In: *Nickel and Its Surprising Impact in Nature* (A. Sigel, H. Sigel and R.K.O. Sigel, eds). Vol. 2. John Wiley & Sons, Chichester.

Nimmo, M., Van den Berg, C. M. G., and Brown, J. (1989). The chemical speciation of dissolved nickel, copper, vanadium and iron in Liverpool Bay, Irish Sea. *Estuar. Coast. Shelf Sci.* **29**, 57–74.

Niyogi, S., and Wood, C. M. (2004). Biotic ligand model, a flexible tool for developing site-specific water quality guidelines for metals. *Environ. Sci. Technol.* **38**, 6177–6192.

Nriagu, J. (1980). Global cycle and properties of nickel. In: *Nickel in the Environment* (J.O. Nriagu, ed.), pp. 1–26. John Wiley & Sons, New York.

Ojo, A. A., and Wood, C. M. (2007). *In vitro* analysis of the bioavailability of six metals via the gastro-intestinal tract of the rainbow trout (*Oncorhynchus mykiss*). *Aquat. Toxicol.* **83**, 10–23.

Pane, E. F., Richards, J. G., and Wood, C. M. (2003). Acute waterborne nickel toxicity in the rainbow trout (*Oncorhynchus mykiss*) occurs by a respiratory rather than ionoregulatory mechanism. *Aquat. Toxicol.* **63**, 65–82.

Pane, E. F., Haque, A., Goss, G. G., and Wood, C. M. (2004a). The physiological consequences of exposure to chronic, sublethal waterborne nickel in rainbow trout (*Oncorhynchus mykiss*): exercise vs resting physiology. *J. Exp. Biol.* **207**, 1249–1261.

Pane, E. F., Haque, A., and Wood, C. M. (2004b). Mechanistic analysis of acute, Ni-induced respiratory toxicity in the rainbow trout (*Oncorhynchus mykiss*): an exclusively branchial phenomenon. *Aquat. Toxicol.* **69**, 11–24.

Pane, E. F., Bucking, C., Patel, M., and Wood, C. M. (2005). Renal function in the freshwater rainbow trout (*Oncorhynchus mykiss*) following acute and prolonged exposure to waterborne nickel. *Aquat. Toxicol.* **72**, 119–133.

Pane, E. F., Glover, C. N., Patel, M., and Wood, C. M. (2006a). Characterization of Ni transport into brush border membrane vesicles (BBMVs) isolated from the kidney of the freshwater rainbow trout (*Oncorhynchus mykiss*). *Biochim. Biophys. Acta Biomemb.* **1758**, 74–84.

Pane, E. F., McDonald, M. D., Curry, H. N., Blanchard, J., Wood, C. M., and Grosell, M. (2006b). Hydromineral balance in the marine gulf toadfish (*Opsanus beta*) exposed to waterborne or infused nickel. *Aquat. Toxicol.* **80**, 70–81.

Pane, E. F., Patel, M., and Wood, C. M. (2006c). Chronic, sublethal nickel acclimation alters the diffusive properties of renal brush border membrane vesicles (BBMVs) prepared from the freshwater rainbow trout. *Comp. Biochem. Physiol. C* **143**, 78–85.

Perez-Coll, C. S., Sztrum, A. A., and Herkovits, J. (2008). Nickel tissue residue as a biomarker of sub-toxic exposure and susceptibility in amphibian embryos. *Chemosphere* **74**, 78–83.

Phipps, T., Tank, S. L., Wirtz, J., Brewer, L., Coyner, A., Ortego, L. S., and Fairbrother, A. (2002). Essentiality of nickel and homeostatic mechanisms for its regulation in terrestrial organisms. *Environ. Rev.* **10**, 209–261.

Pickering, Q. H. (1974). Chronic toxicity of nickel to the fathead minnow. *Wat. Pollut. Cont. Fed.* **46**, 760–765.

Pickering, Q. H., and Henderson, C. (1966). The acute toxicity of some heavy metals to different species of warmwater fishes. *Air Wat. Pollut. Int. J.* **10**, 453–463.

Pierron, F., Bourret, V., St-Cyr, J., Campbell, P. G., Bernatchez, L., and Couture, P. (2009). Transcriptional responses to environmental metal exposure in wild yellow perch (*Perca flavescens*) collected in lakes with differing environmental metal concentrations (Cd, Cu, Ni). *Ecotoxicology* **18**, 620–631.

Playle, R. C., Gensemer, R. W., and Dixon, D. G. (1992). Copper accumulation on gills of fathead minnows: influence of water hardness, complexation and pH of the gill micro-environment. *Environ. Toxicol. Chem.* **11**, 381–391.

Ptashynski, M. D., and Klaverkamp, J. F. (2002). Accumulation and distribution of dietary nickel in lake whitefish (*Coregonus clupeaformis*). *Aquat. Toxicol.* **58**, 249–264.

Ptashynski, M. D., Pedlar, R. M., and Evans, R. E. (2001). Accumulation, distribution and toxicology of dietary nickel in lake whitefish (*Coregonus clupeaformis*) and lake trout (*Salvelinus namaycush*). *Comp. Biochem. Physiol. C* **130**, 145–162.

Ptashynski, M. D., Pedlar, R. M., Evans, R. E., Baron, C. L., and Klaverkamp, J. F. (2002). Toxicology of dietary nickel in lake whitefish (*Coregonus clupeaformis*). *Aquat. Toxicol.* **58**, 229–247.

Pyle, G. G. (2000). *The toxicity and bioavailability of nickel and molybdenum to standard toxicity-test fish species and fish species found in northern Canadian lakes.* PhD thesis, Department of Biology, University of Saskatchewan, Saskatoon, p. 273.

Pyle, G. G., and Mirza, R. S. (2007). Copper-impaired chemosensory function and behavior in aquatic animals. *Hum. Ecol. Risk Assess.* **13**, 492–505.

Pyle, G., and Wood, C. (2007). Predicting "non-scents": rationale for a chemosensory-based biotic ligand model. *Austral. J. Ecotoxicol.* **13**, 47–51.

Pyle, G. G., Swanson, S. M., and Lehmkuhl, D. M. (2002a). Toxicity of uranium mine receiving waters to early life stage fathead minnows (*Pimephales promelas*) in the laboratory. *Environ. Pollut.* **116**, 243–255.

Pyle, G. G., Swanson, S. M., and Lehmkuhl, D. M. (2002b). The influence of water hardness, pH, and suspended solids on nickel toxicity to larval fathead minnows (*Pimephales promelas*). *Wat. Air Soil Pollut.* **133**, 215–226.

Pyle, G. G., Rajotte, J. W., and Couture, P. (2005). Effects of industrial metals on wild fish populations along a metal contamination gradient. *Ecotoxicol. Environ. Saf.* **61**, 287–312.

Ragsdale, S. W. (1998). Nickel biochemistry. *Curr. Opin. Chem. Biol.* **2**, 208–215.

Ragsdale, S. W. (2007). Nickel and the carbon cycle. *J. Inorg. Biochem.* **101**, 1657–1666.

Ragsdale, S. W. (2009). Nickel-based enzyme systems. *J. Biol. Chem.* **284**, 18571–18575.

Ray, D., Banerjee, S. K., and Chatterjee, M. (1990). Bioaccumulation of nickel and vanadium in tissues of the catfish *Clarias batrachus. J. Inorg. Biochem.* **38**, 169–173.

Reck, B. K., Müller, D. B., Rostkowski, K., and Graedel, T. E. (2008). Anthropogenic nickel cycle: Insights into use, trade, and recycling. *Environ. Sci. Technol.* **42**, 3394–3400.

Rehwoldt, R., Bida, G., and Nerrie, B. (1971). Acute toxicity of copper, nickel and zinc ions to some Hudson River fish species. *Bull. Environ. Contam. Toxicol.* **6**, 445–448.

Richter, R. O., and Theis, T. L. (1980). Nickel speciation in a soil/water system. In: *Nickel in the Environment* (J.O. Nriagu, ed.), pp. 189–202. John Wiley & Sons, New York.

Roesijadi, G. (1992). Metallothioneins in metal regulation and toxicity in aquatic animals. *Aquat. Toxicol.* **22**, 81–114.

Roesijadi, G. (1994). Metallothionein induction as a measure of response to metal exposure in aquatic animals. *Environ. Health Perspect.* **102**, 91–96.

Sayer, M. D., Reader, J. P., and Morris, R. (1991). Effects of six trace metals on calcium fluxes in brown trout (*Salmo trutta* L.) in soft water. *J. Comp. Physiol. B* **161**, 537–542.

Schaumloffel, D. (2005). Speciation of nickel. In: *Handbook of Elemental Speciation II: Species in the Environment, Food, Medicine, and Occupational Health* (R. Cornelis, H. Crews, J. Caruso and K.G. Heumann, eds), pp. 310–326. John Wiley and Sons, New York.

Scheil, V., and Köhler, H. R. (2009). Influence of nickel chloride, chlorpyrifos, and imidacloprid in combination with different temperatures on the embryogenesis of the zebrafish *Danio rerio. Arch. Environ. Contam. Toxicol.* **56**, 238–243.

Schubauer-Berigan, M. K., Dierkes, J. R., Monson, P. D., and Ankley, G. T. (1993). pH-dependent toxicity of Cd, Cu, Ni, Pb and Zn to *Ceriodaphnia dubia, Pimephales promelas, Hyalella azteca* and *Lumbriculus variegatus. Environ. Toxicol. Chem.* **12**, 1261–1266.

Scott, G. R., and Sloman, K. A. (2004). The effects of environmental pollutants on complex fish behaviour: integrating behavioural and physiological indicators of toxicity. *Aquat. Toxicol.* **68**, 369–392.

Sreedevi, P., Sivaramakrishna, B., Suresh, A., and Radhakrishnaiah, K. (1992a). Effect of nickel on some aspects of protein metabolism in the gill and kidney of the freshwater fish, *Cyprinus carpio* L. *Environ. Pollut.* **77**, 59–63.

Sreedevi, P., Suresh, A., Sivaramakrishna, B., Prabhavathi, B., and Radhakrishnaiah, K. (1992b). Bioaccumulation of nickel in the organs of the freshwater fish, *Cyprinus carpio*, and the freshwater mussel, *Lamellidens marginalis*, under lethal and sublethal nickel stress. *Chemosphere* **24**, 29–36.

Stohs, S. J., and Bagchi, D. (1995). Oxidative mechanisms in the toxicity of metal ions. *Free Radic. Biol. Med.* **18**, 321–336.

Takumi, S., Kimura, H., Matsusaki, H., Kawazoe, S., Tominaga, N., and Arizono, K. (2010). DNA microarray analysis of genomic responses of yeast *Saccharomyces cerevisiae* to nickel chloride. *J. Toxicol. Sci.* **35**, 125–129.

Tallkvist, J., and Tjalve, H. (1997). Effect of dietary iron-deficiency on the disposition of nickel in rats. *Toxicol. Lett.* **92**, 131–138.

Tallkvist, J., Henriksson, J., d'Argy, R., and Tjälve, H. (1998). Transport and subcellular distribution of nickel in the olfactory system of pikes and rats. *Toxicol. Sci.* **43**, 196–203.

Taylor, L. N., Baker, D. W., Wood, C. M., and Gordon McDonald, D. (2002). An *in vitro* approach for modelling branchial copper binding in rainbow trout. *Comp. Biochem. Physiol. C* **133**, 111–124.

Terreros, D. A., Knight, J. A., and Ashwood, E. R. (1988). Nickel inhibition of the osmotic-sensitive ionic cellular channels. *Ann. Clin. Lab. Sci.* **18**, 444–450.

Tjälve, H., and Borg-Neczak, K. (1994). Effects of lipophilic complex formation on the disposition of nickel in experimental animals. *Sci. Tot. Environ.* **148**, 217–242.

Tjälve, H., Gottofrey, J., and Borg, K. (1988). Bioaccumulation, distribution and retention of $^{63}Ni^{2+}$ in the brown trout (*Salmo trutta*). *Wat. Res.* **22**, 1129–1136.

Tjälve, H., Mejàre, C., and Borg-Neczak, K. (1995). Uptake and transport of manganese in primary and secondary olfactory neurones in pike. *Pharmacol. Toxicol.* **77**, 23–31.

Turner, A., Nimmo, M., and Thuresson, K. A. (1998). Speciation and sorptive behaviour of nickel in an organic-rich estuary (Beaulieu, UK). *Mar. Chem.* **63**, 105–118.

USEPA 2005. *National Recommended Water Quality Criteria.* United States Environmental Protection Agency. http://www.epa.gov/waterscience/criteria/wqctable/

Van den Berg, C. M. G., and Nimmo, M. (1987). Determination of interactions of nickel with dissolved organic material in seawater using cathodic stripping voltammetry. *Sci. Tot. Environ.* **60**, 185–195.

Winterhalder, K. (1995). Early history of human activities in the Sudbury area and ecological damage to the landscape. In: *Restoration and Recovery of an Industrial Region* (J.M. Gunn, ed.), pp. 17–32. Springer, New York.

Wood, C. M. (2001). Toxic responses of the gill. In: *Target Organ Toxicity in Marine and Freshwater Teleosts* (D. Schlenk and W.H. Benson, eds), pp. 1–89. Taylor and Francis, New York.

Xue, H. B., Jansen, S., Prasch, A., and Sigg, L. (2001). Nickel speciation and complexation kinetics in freshwater by ligand exchange and DPCSV. *Environ. Sci. Technol.* **35**, 539–546.

6

COBALT

RONNY BLUST

Homeostasis and Toxicology of Essential Metals: Volume 31A
FISH PHYSIOLOGY

Cobalt (Co) is an essential element to fish and other organisms. Its main role is as an intrinsic part of vitamin B_{12} or cobalamin. Fish and all other animals are not capable of synthesizing this vitamin and are therefore dependent on bacterial production of this essential compound. The essentiality of Co makes it of importance in fish nutrition and aquaculture. From an environmental perspective, Co is especially important as a radioactive waste product from the nuclear industry in the form of the radionuclide ^{60}Co. Most studies on the uptake and accumulation of Co by fish have used this radionuclide. The uptake, accumulation, and toxicity of stable Co in fish have received less attention, although the acute toxicity of Co in freshwater fish is reasonably well documented. The relative importance of water and food as sources of exposure has also been documented in the framework of radionuclide risk assessment. The uptake of Co in freshwater fish strongly depends on the speciation of Co and the calcium concentration in the water, but overall the effects of environmental conditions, including pH, remain poorly documented. The chronic toxicity of Co is also poorly documented, especially for the marine environment. Very little is known concerning the molecular mechanisms of Co uptake and toxicity in fish since no specific studies into this area have been conducted so far and possible mechanisms are inferred from mammalian studies.

1. CHEMICAL SPECIATION IN FRESHWATER AND SEAWATER

1.1. The Element Cobalt

Cobalt (Co) is a transition group metal with atomic number 27 and standard atomic weight of 58.93. It is a silvery grey solid at 20°C and has ferromagnetic properties. Pure metallic Co does not occur in the natural environment but the metal is present in different mineral phases. Cobalt occurs in the 0, +2, and +3 valence states. Cobalt(II) is more stable than Co(III), which is a powerful oxidizing agent (Greenwood and Earnshaw, 1997). Cobalt is one of the least common metals and with a crustal abundance of only 25 mg kg^{-1} it is the 33rd most abundant metal in the Earth's crust. Cobalt-59 is the only stable isotope of Co in the natural environment. Over 20 radionuclides have been identified, of which ^{60}Co is the most stable with a half-life of 5.27 years, ^{57}Co with a half-life of 271.79 days, ^{55}Co with a half-life of 77.27 days, and ^{58}Co with a half-life of 70.86 days. All other radioisotopes have half-lives shorter than 24 h and most of them even less than 1 s (Smith and Carson, 1981).

1.2. Partitioning and Speciation in Natural Waters

Released into water, Co will form a variety of inorganic and organic complexes and adsorb to particles and settle into the sediment (Hamilton, 1994). Cobalt binds relatively strongly to humic and fulvic substances which are naturally present in aquatic environments. However, humic and fulvic complexes with Co are not as stable as those of Cu, Fe, Ni, and Pb (Burba et al., 1994). The partitioning of Co between water and sediment phases strongly depends on the physicochemical conditions and complexation of Co to dissolved organic matter can reduce sediment sorption. Garnier et al. (1997) and Albrecht (2003) have reported K_d values for Co of 10^4–10^6 L kg^{-1}, between the dissolved and colloidal particulate phase, for river-water samples. Concentration profiles of Co indicate that dissolved concentrations decrease with increasing depth and that dissolved Co is precipitated in the adsorbed state with oxides of Fe and Mn and with crystalline sediments such as aluminosilicate and goethite. In the deep sea, formation of Mn nodules removes Co by interaction with MnO_2 (Barceloux, 1999).

Total dissolved concentrations of Co in aquatic environments show strong variation depending on input and sediment binding. Under natural conditions the speciation of Co is controlled by the Co^{2+} oxidation state (Collins and Kinsela, 2010). This is related to the redox potential of the relevant Co^{2+} redox couples ($E_o = -0.28 + 0.03$ log $[Co^{2+}]$ for the Co/Co^{2+} couple and $E_o = 1.81 + 0.06$ log $[Co^{3+}]/[Co^{2+}]$ for the Co^{2+}/Co^{3+} couple). Owing to the extremely low solubility of Co^{3+} [K_{sp} Co(OH)$_3$ = $10^{-44.5}$, K_{sp} CoOOH = $10^{-50.0}$], this oxidation state will only be detected in solution if it is complexed by a strong chelating organic molecule, such as the trihydroxamate siderophore–desferrioxamine B (Duckworth et al., 2009). The inorganic speciation of Co^{2+} is controlled by the pH of the water and complexation with carbonate, hydroxide, chloride, and sulfate ions. Total and dissolved concentrations of Co in natural waters are generally in the ng L^{-1} to μg L^{-1} range but show large variation. Reported total dissolved Co concentrations in different freshwater water bodies are generally low, showing a range from 1.8 ng L^{-1} (0.03 nM) to 5.8 μg L^{-1} (98 nM) in rivers and 1 ng L^{-1} (0.02 nM) to 0.35 μg L^{-1}) (5.9 nM) in freshwater lakes, but more extreme situations with concentrations up to 3.3 mg L^{-1} (56,000 nM) are documented for acid mine drainage sites (Collins and Kinsela, 2010).

In fresh and marine waters Co is strongly associated with dissolved and colloidal organic carbon (Zhang et al., 1990; Tanizaki et al., 1992; Garnier et al., 1997; Pham and Garnier, 1998; Qian et al., 1998; Ellwood and van den Berg, 2001; Saito and Moffett, 2001; Saito et al., 2005; Sekaly et al., 2003; Fasfous et al., 2004; Ellwood et al., 2005; Pokrovsky et al., 2006; Warnken

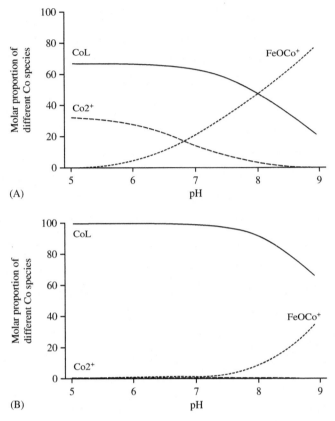

Fig. 6.1. Speciation modeling of Co in a system with competition between adsorption and organic complexation. The total concentrations for Co^{2+} and ligand (L) were taken from Qian et al. (1998). The Co concentration is 2 nM and FeOH 400 nM. The ligand concentration in case A = 1.4 nM and in case B = 7.6 nM. The log K for the formation of the CoL organic complex is 10.55 and the log K for the adsorption of Co on FeOH is −0.46. The concentration of surface sites is calculated based on a particle concentration of 2 mg L^{-1} and a sorption site density of 2×10^{-4} mol g^{-1} (from Albrecht, 2003).

et al., 2007). The effect of complexation and adsorption on Co speciation is illustrated in Fig. 6.1 based on a modeling exercise by Albrecht (2003).

Several studies examining Co complexation by fulvic and humic acids have reported conditional thermodynamic stability constants between $10^{2.7}$ and $10^{8.3}$ M (van Loon et al., 1992; Higgo et al., 1993; Westall et al., 1995; Glaus et al., 2000; Kurk and Choppin, 2000; Hamilton-Taylor et al., 2002; Chang et al., 2006). Qian et al. (1998) have estimated a Co-ligand

conditional thermodynamic stability constant of $10^{9.5}$–$10^{11.6}$ M. High
conditional thermodynamic stability constants have also been determined
for Co complexes in estuarine and marine waters (K in the range
10^{15}–10^{16} M) and although the ligands involved in complexation were not
identified it has been postulated that Co was present as part of cobalamin
where Co^+ is chelated in the corrin ring of the molecule. However, all of
these results were obtained with one analytical methodology, competitive
ligand exchange-differential pulse cathodic stripping voltammetry (CLE-
DPCSV). The limitations of this methodology for measuring such
thermodynamically stable metal complexes have recently been highlighted
and it has been suggested, based on kinetic aspects, that this methodology
has an upper detection limit of Co complexes with stability constants of
about 10^{11} M (van Leeuwen and Town, 2005). Therefore, there is still
considerable uncertainty concerning the stability and kinetics of Co
complexation in natural waters. This may explain why Co complexation
data for freshwater systems estimated by computer speciation programs
(Balistrieri et al., 1994) have differed so widely compared to values that have
been empirically determined (Qian et al., 1998).

2. SOURCES (NATURAL AND ANTHROPOGENIC) OF COBALT AND ECONOMIC IMPORTANCE

2.1. Sources of Minerals

Thirty-four Co minerals have been recognized, principally sulfides,
selenides, arsenides, sulfarsenides, carbonates, sulfates, and arsenates. The
main ore minerals of Co are the sulfides cobaltite, linnaeite, and carrollite,
and the hydrated oxide asbolane. Cobaltiferous pyrite is a further source
(Smith, 2001). Workable deposits generally contain 0.1–0.4% Co and belong
to one of four geologically distinct types: (1) sediment-hosted, largely
Precambrian, typified by the copperbelts of the Congo Democratic Republic
and Zambia, which since the 1970s have contributed between 25 and 50% of
the world's mine production; (2) mid-Tertiary to recent Ni-rich lateritic
deposits generated by weathering of peridotitic rocks, most notably in New
Caledonia, Cuba, and Australia; (3) primary magmatic Ni–Cu sulfide
concentrations, such as Sudbury, Noril'sk, Voisey's Bay, and Bushveld; and
(4) a more diverse group attributable to hydrothermal and volcanogenic
processes, of which the most important are the ophiolite-hosted Co–As
deposit at Bou Azzer, Morocco, and the epigenetic Cu–Au–Co concentrations of the Idaho Cobalt Belt, USA.

2.2. Applications of Cobalt

An early application of Co was as a pigment in the production of Co blue glass and certain paints. Most of the Co produced today is used in the production of superalloys which have high temperature stability and are also corrosion and wear resistant. Lithium cobalt oxide is widely used in Li ion battery electrodes, and Ni–Cd and Ni–metal hydride batteries contain Co. Organocobalt compounds play an important role in several organic reactions as oxidation catalysts. Another application is the use of a Co–Mb catalyst in the removal of sulfur impurities from petroleum that interfere with the refining of fuels (Wang, 2006).

Cobalt-60 is an important source of gamma radiation as it is produced by exposing stable Co to thermal neutrons in a nuclear reactor. Its industrial uses include sterilization of foods and medical materials, industrial radiography, and density measurements, and the isotope is also used in radiotherapy to treat cancer. The radioactive ^{60}Co from radiation instruments requires proper handling and disposal. Cobalt-57 is most used in medical tests as a radiolabel for cobalamin and in the Schilling test to determine whether a patient has pernicious anemia. Cobalt-57 is also used as a source in Mössbauer spectroscopy and as a radiation source in X-ray fluorescence spectroscopy (Mayles and Nahum, 2007).

3. ENVIRONMENTAL SITUATIONS OF CONCERN

Cobalt-60 can be released to the environment through leaks or spills at nuclear power plants, and in solid waste originating from nuclear power plants. Nuclear regulations allow small amounts of ^{58}Co and ^{60}Co to be released into the air or liquid waste streams (Leonard et al., 1993). Nuclear weapons may intentionally contain ^{60}Co and the radionuclide may also be formed in a nuclear explosion and be part of nuclear fallout (Payne, 1977; Hu et al., 2010). Stable Co is unlikely to cause harmful effects unless exposure is very high such as may occur in accidental releases.

4. A SURVEY OF ACUTE AND CHRONIC AMBIENT WATER QUALITY CRITERIA IN VARIOUS JURISDICTIONS IN FRESHWATER AND SEAWATER

The World Health Organization (Kim et al., 2006) has derived criteria using a probabilistic approach. For the freshwater environment, 28 data

Table 6.1
Ambient water quality guidelines from various sources

Guideline	Freshwater (µg L^{-1})	Saltwater (µg L^{-1})
WHO (2006)^a	8 (chronic)	20 (chronic)
British Columbia (2004)^b	4 (30 d average)/110 (max)	
The Netherlands (2000)^c	2.8 (chronic dissolved)	
Australia/New Zealand (2000)^d		1 (95% protection)

^{a}Kim et al. (2006).
^{b}Nagpal (2004).
^{c}van de Guchte et al. (2000).
^{d}ANZECC (2000).

points were used in the derivation and acute values were converted to chronic values by applying a ratio of 10. A guidance value for the protection of 95% of freshwater species with 50% confidence [HC5(50)] of 8 µg L^{-1} was obtained. A comparison of this value with environmental concentrations suggests that effects are likely only in the vicinity of major anthropogenic releases. The greatest risk to aquatic organisms would be in very soft water areas where the Ca^{2+} ion concentration is <10 mg L^{-1}.

For the marine environment, 12 acute toxicity values were used in the analysis and converted to chronic estimates as for freshwater data. Using the calculated chronic no observed effect concentrations (NOECs), the HC5(50) was 140 µg L^{-1}. However, this guidance value is not sufficiently protective of the most sensitive marine species. Therefore these authors also calculated the 99% protection level [HC1(50)] resulting in a guidance value of 20 µg L^{-1}. Ambient water quality guidelines have also been developed by a number of other authorities and are summarized in Table 6.1.

5. MECHANISMS OF TOXICITY

5.1. Toxicological Effects of Cobalt

The acute and chronic toxicity of Co to freshwater and marine fish has not been extensively documented and studies concerning the impact of environmental conditions on the toxicity are even more limited. Studies on the effects of Co in fish have focused on mortality, metabolism, effects on gill epithelium, ion regulation, hematology, rheotaxis, hypoxia, and general stress (Hertz et al., 1989; Montgomery et al., 1997; Marr et al., 1998; Janssen, 2000; Majmudar and Burleson, 2006; Atamanalp et al., 2010). In humans, chronic Co exposure has been shown to induce pathologies similar to chronic hypoxia and chronic mountain sickness. These include cardiomyopathy, elevated hemoglobin concentration, and polycythemia (Seghizzi et al., 1994; Jefferson et al., 2002). Mammalian studies have also shown that Co stimulates cardioventilatory reflexes, production of hypoxia inducible factor 1 (HIF-1), erythropoiesis, hypertrophy of carotid body glomus cells, and increased carotid body neural discharge in a manner consistent with the effects of hypoxia (Morelli et al., 1994). Studies using Co to mimic hypoxia have identified one or more subtypes of cytochrome b as a putative oxygen sensor (Ehleben et al., 1998; Lutz and Prentice, 2002).

Atamanalp et al. (2010) studied the effect of $CoCl_2$ on hematology in brown trout (*Salmo trutta fario*) at an exposure level of 180 μg L^{-1} $CoCl_2$ at 12 h intervals for 28 days. Increases were observed in red blood cell count, thrombocyte count, erythrocyte sedimentation rate, and hematocrit, whereas white blood cell count, hemoglobin, mean corpuscular volume, mean corpuscular hemoglobin, and mean corpuscular hemoglobin concentration values were decreased because of the $CoCl_2$ exposure. The effects of $CoCl_2$ on heart rate, blood pressure, ventilatory frequency, and opercular pressure amplitude in channel catfish (*Ictalurus punctatus*) were studied by Majmudar and Burleson (2006) to evaluate the potential of Co as a histochemical probe to study mechanisms of oxygen (O_2) chemoreception, as well as to assess the general effects of Co on the cardioventilatory physiology of fish. Cobalt, like cyanide, has been previously used to stimulate oxygen chemoreceptors and hypoxic reflexes in mammals. Although Co stimulated some cardioventilatory reflexes, the pattern and magnitude of the responses were noticeably different from those of cyanide and hypoxia. The authors concluded that the results suggest that the cardioventilatory reflexes stimulated by Co were not mediated by O_2-sensitive chemoreceptors and that Co is not an effective O_2 receptor stimulant in fish.

5.2. Acute and Chronic Toxicity of Cobalt

A screening of available toxicity data for fish using the USEPA ECOTOX database provided toxicity results in terms of LC50 for 10 different fish species. Most of the results refer to acute exposure scenarios (<10 days) and only a few of the reported LC50 values refer to chronic scenarios (Table 6.2). Acute or chronic NOEC, lowest observed effect concentrations (LOECs), or maximally allowable toxicant concentrations (MATCs) were available for only three species (Table 6.3). The LC50 values are reported for 1–28 days of exposure and range from 470 μg L^{-1} (*Oncorhynchus mykiss*, 28 days exposure) to 1,875,000 μg L^{-1} (*Poecilia reticulata*, 1 day exposure). The NOEC values are reported for 2–28 days and range from 60 μg L^{-1} (*Danio rerio*, 13 days) to 13,733 μg L^{-1}

Table 6.2
Four to twenty-eight day lethal concentrations (LC50) of cobalt for different fish species
reported in the ECOTOX database

Species name	Exposure time (days)	LC50 (μg L^{-1})	Exposure medium
Oncorhynchus mykiss	4	1,406	FW
Pimephales promelas	4	3,460	FW
Pimephales promelas	4	3,750	FW
Terapon jarbua	4	52,500	SW
Carassius auratus	4	66,800	FW
Cyprinus carpio	4	82,700	FW
Fundulus heteroclitus	4	275,000	SW
Fundulus heteroclitus	4	275,000	SW
Cyprinus carpio	4	332,980	FW
Fundulus heteroclitus	4	1,000,000	SW
Fundulus heteroclitus	4	1,000,000	SW
Colisa fasciata	4	102,000	FW
Pimephales promelas	4	22,000	FW
Blennius pholis	4	454,000–681,000	SW
Pleuronectes platessa	4	454,000–681,000	SW
Oncorhynchus mykiss	6	520	FW
Carassius auratus	7	810	FW
Carassius auratus	7	810	FW
Pimephales promelas	8	2,720	FW
Pimephales promelas	8	2,760	FW
Blennius pholis	9	227,000	SW
Oncorhynchus mykiss	28	470	FW
Oncorhynchus mykiss	28	490	FW

FW: freshwater; SW: saltwater.

(*Pimephales promelas*, 2 days). On the basis of the LC50 values, freshwater fish appear more sensitive than saltwater fish. The 96 h LC50 values for freshwater fish range from 1406 μg L^{-1} in *Oncorhynchus mykiss* to 332,980 μg L^{-1} in *Cyprinus carpio* and for saltwater fish from 52,500 μg L^{-1} in *Terapon jarbua* to 1000,000 μg L^{-1} *Fundulus heteroclitus*. A comparison on the basis of NOEC values was not possible since no values are reported for saltwater fish. A comparison of 96 h LC50 with 28 day NOEC values for *Pimephales promelas* obtained within a single study gives an acute to chronic ratio between 5 and 20. The lowest reported NOEC are reported by Dave and Xiu (1991) for embryos and larvae of the zebrafish (*Danio rerio*), at a hardness of 100 mg L^{-1} (as CaCO$_3$) and a pH of 7.5–7.7. The NOEC determined from the dose–response relationships was 3840 μg L^{-1} for effect on hatching time and 60 μg L^{-1} for effect on survival time.

Table 6.3

No observed effect concentration (NOEC), lowest observed effect concentration (LOEC), and maximally allowable toxicant concentration (MATC) of cobalt for different fish species reported in the ECOTOX database

Species name	Criterion	Effect	Exposure time (days)	Concentration ($\mu g\ L^{-1}$)
Pimephales promelas	NOEC	Mortality	2	6,200
Pimephales promelas	NOEC	Mortality	2	13,733
Danio rerio	NOEC	Hatching	4	3,840
Pimephales promelas	NOEC	Mortality	7	1,232
Pimephales promelas	NOEC	Mortality	7	1,932
Pimephales promelas	NOEC	Mortality	7	3,833
Danio rerio	NOEC	Mortality	13	60
Danio rerio	MATC	Mortality	14	340
Oncorhynchus mykiss	LOEC	Mortality	14	346
Danio rerio	MATC	Hatching	14	10,840
Pimephales promelas	NOEC	Length	28	210
Pimephales promelas	MATC	Length	28	290
Pimephales promelas	LOEC	Length	28	390
Pimephales promelas	NOEC	Weight	28	390
Pimephales promelas	MATC	Weight	28	560
Pimephales promelas	LOEC	Weight	28	810
Pimephales promelas	NOEC	Mortality	28	810
Pimephales promelas	MATC	Mortality	28	1,140
Pimephales promelas	LOEC	Mortality	28	1,610

All experiments refer to freshwater as the exposure medium.

6. ESSENTIALITY OR NON-ESSENTIALITY OF COBALT: EVIDENCE FOR AND AGAINST

6.1. Essentiality of Cobalt for Biological Systems

Cobalt is essential to all living organisms and is a key component of cobalamin. It is involved in the physiology of every cell of the body and has an important role in the regulation of DNA synthesis, fatty acid synthesis, and energy metabolism (Combs, 2008). Cobalamin is not a single molecule but consists of a family of closely related compounds of which the basic structure can only be synthesized by bacteria and archaea. The structure of cobalamin is based on a corrin, which is similar to the porphyrin ring found in heme, chlorophyll, and cytochrome (Fig. 6.2). Four of the six coordination sites for Co are provided by the corrin ring, and a fifth is provided by a dimethylbenzimidazole group (Ludwig and Matthews, 1997).

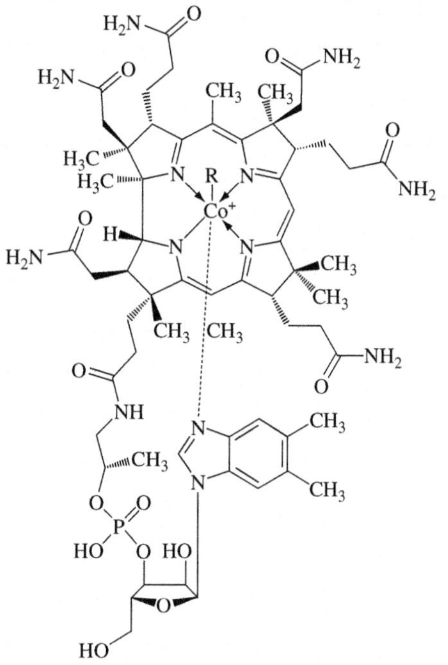

Fig. 6.2. Molecular structure of vitamin B_{12} or cobalamin. Four of the six coordination sites for Co are provided by the corrin ring, and a fifth by a dimethylbenzimidazole group. The sixth coordination site, the center of reactivity, is variable, being a cyano group (-CN), a hydroxyl group (-OH), a methyl group (-CH$_3$), or a 5'-deoxyadenosyl group, of which the C5' atom of the deoxyribose forms the covalent bond with Co.

Cobalt also occurs in some other proteins, including methionine aminopeptidase 2 and nitrile hydratase (Kobayashi and Shimizu, 1999). Methionine aminopeptidase 2 binds two Co or Mn ions and the enzyme functions by protecting the alpha unit of eukaryotic initiation factor 2 from inhibitory phosphorylation and by removing the amino-terminal methionine residue from nascent protein. Nitrile hydratases are single Fe or Co enzymes present in bacteria that catalyze the hydration of a variety of nitriles to their corresponding amides. There is recent evidence that they may also play a role in eukaryotic systems (Foerstner et al., 2008).

6.2. Cobalt in Fish Nutrition

Cobalt is an essential component in the diet of fish (Davis and Gatlin, 1991; Watanabe et al., 1997). The main nutritional requirement for Co is cobalamin. Intestinal microorganisms help in the synthesis of cobalamin (Kashiwada et al., 1970) and the cobalamin production in the gut depends on the dietary supply of Co (Limsuwan and Lovell, 1981). Deficiency of cobalamin results in a poor general condition and is reflected in a loss of appetite, limited growth, low hemoglobin, and anemia (Stickney, 1994).

Dietary Co has a positive effect on hematology, growth, and survival in different fish species (Frolova, 1960; Sabalina, 1964, cited in Steffens, 1989; Khan and Mukhopadhay, 1971; Bhanot and Gopalakrishnan, 1973; Castell et al., 1986; Anadu et al., 1990; Hossein et al., 2008). For example, gold-spot mullet supplemented with dietary $CoCl_2$ showed improved survival and growth compared to non-supplemented fish, with a net percent increase of 229% in weight and 108% in length in the group supplemented with $1.0 \, mg \, kg^{-1}$ dietary Co (Ghosh, 1975).

Carp fingerlings (*Cyprinus carpio*) fed $CoCl_2$-supplemented diets (0.05, 0.10, or 1.0%) showed significantly higher growth than the control. In addition, a significantly increased accumulation of Co was observed in the liver and kidneys of the 0.1% and 1.0% groups and in the gills, gut, and caudal trunk of the 1.0% dietary Co group. The fact that no difference in accumulation between the control and 0.05% group was observed indicates that Co is under homeostatic control (Mukherjee and Kaviraj, 2009).

Lin et al. (2010) performed growth trials to determine the dietary essentiality of Co and cobalamin for grouper (*Epinephelus malabaricus*). In a first experiment, a cobalamin-free basal diet was supplemented with 0.05, 6.5, 10.7, 20.7, 41.6, 82.7, and $153.1 \, mg \, Co \, kg^{-1}$ diet. In a second experiment, a cobalamin-free basal diet with $10 \, mg \, Co \, kg^{-1}$ was supplemented with 0, 8.3, 18.1, 54.0, 87.1, and $193.2 \, \mu g \, cobalamin \, kg^{-1}$ diet. In the first experiment weight gain and feed efficiency were significantly higher in the 10.7 and $20.7 \, mg \, Co \, kg^{-1}$ treatment compared to the $6.5 \, mg \, Co \, kg^{-1}$ dietary group. Further analysis indicated that 10 and $20 \, mg \, Co \, kg^{-1}$ diet are required for optimal growth and maximum hepatic and plasma cobalamin concentrations in grouper, respectively (Fig. 6.3). In the second experiment, the dietary treatments had no significant effects on weight gain, feed efficiency, or survival. These results indicate that grouper require $10 \, mg \, Co \, kg^{-1}$ diet for optimal growth and that no cobalamin is needed in the diet containing an adequate supply of Co.

6.3. Cobalt-dependent Enzymes

Two major coenzyme B_{12}-dependent enzyme families are known in vertebrates. These are typified by the following two enzymes: methylmalonyl coenzyme A mutase (MUT) and 5-methyltetrahydrofolate-homocysteine methyltransferase (MTR) (Banerjee and Ragsdale, 2003). MUT's reaction

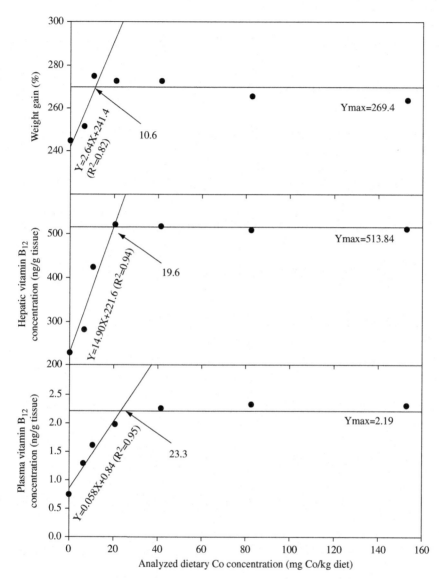

Fig. 6.3. Broken-line analysis of weight gain, hepatic and plasma vitamin B_{12} concentrations in juvenile grouper as a function of dietary Co concentration, indicating that the optimal Co levels are 10.6, 19.6, and 23.3 mg Co kg^{-1} diet, respectively. The break-points refer to the dietary Co concentration where the maximum plasma or liver vitamin B_{12} level or fish weight gain is reached. The Y_{max} is the maximum vitamin B_{12} level or weight gain reached. Each point represents the mean of three aquaria within a treatment with 10 fish per aquarium for weight gain and with six fish per aquarium for hepatic and plasma vitamin B_{12} concentration (from Lin et al., 2010).

converts MMl-CoA to Su-CoA, an important step in the extraction of energy from proteins and fats. This functionality is lost in cobalamin deficiency. The enzyme MTR, also known as methionine synthase, is a methyltransferase that uses MeB_{12} to catalyze the conversion of the amino acid homocysteine (Hcy) back into methionine (Met) (Banerjee and Matthews, 1990). Again, this functionality is lost in cobalamin deficiency. Increased homocysteine can also be caused by a folic acid deficiency, since cobalamin helps to regenerate the tetrahydrofolate (THF) active form of folic acid. This results in problems with DNA synthesis, in particular in blood cells and also intestinal wall cells which are responsible for absorption. The failure of blood cell production results in pernicious anemia which is a form of megaloblastic anemia (Lahner and Annibale, 2009). A lack of cobalamin will not lead to the anemia in the presence of a sufficient supply of folic acid for the production of thymine. When sufficient folic acid is available, all known cobalamin-related deficiency syndromes normalize, except those narrowly connected with the cobalamin-dependent enzymes MUT and MTR.

7. POTENTIAL FOR BIOCONCENTRATION AND/OR BIOMAGNIFICATION OF COBALT

7.1. Bioconcentration and Bioaccumulation Factors of Cobalt

Depending on the metal exposure and uptake pathway being considered, a number of different concentration or accumulation factors can be defined. These include: (1) the bioconcentration factor (BCF), which is the ratio of the metal concentration in an organism (C_b) on a per unit tissue fresh weight basis to that in water (C_w) being the only exposure phase; (2) the bioaccumulation factor (BAF), which is the ratio of the metal concentration in an organism (C_b) from all exposure pathways (including water, sediment, and dietary pathways) on a per unit tissue fresh weight basis to that in water (C_w); and (3) the biota sediment accumulation factor (BSAF), which is the ratio of the concentration of a metal in an organism (C_b) to the metal concentration measured in the sediment (C_s). BAF and BSAF values have been compiled from various sources by the International Atomic Energy Agency (IAEA, 2009, 2010) and are summarized in Table 6.4 for Co in freshwater fish. The results show a relatively broad range, even within one species, reflecting differences in exposure conditions and routes. The reported values show that Co is not strongly accumulated in most fish species and as is the case with several other metals, Co does not appear to biomagnify in the food web extensively. Smith (2006) has developed a

Table 6.4
Cobalt bioaccumulation factors (BAFs) for freshwater fish on a whole-body and muscle-tissue basis, and sediment to fish bioaccumulation factors (BSAFs) for freshwater fish on a whole-body, liver, and muscle-tissue basis

	Sample	GM	GMSD	Min	Max
BAF (L kg^{-1} FW)					
Whole-body	118	4.0×10^2	1.6×10^0	2.3×10^1	2.4×10^3
Muscle-tissue	65	7.6×10^1	2.4×10^0	9.0×10^0	5.6×10^2
BSAF (L kg^{-1} FW)					
Whole-body	751	2.8×10^{-1}	4.3×10^0	7.1×10^{-4}	2.9×10^1
Liver-tissue	133	6.1×10^{-2}	1.1×10^1	1.3×10^{-3}	9.7×10^1
Muscle-tissue	206	2.0×10^{-1}	4.4×10^0	3.0×10^{-4}	5.9×10^0

GM: geometric mean; GMSD: standard deviation of the geometric mean; FW: fresh weight.
Source: IAEA (2009).

dynamic model to estimate the concentration of ^{60}Co in fish, which indicated that differences in uptake rates (i.e. water chemistry and temperature-dependent feeding rate) and body size (i.e. metabolic activity), in particular, had an important influence on the concentration factors obtained.

In general, it appears that the bioconcentration of ^{60}Co is proportionally lower when the species belongs to a higher trophic level (Kimura and Ichikawa, 1972; Coughtrey and Thorne, 1983). A comparison of the mean and ranges of tissue-specific Co accumulation shows that the gills and the kidneys are the strongest accumulators of Co, followed by the viscera (including the gut) and the liver, and that muscle tissue is not a strong accumulator of Co (Table 6.5).

7.2. Food Web Transfer and Cobalt Exposure of Fish

Harrison et al. (1990) introduced six gamma-emitting isotopes including ^{60}Co, in a single addition to an oligotrophic lake on the Precambrian Shield of north-western Ontario, Canada. Accumulation of each isotope was monitored in fathead minnow (*Pimephales promelas*) and lake trout (*Salvelinus namaycush*). Fathead minnow accumulated higher concentrations of all isotopes than lake trout. In lake trout, the highest concentrations of ^{60}Co were found in the kidneys. Food seemed to be the primary source of all isotopes to fish. Ratios of isotope concentrations in fish to those in water were higher for both fathead minnow and lake trout than ratios reported from laboratory studies in which fish were exposed via the water only. Bird et al. (1998a,b, 1999) studied the bioaccumulation of different radionuclides including ^{60}Co in Canadian Shield lake basins. The two basins were part of

Table 6.5
Cobalt mean organ to reference organ (muscle) concentration ratios

Organ	Sample	AM	SD	Min.	Max.
Bone	8	4.1	5.2	0.39	17
Blood	2	1.7	1.5	0.69	2.8
Gills	4	130	250	0.55	500
Kidney	5	17	23	1.6	57
Liver	5	5.7	7.6	0.37	19
Muscle	11	1.0	0.0	1.0	1.0
Skin	3	1.7	1.3	0.4	3.1
Spleen	1	0.0022			
Viscera	2	6.2	0.47	5.8	6.5
Body	4	12	22	1.4	45

AM: arithmetic mean; SD: standard deviation.
Source: IAEA (2009).

an eutrophication experiment and one was eutrophic whereas the other was mesotrophic. The trophic status of the lakes did not appear to have a marked effect on the accumulation of radionuclides by the biota. Concentrations in the biota generally decreased in the order: fathead minnow > pearl dace > tadpoles > slimy sculpin > leeches. Concentrations in biota generally decreased in the order: ^{65}Zn > ^{203}Hg > ^{75}Se > ^{134}Cs > ^{60}Co > ^{85}Sr > ^{59}Fe. Cobalt-60 concentrations in tadpoles were greater than in the other biota. Radionuclide concentrations in the tissues of lake whitefish indicated that uptake was predominantly from food. Radionuclide concentrations were usually higher in the posterior gut, liver, and kidney than in other tissues, whereas body burdens were generally higher in kidney and gut for ^{60}Co.

Garnier-Laplace et al. (2000) proposed a procedure to calibrate dynamic models for radionuclide uptake and elimination in fish from water and food. The model takes into account the food-web effect, the feeding rate, and the growth of organisms. One of the examples considered the transfer dynamics of ^{60}Co through a simple pelagic food-chain (phytoplankton, zooplankton, prey fish, and predator fish). Concentration factors for the transfer of ^{60}Co from water were 8.7 L kg^{-1} wet weight for prey fish and 4.14 for predator fish. For the trophic route, a trophic transfer factor at equilibrium (TTF$_{eq}$) of 0.0134 was reached in 310 days for prey fish and a TTF$_{eq}$ of 0.077 reached in 3.5 years for the predator fish. A sensitivity analysis adapted to a chronic contamination scenario of a watercourse showed that the most influential factors with regard to the concentration in fish for both trophic levels were the phytoplankton biomass, the contact time of the particles with the radionuclide, and the feeding rates of the fish.

Nolan et al. (1992) performed a comparative study, using double-radiolabeling techniques, on the uptake and retention of Co species in a simple marine food chain which included phytoplankton (*Dunaliella tertiolecta* and *Chaetoceros pseudocurviseturn*), mixed copepods (mainly *Centropages* sp.), and fish (the sea perch, *Serranus scriba*). Retention of Co from the diet was also studied for the latter two organisms. Phytoplankton accumulated more than 60 times as much cobalamin compared to $CoCl_2$ from the water, and retained the metal for a significantly longer period (a retention half-time of 4.4 vs 0.6 days). Accumulation of $CoCl_2$ after ingestion of radiolabeled phytoplankton by copepods was not measurable, whereas retention of cobalamin reached 42% of the quantities ingested. Fish accumulated cobalamin 21 times more rapidly from seawater than $CoCl_2$ and retained ingested cobalamin 20 times more efficiently (100%) than ingested $CoCl_2$ (5%). Two-thirds of the ingested cobalamin was retained in the fish with a retention half-time of 8 days. The remaining one-third of the organic form was retained with a half-time of 54 days, a value that was not significantly different from that of $CoCl_2$ (47 days). A simple biokinetic model showed that preferential accumulation of the cobalamin complex over inorganic Co species in the food web could explain the Co concentrations measured in marine organisms including fish.

8. CHARACTERIZATION OF UPTAKE ROUTES

8.1. Uptake of Cobalt Across the Gills

Cobalt is taken up by fish via the gills and gut as main routes of uptake, although direct uptake via other routes such as the skin is not excluded. Studies on Co uptake have usually used ^{57}Co or ^{60}Co as a tracer. Baudin et al. (1997) studied the uptake and elimination kinetics of ^{60}Co in rainbow trout (*Onchorhynchus mykiss*) from water. The BCF value, calculated from the ratio of the radionuclide concentration in fish and filtered water, reached a maximum of 4.6 on a wet weight basis after 30 days' exposure. After the 42 day depuration phase, the fish retained about 29% of the accumulated radionuclide. A single-compartment exponential model was fitted to the ^{60}Co elimination data, and the corresponding radionuclide half-life was 21 days. At the end of the exposure phase, tissue counting showed ^{60}Co accumulation by the gills, viscera (air bladder, heart, and spleen), and kidneys to be the highest. At the end of the depuration phase, the kidney was the most contaminated organ, followed by the viscera, head, gills, and liver. In both cases, ^{60}Co concentration was by far the lowest in the muscle, which accounted for about 45% of the total body weight and only 20% of the total radionuclide body load.

8.2. Effect of Cobalt Complexation on Uptake

Complexation of dissolved Co results in a decreased free Co ion activity and this has a direct effect on the uptake of waterborne Co by fish, as illustrated for the common carp (*Cyprinus carpio*) by Blust et al. (1997). They studied the uptake of Co in carp using ^{57}Co as a tracer in the presence of different types and concentrations of ligands generating a wide range of free Co ion levels. The results showed that complexation decreases Co uptake and that the effect appears to depend only on the free metal ion activity and not on the complexing ligand (Fig. 6.4). However, with increasing free Co ion activities the rate of Co uptake levels off (note the logarithmic scale on the *x*-axis in Fig. 6.4) and could be analyzed using a Michaelis–Menten model for facilitated transport (Weiss, 1996). The results could only be explained in a reasonable way when two uptake sites for Co were assumed: one site with a high affinity for Co and a relatively low transport rate and a second site with a much lower affinity for Co but higher

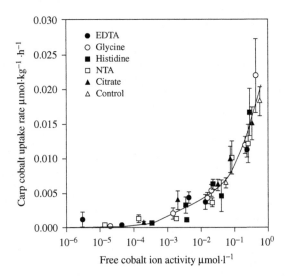

Fig. 6.4. Effect of complexation by organic ligands (glycine, citrate, histidine, NTA, EDTA) with different Co affinities [glycine (log $K = 5.1$), citrate (6.6), histidine (7.6), NTA (11.8), EDTA (18.2)] on the rate of Co uptake in common carp, measured over a 3 h exposure period using ^{57}Co as tracer, as a function of the free Co ion activity in medium hard water containing either 0.1 or 1.0 μM of dissolved Co and ligand concentrations ranging from 10^{-8} to 10^{-2} M at a pH of 8.0 and temperature of 25°C. Data points represent means with standard deviations for six to seven fish. The solid line is the fit of a two binding-site Michaelis–Menten-type equation to the influx data (from Blust et al., 1997).

transport rate. The first binding site becomes saturated at relatively low Co levels in the environment, while the second binding site only saturates at rather high Co exposure levels.

8.3. Effect of Calcium on Cobalt Uptake

Comhaire et al. (1994, 1997) studied the effect of increasing Co concentrations and different Ca concentrations on the uptake of Co across the gills of the common carp (*C. carpio*) in chemically defined freshwater. Before the Co uptake experiments, fish were acclimated for 16 days to different Ca concentrations ranging from 0.1 to 10 mM Ca. Cobalt and Ca influx were measured, over a 3 h period at the same range of Ca concentrations, using ^{57}Co and ^{45}Ca as tracers, and uptake was determined in the whole body, gills, and blood. The uptake of Co increased with increasing Co concentration in the exposure water in a biphasic manner. At relatively low Co concentrations Co uptake displayed saturable kinetics but at higher exposure concentrations a linear relation between exposure concentration and influx rate was observed (Fig. 6.5). The results showed a strong decrease in the uptake of Co with increasing Ca concentrations in the exposure water. Vice versa, Co had a strong inhibitory effect on Ca uptake

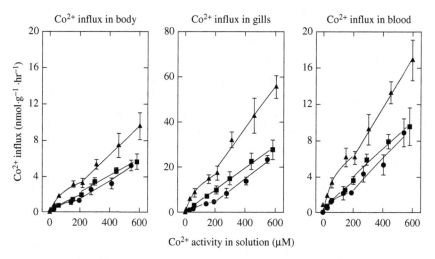

Fig. 6.5. Effect of Co^{2+} ion activity in exposure water on Co influx in the whole body, gills, and blood of carp over a 3 h exposure period using ^{57}Co as a tracer at three different calcium concentrations (circles: 1000 μM Ca; squares: 348 μM Ca; triangles: 100 μM Ca) at a pH of 8.0 and temperature of 25°C. Data points represent means with standard deviations for five to 14 fish (from Comhaire et al., 1997).

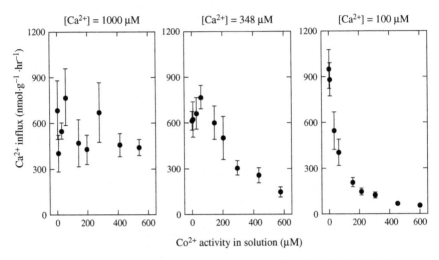

Fig. 6.6. Effect of Co ion activity in exposure water on calcium influx in the gills of common carp over a 3 h exposure period using ^{45}Ca as tracer at three different calcium concentrations at a pH of 8.0 and temperature of 25°C. Data points represent means with standard deviations for five to 14 fish (from Comhaire et al., 1997).

(Fig. 6.6). Above a Ca concentration of 1000 μM, no further effect on Co uptake was observed. In addition to the direct effect of Ca on Co uptake, the effect of the Ca concentration in the acclimation water was significant, but of less importance. In another study, Comhaire et al. (1998) investigated whether the branchial uptake of Co occurs via selective, inhibitable, calcium uptake routes. Modulation of the Ca transport system was performed using general Ca channel blockers (Cd^{2+}, La^{3+}, Mg^{2+}), a voltage-dependent Ca channel blocker (diltiazem), and an intraperitoneal $CaCl_2$ injection. Adding Cd^{2+} or Mg^{2+} to the water or injecting the fish with the Ca^{2+} solution resulted in decreased Co and Ca uptake rates. Addition of La^{3+} reduced Ca and Co uptake rates except for Co uptake in the gills. Diltiazem did not show a clear effect on either Co or Ca uptake. All inhibitors which inhibited Ca uptake also inhibited Co uptake, but the degree of inhibition was not always the same. Taken together, these results provide strong evidence for a direct competitive type of interaction between Co and Ca at the site of uptake in the gills of fish.

8.4. Cobalt Gill Binding Model

To characterize the binding of Co to fish gills, Richards and Playle (1998) exposed rainbow trout (*Oncorhynchus mykiss*) to approximately 442 μg L^{-1} (7.5 μM) Co in soft water for 2–3 h at pH 6.5. The water contained complexing ligands such as nitrilotriacetic acid (NTA) and natural dissolved organic matter (DOM) and competing cations such as Ca^{2+}, Na^+, or H^+. A Langmuir isotherm was used to calculate the conditional equilibrium binding constant (K) for Co binding to trout gills plus the concentrations of gill Co binding sites. The calculated binding constant for Co to trout gills was log $K_{gill} = 5.1$, with 88 nmol Co binding sites per 1 g of wet gill tissue. Conditional equilibrium binding constants were also calculated for Ca^{2+}, Na^+, and H^+ binding to the gill Co sites and for Co binding to DOM. The experimentally determined binding constants were entered into an aquatic equilibrium speciation program, MINEQL(+), to predict Co binding by fish gills. Predicted and observed results indicate that Co would not accumulate on or in gills of trout held in a series of natural and 1:1 diluted natural waters supplemented with approximately 8.7 μM Co. Model analysis of the reasons for Co being kept off gills of trout held in natural waters spanning a wide range of Ca^{2+} and Na^+, pH (4.2–7.6), and DOM levels, indicated that Ca^{2+} competition and DOM complexation were the most important factors. However, the model significantly underpredicted the gill Co accumulations at the highly alkaline range (pH > 9) in the laboratory soft water. Considering the fact that all Co in the media was in dissolved form, this finding raised the possibility of $CoOH^+$ and/or $CoCO_3(aq)$ (the two dominant Co species under the existing conditions) also binding to the gill sites.

The obtained binding constants were used to develop a Co gill binding model that takes into account effects of environmental conditions and metal speciation on metal interaction with the gills. These models can then be used to develop a biotic ligand model (BLM) (see Paquin et al. Chapter 9, Vol. 31B) to predict the acute or chronic toxicity of metal exposure in fish and other aquatic organisms. Niyogi and Wood (2004) included these results in a multimetal gill binding model which also considered metals such as Cd, Pb, and Zn, which have been shown to interact with Ca uptake sites in the fish gills. Comparison of the binding constants of Co with other metals indicates that Co^{2+} binds to gill sites over 1000 times more weakly than Cd^{2+}, 10 times more weakly than Pb^{2+}, and about six times more weakly than Zn^{2+}. However, the binding strengths of Ca^{2+} and H^+ at the same sites are more or less similar for the three metals. While this gill–Co binding model provides a basic framework, coupling to acute and chronic toxicity data will be needed to transform it into a BLM that can be used for environmental regulation.

8.5. Cobalt Uptake via the Gut

Several studies have explored the uptake of Co in fish via the dietary route and determined the relative importance of water and food as sources of Co. The accumulation of ^{59}Fe and [58]Co from water by plaice (*Pleuronectes platessa* L.) in tissues, and their levels of accumulation relative to stable element concentrations was studied by Pentreath (1973a). The retention of both radionuclides from labeled *Nereis* and labeled pellets was also examined, and the loss of activity from intraperitoneal injections followed over a 3 month period. Biological half-times of the exchange with water were calculated for different organs and for both radionuclides the direct accumulation from water resulted in concentration factors of less than 1% of the values inferred from stable metal analysis, indicating that Co is mainly taken up via the dietary route. In another study, Pentreath (1973b) investigated the accumulation of ^{59}Fe and ^{58}Co by the thornback ray (*Raja clavata* L.) on a whole-body basis. Lower values of accumulation of both radionuclides were obtained than in the case of the teleost (*Pleuronectes platessa* L.). It was concluded that for both species direct input from water plays only a minor role in the accumulation of the radionuclides studied. In contrast, Suzuki et al. (1979) showed that water and food were about equally important in the accumulation of ^{60}Co in the Japanese amberjack (*Seriola quinqueradiata*).

Baudin and Fritsch (1989) studied the relative importance of food and water as sources of ^{60}Co to the carp (*Cyprinus carpio*). The fish were exposed to Co via water, the diet, and a combination of both routes. The individuals of the first group were offered 45 daily rations of labeled food over a 63 day period. The carp of the two other groups were exposed to 6.8×10^4 Bq ^{60}Co L^{-1}. One group was fed with radioactive food, the other with non-radioactive food. The contamination kinetics showed that the steady state should be reached after 165 days for fish exposed to ^{60}Co in food, 92 days for fish exposed to ^{60}Co in water, and 120 days for fish exposed to ^{60}Co in both sources. According to the ^{60}Co concentration in the fish of the three treatments, the accumulation from water accounted for 75% of the total radioactivity and the accumulation from both water and food was additive. Depuration of ^{60}Co from carp was a relatively fast process, reflecting a high Co turnover. Biological half-lives for loss from the long-lived compartment ranged from 53 days in fish previously contaminated from food to 87 days in fish previously contaminated from water.

9. CHARACTERIZATION OF INTERNAL HANDLING

The internal handling of Co is not documented for fish except for the accumulation in different tissues as presented in Section 9.1. A model for the biokinetics of inorganic Co has been developed for humans by the International Commission on Radiological Protection (ICRP) (Leggett, 2008). According to the ICRP's systemic model, Co entering blood moves to excretion pathways (50%), liver (5%), and remaining tissues (45%). According to the compartmental model, fractions of 0.6, 0.2, and 0.2 deposited in tissues are removed to excretion pathways with half-times of 6, 60, and 800 days, respectively, with six-sevenths going to urinary pathways and one-seventh to fecal pathways. However, these results cannot be compared directly since there are important differences between fish and humans. Nonetheless, this model may provide a basis for the future development of biokinetic models for metals in fish, including Co.

There are no studies concerning the metabolism and biotransformation of Co or cobalamin in fish but the processes resemble, at least partially, those described in mammalian systems. These studies are largely confined to the handling of cobalamin, which is the biologically most important form of Co. The absorption of cobalamin from the gut is a receptor-mediated process (Quadros, 2010). The gastric uptake is characterized by a number of steps, of which the first is the release from proteins in the food by the action of pepsins. Secondly, cobalamin is bound to R-proteins, which have a high affinity for cobalamin intrinsic factor. Intrinsic factor is secreted with HCl by the parietal cells of the stomach. Cobalamin bound to R-proteins is transferred to intrinsic factor, forming a cobalamin–intrinsic factor complex after the R-proteins have been degraded by proteases. In a third phase, the cobalamin–intrinsic factor complex binds to a mucosal receptor in the thin gut (Banerjee, 2006). After internalization, the complex enters the lysosomes, where the intrinsic factor is degraded and cobalamin released to the intracellular environment. It is exported from the mucosa in association with transcobalamin II, which is the main plasma transport protein for cobalamin and transfer to different tissues. Transcobalamin I and III also bind cobalamin in plasma but are less specific and have a lower turnover rate. The receptors on the target cells recognize the complex, which enters the lysosomes and again releases the cobalamin.

9.1. Accumulation in Specific Organs

As discussed in Section 7, Co appears to accumulate in different tissues and results show large variation (IAEA, 2009). In an experimental dietary exposure study by Pentreath (1973a) with the marine teleost *Pleuronectes platessa*, between 11.2 and 48.1% of the radionuclide ^{60}Co was recovered in the gut wall 4 days after the fish were force fed with pellets of live *Nereis*, while gills, liver, and kidney only retained less than 1 or 2% of the total amount of ^{60}Co that was accumulated. Baudin and Fritsch (1989) observed a retention of ^{60}Co in the gut of common carp (*Cyprinus carpio*), ranging between 19.6 and 26.5% of the total amount, with no clear difference between waterborne and dietary exposure. In this case, the gills accounted for 2.8–4.9%, the viscera including the liver for 2.3–3.2%, and the kidneys for 9.0–13.3% of the total amount. Baudin et al. (1997, 2000) also found the kidneys to be the main ^{60}Co-sequestering organ in water or dietary exposure studies with trout (*Oncorynchus mykiss*). The relative higher accumulation in certain tissues has been related to the exposure pathway (i.e. gills and gut), the physiological role of the organ (i.e. hematopoietic function), and the excretion pathway (i.e. gut and kidneys). In a Co dietary supplementation study on carp, Mukherjee and Kaviraj (2009) found that under normal non-Co-supplemented conditions, the highest concentration of Co was found in the kidneys, followed by the gut, liver, and gills. With increasing dietary Co, the role of the kidneys and liver became more important and at the highest Co supplementation the situation reversed and strong accumulation in the gut, kidneys, gills, and liver occurred. These results also indicate that Co is under homeostatic control and that the highest dietary exposure level was clearly above the exposure range of regulation at the tissue level.

10. CHARACTERIZATION OF EXCRETION ROUTES

In their study on the relative importance of waterborne and dietary ^{60}Co to carp, Baudin and Fritsch (1989) concluded that regardless of the exposure route, the excretion pathways were similar. Using a two-compartmental model, the elimination of ^{60}Co was characterized by two biological half-lives of 1.2 and 87 days. Branchial and renal excretions were more important than fecal elimination. In the water-exposed group, fecal elimination accounted for $34 \pm 7\%$ of the elimination, which was not significantly different from the $38 \pm 7\%$ and $37 \pm 8\%$ obtained for the food-only and food- and water-exposure groups. Monitoring of the ^{60}Co levels in trout during the water exposure phase and subsequent depuration phase resulted in an estimated half-life of the radionuclide of 21 days using a one-compartment model (Baudin et al., 1997). Comparison of the tissue levels at the end of the exposure and depuration phase showed that after the kidneys and the liver it was the skeleton that showed least elimination, followed by the head, viscera, and muscle. In a similar dietary exposure study with trout, Baudin et al. (2000) obtained biological half-lives of 4 and 170 days, with the kidneys showing again the highest ^{60}Co accumulation.

11. BEHAVIORAL EFFECTS OF COBALT

Very few studies have considered the potential behavioral effects of Co in fish. Hansen et al. (1999) showed that the behavioral avoidance of Cu, Co, and a Cu and Co mixture in soft water differed between rainbow trout (*Oncorhynchus mykiss*) and chinook salmon (*O. tshawytscha*). Chinook salmon avoided $\geq 0.7\,\mu g$ Cu L^{-1} and $\geq 24\,\mu g$ Co L^{-1}, and a mixture of $1.0\,\mu g$ Cu L^{-1} and $0.9\,\mu g$ Co L^{-1}, whereas rainbow trout avoided $\geq 1.6\,\mu g$ Cu L^{-1} and $\geq 180\,\mu g$ Co L^{-1}, and a mixture of $2.6\,\mu g$ Cu L^{-1} and $2.4\,\mu g$ Co L^{-1}. These values are lower than most of the Co NOEC or LOEC values reported for fish.

Although not of immediate ecotoxicological relevance, Co is also used to block the lateral line of fish. This is because of the blocking effect of Co^{2+} on the mechanosensory hair cells of the superficial and canal neuromasts. These studies are conducted in relation to the involvement of the lateral line and the inner ear in the detection of infrasound, in directional hearing in the near field, and in detection and attack of swimming prey below the surface (Karlsen and Sand, 1987; Hassan et al., 1992; Liao, 2006; Ayali et al., 2009) and rheotaxis (Montgomery et al., 1997; Baker and Montgomery, 1999).

By recording multiunit activity from the lateral line nerve and microphonic potentials from the inner ear in the roach (*Rutilus rutilus*), Karlsen and Sand (1987) have shown that Co^{2+} ions in the external water may completely block the mechanosensitivity of the lateral line without affecting the utricular microphonic activity. This inhibitory effect of Co^{2+} is antagonized by Ca^{2+}, which is indicative of a competitive interaction between these two ions for binding sites. For practical work, they recommended 12–24 h exposure to 0.1 mmol L^{-1} Co^{2+} ($5900\,\mu g\,L^{-1}$) at a Ca^{2+} concentration of less than 0.1 mmol L^{-1}. The fish showed no sign of general behavioral disorders even after 1 week in this solution, and the microphonic sensitivity of the inner ear was not reduced. The blocking effect of Co^{2+} was clearly reversible, and the recovery was dependent on both the duration of the Co^{2+} exposure and the Ca^{2+} concentration of the recovery solution. In these studies Co concentrations are high compared to environmental exposure conditions and exposure times should be kept short to avoid toxicological effects interfering with the pharmacological use of Co as blocking agent (Janssen, 2000).

12. MOLECULAR CHARACTERIZATION OF COBALT TRANSPORTERS, STORAGE PROTEINS, AND CHAPERONES

There are no studies on specific Co proteins involved in transmembrane, extracellular, or intracellular transport and storage in fish. However, it has been shown that waterborne Co uptake decreases with increasing Ca levels and that at least part of the Co is taken up across the gills via Ca gateways

(Comhaire et al., 1998). Across a diversity of biological systems several transport system have been implicated in the transport of Co which in general were shown to transport certain other metals as well. The transport may be by passive channeling or active using ATP or other solutes or ions (Paulsen and Saier, 1997; Ballatori, 2002).

Ion channels and active ion pumps are, in addition to the transport of essential alkaline and alkali-earth metal, often involved in the transport of other essential and non-essential metals (Nelson, 1999). Ligand-sensitive and voltage-sensitive Ca channels of excitable tissues provide a gateway for the uptake of Co (Diaz et al., 2005). Such Ca channels may also allow other divalent cations to enter the cell, as supported by the fact that these Ca channels are inhibited by Zn^{2+}, Cd^{2+}, Ni^{2+}, Co^{2+}, and Mn^{2+}. Metals also interact with plasma membrane Ca^{2+}-ATPases, which are the most important proteins for Ca^{2+} efflux from cells and may also be involved in the transport of other metal ions including Co (Ballatori, 2002).

Another important group is the natural resistance-associated macro-phage protein (NRAMP) family of transport proteins to which belong the divalent cation transporter (DCT), several putative zinc transporters (ZIP and ZNT families), and iron-regulatory protein-1 (IREG1) (Nevo and Nelson, 2006). Some of these transporters have a high specificity towards one single metal ion, but others appear less selective. DCT1 is such a multimetal transporter and a major mechanism for the cellular uptake of Fe^{2+} and other divalent metal ions including Co, and is expressed in a wide number of tissues (Andrews et al., 1999) (see also Bury et al., Chapter 4, on Fe). Gaither and Eide (2001) have described a family of putative zinc transporters, the ZIP family, that may function for uptake of Zn and possibly other divalent metal ions from the extracellular medium into the cytoplasm (see also Hogstrand, Chapter 3, on Zn). The activity of the CorA proteins has generally been associated with the transport of Mg ions but the members of the CorA family can also transport other ions such as Co and Ni (Forbes and Gros, 2003; Niegowski and Eshaghi, 2007). P_{1B}-type ATPases also transport metals in which the P_{1-4B} subgroup has a distinct sensitivity for Co^{2+} ions (Argüello et al., 2007).

Another important class of transporters that may be involved in the export of metal complexes is the ATP-binding cassette (ABC) group of proteins, which includes multidrug resistance (MDR) families. These transport proteins have a broad specificity and are involved in the transport of a variety of structurally different molecules including metal complexes. For example, there is evidence that the multidrug resistance protein 2 (MRP2) mediates biliary transport of metals complexed with glutathione (Borst and Elferink, 2002). In prokaryotes, members of the ABC group of transporters have been implicated in the specific uptake of Ni and Co (Rodionov et al., 2006).

Although almost nothing is known about the specific transport and handling of Co in fish it is likely that the different systems mentioned above also play a more or less important role in the transport of Co in fish.

13. GENOMIC AND PROTEOMIC STUDIES

Full-scale genomic, transcriptomic, or proteomic studies addressing the physiology or toxicology of Co in fish have not been performed so far. However, exposure to or injection of fish with $CoCl_2$ has been used to activate certain genes. Chu et al. (2010) investigated whether the gene that encodes the appetite-suppressing hormone leptin is regulated by hypoxia in zebrafish (*Danio rerio*). Exposure of adult zebrafish to hypoxic conditions $(1 \pm 0.2 \text{ mg O}_2 \text{ L}^{-1})$ for 4 and 10 days significantly increased leptin-a (zlep-a) mRNA levels in the liver. To evaluate the role of hypoxia-inducible factor-1 (HIF-1) in regulating zlep-a expression, zebrafish embryos were exposed to $CoCl_2$, which is a known HIF-1 inducer and overexpressed with HIF-1α mRNA. Both $CoCl_2$ treatment and HIF-1α markedly increased zlep-a expression in developing embryos, indicating the possible involvement of HIF-1 in zlep-a regulation. *In vivo* promoter analysis indicated that zlep-a promoter activity is found in the muscle fibers of zebrafish embryos and enhanced by $CoCl_2$.

14. INTERACTIONS WITH OTHER METALS

The effects of Co on Ca uptake and vice versa are well documented, as discussed earlier. There is strong evidence that Co and Ca use the same uptake sites in the gills of fish and interact in a competitive manner (Comhaire et al., 1994, 1997, 1998). Diamond et al. (1992) investigated whether acute Co toxicity in freshwater was dependent on dilution water hardness. Toxicity tests using the fathead minnow (*Pimephales promelas*) were performed in synthetic waters of 50, 200, 400, and 800 mg L^{-1} hardness as $CaCO_3$. Regression analysis of *P. promelas* acute NOEC values and test water hardness showed a direct relationship between water hardness and acute Co toxicity. The 7 day NOEC for *P. promelas* increased from 1232 μg L^{-1} Co (at 50 mg L^{-1} hardness) to ≥ 3833 μg L^{-1} (at 800 mg L^{-1} hardness). Effects were on fish survival but not on growth. Regression analysis yielded a linear relationship based on fish survival NOECs and the natural log of hardness. The results suggest that within the 50–200 mg L^{-1} water hardness range, Co acute toxicity is inversely related to water hardness. Chronic endpoints for *P. promelus* in soft and hard water (≤ 400 mg L^{-1} hardness) were similar, suggesting that chronic toxicity is not hardness dependent over this range. However, at 800 mg L^{-1} hardness, the fish were less sensitive than under more soft water conditions, suggesting that very hard waters significantly reduce the chronic toxicity potential of Co.

Interaction with Ca and a number of other metals has also been incorporated in the Co gill binding model developed by Richards and Playle (1998) and the BLM, of which the various implementations have been reviewed by Niyogi and Wood (2004). This model considers the site of calcium uptake in the gills as the main site of Cd, Co, Pb, and Zn uptake (see Section 8.4).

Marr et al. (1998) conducted 14 day laboratory toxicity tests with rainbow trout to evaluate individual and combined effects of Co and Cu administered as chloride salts. They made a comparison of mortality percentages observed in Co/Cu mixtures to those predicted from two models of joint toxicity. Cobalt was a slower acting and less potent toxicant than was Cu. For Co, the incipient lethal level (ILL50) for 50% mortality was 346 μg L^{-1}, whereas for Cu, the ILL50 was 14 μg L^{-1}. Moreover, in Co/Cu mixtures, Co acted as an antagonist during the first 48–96 h, but later acted as an additive or slightly synergistic toxicant, making it difficult to predict short-term mortality of fish in Co/Cu mixtures.

Kwong and Niyogi (2009) examined the concentration-dependent interactive effects of four essential (Cu, Zn, Ni, Co) and two non-essential (Pb, Cd) divalent metals on intestinal iron (Fe^{2+}) absorption in freshwater rainbow trout (*Oncorhynchus mykiss*) using an *in vitro* gut sac technique. All of the divalent metals except Co inhibited the intestinal Fe^{2+} absorption in fish. The mucosal epithelium of the intestine was found to be the most sensitive to inhibition in comparison to the mucus layer or blood compartment, suggesting that these interactions are likely to occur via the divalent metal transporter-1 (DMT1). The absence of an inhibitory effect of Co on Fe uptake implies that Co uptake across the gut in fish does not involve the DMT1 transporter. Bury and Grosell (2003) also showed that branchial Fe uptake in zebrafish (*Danio rerio*) was not affected by other divalent metals, including Co, except Cd.

15. KNOWLEDGE GAPS AND FUTURE DIRECTIONS

In contrast with those of other essential metals such as Cu or Zn, the physiology and toxicology of Co in fish are poorly documented and understood. The physiological role of Co in fish is as an intrinsic part of cobalamin. Several studies have studied the need for Co or cobalamin supplementation in fish aquaculture. All these studies focus on the effects of Co supplementation on fish production and do not consider the internal handling or biochemistry of Co. In fact, almost all information on the biochemistry of Co comes from studies in mammalian models.

Studies on the accumulation and toxicity of Co are scattered and provide an incomplete picture. Most accumulation studies have been performed in relation to biomonitoring and experimental exposure studies concerning the release of the radionuclide ^{60}Co in the aquatic environment.

The ecotoxicological literature contains a number of toxicity studies with fish but most have only considered acute scenarios and the number of chronic studies is very limited. Estuarine and marine species, in particular, are very poorly covered in relation to Co toxicity.

Studies concerning the effects of environmental conditions on uptake and toxicity are limited in number, despite the importance of these in relation to the development of models to predict accumulation and toxicity under different exposure scenarios. Although it has been clearly shown that, as for other metals, changes in chemical speciation, water hardness, and pH have an important impact on Co availability and toxicity, these studies are confined to only a few species and cases.

Almost nothing is known concerning the interactions of Co with the biological machinery in fish and the toxicological modes of action. Studies concerning the molecular aspects of Co uptake, its internal processing, and mechanisms of toxicity are largely lacking. Genomic and proteomic studies on Co in fish are very limited and the few available are related to specific genes and gene products.

As such, different knowledge gaps and future directions can be identified in the case of Co. Future studies could: (1) provide a more extensive documentation of the chronic toxicity of Co in freshwater, estuarine, and marine fish, (2) characterize the effects of environmental conditions on Co uptake and toxicity under environmentally realistic scenarios, (3) identify the transporters involved in Co uptake across the gills and the digestive system, (4) characterize the internal handling of Co in fish including excretion pathways, and (5) characterize the molecular mechanisms of toxicity and resulting physiological effects.

ACKNOWLEDGMENTS

The author wishes to thank Andrea Vlaeminck for invaluable assistance with database searches, literature compilations, and critical reading of the manuscript. Also thanks to Chris Wood, Colin Brauner, and two anonymous reviewers for their critical and constructive comments on draft versions of the manuscript.

REFERENCES

Albrecht, A. (2003). Validating riverine transport and speciation models using nuclear reactor-derived radiocobalt. *J. Environ. Radioactiv.* **66**, 295–307.

Anadu, D. I., Anozie, O. C., and Anthony, A. D. (1990). Growth responses of *Tilapia zillii* fed diets containing various levels of ascorbic acid and cobalt chloride. *Aquaculture* **88**, 329–336.

Andrews, N. C., Fleming, M. D., and Gunshin, H. (1999). Iron transport across biological membranes. *Nutr. Rev.* **57**, 114–123.

ANZECC (2000). *Australian and New Zealand Guidelines for Fresh and Marine Water Quality.* Australian and New Zealand Environment and Conservation Council and the Agriculture and Resource Management Council of Australia and New Zealand.

Argüello, J. M., Eren, E., and Gonzalez-Guerrero, M. (2007). The structure and function of heavy metal transport P-1B-ATPases. *Biometals* **20**, 233–248.

Atamanalp, M., Kocaman, E. M., Ucar, A., and Alak, G. (2010). The alterations in the hematological parameters of brown trout *Salmo trutta fario*, exposed to cobalt chloride. *J. Anim. Vet. Adv.* **9**, 2167–2170.

Ayali, A., Gelman, S., Tytell, E. D., and Cohen, A. H. (2009). Lateral-line activity during undulatory body motions suggests a feedback link in closed-loop control of sea lamprey swimming. *Can. J. Zool.* **87**, 671–683.

Baker, C. F., and Montgomery, J. (1999). The sensory basis of rheotaxis in the blind Mexican cavefish, *Astyanax fasciatus*. *J. Comp. Physiol. A* **184**, 519–527.

Balistrieri, L. S., Murray, J. W., and Paul, B. (1994). The geochemical cycling of trace elements in a biogenic meromictic lake. *Geochim. Cosmochim. Acta* **58**, 3993–4008.

Ballatori, N. (2002). Transport of toxic metals by molecular mimicry. *Environ. Health Perspect.* **110**, 689–694.

Banerjee, R. (2006). B-12 trafficking in mammals: a case for coenzyme escort service. *ACS Chem. Biol.* **1**, 149–159.

Banerjee, R. V., and Matthews, R. G. (1990). Cobalamin-dependent methionine synthase. *FASEB J.* **4**, 1450–1459.

Banerjee, R., and Ragsdale, S. W. (2003). The many faces of vitamin B_{12}: catalysis by cobalamin-dependent enzymes. *Annu. Rev. Biochem.* **72**, 209–247.

Barceloux, D. G. (1999). Cobalt. *Clin. Toxicol.* **37**, 201–216.

Baudin, J. P., and Fritsch, A. F. (1989). Relative contributions of food and water in the accumulation of ^{60}Co by a freshwater fish. *Water Res.* **23**, 817–823.

Baudin, J. P., Veran, M. P., Adam, C., and Garnier-Laplace, J. (1997). Co-60 transfer from water to the rainbow trout (*Oncorhynchus mykiss* Walbaum). *Arch. Environ. Contam. Toxicol.* **33**, 230–237.

Baudin, J. P., Adam, C., and Garnier-Laplace, J. (2000). Dietary uptake, retention and tissue distribution of ^{54}Mn, ^{60}Co, and ^{137}Cs in the rainbow trout (*Oncorhynchus mykiss* Walbaum). *Water Res.* **34**, 2869–2878.

Bhanot, K. K., and Gopalakrishnan, V. (1973). Priorities in nutritional research for formulating artificial feeds for fishes. *J. Inland Fish Soc. India* **5**, 162–170.

Bird, G. A., Hesslein, R. H., Mills, K. H., Schwartz, W. J., and Turner, W. A. (1998a). Bioaccumulation of radionuclides in fertilized Canadian Shield Lake basins. *Sci. Total Environ.* **218**, 67–83.

Bird, G. A., Schwartz, W. J., Motycka, M., and Rosentreter, J. (1998b). Behavior of ^{60}Co and ^{134}Cs in a Canadian shield lake over 5 years. *Sci. Total Environ.* **212**, 115–135.

Bird, G. A., Mills, K. H., and Schwartz, W. J. (1999). Accumulation of ^{60}Co and [134]Cs in lake whitefish in a Canadian shield lake. *Water Air Soil Pollut.* **114**, 303–322.

Blust, R., Van Ginneken, L. and Comhaire, S. (1997). Bioavailability of radiocobalt to the common carp, *Cyprinus carpio*, in complexing environments. International Seminar on Freshwater and Estuarine Radioecology, March 21–25, 1994, Lisbon, Portugal. Freshwater and Estuarine Radioecology. Book Series: *Studies in Environmental Science*, Vol. 68, pp. 299–305.

Borst, P., and Elferink, R. O. (2002). Mammalian ABC transporters in health and disease. *Annu. Rev. Biochem.* **71**, 537–592.

Burba, P., Rocha, J., and Klockow, D. (1994). Labile complexes of trace metals in aquatic humic substances: investigations by means of an ion exchange-based flow procedure. *Fresenius J. Anal. Chem.* **349**, 800–807.

Bury, N. R., and Grosell, M. (2003). Waterborne iron acquisition by a freshwater teleost fish, zebrafish *Danio rerio*. *J. Exp. Biol.* **206**, 3529–3535.

Castell, J. D., Conklin, D. E., Craigie, J. S., Lall, S. P., and Norman-Boudreau, K. (1986). In *Realism in Aquaculture: Achievements, Constraints, Perspectives* (M. Bile, H. Rosenthal and C. Sindermann, eds), pp. 251–308. European Aquaculture Society, Belgium.

Chang, Z., Ambe, S., Takahashi, K., and Ambe, F. (2006). A study on the metal binding of humic acid by multitracer technique. *Radiochim. Acta* **94**, 37–46.

Chu, D. L. H., Li, V. W., and Yu, R. M. K. (2010). Leptin: clue to poor appetite in oxygen-starved fish. *Mol. Cell. Endocrinol.* **319**, 143–146.

Collins, R. N., and Kinsela, A. S. (2010). The aqueous phase speciation and chemistry of cobalt in terrestrial environments. *Chemosphere* **79**, 763–771.

Combs, G. F. (2008). *The Vitamins: Fundamental Aspects in Nutrition and Health* (3rd edn.). Elsevier, Amsterdam.

Comhaire, S., Blust, R., Van Ginneken, L., and Vanderborght, O. L. J. (1994). Cobalt uptake across the gills of the common carp, *Cyprinus carpio*, as a function of calcium-concentration in the water of acclimation and exposure. *Comp. Biochem. Physiol. C* **109**, 63–76.

Comhaire, S., Blust, R., Van Ginneken, L., D'haeseleer, F. and Vanderborght, O. (1997). Calcium influences radio-cobalt uptake by the common carp, *Cyprinus carpio*. International Seminar on Freshwater and Estuarine Radioecology, March 21–25, 1994, Lisbon, Portugal. Freshwater and Estuarine Radioecology. Book Series: *Studies in Environmental Science*, Vol. 68, pp. 321–328.

Comhaire, S., Blust, R., Van Ginneken, L., Verbost, P. M., and Vanderborght, O. L. J. (1998). Branchial cobalt uptake in the carp, *Cyprinus carpio*: effect of calcium channel blockers and calcium injection. *Fish Physiol. Biochem.* **18**, 1–13.

Coughtrey, P. J. and Thorne, M. C. (1983). Cobalt. In *Radionuclide Distribution and Transport in Terrestrial and Aquatic Ecosystems*, Vol. 2. pp. 41–94. A.A. Balkema, Rotterdam.

Dave, G., and Xiu, R. Q. (1991). Toxicity of mercury, copper, nickel, lead, and cobalt to embryos and larvae of zebrafish, *Brachydanio rerio*. *Arch. Environ. Contam. Toxicol.* **21**, 126–134.

Davis, A. D. and Gatlin, D. M. III (1991). In *Dietary Mineral Requirements of Fish and Shrimp* (eds D. M. Akiyama and K. H. Tan), pp. 49–67. Proceedings of the Aquaculture Feed Processing and Nutrition Workshop, Singapore. American Soybean Association.

Diamond, J. M., Winchester, E. L., Mackler, D. G., Rasnake, W. J., Fanelli, J. K., and Gruber, D. (1992). Toxicity of cobalt to fresh-water indicator species as a function of water hardness. *Aquat. Toxicol.* **22**, 163–179.

Diaz, D., Bartolo, R., Delgadillo, D. M., Higueldo, F., and Gomora, J. C. (2005). Contrasting effects of Cd^{2+} and Co^{2+} on the blocking/unblocking of human Ca(v)3 channels. *J. Membr. Biol.* **207**, 91–105.

Duckworth, O. W., Bargar, J. R., Jarzecki, A. A., Oyerinde, O., Spiro, T. G., and Sposito, G. (2009). The exceptionally stable cobalt(III)–desferrioxamine B complex. *Mar. Chem.* **113**, 114–122.

Ehleben, W., Bölling, B., Merten, E., Porwol, T., Strohmaier, A. R., and Acker, H. (1998). Cytochromes and oxygen radicals as putative members of the oxygen sensing pathway. *Respir. Physiol.* **114**, 25–36.

Ellwood, M. J., and van den Berg, C. M. G. (2001). Determination of organic complexation of cobalt in seawater by cathodic stripping voltammetry. *Mar. Chem.* **75**, 33–47.

Ellwood, M. J., van den Berg, C. M. G., Boye, M., Veldhuis, M., de Jong, J. T. M., de Baar, E. J. W., Croot, P. L., and Kattner, G. (2005). Organic complexation of cobalt across the Antarctic polar front in the southern ocean. *Mar. Freshwat. Res.* **56**, 1069–1075.

Fasfous, I., Yapici, T., Murimboh, J., Hassan, I. M., Chakrabarti, C. L., Back, M. H., Lean, D. R. S., and Gregoire, D. C. (2004). Kinetics of trace metal competition in the freshwater environment: some fundamental characteristics. *Environ. Sci. Technol.* **38**, 4979–4986.

Foerstner, K. U., Doerks, T., Muller, J., Raes, J., and Bork, P. (2008). A nitrile hydratase in the eukaryote *Monosiga brevicollis*. *PLoS ONE* **3**, e3976.

Forbes, J. R., and Gros, P. (2003). Iron, manganese, and cobalt transport by Nramp1 (Slc11a1) and Nramp2 (Slc11a2) expressed at the plasma membrane. *Blood* **102**, 1884–1892.

Frolova, L. K. (1960). Vlijanie kobal'ta na morfologiEeskuju kartinu krovi karpa. *Dokl. Akad. Nauk. USSR* **131**, 983–984.

Garnier, J.-M., Pham, M. K., Ciffroy, P., and Martin, J.-M. (1997). Kinetics of trace element complexation with suspended matter and with filterable ligands in freshwater. *Environ. Sci. Technol.* **31**, 1597–1606.

Garnier-Laplace, J., Adam, C., and Baudin, J. P. (2000). Experimental kinetic rates of food-chain and waterborne radionuclide transfer to freshwater fish: a basis for the construction of fish contamination charts. *Arch. Environ. Contam. Toxicol.* **39**, 133–144.

Gaither, L. A., and Eide, D. J. (2001). Eukaryotic zinc transporters and their regulation. *Biometals* **14**, 251–270.

Ghosh, S. R. (1975). Preliminary observation on the effect of cobalt on the survival and growth of *Mugil parsia*. *Bamidgeh* **27**, 110–111.

Glaus, M. A., Hummel, W., and van Loon, L. R. (2000). Trace metal–humate interactions. I. Experimental determination of conditional stability constants. *Appl. Geochem.* **15**, 953–973.

Greenwood, N. N., and Earnshaw, A. (1997). *Chemistry of the Elements* (2nd edn.). Butterworth-Heinemann, Oxford.

Hamilton, E. I. (1994). The geobiochemistry of cobalt. *Sci. Total Environ.* **150**, 7–39.

Hamilton-Taylor, J., Postill, A. S., Tipping, E., and Harper, M. P. (2002). Laboratory measurements and modeling of metal–humic interactions under estuarine conditions. *Geochim. Cosmochim. Acta* **66**, 403–415.

Hansen, J. A., Marr, J. C. A., Lipton, J., Cacela, D., and Bergman, H. L. (1999). Differences in neurobehavioral responses of chinook salmon (*Oncorhynchus tshawytscha*) and rainbow trout (*Oncorhynchus mykiss*) exposed to copper and cobalt: behavioral avoidance. *Environ. Toxicol. Chem.* **18**, 1972–1978.

Harrison, S. E., Klaverkamp, J. F., and Hesslein, R. H. (1990). Fates of metal radiotracers added to a whole lake – accumulation in fathead minnow (*Pimephales promelas*) and lake trout (*Salvelinus namaycush*). *Water Air Soil Pollut.* **52**, 277–293.

Hassan, E. S., Abdellatif, H., and Biebricher, R. (1992). Studies on the effects of Ca^{++} and Co^{++} on the swimming behavior of the blind Mexican cave fish. *J. Comp. Physiol. A* **171**, 413–419.

Hertz, Y., Madar, Z., Hepper, B., and Gertler, A. (1989). Glucose metabolism in the common carp (*Cyprinus carpio* L.): the effects of cobalt and chromium. *Aquaculture* **76**, 255–267.

Higgo, J. J. W., Kinniburgh, D., Smith, B., and Tipping, E. (1993). Complexation of Co^{2+}, Ni^{2+}, UO_2^{2+} and Ca^{2+} by humic substances in groundwaters. *Radiochim. Acta* **61**, 91–103.

Hossein, E., Abbas, M. and Shohreh, B. (2008). Effects of cobalt as growth promotant on the growth of rainbow trout, *Oncorhynchus mykiss*. World Aquaculture 2008: Aquaculture for Human Wellbeing – The Asian Perspective, 19–23 May 2008. Busan, Korea.

Hu, Q. H., Weng, J. Q. and Wang, J. S. (2010). Sources of anthropogenic radionuclides in the environment: a review. *J. Environ. Radiat.* **101**, 426–437. Special issue presented at the 9th International Conference on Biogeochemistry of Trace Elements, Beijing, P.R. China, July 2007.

IAEA (2009). *Quantification of Radionuclide Transfer in Terrestrial and Freshwater Environments for Radiological Assessments*. IAEA-TECDOC-1616. Vienna: International Atomic Energy Agency.

IAEA (2010). *Handbook of Parameter Values for the Prediction of Radionuclide Transfer in Terrestrial and Freshwater Environments*. IAEA Technical Report Series No. 472. Vienna: International Atomic Energy Agency.

Janssen, J. (2000). Toxicity of Co^{2+}: implications for lateral line studies. *J. Comp. Physiol. A* **186**, 957–960.

Jefferson, J. A., Escudero, E., Hurtado, M.-E., Pando, J., Tapia, R., Swenson, E. R., Prchal, J., Schreiner, G. F., Schoene, R. B., Hurtado, A., and Johnson, R. J. (2002). Excessive erythrocytosis, chronic mountain sickness and serum cobalt levels. *Lancet* **359**, 407–408.

Karlsen, H. E., and Sand, O. (1987). Selective and reversible blocking of the lateral line in freshwater fish. *J. Exp. Biol.* **133**, 249–262.

Kashiwada, K., Teshima, S., and Kanazawa, A. (1970). Studies on the production of B vitamins by intestinal bacteria of fish – V. Evidence of the production of vitamin B_{12} by microorganisms in the intestinal canal of carp *Cyprinus carpio*. *Nippon Suisan Gakk.* **36**, 421–424.

Khan, H. A. and Mukhopadhay, S. K. (1971). Observation on the effects of yeast and cobalt chloride in increasing the survival rate of the hatchlings of *Heteropneustesfossilis* (Bloch). *Symposium on Trends of Research in Zoology*. Zoological Society, pp. 1, 1–12. Calcutta.

Kim, J. H., Gibb, J. H. and Howe P. D. (2006). *Cobalt and Inorganic Cobalt Compounds*. Concise international chemical assessment document 69. Geneva: World Health Organization.

Kimura, K., and Ichikawa, R. (1972). Accumulation and retention of ingested ^{60}Co by the common goby. *Bull. Jpn. Soc. Scient. Fish.* **38**, 1087–1103.

Kobayashi, M., and Shimizu, S. (1999). Cobalt proteins. *Eur. J. Biochem.* **261**, 1–9.

Kurk, D. N., and Choppin, G. R. (2000). Determination of Co(II) and Ni(II)-humate stability constants at high ionic strength NaCl solutions. *Radiochim. Acta* **88**, 583–586.

Kwong, R. W. M., and Niyogi, S. (2009). The interactions of iron with other divalent metals in the intestinal tract of a freshwater teleost, rainbow trout (*Oncorhynchus mykiss*). *Comp. Biochem. Physiol. C* **150**, 442–449.

Lahner, E., and Annibale, B. (2009). Pernicious anemia: new insights from a gastroenterological point of view. *World J. Gastroenterol.* **15**, 5121–5128.

Leggett, R. W. (2008). The biokinetics of inorganic cobalt in the human body. *Sci. Total Environ.* **389**, 259–269.

Leonard, K. S., McCubbin, D., and Harvey, B. R. (1993). Chemical speciation and environmental behaviour of ^{60}Co discharged from a nuclear establishment. *J. Environ. Radioactiv.* **20**, 1–21.

Liao, J. C. (2006). The role of the lateral line and vision on body kinematics and hydrodynamic preference of rainbow trout in turbulent flow. *J. Exp. Biol.* **209**, 4077–4090.

Limsuwan, T., and Lovell, R. T. (1981). Intestinal synthesis and absorption of vitamin B, in channel catfish. *J. Nutr.* **111**, 2125–2132.

Lin, Y. H., Wu, J. Y., and Shiau, S. Y. (2010). Dietary cobalt can promote gastrointestinal bacterial production of vitamin B-12 in sufficient amounts to supply growth requirements of grouper, *Epinephelus malabaricus*. *Aquaculture* **302**, 89–93.

Ludwig, M. L., and Matthews, R. G. (1997). Structure-based perspectives on B-12 dependent enzymes. *Annu. Rev. Biochem.* **66**, 269–313.

Lutz, P. L., and Prentice, H. M. (2002). Sensing and responding to hypoxia, molecular and physiological mechanisms. *Integr. Comp. Biol.* **42**, 463–468.

Majmudar, K., and Burleson, M. L. (2006). An evaluation of cobalt chloride as an O_2-sensitive chemoreceptor stimulant in channel catfish. *Comp. Biochem. Physiol. C* **142**, 136–141.

Marr, J. C. A., Hansen, J. A., Meyer, J. S., Cacela, D., Podrabsky, T., Lipton, J., and Bergman, H. L. (1998). Toxicity of cobalt and copper to rainbow trout: application of a mechanistic model for predicting survival. *Aquat. Toxicol.* **43**, 225–238.

Mayles, P. and Nahum, A., eds (2007). *Handbook of Radiotherapy Physics: Theory and Practice.* Liverpool: Clatterbridge Centre for Oncology; Paris: J.C. Rosenwald, Institute Curie.

Montgomery, J., Baker, C., and Carton, A. (1997). The lateral line can mediate rheotaxis in fish. *Nature* **389**, 960–963.

Morelli, L., Di Giulio, C., Iezzi, M., and Data, P. G. (1994). Effect of acute and chronic cobalt administration on carotid body chemoreceptor responses. *Sci. Total Environ.* **150**, 215–216.

Mukherjee, S., and Kaviraj, A. (2009). Evaluation of growth and bioaccumulation of cobalt in different tissues of common carp, *Cyprinus carpio* (Actinopterygii: Cypriniformes: Cyprinidae), fed cobalt-supplemented diets. *Acta Ichthyol. Pisc.* **39**, 87–93.

Nagpal, N. A. (2004). *Water Quality Guidelines for Cobalt.* Technical Report. British Columbia: Water Protection Section; Water, Air and Climate Change Branch; Ministry of Water, Land and Air Protection.

Nelson, N. (1999). Metal ion transporters and homeostasis. *EMBO J.* **18**, 4361–4371.

Nevo, Y., and Nelson, N. (2006). The NRAMP family of metal-ion transporters. *Biochim. Biophys. Acta Mol. Cell. Res.* **1763**, 609–620.

Niegowski, D., and Eshaghi, S. (2007). The CorA family: structure and function revisited. *Cell. Mol. Life Sci.* **64**, 2564–2574.

Niyogi, S., and Wood, C. M. (2004). Biotic ligand model, a flexible tool for developing site-specific water quality guidelines for metals. *Environ. Sci. Technol.* **38**, 6177–6192.

Nolan, C. V., Fowler, S. W., and Teyssie, J. L. (1992). Cobalt speciation and bioavailability in marine organisms. *Mar. Ecol. Prog. Ser.* **88**, 105–116.

Paulsen, I. T., and Saier, M. H. (1997). A novel family of ubiquitous heavy metal ion transport proteins. *J. Membr. Biol.* **156**, 99–103.

Payne, L. R. (1977). The hazards of cobalt. *Occup. Med.* **27**, 20–25.

Pentreath, R. J. (1973a). The accumulation and retention of ^{59}Fe and ^{58}Co by the plaice, *Pleuronectes platessa* L.. *J. Exp. Mar. Biol. Ecol.* **12**, 315–326.

Pentreath, R. J. (1973b). The accumulation from sea water of ^{65}Zn, ^{54}Mn, ^{58}Co and ^{59}Fe by the thornback ray, *Raja clavata* L.. *J. Exp. Mar. Biol. Ecol.* **12**, 327–334.

Pham, M. K., and Garnier, J.-M. (1998). Distribution of trace elements associated with dissolved compounds ($< 0.45\ \mu m$–1 nm) in freshwater using coupled (frontal cascade) ultrafiltration and chromatographic separations. *Environ. Sci. Technol.* **32**, 440–449.

Pokrovsky, O. S., Schott, J., and Dupre, B. (2006). Trace element fractionation and transport in boreal rivers and soil porewaters of permafrost-dominated basaltic terrain in Central Siberia. *Geochim. Cosmochim. Acta* **70**, 3239–3260.

Qian, J., Xue, H. B., Sigg, L., and Albrecht, A. (1998). Complexation of cobalt by natural ligands in freshwater. *Environ. Sci. Technol.* **32**, 2043–2050.

Quadros, E. V. (2010). Advances in the understanding of cobalamin assimilation and metabolism. *Br. J. Haematol.* **148**, 195–204.

Richards, J. G., and Playle, R. C. (1998). Cobalt binding to gills of rainbow trout (*Oncorhynchus mykiss*): an equilibrium model. *Comp. Biochem. Physiol. C* **119**, 185–197.

Rodionov, D. A., Hebbeln, P., Gelfand, M. S., and Eitinger, T. (2006). Comparative and functional genomic analysis of prokaryotic nickel and cobalt uptake transporters: evidence for a novel group of ATP-binding cassette transporters. *J. Bacteriol.* **188**, 317–327.

Saito, M. A., and Moffett, J. W. (2001). Complexation of cobalt by natural organic ligands in the Sargasso Sea as determined by a new high-sensitivity electrochemical cobalt speciation method suitable for open ocean work. *Mar. Chem.* **75**, 49–68.

Saito, M. A., Rocap, G., and Moffett, J. W. (2005). Production of cobalt binding ligands in a *Synechococcus* feature at the Costa Rica upwelling dome. *Limnol. Oceanogr.* **50**, 279–290.

Seghizzi, P., D'Adda, F., Borleri, D., Barbic, F., and Mosconi, G. (1994). Cobalt myocardiopathy. A critical review of literature. *Sci. Total Environ.* **150**, 105–109.

Sekaly, A. L. R., Murimboh, J., Hassan, N. M., Mandal, R., Younes, M. E. B., Chakrabarti, C. L., and Back, M. H. (2003). Kinetic speciation of Co(II), Ni(II), Cu(II), and Zn(II) in model solutions and freshwaters: lability and the d electron configuration. *Environ. Sci. Technol.* **37**, 68–74.

Smith, C. G. (2001). Always the bridesmaid, never the bride: cobalt geology and resources. *Trans. Inst. Mining Metallurgy B Appl. Earth Sci.* **110**, 75–80.

Smith, I. C., and Carson, B. L. (1981). *Trace Metals in the Environment*. Ann Arbor Science Publishers, Ann Arbor, MI.

Smith, J. T. (2006). Modelling the dispersion of radionuclides following short duration releases to rivers. Part 2: Uptake by fish. *Sci. Total. Environ.* **368**, 502–518.

Steffens, W. (1989). *Principles of Fish Nutrition*. Ellis Horwood, Chichester.

Stickney, R. R. (1994). *Principles of Aquaculture*. John Wiley and Sons, New York.

Suzuki, Y., Nakahara, M., Nakamure, R., and Ueda, T. (1979). Roles of food and sea water in the accumulation of radionuclides by marine fish. *Bull. Jpn. Soc. Scient. Fish.* **45**, 1409–1416.

Tanizaki, Y., Shimokawa, T., and Nakamura, M. (1992). Physicochemical speciation of trace elements in river waters by size fractionation. *Environ. Sci. Technol.* **26**, 1433–1444.

van Leeuwen, H. P., and Town, R. M. (2005). Kinetic limitations in measuring stabilities of metal complexes by competitive ligand exchange-adsorptive stripping voltammetry (CLE-AdSV). *Environ. Sci. Technol.* **39**, 7217–7225.

van Loon, L. R., Granacher, S., and Harduf, H. (1992). Equilibrium dialysis-ligand exchange: a novel method for determining conditional stability constants of radionuclide–humic acid complexes. *Anal. Chim. Acta* **268**, 235–246.

van de Guchte, C., Beek, M., Tuinstra, J., and van Rossenberg, M. (2000). *Normen voor het Waterbeheer Achtergronddocument NW4 CIW*. Commissle Integraal Waterbeheer, Rijks-waterstaat, Nederland

Wang, S. (2006). Cobalt – Its recovery, recycling, and application. *J. Miner. Metals Mater. Soc.* **58**, 47–50.

Warnken, K. W., Davison, W., Zhang, H., Galceran, J., and Puy, J. (2007). *In situ* measurements of metal complex exchange kinetics in freshwater. *Environ. Sci. Technol.* **41**, 3179–3185.

Watanabe, T., Kiron, V. and Satoh, S. (1997). Trace minerals in fish nutrition. 6th International Symposium on Feeding and Nutrition in Fish: Nutrition and the Production of Fish and Shellfish, October 04–07, 1993 Hobart, Australia. *Aquaculture* **151**, 185–207.

Weiss, T. F. (1996). *Cellular Biophysics*. Vol. 1. *Transport*. MIT Press, Cambridge, MA.

Westall, J. C., Jones, J. D., Turner, G. D., and Zachara, J. M. (1995). Models for association of metal-ions with heterogeneous environmental sorbents. 1. Complexation of Co(II) by Leonardite humic-acid as a function of pH and $NaClO_4$ concentration. *Environ. Sci. Technol.* **29**, 951–959.

Zhang, H., van den Berg, C. M. G., and Wollast, R. (1990). The determination of interactions of cobalt(II) with organic compounds in seawater using cathodic stripping voltammetry. *Mar. Chem.* **28**, 285–300.

7

SELENIUM

DAVID M. JANZ

Homeostasis and Toxicology of Essential Metals: Volume 31A
FISH PHYSIOLOGY

The physiological and toxicological importance of selenium (Se) has long intrigued scientists, and fish have played an important role in this regard. Selenium is an essential trace element in fish that is specifically incorporated into proteins as selenocysteine. The best characterized selenoproteins in fish are enzymes involved in antioxidant defense (glutathione peroxidases and thioredoxin reductases) and thyroid hormone metabolism (iodothyronine deiodinases). Fish appear to possess the largest number of selenoproteins among all biota, including several not found in other vertebrates; however, the molecular, biochemical, and physiological characterization of the majority of selenoproteins in fish remains to be completed. The natural Se concentrations in most freshwater and saltwater environments are low $(0.01–0.1 \, \mu g \, L^{-1})$, but certain anthropogenic activities can increase Se loading into aquatic ecosystems. Selenium is extremely toxic to fish, with a narrow margin between essentiality and toxicity. The major toxicological hazard associated with Se is its efficient incorporation into aquatic food webs by primary producers and subsequent trophic transfer to fish via dietary pathways. The toxicity of Se is related to its non-specific incorporation into the other sulfur-containing amino acid (methionine), producing selenomethionine. The most sensitive toxicological effects in fish are a characteristic and diagnostic suite of early life-stage deformities resulting from maternal transfer of selenomethionine to yolk proteins during vitellogenesis, and subsequent Se exposure during yolk resorption by developing larvae. In several instances, increased frequencies of larval deformities in wild fish exposed to elevated Se have been associated with impaired recruitment and extirpation of fish populations. As a result, Se is recognized as a contaminant of global concern with many knowledge gaps and research needs, making investigation of the homeostasis and toxicology of Se in fish an area of intense future research.

1. INTRODUCTION

Selenium (Se) was discovered by the Swedish chemist Berzelius in 1817 and is named after Selene, the Greek goddess of the moon. In 1957, Se was first identified as an essential trace element in mammals (Schwarz and Foltz, 1957), and is now known to be required for a variety of functional Se-dependent proteins (selenoproteins) in most living organisms (Mayland, 1994; Hesketh, 2008). Our knowledge of the physiological roles of Se has increased dramatically in recent years, with several annual scientific conferences now dedicated solely to research on the essentiality and toxicity of this element. Indeed, it is the relatively narrow range between essentiality and toxicity in vertebrates that generates tremendous scientific and conservation interest in Se. The majority of Se research historically focused on aspects of nutritional Se deficiency and toxicological excess in humans and domestic animals. However, a recent surge in research is focused on the homeostasis and especially toxicity of Se in fish due to the realization that Se is exquisitely toxic to oviparous fish species and can negatively impact the sustainability of wild fish populations (Lemly, 1993b, 2004; Maier and Knight, 1994; Skorupa, 1998; Janz et al., 2010).

In aquatic ecotoxicology, Se is one of the most hazardous trace elements for fish, following mercury (Luoma and Presser, 2009; Janz et al., 2010). Although inorganic selenate and selenite are the dominant forms of Se in surface waters, Se toxicity to fish is governed by the bioavailability of organic Se in food webs (Maher et al., 2010; Stewart et al., 2010). Several human activities can increase loading of Se into aquatic systems, such that Se is now considered an aquatic contaminant of global concern (Lemly, 2004; Chapman, 2009). From a regulatory and risk assessment perspective, strong debate has focused on developing guidelines that protect fish populations. This lack of regulatory harmonization is due largely to a failure to link Se biogeochemistry to physiological and ecological processes (Luoma and Presser, 2009). The goal of this chapter is to provide a comprehensive review of the current knowledge of Se homeostasis and toxicity in fish. Important discoveries are needed to bridge the many knowledge gaps. For further details on broader perspectives of Se in aquatic ecosystems, the reader is referred to a recent book focused solely on this topic (Chapman et al., 2010).

2. CHEMICAL SPECIATION IN FRESHWATER AND SEAWATER

2.1. Freshwater

Selenium in aquatic environments displays a diverse array of chemical forms, or species, which are governed by biogeochemical transformation reactions. The speciation, biotransformation, and cycling of Se in freshwater are more complex than most trace elements (Fig. 7.1; see Section 8), and dictate its fate and toxicological effects in aquatic ecosystems (Luoma and Presser, 2009; Stewart et al., 2010). It is beyond the scope of this chapter to provide a thorough discussion of all Se species present in abiotic and biotic components of aquatic ecosystems, and thus the focus will be on the inorganic and organic forms that have greatest physiological and toxicological relevance to fish. A detailed review of Se speciation in aquatic systems has recently been published (Maher et al., 2010).

Selenium (atomic number 34, atomic mass 78.96) is the third member of group 16 of the periodic table. It shares fundamental aspects of chemical behavior with oxygen and sulfur, the lighter elements of group 16, as well as the heavier and less abundant tellurium and polonium. The unique physical, chemical, and biological properties of Se have long fascinated scientists from a variety of perspectives. Selenium is a non-metal or metalloid, and consequently non-metallic behavior governs its geochemical cycling. However, biologically mediated reactions also play critical roles in Se cycling (Fig. 7.1). In contrast to most trace elements, Se speciation in aquatic ecosystems is poorly predicted by thermodynamic models, since biological (kinetic) processes are just as important in determining Se speciation (Luoma and Presser, 2009; Stewart et al., 2010). Fortunately, environmental levels of different Se species can be determined reliably using analytical techniques, so that the distribution and cycling of Se species in aquatic ecosystems are well characterized (Cutter, 1989; Tamari, 1998; Fan et al., 2002; Maher et al., 2010).

Selenium exists in four oxidation states (VI, IV, 0, −II) and as inorganic or organic compounds. Unlike metals or transition metals, which typically exist as dissolved cations in water, Se is hydrolyzed in aqueous solution to form the oxyanions selenate (SeO_4^{2-}, or Se[VI]) and selenite (SeO_3^{2-}, or Se[IV]). As a result, selenate and selenite display increased solubility and mobility with increasing pH, in contrast to most metals where the opposite effect occurs. The other primary Se species in water are organoselenides (org-Se[−II]). A key step in the biogeochemical cycling of Se occurs when primary and secondary producers (bacteria, algae, and higher plants) take up dissolved inorganic Se (selenate and selenite) and convert it into organoselenides, predominantly selenomethionine (Maher et al., 2010; Stewart et al., 2010). This step has important ramifications for bioaccumulation and toxicity at higher trophic levels, as will be discussed in Section 8. Elemental Se (Se[0]) is predominantly found in anoxic environments such as sediments, where chemical, physical, and biological factors promote the formation of this reduced Se species (Maher et al., 2010).

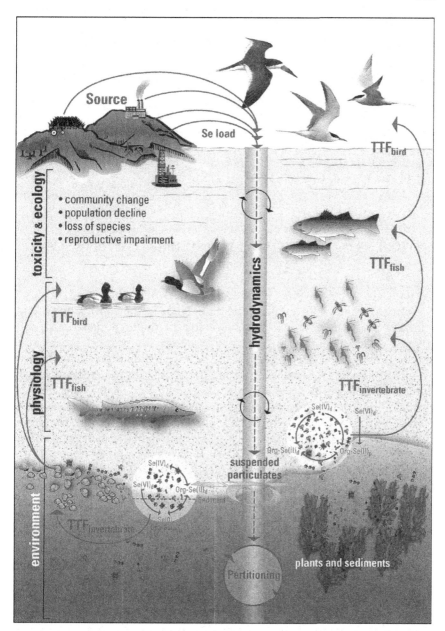

Fig. 7.1. A conceptual model illustrating the roles of Se speciation, biogeochemical transformation, and trophic transfer functions (TTFs) in the biogeochemical cycling within freshwater and marine aquatic ecosystems. A representative benthic food web (left) and water column food web (right) are shown. Particulate and dissolved Se are indicated by subscripts p and d, respectively. Reprinted with permission from Luoma and Presser (2009).

2.2. Seawater

Similar to freshwater environments, the dissolved Se species in seawater are selenate, selenite, and various organoselenides. Dissolved organoselenides, existing primarily as seleno-amino acids such as selenomethionine, are generally highest in surface ocean waters and decrease rapidly with depth (Cutter and Bruland, 1984; Cutter, 1992). In areas with high productivity, organoselenides may represent up to 80% of total dissolved Se at the surface (Cutter and Bruland, 1984; Cutter and Cutter, 1998). In contrast, selenate becomes the dominant Se species at depth in marine environments (Cutter and Bruland, 1984; Cutter, 1992).

3. SOURCES OF SELENIUM AND ECONOMIC IMPORTANCE

3.1. Natural Sources

Selenium ranks 69th in elemental abundance in the Earth's crust, and generally occurs in low amounts in geological raw materials, soils, and sediments (i.e. low mg kg^{-1}). Coals and crude oils can be much greater (i.e. hundreds of mg kg^{-1}) depending on the geological evolution of the deposit (Maher et al., 2010). Selenium concentrations in soils and sediments vary throughout the world depending on the parent rock, with most regions characterized by low to moderate soil levels. A notable exception is the central region of North America, where Se originates largely from Cretaceous marine sedimentary rocks, particularly black shale, phosphate rocks (phosphorites), and coal (Presser et al., 2004; Maher et al., 2010). Natural processes that redistribute Se include terrestrial weathering of rocks and soils, volcanic activity, wildfires, and volatilization from water bodies and plants (Nriagu, 1989).

Variation in Se sources and associated speciation are important considerations when managing Se risks to aquatic species such as fish, since background concentrations of Se can vary greatly among regions (Maher et al., 2010). Background Se concentrations in freshwater and seawater are usually between 0.01 and 0.1 µg L^{-1}, but can be as high as 5–50 µg L^{-1} in certain locations when surface water is in contact with highly seleniferous exposed shale deposits (Canton and Van Derveer, 1997; Canton, 2010; Maher et al., 2010). Although natural weathering of rocks and soils, and atmospheric deposition, contribute to regional Se mass balances (Nriagu, 1989), anthropogenic activities can greatly accelerate these processes. As a result, the largest Se sources on a regional basis are mediated by human activities and not natural processes (Maher et al., 2010; Young et al., 2010a).

3.2. Anthropogenic Sources

Selenium is mobilized from a wide variety of human activities that generally involve association of the natural seleniferous matrices described above with water (Young et al., 2010a). Coal, phosphate, uranium, and certain base metal and precious metal mining operations can increase loading of Se into aquatic ecosystems via runoff from waste rock (overburden), tailings impoundments, ore-milling waste waters, and reclaimed mining areas (Lemly, 2004; Maher et al., 2010). Ore smelting and coal combustion (particularly in coal-fired electricity generating plants) can release particulate and volatile forms of Se into the atmosphere that are then transported and deposited elsewhere via dry or wet deposition (Nriagu and Wong, 1983). Coal-fired power plants also produce fly ash containing Se, which is commonly allowed to settle in ponds whereby effluent is released into local water bodies (Maher et al., 2010). Crude oil originating from marine shale deposits is rich in Se, and refining effluents can thus be a major source of Se loading to aquatic environments (Cutter, 1989). Agricultural irrigation and drainage of seleniferous soils can cause significant Se contamination of ecosystems, as reported in the Central Valley of California (Ohlendorf, 2003). Other agricultural activities such as field application of biosolids from sewage treatment plants, manure, and chemical fertilizers also contribute to Se loading in aquatic environments (Bouwer, 1989). Selenate is usually the dominant Se species in anthropogenic inputs to aquatic ecosystems, although selenite may predominate in certain discharges such as petroleum refining waste waters and fly ash from coal-fired power plants (Maher et al., 2010).

3.3. Economic Importance

The most common use of Se is as a pigment in the manufacture of glass, plastics, ceramics, and paints (Haygarth, 1994; George, 2009). Selenium is a photovoltaic substance used in photoelectric cells such as solar energy panels, photographic exposure meters, photometers, and light-controlled switches. The electrical properties of Se also result in its use in rectifiers, which convert alternating to direct current (AC to DC). Selenium is used in the rubber industry as an accelerant and vulcanizing agent. Dietary supplements and anti-dandruff shampoos used by humans commonly contain Se. In diagnostic medicine, ^{75}Se aids in visualization of certain malignant tumors. A more recent use of Se is as a component of the inner core of semiconductor nanocrystals (quantum dots), which are increasingly used in photovoltaic devices and as replacements for fluorescent dyes in medical research and diagnostics (Bouldin et al., 2008).

4. ENVIRONMENTAL SITUATIONS OF CONCERN

Hazardous Se exposures for fish stem primarily from the anthropogenic sources described in Section 3.2 (Maier and Knight, 1994; Lemly, 2004). Selenium contamination is a global issue that results from basic agricultural practices through to highly technical industrial processes, including nanotechnology (Lemly, 2004; Young et al., 2010a, b). Several instances of unexpected anthropogenic Se contamination have negatively impacted wild fish populations. For ecotoxicologists, these cases provide perhaps the clearest connection between exposure to a toxic substance and resulting adverse effects on fish population dynamics (Lemly, 1985; Skorupa, 1998; Janz et al., 2010). This section briefly summarizes current environmental concerns. For detailed case studies in which Se negatively affected fish populations, see Skorupa (1998), Lemly (2004), and Young et al. (2010b). These case studies represent instances where significant effects of Se on fish populations have been recognized; however, there remain many similar situations where cause–effect relationships between Se exposure and negative effects in fish have not been established that require similar ecotoxicological research (Lemly, 2004).

4.1. Power Plant Coal Combustion Wastes

Belews Lake is a reservoir constructed to provide cooling water for a large coal-fired electric power plant in North Carolina, USA, which commenced operation in 1974. Water entered Belews Lake from a fly ash settling pond that contained 150–200 µg Se L^{-1}. Of 16 trace elements analyzed, only Se was elevated in lake water (averaging 10 µg Se L^{-1}) and tissues of fish inhabiting the lake (Lemly, 1985). By 1977, only three of the 29 resident fish species remained, providing clear evidence of a severe episode of Se poisoning (Cumbie and Van Horn, 1978; Lemly, 1985). This classic case study in aquatic ecotoxicology was among the first to demonstrate the insidious nature of Se contamination, and has provided much useful information on the symptoms, pathology, and ecological consequences of aquatic Se contamination in fish (Lemly, 1985, 2002; Skorupa, 1998). In 1986, Se discharge to the lake stopped, and a period of slow natural fish population recovery began (Lemly, 1997b), although Se in fish tissues remains elevated (Finley and Garrett, 2007). Similar cases involving coal combustion wastes have been documented in North America (Skorupa, 1998) and Australia (Kirby et al., 2001; Barwick and Maher, 2003). A recent large spill of fly ash waste in Tennessee released approximately 24,000 kg of Se into a local watershed (Tennessee Valley Authority, 2009), illustrating the ongoing importance of this source of Se to aquatic environments.

4.2. Agricultural Irrigation

Elevated Se in irrigation waters is a major environmental concern in arid and semi-arid regions with seleniferous soils (Outridge et al., 1999; Seiler et al., 2003; Hu et al., 2009). This was most evident in a large reservoir within the Kesterson National Wildlife Refuge during the 1980s that received runoff from the San Joaquin Valley of California (Presser and Ohlendorf, 1987). There, eight warm water fish species were extirpated, with only the tolerant mosquitofish (*Gambusia affinis*) persisting (Skorupa, 1998). Thus, other situations involving extensive agricultural irrigation of seleniferous soils in arid regions are of concern, such as in the prairie region of Canada (Outridge et al., 1999; Hu et al., 2009).

4.3. Petroleum Refinery Waste Water

Crude oil commonly contains high Se levels, and disposal of waste waters from various steps in the refining process can release Se into local aquatic ecosystems (Young et al., 2010a). An excellent example of this occurred from 1982 to 1995 as a result of oil refining at San Francisco Bay, CA, where increased Se loading has impacted local estuarine and coastal marine ecosystems (Cutter and San Diego-McGlone, 1990; Cutter and Cutter, 2004).

4.4. Coal Mining

Open pit coal mining, such as in the Elk River Valley, BC, Canada, and mountain-top coal mining, such as in the central and southern Appalachian Mountains, USA, release significant quantities of Se into local watersheds (Young et al., 2010a). In both cases, waste rock is transported to large dump areas, often valley fills, and the resulting leachate enters aquatic ecosystems. In the Elk Valley, aqueous Se concentrations (predominantly selenate) are as high as 300 μg L^{-1} in leachates, between 50 and 80 μg L^{-1} in nearby ponds and marshes, and remain elevated (5–10 μg L^{-1}) in the Elk River as far as 60 km downstream of mines (Young et al., 2010a).

4.5. Metal Mining and Milling

Mining operations that extract base metals (e.g. copper, nickel), precious metals (e.g. gold, silver) and uranium from ores with marine sedimentary origin can increase loading of Se into aquatic ecosystems due to smelting, ore milling waste water effluents, and runoff from waste rock and tailings impoundments (Lemly, 2004; Muscatello et al., 2008; Maher et al., 2010). Selenium has often been overlooked as a pollutant of concern in monitoring such operations, since managers historically focus on toxic trace elements such as mercury, lead, arsenic, and cadmium. As a result, Se has only recently emerged as an aquatic contaminant of concern from metal mining activities (Lemly, 2004).

4.6. Phosphate Mining

Open pit phosphate mining within the Phosphoria Formation in the western USA can cause aquatic Se contamination, and is an emerging environmental issue in other areas of the world (Lemly, 2004; Hamilton and Buhl, 2005). Similar to coal and metal mining, solid wastes from phosphate mines contain elevated concentrations of Se that enter surrounding aquatic ecosystems depending on local climate and hydrology (Lemly, 2004).

4.7. Emerging Concerns

Although not studied extensively, several environmental scenarios have the potential to increase Se loading into aquatic ecosystems. Municipal landfills can create leachates with elevated Se concentrations (5–$50 \ \mu g \ L^{-1}$), largely from photoelectronic components (Lemly, 2004). In areas with low soil Se levels, supplementation of livestock diets with Se to enhance growth is of concern owing to the subsequent excretion and transport via runoff into aquatic ecosystems. This is of particular concern since the common form of Se in diets (selenomethionine) is efficiently bioaccumulated in aquatic food webs (Young et al., 2010a). Manure application to fields can result in non-point source inputs of Se to aquatic systems, a problem accentuated by the large scale of intensive livestock operations. Commercial fertilizers are often amended with selenate to address soil deficiencies, and this practice can cause pulsed inputs via runoff (Bouwer, 1989; Young et al., 2010a). An emerging concern stems from nanotechnology, where Se is a major component of quantum dots (as CdSe or PbSe). Selenium-containing nanomaterials may differ in aquatic ecotoxicological consequences because uptake routes and bioavailable forms are undefined (Bouldin et al., 2008; Young et al., 2010a).

5. SURVEY OF WATER QUALITY GUIDELINES

Ambient water quality guidelines for Se (as total aqueous Se) have been developed for the USA, Canada, Australia/New Zealand, and South Africa (Table 7.1). These guidelines were developed to protect populations of aquatic biota, including fish, from chronic exposure to Se. The United States Environmental Protection Agency (EPA) currently has chronic water quality criteria for the protection of aquatic organisms set at 5 μg Se L^{-1} for freshwater and 71 μg Se L^{-1} (continuous exposure) for saltwater (USEPA, 1987). The Canadian Council of Ministers of the Environment (CCME) guideline for protection of aquatic life is 1 μg Se L^{-1} for freshwater, with no saltwater guideline (CCME, 2003). The Australian and New Zealand Environment and Conservation Council (ANZECC) guidelines (termed "trigger" values) are based on protection of a percentage of aquatic species, and are 5 μg Se L^{-1} for protection of 99% of freshwater aquatic species (the most conservative value), increasing to 34 μg Se L^{-1} for protection of 80% of species, with no saltwater guideline (ANZECC, 2000). South Africa has a freshwater guideline of 5 μg Se L^{-1} (based on dissolved Se), with no saltwater guideline (Department of Water Affairs and Forestry, 1996). Ambient water quality guidelines for protection of aquatic life have not been developed for other countries, including the European Union. In Canada, certain provincial governments have developed their own

Table 7.1

Summary of chronic ambient water quality guidelines for the protection of aquatic life for selenium

Country	Chronic water quality guideline for Se (μg L^{-1})[a]	
	Freshwater	Seawater
USA[b]	5	71
Canada[c]	1	N/A
Australia/New Zealand[d]	5	N/A
South Africa[e]	5	N/A

N/A: Not available.
[a]Expressed as total Se concentration in water unless otherwise indicated.
[b]United States Environmental Protection Agency (EPA) chronic criteria for the protection of aquatic organisms (USEPA, 1987).
[c]Canadian Council of Ministers of the Environment (CCME) water quality guideline for the protection of aquatic life (CCME, 2003).
[d]Australian and New Zealand Environment and Conservation Council (ANZECC) trigger value for the protection of 99% of aquatic organisms (ANZECC, 2000).
[e]South African Department of Water Affairs and Forestry (1996) chronic guideline (based on dissolved Se).

guidelines, which vary greatly. For example, British Columbia has set both freshwater and marine water quality guidelines at $2\,\mu g$ Se L^{-1} (Nagpal, 2001). In Ontario, the ambient provincial water quality objective (PWQO) for the protection of aquatic life is $100\,\mu g$ Se L^{-1} for freshwater (Ontario Ministry of Environment and Energy, 1999).

There is ongoing debate concerning regulatory guidelines for Se in North America owing to the prevalence of anthropogenic inputs and increasing concern over protection of aquatic species such as fish. The water quality guidelines for the USA and Canada are currently under review, as both countries realize that the current guidelines based on aqueous Se concentrations create uncertainty in ecological risk assessments. A universal dissolved Se concentration cannot predict toxicity in fish (see Section 8). This is because differences in feeding preferences, geochemistry, hydrology, and ecology among aquatic ecosystems dictate the speciation, bioavailability, and toxicity of Se to fish (Luoma and Presser, 2009; Stewart et al., 2010) (Fig. 7.1). A tissue-based aquatic criterion for Se has been proposed for some time (DeForest et al., 1999; Hamilton, 2002, 2003). In 2004, the US EPA proposed a whole body Se concentration of $7.91\,\mu g\,g^{-1}$ dry weight as a chronic criterion for the protection of aquatic organisms (USEPA, 2004), which is currently under revision. In British Columbia, Canada, an interim tissue guideline for fish of $1\,\mu g$ Se g^{-1} whole body wet weight (which corresponds to $5\,\mu g$ Se g^{-1} dry weight assuming 80% moisture) has been proposed (Nagpal, 2001). Both Canada (CCME) and the USA (EPA) will likely set tissue-based aquatic life chronic criteria in the near future.

6. MECHANISMS OF TOXICITY

6.1. Acute Toxicity

The acute toxicity of Se to fish has been widely studied, with the first published report dating to the 1930s (Ellis et al., 1937). Acute selenosis follows aqueous exposure to selenate or selenite, with the latter consistently more acutely toxic to a variety of teleosts. For example, 96 h LC50 values for sodium selenite and sodium selenate in juvenile rainbow trout (*Oncorhynchus mykiss*) range from 4.2 to 9.0 mg Se L^{-1} and 32 to 47 mg Se L^{-1}, respectively (Hodson et al., 1980; Buhl and Hamilton, 1991). It should be noted that these LC50 values are two to three orders of magnitude greater than total Se concentrations observed in even the most severely Se-contaminated aquatic ecosystems; thus acute Se toxicity is not generally considered ecotoxicologically relevant. The mechanism(s) of acute toxicity have not been fully elucidated, although oxidative stress may play a role in the acute toxicity of selenite to rainbow trout (Miller et al., 2007; Misra and Niyogi, 2009). Dietary exposure of fish to organoselenium compounds such as selenomethionine is not considered an acute toxicity hazard (Cleveland et al., 1993), although this is the major exposure route leading to chronic toxicity, as discussed below.

6.2. Chronic Toxicity

The major ecological concern related to Se exposure is its teratogenic effects on early life stages (i.e. embryo–alevin–fry) of fish, which can negatively affect recruitment of young individuals into fish populations (Lemly, 2004; Janz et al., 2010). Similar to most toxic agents, there are both toxicokinetic (transport and delivery to site of toxic action) and toxicodynamic (mode of toxic effect) mechanisms determining the chronic toxicity of Se to fish. As discussed in detail below, the toxicokinetic mechanism for Se is well characterized, but there is uncertainty surrounding the toxicodynamic mechanism of teratogenicity. Unlike most trace elements, aqueous exposure to dissolved Se species is a minor exposure route, and it is well recognized that dietary exposure to organoselenium compounds is the dominant exposure route in fish (Luoma and Presser, 2009; Stewart et al., 2010). Although adult fish are relatively tolerant of elevated dietary organoselenium exposure, early life stages are extremely sensitive. The following sections will describe toxicokinetic and toxicodynamic mechanisms of action in detail.

6.2.1. TOXICOKINETIC MECHANISMS

Maternal deposition of Se into eggs during vitellogenesis and subsequent assimilation of yolk by the developing embryo is the toxicokinetic mechanism that results in larval deformities (Janz et al., 2010). Biochemically, the dose-dependent transfer of Se to yolk occurs because Se readily substitutes for sulfur in the amino acid methionine, producing selenomethionine, which is indiscriminately substituted for methionine during protein synthesis in direct proportion to dose. Conversely, incorporation of Se into the other sulfur-containing amino acid, cysteine, is a regulated process for production of selenocysteine (the 21st amino acid), associated with Se essentiality (see Section 7) (Schrauzer, 2000). When the yolk protein precursor, vitellogenin, is synthesized in the liver of female fish, greater dietary exposure of Se results in greater selenomethionine incorporation into vitellogenin in place of methionine. Transport and uptake of vitellogenin enriched with selenomethionine from the liver to oocytes thus determine the ultimate Se dose received by eggs (Janz et al., 2010). As discussed below in Section 6.2.2, assimilation of yolk proteins during the embryo–alevin–fry developmental stages produces a suite of larval malformations that are diagnostic of Se exposure.

The other major route of Se exposure in juvenile and adult fish is directly via organoselenium compounds in the diet. There have been numerous earlier laboratory studies investigating the chronic toxicity of Se in fish (reviewed in Jarvinen and Ankley, 1999). These laboratory studies have used a myriad of exposure conditions involving aqueous and/or Se-spiked diets, inorganic, and/or organic forms of Se, alone and in combination, and for varying durations. Field studies have mostly involved chronic exposure of wild fish to contaminant mixtures containing elevated Se (reviewed in Janz et al., 2010). As discussed below, it appears that toxicological effects on larval fish (i.e. deformities) are considerably more sensitive responses than chronic effects in juveniles or adults exposed solely via the diet.

6.2.2. Toxicodynamic Mechanisms

As mentioned above, the most sensitive and ecologically relevant toxic effects are a diagnostic suite of skeletal, craniofacial, and fin malformations that arise in larval fish following maternal transfer of elevated Se concentrations to egg yolk during vitellogenesis (Fig. 7.2). Exposure of larval fish to elevated Se occurs during yolk resorption between hatch and swim-up life stages (Janz et al., 2010). Skeletal deformities include spinal curvatures of lateral (scoliosis) and dorsoventral (kyphosis and lordosis) aspects. Craniofacial deformities include improper development of the mandible and operculum, shortened facial (oral–nasal) region (due mainly to improper development of the maxilla), reduced forebrain, and micro-phthalmia (i.e. abnormally small eyes). Fin deformities include shortened or missing fins. In addition, various forms of edema (pericardial, yolk sac, and cranial) are commonly associated with skeletal, craniofacial, and fin deformities. Figure 7.2 shows representative deformities in each of these categories. This suite of malformations is remarkably consistent among fish species and is an extremely useful diagnostic tool to identify Se toxicity in field studies where fish are exposed to complex mixtures of contaminants and other stressors. Importantly, an increasing frequency and severity of these larval malformations can impact juvenile recruitment in fish populations exposed to elevated Se, and an index has been proposed (based on the Belews Lake studies described in Section 4.1) that relates the frequency of deformities in fry to negative impacts on fish population dynamics (Lemly, 1997a).

Until recently, it was thought that the primary mechanism responsible for teratogenic effects was the substitution of Se for sulfur in cysteine and methionine, affecting the tertiary structure, and thus functions, of proteins by altering disulfide linkages (Maier and Knight, 1994; Lemly, 2004).

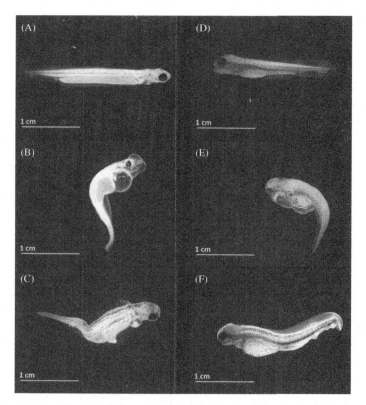

Fig. 7.2. Characteristic Se-induced larval deformities in white sucker (A–C) and northern pike (D–F). (A, D) Larvae originating from adult female fish collected at reference sites showing normal development following yolk sac resorption; (B) craniofacial and pericardial edema, and a lateral spinal curvature (scoliosis); (C) a shortened pelvic fin and scoliosis; (E) a craniofacial deformity, microphthalmia, pericardial edema, yolk sac edema, and a dorsoventral spinal curvature (kyphosis); (F) a craniofacial deformity and a dorsoventral spinal curvature (lordosis). Reprinted with permission from Muscatello (2009).

However, this proposed mechanism has been questioned because of the different nature of Se incorporation into cysteine and methionine (see Section 6.2.1), and the relative lack of importance of methionine in coordinating disulfide linkages and thus tertiary protein structure (Yuan et al., 1998; Mechaly et al., 2000). More recently, oxidative stress has been proposed as a key mechanism responsible for Se-induced larval fish deformities (Palace et al., 2004), in an etiology similar to certain organic toxicants (Bauder et al., 2005; Wells et al., 2009).

Selenide (Se^{2-}) is believed to be the common metabolic intermediate involved in essential selenoprotein synthesis or Se excretion (Birringer et al.,

2002; Suzuki and Ogra, 2002; see Section 10.1). The pro-oxidant nature of Se is due to production of metabolic intermediates such as methylselenol and dimethylselenide, which paradoxically react with glutathione (GSH) to produce the reactive oxygen species (ROS), superoxide anion (Spallholz, 1994; Spallholz et al., 2001, 2004). At sufficiently high Se exposures, production of superoxide anion and further reduced ROS (i.e. hydrogen peroxide and hydroxyl radical) overwhelms antioxidant defenses and causes oxidative damage to proteins, lipids, and DNA (Spallholz, 1994, Spallholz et al., 2001), which has been postulated to cause many of the Se-induced deformities observed in fish (Palace et al., 2004) and aquatic birds (Spallholz and Hoffman, 2002). Further research is needed to determine the importance of oxidative stress in the etiology of Se-induced deformities in fish.

6.2.3. WEIGHT OF EVIDENCE CASE STUDY

The most common approach used to assess potential effects of Se on fish populations in aquatic ecosystems receiving increased Se inputs is to conduct field research using resident fish species. In such field studies, fish are usually exposed to complex mixtures of contaminants, so that determining cause–effect relationships for Se toxicity is challenging. However, by using a weight of evidence experimental design, identification of Se as a causal toxicant can be inferred. In these studies, gametes (eggs and milt) are collected from wild-caught adult fish during the spawning period from sites representing varying levels of Se contamination, in addition to ecologically similar reference site(s). Eggs are fertilized and embryos are incubated in the laboratory (or less commonly, in the field). After hatch, larval fish are raised until the swim-up stage, when yolk resorption is complete and Se exposure from maternally transferred Se has occurred. Fry are then euthanized and preserved for determination of the frequency (and, less commonly, severity) of spinal, craniofacial and fin deformities, and edema. Selenium analysis of eggs from the same females that produced embryos can be used to derive a site-specific dose–response relationship between egg Se concentration and deformity frequency. A comparison of threshold egg Se concentrations causing an increase in larval deformities or mortality above control (i.e. EC10 values) has been performed in eight fish species where adequate data exist (Fig. 7.3). Although a limited number of species has been studied, there is a remarkably narrow range (17–24 μg Se g^{-1} dry weight) in threshold values among species representing salmonids, centrarchids, cyprinids, and esocids (Janz et al., 2010). Thus, initial assessments of a potential Se hazard should collect Se values for egg or ovary samples from resident fish species. If egg or ovary total Se

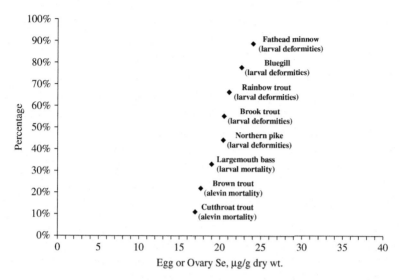

Fig. 7.3. Species sensitivity distribution for threshold (EC10 or equivalent) egg- or ovary-based Se concentrations causing an increased incidence of larval deformities or alevin mortality for eight species of freshwater fish. Reprinted with permission from Janz et al. (2010).

concentration is observed to be within this range (Fig. 7.3), it can be inferred that an Se hazard exists, prompting further field research.

The following case study provides an example of the use of a weight of evidence approach in field studies where Se is a contaminant of potential concern. Field studies were conducted near a uranium milling operation in northern Saskatchewan, Canada (Muscatello et al., 2006, 2008). Spawning northern pike (*Esox lucius*) were collected from three exposure sites downstream of mill effluent release representing a gradient of Se contamination, and from two ecologically similar reference sites. There were no significant differences in the age of adult female pike. Eggs were stripped and fertilized in the field then transported to the laboratory within 24 h of fertilization. Embryos were incubated in a photoperiod- and temperature-controlled environmental chamber using a two-way analysis of variance (ANOVA) (cross-over) experimental design, where embryos originating from exposure and reference sites were incubated in either reference site or exposure site water (Muscatello et al., 2006). Water from each fish collection site was transported weekly for use in embryo incubations. The experiment was terminated when fry reached the swim-up stage.

Selenium concentrations in pike eggs ranged from 3.2 to 48.2 μg Se g^{-1} dry weight. In this study, there were no differences in fertilization success, egg diameter, time to eyed embryo stage, time to hatch, time to swim-up, cumulative mortality, or fry condition factor among larval fish originating from study sites (Muscatello et al., 2006). Lack of impairment of these classic developmental endpoints was similar to previous studies that showed no effects of maternally derived Se exposure (Janz et al., 2010). However, there were statistically significant increases in frequencies of each deformity category, and total deformities, in pike collected from the high exposure site. Although sample sizes of adult females were small ($n = 3$–5), there was sufficient power ($1-\beta$) to reliably detect differences (i.e. power > 0.8). Importantly, the two-way ANOVA experimental design for embryo incubations indicated that this difference in deformity frequencies was due to maternal transfer and not aqueous exposure of larvae to site water. Among 19 trace elements analyzed, only Se was significantly greater in eggs from female pike originating from the high exposure site compared to eggs from reference fish. An EC10 value of 20.4 µg Se g^{-1} egg (dry weight) was calculated as an effects threshold (Muscatello et al., 2006) (Fig. 7.3). Overall, the weight of evidence from this study identified Se as the cause of an elevated incidence of larval deformities in fish from the exposure site. This approach can derive site-specific tissue residue guidelines for Se management purposes (McDonald and Chapman, 2007). However, further research is needed demonstrating potential linkages between increased frequencies/ severities of fish deformities and impacts on population dynamics (Janz et al., 2010).

7. SELENIUM ESSENTIALITY

7.1. History and Background

Selenium was first recognized as an essential micronutrient in animals by the pioneering work of Schwarz and Foltz (1957), which was further supported by the discovery that Se was essential for glutathione peroxidase activity (Rotruck et al., 1973). Another decade passed before a second mammalian selenoprotein was identified as selenoprotein P (SelP) (Motsenbocker and Tappel, 1982). Unlike the majority of essential trace elements, Se is not coordinated to proteins, but is covalently incorporated as a selenocysteine residue. Selenium is integrated into essential selenoproteins via a selenocysteinyl-tRNA that is directed by a UGA codon reprogrammed by a selenocysteine insertion sequence (SECIS) (Zinoni et al., 1987). As a result, selenocysteine was designated the 21st amino acid (Sec, U) in 1991 (Bock et al., 1991). Details of the biosynthesis of selenocysteine and the translational machinery governing its incorporation into proteins can be found in recent reviews (Reeves and Hoffmann, 2009; Gladyshev and Hatfield, 2010).

Proteins containing selenocysteine occur in bacteria, archaea and eukaryotes (Lobanov et al., 2009; Gladyshev and Hatfield, 2010). Compared to other amino acids, selenocysteine is rarely used in protein synthesis. With the exception of two known examples (SelP and selenoprotein L; see Section 7.3), only one selenocysteine is used per polypeptide. Selenocysteine is known to be present in diverse taxa, including fish, at the active site of glutathione peroxidases, thioredoxin reductases, iodothyronine deiodinases, and selenophosphate synthetases, as well as being an essential component of other selenoproteins, most of which have unknown functions (Hesketh, 2008; Reeves and Hoffmann, 2009; Gladyshev and Hatfield, 2010). There are significant differences in the total number of selenoproteins among taxa, with no functional selenoproteins definitively found in yeast or terrestrial plants, only one in the nematode *Caenorhabditis elegans*, three in *Escherichia coli* and *Drosophila melanogaster*, and about 23–25 in mammals including humans (Kryukov et al., 2003). Many selenoprotein enzyme families have mammalian homologues where selenocysteine is replaced by cysteine; however, their catalytic efficiency is reduced by 100–1000-fold (Gladyshev and Hatfield, 2010).

7.2. Selenium Essentiality in Fish

Selenium was first identified as an essential trace element in fish over three decades ago in Atlantic salmon (*Salmo salar*) (Poston et al., 1976) and rainbow trout (Hilton et al., 1980). Subsequent studies demonstrated Se essentiality in channel catfish (*Ictalurus punctatus*) (Gatlin and Wilson, 1984) and grouper (*Epinephelus malabarious*) (Lin and Shiau, 2005). Although different forms of Se (inorganic or organic) were used, collectively these nutritional studies suggest that dietary Se requirements for normal physiological function in fish are between 5 and 25 μg Se kg^{-1} body weight day^{-1}, assuming a feeding ration of 5% body weight day^{-1}. The studies in juvenile rainbow trout were the first to identify the narrow margin between essentiality and toxicity of Se, with toxicity occurring at between seven and 30 times greater dietary exposure than essential levels (Hodson et al., 1980; Hilton and Hodson, 1983). This finding, combined with the realization that certain anthropogenic activities result in dietary Se concentrations exceeding this toxic threshold (Lemly, 1993b), stimulated research that continues to this day investigating potential adverse effects of Se in wild fish populations (Janz et al., 2010). Although investigation of the normal physiological roles of Se in fish has not received high priority, several recent studies have advanced our understanding of Se essentiality in fish, particularly with respect to the identity and function of selenoproteins.

7.3. The Piscine Selenoproteome

Since early nutritional studies demonstrating that Se was required for maximal glutathione peroxidase activity (Poston et al., 1976; Hilton et al., 1980), our knowledge of selenoprotein families in fish has increased significantly. The majority of studies have been conducted in zebrafish (*Danio rerio*) owing to their abundant genetic information and the ease of experimental manipulation in this model species. Since selenoproteomes (the full set of selenoproteins in an organism) apparently expanded from prokaryotes to eukaryotes, the human selenoproteome was initially believed to encompass all known eukaryotic selenoprotein families (Kryukov et al., 2003). The zebrafish expressed sequence tag (EST) database initially identified 18 genes for selenocysteine-containing seleno-proteins (Kryukov and Gladyshev, 2000), most of which have known human homologues.

Subsequent studies identified a fish-specific selenoprotein (SelU) in the green spotted pufferfish (*Tetraodon nigroviridis*), which has cysteine-containing homologues among mammals (Castellano et al., 2004). Selenoprotein J (SelJ), which is expressed during eye lens development in zebrafish, was then identified in the genome of the fugu fish (*Takifugu rubripes*; another species of pufferfish) (Castellano et al., 2005). Thus, SelJ appears to have a structural role, unique for selenoproteins. Moreover, in contrast to all known eukaryotic selenoproteins SelJ is not present in mammalian genomes, even as a cysteine homologue (Castellano et al., 2005). Another fish-specific selenoprotein (Fep15; fish 15 kDa selenoprotein-like protein) was identified in zebrafish, *Takifugu* and *Tetraodon* (Novoselov et al., 2006). Similarly, selenoprotein L (SelL) was subsequently identified in fish and does not appear to be present in mammals (Shchedrina et al., 2007).

These recent studies identifying selenoprotein families in zebrafish and pufferfish suggest that evolution of selenoproteins displays a mosaic of evolutionary histories among taxa, likely due to varied selection pressures (Castellano et al., 2004; Lobanov et al., 2008). Fish appear to have the largest selenoproteomes, with genes coding for 32–37 selenoproteins (36 in zebrafish) compared to 25 in humans (Lobanov et al., 2008, 2009; Reeves and Hoffmann, 2009). In eukaryotes, it is hypothesized that aquatic environments favor greater reliance on selenoproteins than in terrestrial environments (Lobanov et al., 2007), although the environmental factors influencing this trend remain unknown. Indeed, it appears that most of the rare selenoproteins occur only in aquatic eukaryotes (Lobanov et al., 2007,

2008). As noted above, several piscine selenoproteins have mammalian homologues in which cysteine is present in place of selenocysteine, suggesting reduced reliance on Se during vertebrate evolution (Lobanov et al., 2008). What has become apparent from genomic comparisons is that many other taxa-specific selenoproteins likely exist, and that we currently comprehend only a small portion of selenoprotein dependence in biota. In fish, only a handful of the more than 25,000 extant species have been investigated, making this an area where advancement is likely in the future, including the discovery of additional selenoproteins.

Identification of selenoprotein families in fish has been deduced largely using bioinformatic approaches (Kryukov and Gladyshev, 2000; Castellano et al., 2004), and tissue-specific expression has only been conducted in zebrafish using whole-mount *in situ* hybridization (Tujebajeva et al., 2000; Thisse et al., 2003; Castellano et al., 2005). Functional studies in fish are lacking, and in fact the physiological roles of most selenoproteins are unknown even in mammals, although recent studies using transgenic (knockout) mouse models are making progress in this area (Gladyshev and Hatfield, 2010).

The best characterized selenoproteins in fish are oxidoreductase enzymes incorporating selenocysteine in the active site. These include glutathione peroxidases (GPx), which function primarily to reduce hydroperoxides to corresponding alcohols at the expense of glutathione. There are five known GPx in vertebrates, and four have been identified in fish (Kryukov and Gladyshev, 2000; Thisse et al., 2003). Glutathione peroxidase activity, ideally in combination with measures of membrane lipid peroxidation, is a sensitive biochemical marker of oxidative stress in ecotoxicological studies of fish (e.g. Kelly and Janz, 2009). Thioredoxin reductases (TRs) are also important antioxidant enzymes that maintain cellular redox status, primarily by maintaining reduced cysteine (Reeves and Hoffmann, 2009). Three forms of TR are identified in vertebrates, and two of these (TR2 and TR3) are expressed in zebrafish (Kryukov and Gladyshev, 2000; Thisse et al., 2003). The third group of selenoenzymes well characterized in vertebrates is the iodothyronine deiodinases, which activate the prohormone thyroxine (T_4) to the active thyroid hormone triiodothyronine (T_3), and catalyze the inactivation of T_4 to reverse T_3 (rT_3) and T_3 to diiodothyronine (T_2). Each of these steps is catalyzed by one or more deiodinases, and all three forms of human deiodinase (i.e. types I, II, and III) occur in fish (Valverde et al., 1997; Sanders et al., 1999; Kryukov and Gladyshev, 2000; Thisse et al., 2003). Finally, selenophosphate synthetase-2, which synthesizes selenoproteins by producing the active (phosphorylated) form of Se that is transferred to the selenocysteinyl-tRNA (Reeves and Hoffmann, 2009), has been identified in zebrafish (Thisse et al., 2003).

Of the remaining selenoproteins, most are poorly characterized in vertebrates with the exception of SelP. In human, rat and mouse, the single SelP has 10 selenocysteine residues, suggesting a hypothetical role as a transport protein delivering Se to organs and tissues (Motsenbocker and Tappel, 1982; Burk and Hill, 2009). In zebrafish, there are two SelP forms, with SelPa containing 17 selenocysteines and SelPb containing just one selenocysteine residue (Kryukov and Gladyshev, 2000; Tujebajeva et al., 2000). Of the remaining selenoprotein families identified, including the fish-specific SelJ, SelL, SelU, and Fep15 described above, very little information exists on their function, although many contain a thioredoxin fold motif (Lobanov et al., 2008) indicating potential regulation of cellular antioxidant status. Additional selenoproteins identified in the human selenoproteome (consisting of 25 selenoprotein families) include SelH, SelI, SelK, SelM, Sep15 (a 15 kDa selenoprotein showing some sequence homology to Fep15), SelN, SelO, SelS, SelT, SelV, and SelW, most of which are identified in zebrafish (Kryukov and Gladyshev, 2000; Thisse et al., 2003). Clearly, investigation of the piscine selenoproteome is at a very early stage, and many surprises remain to be discovered.

8. POTENTIAL FOR BIOACCUMULATION AND BIOMAGNIFICATION OF SELENIUM

Selenium differs from the majority of trace elements in that it behaves more like a persistent organic contaminant with respect to its high potential to bioaccumulate, and in certain cases to biomagnify, in aquatic food webs. Understanding the bioaccumulation and trophic transfer of Se is central to properly managing ecological risks to fish populations (Luoma and Presser, 2009; Hodson et al., 2010; Stewart et al., 2010). Selenium bioaccumulation and toxicity in fish cannot be predicted solely on concentrations of Se in water, but require integrated knowledge of the site-specific hydrodynamics and ecology of an aquatic ecosystem (Luoma and Presser, 2009; Stewart et al., 2010) (Fig. 7.1). A single, universal water quality guideline cannot protect fish populations in all aquatic ecosystems with any degree of certainty (Stewart et al., 2010). This is a primary reason underlying recent efforts in the USA and Canada towards tissue-based Se guidelines for protecting fish populations, as described in Section 5. More than perhaps any other contaminant, Se may require some degree of site-specific ecotoxico-logical research to be conducted. This section provides a brief overview of Se bioaccumulation in fish; for more detailed aspects on this topic the reader is referred to Luoma and Presser (2009) and Stewart et al. (2010).

Since diet comprises the predominant source of Se for fish, under-standing the bioaccumulation of Se in aquatic ecosystems requires knowledge of Se uptake and biotransformation at the base of the food web (see Fig. 7.1). Selenium is an essential trace element for algae, and inorganic Se (selenate and selenite) is taken up by a carrier-mediated process (Fournier et al., 2006). Once in algal cells, Se is readily converted to organoselenium species such as selenomethionine and selenocysteine, indicated as Org-Se $(II)_p$ in Fig. 7.1. Organisms that ingest algae (primary consumers) are therefore exposed to organic forms of Se, and this dietary organoselenium then progresses through to secondary and tertiary consumers, including fish (Figs. 7.1 and 7.4). As shown schematically in Fig. 7.4, it is the initial reductive assimilation step from water to primary producers that accounts for the major increase in Se bioconcentration in food webs [termed the enrichment function (EF)], which varies greatly from 100- to 1,000,000-fold depending on the algal community and the hydrodynamics of the aquatic system (Stewart et al., 2010) (see Fig. 7.1). In lentic (standing) waters such as lakes and wetlands, increased residence time causes greater bioconcentration in primary producers, whereas in lotic (flowing) waters such as rivers, bioconcentration is not as great (Simmons and Wallschläger, 2005; Hillwalker et al., 2006; Orr et al., 2006). Reductive assimilation of inorganic

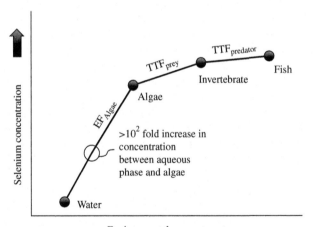

Fig. 7.4. A generalized representation of Se accumulation in freshwater and marine food webs. The major increase in Se concentration (ranging from 100- to 1,000,000-fold depending on hydrology and ecology) occurs between water and algae (termed enrichment function; EF_{algae}). Trophic transfer functions (TTFs) between algae and invertebrates (TTF_{prey}) and between invertebrates and fish ($TTF_{predator}$) are lower. Reprinted with permission from Stewart et al. (2010).

Se proceeds readily in both the water column and sediment (Fig. 7.1). In sediment, biogeochemical processes at the sediment–water interface greatly influence Se bioavailability to primary consumers such as benthic invertebrates (Wiramanaden et al., 2010a, b).

Compared to the large and variable enrichment of Se from water to algae, trophic transfer functions (TTFs; somewhat analogous to bioaccumulation factors) from primary producers to primary, secondary and tertiary consumers, including fish, are much lower, ranging from 0.6 to 23 from algae to invertebrates and 0.5 to 1.8 in fish (Stewart et al., 2010) (Fig. 7.4). In food webs with TTFs >1 between trophic levels, biomagnification occurs (Luoma and Presser, 2009). Importantly, the majority of lower trophic organisms such as algae, zooplankton, and benthic invertebrates are extremely tolerant of high Se exposures, whereas fish are among the most sensitive organisms (Janz et al., 2010). Thus, prey items can act as vectors that deliver high dietary Se concentrations to fish. Herein lays a key factor that makes Se such a hazard to fish populations: under appropriate hydrological and ecological conditions, relatively low aqueous Se concentrations (1–5 µg Se L^{-1}) can lead to toxicity in fish via bioaccumulation in their prey (Muscatello et al., 2006; Janz et al., 2010; Stewart et al., 2010).

The unique behavior of Se in aquatic systems compared to other trace elements necessitates new approaches to modeling. Biokinetic or biodynamic models address the central issue of food web relationships in predicting and managing ecological risks of Se to fish and other susceptible animals such as aquatic birds (Luoma and Rainbow, 2005; Luoma and Presser, 2009). Such models consider the combined influence of aquatic Se concentrations and physiological processes on bioaccumulation. Assuming diet is the dominant source of Se uptake, TTFs (Luoma and Rainbow, 2005) can be calculated based on ingestion rate, assimilation efficiency, elimination rate, and growth rate (see Figs. 7.1 and 7.4). Although these parameters can be determined experimentally, there are few data currently available for fish (Luoma and Presser, 2009). However, when sufficient data are available, this modeling approach is highly predictive of Se bioaccumulation in fish (Stewart et al., 2004), which is closely related to toxicity (Janz et al., 2010). Luoma and Presser (2009) examined the relationship between predicted (from food alone) and measured Se body burdens in 35 species of freshwater and marine fish from field studies, and observed a strong 1:1 relationship ($r^2 = 0.95$) between model predictions and actual Se body burdens. Overall, biodynamic models show considerable promise in predicting and managing Se risks to fish, although they require site- and species-specific information on biogeochemical, ecological, and physiological processes that are often lacking.

9. CHARACTERIZATION OF UPTAKE ROUTES

9.1. Gills

Dissolved inorganic forms of Se (selenate and selenite) are oxyanions that are not absorbed appreciably through gill membranes (Pedersen et al., 1998) or adsorbed to particulate matter (Stewart et al., 2010). In mangrove snapper (*Lutjanus argentimaculatus*), virtually the entire Se body burden comes from dietary uptake because the aqueous uptake rate (k_u) is extremely low (0.0008 L g^{-1} day^{-1}) (Xu and Wang, 2002). However, studies in freshwater fish are lacking, and further work is needed to investigate potential aqueous exposure of dissolved Se via gills. Although organoselenium compounds are present in solution and are potentially bioavailable to fish through gill membranes, even in the most contaminated aquatic systems concentrations are inadequate to cause significant bioaccumulation (Stewart et al., 2010).

9.2. Gut

Since diet is the dominant route of Se exposure in fish, gastrointestinal uptake is central to both nutritional requirements and toxicity. L-Selenomethionine is the major (>80%) form of Se found in organisms at all levels of food webs (Fan et al., 2002; Maher et al., 2010). In mammals, gastrointestinal absorption of selenomethionine occurs mainly by the epithelial Na^+-dependent neutral amino acid transporter (Schrauzer, 2000). Subsequently, the symport system B^0 and exchanger (antiport) system b^{0+} of neutral amino acid transporters apparently dominate selenomethionine absorption at the apical membrane, with similar kinetics between methionine and selenomethionine (Nickel et al., 2009). In green sturgeon (*Acipenser medirostris*), selenomethionine and methionine absorption are mediated by the same apical membrane transporter in the gut, although the identity and number of transporters involved are unknown (Bakke et al., 2010). Unlike in mammals, selenomethionine and methionine absorption rates in sturgeon follow an increasing gradient from proximal to distal portions of the gastrointestinal tract (Bakke et al., 2010).

A small portion of Se exposure in fish may arise from ingesting dissolved inorganic or organic Se in water, a phenomenon that may become greater in marine teleosts. Although studies in fish have not been conducted, selenate absorption from the mammalian intestine occurs via a Na^+-dependent sulfate transporter, whereas selenite is absorbed by passive diffusion (Vendeland et al., 1992). However, these uptake routes are considered insignificant compared to the dominant role of dietary organoselenium compounds in nutrition and toxicity.

9.3. Other Routes

Uptake of Se from routes other than the gut is apparently minimal, although studies in fish are lacking. However, the presence of Se in manufactured nanomaterials such as quantum dots (Bouldin et al., 2008) may present unique exposure scenarios, since these materials have been reported to translocate to the brain via the olfactory bulb in largemouth bass (*Micropterus salmoides*) (Oberdorster, 2004).

10. CHARACTERIZATION OF INTERNAL HANDLING

10.1. Biotransformation

Selenium undergoes a complex array of biotransformation reactions *in vivo*, and this section will focus on basic pathways of known relevance to homeostasis and toxicity in fish. Since the recognition that Se displays anticancer properties, there has been an explosion of human health research investigating the identification and biological activity of different Se forms. Although very little is known regarding the biotransformation (or indeed the transport, storage and excretion) of Se in fish, presumably many pathways are evolutionarily conserved and inferences can be made from the mammalian literature. More research is needed investigating the biotransformation of Se in fish, since these processes influence toxicity.

The majority of Se is present in the body as either selenocysteine or selenomethionine residues incorporated into proteins (Mayland, 1994; Schrauzer, 2000; Suzuki and Ogra, 2002). As described in Sections 6.2.1 and 7, Se is specifically incorporated into essential selenoproteins as selenocysteine. In all other proteins, selenomethionine replaces methionine in an unregulated and dose-dependent manner. From a nutritional standpoint, selenocysteine and selenomethionine obtained from the diet are transported to the liver via the hepatic portal venous system. Selenocysteine synthesis only occurs *de novo* after the biotransformation of dietary selenocysteine or selenomethionine to the common metabolic intermediate, selenide (Se^{2-}) (Suzuki and Ogra, 2002) (Fig. 7.5A and B). Selenate and selenite, although not a major portion of Se uptake in fish (see Section 9.2), can be reduced by GSH to selenide for subsequent selenocysteine synthesis (Fig. 7.5A and B). Selenide is phosphorylated by selenophosphate synthetase (a selenoprotein) before being incorporated into protein via the selenocysteinyl-tRNA insertion sequence (SECIS) and the UGA codon (Gladyshev and Hatfield, 2010; see Section 7.1). Dietary selenomethionine can be either biotransformed to selenide as described above or incorporated

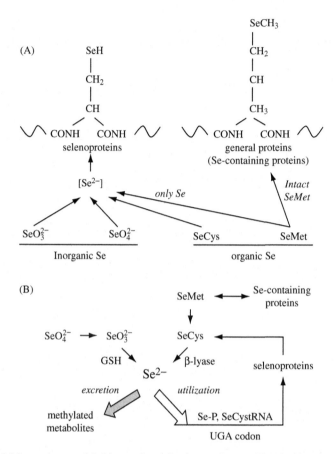

Fig. 7.5. Major pathways of Se biotransformation in vertebrates. (A) Nutritional sources of Se resulting in selenoprotein biosynthesis. Selenocysteine (SeCys) and selenomethionine (SeMet) obtained from dietary sources, or inorganic Se (selenite, SeO_3^{2-} and selenate, SeO_4^{2-}) obtained primarily from the gut in fish, are biotransformed to selenide (Se^{2-}) before being incorporated into essential selenoproteins (as selenocysteine). Selenomethionine is also incorporated non-specifically into all proteins in place of methionine. (B) Pathways of selenide utilization for essential selenoproteins and excretion as methylated or selenosugar metabolites. See Sections 10.1 and 11.2 for details. Reprinted with permission from Suzuki and Ogra (2002).

intact into methionine-containing proteins, since animals cannot distinguish between methionine and selenomethionine (Fig. 7.5A). At normal dietary uptake, selenomethionine can be considered an unregulated pool of Se for eventual selenocysteine synthesis (Suzuki and Ogra, 2002). However, supraphysiological uptake of selenomethionine can lead to toxicity, as discussed below.

From a toxicological standpoint, excessive selenomethionine uptake via diet is closely linked to toxicity via its biotransformation to selenoxides such as methylselenol that can undergo redox cycling, producing reactive oxygen species and oxidative stress (Spallholz, 1994). Herein lies the paradoxical nature of Se, as it displays both pro-oxidant (via biotransformation of selenomethionine) and antioxidant (via oxidoreductase enzyme synthesis using selenocysteine) properties. The carcinostatic properties of Se are potentially related to its pro-oxidant activity, stimulating cell death (apoptosis or necrosis) in cancer cells (Shen et al., 2000; Weiller et al., 2004). In mammals, the enzyme methionine α,γ-lyase (also known as methioninase) catalyzes the conversion of selenomethionine to methylselenol (Spallholz, 1994). The oxidation of methylselenol to methylseleninic acid, using glutathione as cofactor, results in reduction of O_2 to superoxide anion ($O_2{}^-$), which can lead to oxidative DNA, lipid and protein damage if produced in excess (Spallholz, 1994). This mechanism is hypothesized to be responsible for the early life stage deformities in fish exposed to excess selenomethionine via maternal transfer (Palace et al., 2004; Janz et al., 2010), although further work is needed to test this hypothesis.

10.2. Transport through the Bloodstream

Very little is known regarding Se transport in the bloodstream of fish. The major forms of Se in the bloodstream of mammals are selenocysteine, incorporated specifically into selenoproteins, and selenomethionine, replacing methionine non-specifically in albumin (Burk et al., 2001). Selenoprotein P is the dominant selenoprotein in human and rat plasma (25–30 mg L^{-1} under normal nutritional status), and functions primarily in Se transport and distribution throughout the body (Motsenbocker and Tappel, 1982; Burk and Hill, 2009). Approximately 65% of circulating Se is bound to SelP in the rat as selenocysteine residues (Burk and Hill, 2009). In mammals, SelP is produced primarily in liver, secreted into the bloodstream, and taken up into specific organs (Burk and Hill, 2009). Uptake of circulating SelP into organs such as testis and brain is mediated by apolipoprotein E receptor-2 (apoER2), whereas uptake into kidney proximal convoluted tubular cells is mediated by another member of the lipoprotein receptor family, megalin (Burk and Hill, 2009). Another selenoprotein in the bloodstream is the extracellular glutathione peroxidase (GPx3; see Section 7.2), which has not been identified in fish (Kryukov and Gladyshev, 2000). Although minor, a hypothesized additional mode of Se transport in blood is covalent binding of circulating selenide to cysteine residues of albumin (Suzuki and Ogra, 2002).

Although the functional role of SelP in Se transport has not been investigated in fish, zebrafish possess two isoforms (SelPa and SelPb) that are expressed in newly fertilized eggs, embryonic yolk sac, liver, kidney, heart, and brain (Tujebajeva et al., 2000; Thisse et al., 2003). Zebrafish SelPa contains 17 selenocysteine residues, whereas rat, mouse, and human SelP contains 10 (Kryukov and Gladyshev, 2000; Tujebajeva et al., 2000; Burk and Hill, 2009). This higher number of selenocysteine residues in zebrafish SelPa is consistent with the hypothesis that aquatic animals have a greater reliance on Se than do terrestrial animals (Lobanov et al., 2007; see Section 7.3). It is likely that the greater number of selenocysteine residues in zebrafish is related to a significant role in Se transport throughout the body, although this remains to be verified, and further studies of SelP in other fish species are needed.

10.3. Organ-specific Accumulation

In fish and other vertebrates, the primary organ for Se accumulation is the liver (Sato et al., 1980) because it is the dominant site of selenoprotein synthesis and catabolism (Burk and Hill, 2009). However, all organs will accumulate Se to varying extents owing to its presence as selenomethionine and selenocysteine in proteins. In fish, organ- and tissue-specific Se accumulation has been determined in adult northern pike (*Esox lucius*), where liver and kidney displayed the greatest Se concentrations, followed by ovary (egg), muscle, and bone (Muscatello et al., 2006). Incorporation of Se into otoliths, calcified structures in the inner ear of fish commonly used for age determination, has recently been determined using laser ablation inductively coupled plasma mass spectrometry (LA-ICP-MS) and was reported to provide a sensitive chronological record of fish migration into and out of Se-contaminated aquatic ecosystems (Palace et al., 2007).

Selenium accumulation in organs and tissues is of considerable ecotoxicological importance because of the recent focus on developing tissue-based regulatory guidelines to protect fish populations (see Section 6). The organ with greatest ecotoxicological relevance for Se accumulation is the ovary (ovarian follicles), since yolk proteins provide the ultimate Se dose that leads to larval deformities (Janz et al., 2010; DeForest and Adams, 2011; see Section 6.2.2). Since the yolk protein precursor vitellogenin is heterogeneous among fish in amino acid number and sequence, it is hypothesized that the relative number of methionine residues in yolk proteins (which substitute for selenomethionine in a dose-dependent manner) may predict species sensitivity (Janz et al., 2010).

Collection of eggs for Se analysis is not always possible when sampling fish in monitoring studies, prompting interest in determining relationships between egg/ovary Se and other tissues, especially muscle since it can be collected non-lethally using muscle plugs. However, ratios of egg/ovary to muscle Se concentrations among fish species are highly variable, ranging from 10:1 to 1:1 in a comparison of eight species of freshwater fish (DeBruyn et al., 2008; Janz et al., 2010). Thus, predicting egg/ovary Se concentration from Se levels in other tissues such as muscle may not be a useful approach to derive tissue-based Se guidelines unless species-specific data are available (Holm et al., 2005; DeBruyn et al., 2008; Janz et al., 2010; DeForest and Adams, 2011).

10.4. Subcellular Partitioning

Since available Se within cells is efficiently incorporated into proteins as selenocysteine or selenomethionine, the subcellular distribution of selenoenzymes (containing selenocysteine at the active site) and relative subcellular abundance of all other proteins (containing selenomethionine in place of methionine) will largely dictate the partitioning of Se within subcellular compartments. It is likely that the greatest abundance of Se will be found in the cytoplasm, since many of the characterized selenoproteins are cytosolic enzymes. However, the functions of most selenoproteins remain to be discovered and future research may identify specific subcellular compartmentalization.

10.5. Detoxification and Storage Mechanisms

Selenium compounds from the diet are delivered to the liver via the hepatic portal vein and biotransformed to selenide, which is potentially toxic owing to its involvement in redox cycling and production of ROS (see Section 10.1). However, under normal dietary Se uptake, selenide is incorporated into selenocysteine for selenoprotein synthesis or biotransformed to selenosugars or methylated metabolites for excretion (Fig. 7.5B). Thus, incorporation of selenocysteine into proteins such as SelP can be considered a pathway of dietary Se detoxification (Burk and Hill, 2009). In fact, incorporation of metabolically active selenides into proteins can also be considered a storage mechanism, since endogenous selenoproteins represent a pool of internally stored, yet bioavailable, Se that can be liberated through catabolism of protein.

10.6. Homeostatic Controls

Very little is known regarding homeostatic regulation of Se in fish, and this represents an exciting area of future research. The large piscine selenoproteome and apparent greater reliance on selenoproteins in fish compared to terrestrial vertebrates (Lobanov et al., 2007) suggests that novel insights into physiological, ecological, and evolutionary aspects of Se homeostasis remain to be made. In mammals, our understanding of Se homeostasis involves a balance between SelP synthesis and Se catabolism/ excretion in the liver (Burk and Hill, 2009). Selenoprotein P, unique among selenoproteins owing to its multiple selenocysteine residues, is central to Se storage, transport, and distribution to other organs. Studies using knockout mouse models with inactivated *SelP* gene demonstrate that SelP plays a pivotal role in distributing hepatic Se to target tissues, indicated by greatly reduced Se in plasma and tissues, and elevated hepatic Se (Schomburg et al., 2003). The potential functions of other selenoproteins are unknown, and some may also be involved in Se homeostasis.

11. CHARACTERIZATION OF EXCRETION ROUTES

11.1. Gills

There is no information available on whether excretion of Se metabolites via gills occurs in fish, although given the efficient absorption of organoselenium and incorporation of Se into proteins, this seems unlikely.

11.2. Kidney

Urinary excretion of Se is the dominant route of elimination in mammals (Kobayashi et al., 2002), although its importance in fish is not known. Selenosugars (methylseleno-*N*-acetylgalactosamine, methylseleno-*N*-acetyl-glucosamine, and methylselenogalactosamine) are the major urinary metabolites in mammals at nutritional and low toxic ranges (Gammelgaard et al., 2008). At higher Se exposures, methylated metabolites such as trimethylselenonium appear in mammalian urine and may serve as urinary biomarkers of excess Se intake (Suzuki et al., 2005; Ohta and Suzuki, 2008) (Fig. 7.5B).

11.3. Liver/Bile

Hepatobiliary excretion of Se in mammals is a minor route (Gregus and Klaassen, 1986), involving excretion of Se diglutathione (GS-Se-SG) via the canalicular GSH transporter (Gyurasics et al., 1998). The importance of hepatobiliary Se excretion in fish is not known, but warrants further study.

11.4. Gut

Owing to the efficient absorption of dietary organoselenium from the gastrointestinal tract, the gut is an unlikely route of excretion in fish, although there are no studies investigating this. However, as with most substances, it is likely that a portion of dietary organoselenium is not absorbed and is excreted in feces.

12. BEHAVIORAL EFFECTS OF SELENIUM

In general, fish behavior is a sensitive indicator of low dissolved concentrations of certain trace elements, often occurring before other sublethal effects are evident (Atchison et al., 1987; Scott and Sloman, 2004). In the case of Se, few studies have investigated behavioral effects. Earlier studies exposing fish to high concentrations of dissolved inorganic Se reported no avoidance behavior in fathead minnow (*Pimephales promelas*) exposed to 11.2 mg selenate L^{-1} (Watenpaugh and Beitinger, 1985) or in golden shiner (*Notemigonus crysoleucas*) exposed to 3.5 mg selenite L^{-1} (Hartwell et al., 1989). These test concentrations approached acutely lethal levels, indicating that, unlike some other trace elements, inorganic Se does not influence avoidance behavior. A study investigating potential effects of acute (24 h) selenate exposure on reproductive behaviors in fathead minnow found no reproductive behavioral effects at selenate exposures as high as 36 mg L^{-1} (Pyron and Beitinger, 1989).

Swimming behaviors are known to be sensitive and ecologically relevant sublethal responses to toxicant exposure in fish (Little and Finger, 1990). Swimming behavior in juvenile bluegill sunfish (*Lepomis macrochirus*) was significantly altered by exposure to a 6:1 ratio of dissolved selenate:selenite for 60 days (Cleveland et al., 1993). However, in this study reduced swim duration and frequency of activity, and increased swim speed and swim distance were observed at concentrations exceeding 0.16 mg total Se L^{-1}, indicating behavioral effects at relatively high dissolved Se exposures. In addition, these authors observed no effects of 90 day dietary Se exposure (as L-selenomethionine) on these same swimming behaviors at concentrations up to 25 µg Se g^{-1} diet wet weight (Cleveland et al., 1993). In contrast, a recent study investigated potential effects of 90 day dietary L-selenomethionine exposure (1, 3, 10, and 30 µg Se g^{-1} dry weight) on swim performance in adult zebrafish, and in their F_1 offspring exposed to Se via maternal transfer (Thomas and Janz, 2011). Critical swimming speed (U_{crit}; swimming fatigue water velocity) decreased significantly in adults and in the F_1 generation at 120 days post-hatch, suggesting ecologically relevant effects of dietary Se exposure on swim performance (Thomas and Janz, 2011).

13. MOLECULAR CHARACTERIZATION OF TRANSPORTERS, STORAGE PROTEINS, AND CHAPERONES

Although gut uptake remains uncharacterized at a molecular level, kinetic studies show that gastrointestinal absorption of dietary seleno-methionine is mediated via epithelial neutral amino acid transporters in fish (Bakke et al., 2010; see Section 9.2), similar to mammals (Nickel et al., 2009). Selenoprotein P likely mediates regulated Se transport and storage in fish (see Section 7.3) by acting as a chaperone. However, our knowledge of the roles of SelP in fish is limited to genomic studies in few species. Further molecular studies are needed in fish to characterize SelP and other selenoproteins that may be involved in this aspect of Se homeostasis.

14. GENOMIC AND PROTEOMIC STUDIES

There have been considerable recent advancements in genomic and to a lesser extent proteomic investigations of selenoproteins, all of which contain selenocysteine, in fish. Initial genomic studies in eukaryotes were challenging because of the dual function of the UGA codon, present universally as the stop codon and also "recoded" to direct the insertion of selenocysteine into peptides (Reeves and Hoffmann, 2009; Gladyshev and Hatfield, 2010). Advanced computational methods were subsequently developed to predict SECIS elements in nucleotide sequences (Kryukov et al., 2003), and in combination with comparative sequence analysis methods, allowed the identification of novel selenoproteins (Castellano et al., 2005; Gladyshev and Hatfield, 2010). In fish, such bioinformatic studies have been limited largely to zebrafish and pufferfish (*Tetraodon* and *Takifugu*), because of their abundant genomic information. Since the zebrafish (and likely many other teleosts) genome duplicated after the piscine lineage diverged phylogeneti-cally from mammals (Meyer and Schartl, 1999; Kelly et al., 2000), certain selenoprotein genes were duplicated whereas others were lost during evolution (Castellano et al., 2004; Lobanov et al., 2007, 2008). Proteomic studies in fish are just beginning, and much investigation remains regarding protein expression and functional characterization of the majority of selenoproteins. Among selenoproteins, the iodothyronine deiodinases, glutathione peroxidases, and thioredoxin reductases are better characterized than others, as discussed below.

The best characterized selenoproteins in fish are the type I, II, and III deiodinases involved in thyroid hormone activation and inactivation (Eales, 1985; see Section 7.3). Overall, genomic studies report a high degree of homology between nucleotide and predicted amino acid sequences among fish, amphibians, birds, and mammals, including humans (Valverde et al., 1997; Sanders et al., 1999; Kryukov and Gladyshev, 2000), although in certain instances significant temporal (ontogenic) and spatial (tissue-specific) differences exist among vertebrates (Sambroni et al., 2001; Walpita et al., 2007). Thus, the iodothyronine deiodinases appear to be highly evolutionarily conserved among vertebrates.

There are five mammalian glutathione peroxidases (GPx): the cytosolic GPx1, gastrointestinal GPx2, extracellular/plasma GPx3, phospholipid hydroperoxide GPx4, and epididymis-specific GPx5 (Kryukov et al., 2003). It has proven difficult to clearly separate GPx1 and GPx2 in teleosts. Two forms of GPx1 have been identified in zebrafish, one of which may be homologous to human GPx2 (Kryukov and Gladyshev, 2000). Cytosolic GPx from five fish species shows 78–87% homology in predicted amino acid sequence, with greatest similarity to mammalian GPx2 (Choi et al., 2008). Two forms of GPx4 have been reported in zebrafish (Kryukov and Gladyshev, 2000) and common carp (*Cyprinus carpio*) (Hermesz and Ferencz, 2009), which show predicted amino acid sequence homologies of 78% and 79%, respectively, suggesting they arose from gene duplication. Homologues of the mammalian GPx3 and GPx5 forms have not been identified in fish (Kryukov and Gladyshev, 2000; Thisse et al., 2003).

Little genomic information is available for thioredoxin reductases in fish. Three forms of thioredoxin reductase have been identified in humans (TR1, TR2, and TR3), and only TR2 and TR3 have been identified in zebrafish using EST databases (Kryukov and Gladyshev, 2000). Thioredoxin reductases belong to a family of thioredoxin fold proteins that function primarily as thiol-based oxidoreductases. Recent work has identified mammalian selenoproteins SelH, SelT, SelV, and SelW as belonging to this family (designated Rdx proteins), furthering our knowledge of the previously uncharacterized functions of most selenoproteins (Dikiy et al., 2007). An additional member of the Rdx family (Rdx12) is a cysteine-containing protein in mammals with a selenocysteine homologue in zebrafish and pufferfish. This Rdx12 homologue in fish was previously identified as SelW2 in zebrafish (Kryukov and Gladyshev, 2000), but displays greater sequence alignment with mammalian Rdx12 than SelW (Dikiy et al., 2007). In Senegalese sole (*Solea senegalensis*) SelW2 exhibited the greatest upregulation of 34 mRNA transcripts identified in vitellogenic ovaries, which suggests that SelW2 provides antioxidant protection during this stage of oogenesis (Tingaud-Sequeira et al., 2009).

Another putative member of the thioredoxin family is SelL, which is only present in bony, cartilaginous, and jawless fish, as well as other aquatic organisms including tunicates, mollusks, crustaceans, and bacteria

(Shchedrina et al., 2007). Selenoprotein L contains two selenocysteines, which form the first known diselenide bond in proteins (Shchedrina et al., 2007). This diselenide bond corresponds to the disulfide bond between cysteines at the active site of mammalian thioredoxin reductases, suggesting a redox role for SelL in fish.

As discussed in Section 7.3, there are two forms of SelP in zebrafish, with SelPa containing 17 selenocysteine residues and SelPb containing one selenocysteine residue, compared to 10 selenocysteine residues for the single SelP present in mammals (Kryukov and Gladyshev, 2000; Tujebakeva et al., 2000). The remaining selenoproteins identified in zebrafish using genomic approaches include proteins with mammalian homologues (SelH, SelM, SelN, SelR, SelT, SelW) and four that appear unique to fish (SelJ, SelL, SelU, and Fep15) (Kryukov and Gladyshev, 2000; Thisse et al., 2003; Shchedrina et al., 2007; Lobanov et al., 2008).

Overall, comparative genomics has provided fascinating insights into the evolution of selenoproteins. Selenium utilization appears to be an ancient trait that once was common to all organisms (Zhang and Gladyshev, 2010). The evolution of terrestrial animals coincided with a reduced reliance on Se, likely since the physiological utilization of Se was counter-selected by the presence of oxygen in the atmosphere, possibly due to the pro-oxidant nature of both oxygen and Se (Lobanov et al., 2008). This reduced reliance is evident as the larger selenoproteomes in aquatic animals including fish, which appear to have the largest number of selenoproteins among all biota (Lobanov et al., 2007, 2008). In the future, proteomic and functional characterization studies of selenoproteins using models such as zebrafish will surely provide additional insights into the roles of selenoproteins in vertebrates.

15. INTERACTIONS WITH OTHER METALS

Selenium interacts with a variety of other trace elements, primarily in an antagonistic fashion. From a toxicological perspective, the most notable interactions in mammals involve As, Ag, Au, Cd, Cu, Co, Cr, Hg, Mo, Ni, Pb, and Zn (Schrauzer, 2009), certain of which have been documented in fish (Janz et al., 2010). Although relatively little is known about Se interactions with trace elements in fish, these chemicobiological interactions commonly involve trace element forms (species) and endogenous compounds such as thiols that are highly conserved among vertebrates. Similar interactions likely occur in fish. The following section describes known interactions between Se and selected trace elements in fish.

The best characterized interaction involves protective effects of Se against mercury toxicity, for both inorganic (mercuric mercury, Hg^{2+}) and organic (methylmercury, CH_3Hg^+) forms (Cuvin-Aralar and Furnes, 1991; Yang et al., 2008). Selenite has long been known to antagonize the toxicity of Hg^{2+} in mammals (Parizek and Ostradalova, 1967) by producing inert mercuric selenide (HgSe) compounds in the bloodstream (Burk et al., 1974). Subsequent studies showed that GSH was essential to formation of HgSe complexes (Naganuma and Imura, 1983), and that HgSe was bound to SelP in plasma (Yoneda and Suzuki, 1997; Gailer, 2007). The mechanism(s) involved in antagonism of CH_3Hg^+ by Se is less clear, but the primary interaction likely involves production of bis(methylmercuric) selenide $[(CH_3Hg^+)_2Se]$, which occurs in mercury-contaminated aquatic ecosystems (Yang et al., 2008). Indeed, addition of Se to mercury contaminated lakes significantly reduces Hg body burdens in resident fish species (Paulsson and Lundberg, 1989). Although Se can protect against methylmercury toxicity, methylmercury can also cause the physiological inactivation of essential selenoproteins *in vivo* by binding covalently to selenocysteine at the active site of enzymes such as glutathione peroxidase (Ralston et al., 2008).

Another well-characterized interaction with Se involves arsenic (As). Both arsenite (As[III]) and arsenate (As[V]) interact antagonistically with Se (Gailer, 2007). Although simultaneous administration of selenite and arsenite to animals resulted in the mutual biliary excretion of both species, the mechanism of interaction remains elusive. Potentially, the interaction involves complexation with endogenous thiols such as GSH, producing seleno-bis(S-glutathionyl) arsinium ion $[(GS)_2AsSe]^-$ (Gailer, 2007).

Interactions between copper (Cu) and Se in fish also decrease tissue-specific Se accumulation but apparently not its biological effects (Hilton and Hodson, 1983). Elevated dietary Cu reduced the concentrations of Se in liver of Atlantic salmon (Lorentzen et al., 1998; Berntssen et al., 1999). Similar to the interaction of selenide with Hg^{2+}, it appears that insoluble Se–Cu complexes are formed in the intestinal lumen, reducing the bioavailability of Se from diet (Berntssen et al., 2000).

Selenate also interacts with sulfate in fish acute toxicity studies. In adult fathead minnow, increasing aqueous sulfate concentrations (24 to 1013 mg SO_4^{2-} L^{-1} as Na_2SO_4) significantly reduced the acute toxicity of selenate, as indicated by a six-fold increase in 96 h LC50 values (Brix et al., 2001). This suggests a potential interaction between selenate and sulfate uptake at the fish gill, although there is no evidence for the presence of a sulfate transporter on the fish gill membrane to confirm such interactions. Höbe (1987) observed significant influx of sulfate across gills in rainbow trout and white sucker (*Catostomus commersoni*) exposed to low ambient pH (pH 4.2), but no sulfate influx at pH 6.8, and hypothesized that passive sulfate influx may occur at low ambient pH in these fish species. As discussed in Section 9.2, the interaction between sulfate and selenate may also arise via competition between sulfate and selenate for the intestinal Na^+-dependent sulfate transporter (Vendeland et al., 1992), although this interaction would likely be more important in marine teleosts. However, this interaction is of limited ecotoxicological importance since chronic exposure to dietary organoselenium compounds is the most environmentally relevant route of exposure leading to toxicity in fish.

Most of our knowledge on interactions of Se with other trace elements is derived from human carcinogenicity studies reporting antagonistic effects of Se on organ-specific cancers (Schrauzer, 2009). It is likely that interactions with other transition elements involve complexation of highly reactive selenides in the presence of endogenous electron sources such as thiols. Because of the ability of Se to interact with a range of other trace elements, further research is clearly needed investigating the potential nutritional and toxicological importance of these interactions in fish.

16. INTERACTIONS WITH WATER TEMPERATURE

In aquatic ecotoxicology, the term "winter stress syndrome" has been proposed to describe the potential for contaminants to potentiate overwinter mortality due to increased metabolism resulting from contaminant exposure during periods of reduced food availability (Lemly, 1996). The winter stress syndrome hypothesis is based on a laboratory study in which juvenile bluegill sunfish (*Lepomis macrochirus*) were exposed for 180 days to dietary and waterborne Se under either summer (20°C and 16 h light:8 h dark) or winter (4°C and 8 h light:16 h dark) conditions (Lemly, 1993a). Compared to sunfish exposed to Se under summer conditions, winter conditions exacerbated the toxicity of Se, indicated by increased mortality, decreased condition factor (weight-at-length) and decreased energy (lipid) stores. Lemly's (1993a) laboratory study has recently been replicated, with similar results observed (McIntyre et al., 2008). The juvenile bluegill sunfish studies (Lemly, 1993a; McIntyre et al., 2008) indicate that Se likely causes increased metabolism, which may occur, in part, through oxidative stress (Spallholz, 1994; Palace et al., 2004). Although the concept of winter stress syndrome is a scientifically sound hypothesis, evidence for it occurring under field conditions of elevated Se exposure is lacking. It is possible that winter stress syndrome is most important in fish species at the northern limit of their ranges, and future studies should focus on this aspect (Lemly, 1996; Janz et al., 2010).

17. KNOWLEDGE GAPS AND FUTURE DIRECTIONS

There remains much to be discovered regarding the homeostasis of Se in fish. Although fish appear to possess the largest selenoproteome among all biota, our knowledge of the majority of piscine selenoproteins is limited largely to genomic studies in few species. Tissue-specific mRNA expression has only been determined for the majority of selenoproteins in zebrafish embryos (Kryukov and Gladyshev, 2000; Thisse et al., 2003). Further investigations of selenoprotein expression profiles and characterization of biochemical and physiological functions are greatly needed. With their diversity of life history strategies and range of aquatic environments inhabited, future studies in fish should provide exciting insight into the evolution of selenoprotein functions.

Although toxicokinetic mechanisms of exposure in fish are fairly well understood, the toxicodynamic mechanisms of toxicity remain unclear. The potential role of oxidative stress in mediating toxicity to different life stages remains an important future research direction. Dose–response relationships between dietary Se exposure, maternal transfer to eggs, and resulting teratogenic effects on early life stages have been described in fewer than 10 fish species. Larval deformities arising from maternal transfer of Se to offspring vary in their severity, and an important knowledge gap concerns the ecological importance of more subtle deformities on individual fish fitness. Tissue- and organ-specific distribution of Se varies greatly in the fish species studied to date, and a greater understanding of the factors influencing organ accumulation is needed. Selenium interacts with a wide variety of other trace elements, and with the exception of interactions with Hg, further work is needed investigating the ecotoxicological importance of these interactions. Tolerance in fish exposed to elevated Se over several generations has been hypothesized (Canton and Van Derveer, 1997; Kennedy et al., 2000), but potential mechanisms of physiological acclimation or genetic adaptation have not been investigated.

There is currently general scientific consensus that elevated Se in aquatic ecosystems represents a significant hazard to the sustainability of fish populations, primarily by impacting recruitment. However, there remains uncertainty in ecological risk assessments owing to the unclear associations between early life stage toxicity of maternally derived Se and resulting negative impacts on fish populations. Future research on the physiology and toxicology of Se in fish will provide important information to assist with better management of Se in aquatic ecosystems.

REFERENCES

ANZECC (Australian and New Zealand Environment and Conservation Council) (1987). *Australian and New Zealand Guidelines for Fresh and Marine Water Quality*. Vol. 1: *The Guidelines*. Australian Water Association, Artarmon, NSW.

Atchison, G. J., Henry, M. G., and Sandheinrich, M. B. (1987). Effects of metals on fish behaviour: a review. *Environ. Biol. Fish.* **18**, 11–25.

Bakke, A. M., Tashjian, D. H., Wang, C. F., Lee, S. H., Bai, S. C., and Hung, S. S. O. (2010). Competition between selenomethionine and methionine in the intestinal tract of green sturgeon (*Acipenser medirostris*). *Aquat. Toxicol.* **96**, 62–69.

Barwick, M., and Maher, W. (2003). Biotransference and biomagnification of selenium, copper, cadmium, zinc, arsenic and lead in a temperate seagrass ecosystem from Lake Macquarie Estuary, NSW, Australia. *Mar. Environ. Res.* **56**, 471–502.

Bauder, M. B., Palace, V. P., and Hodson, P. V. (2005). Is oxidative stress the mechanism of blue sac disease in retene-exposed trout larvae? *Environ. Toxicol. Chem.* **24**, 694–702.

Berntssen, M. H. G., Lundebye, A. K., and Hamre, K. (1999). Effects of elevated dietary copper concentrations on growth, feed utilization and nutritional status of Atlantic salmon (*Salmo salar* L.) fry. *Aquaculture* **174**, 167–181.

Berntssen, M. H. G., Lundebye, A. K., and Hamre, K. (2000). Tissue lipid peroxidative responses in Atlantic salmon (*Salmo salar* L.) parr fed high levels of dietary copper and cadmium. *Fish Physiol. Biochem.* **23**, 35–48.

Birringer, M., Pilawa, S., and Flohé, L. (2002). Trends in selenium biochemistry. *Nat. Prod. Rep.* **19**, 693–718.

Bock, A., Forchhammer, K., Heider, J., and Baron, C. (1991). Selenoprotein synthesis: expansion of the genetic code. *Trends Biochem. Sci.* **16**, 463–467.

Bouldin, J. L., Ingle, T. M., Sengupta, A., Alexander, R., Hannigan, R. E., and Buchanan, R. A. (2008). Aqueous toxicity and food chain transfer of quantum dots in freshwater algae and *Ceriodaphnia dubia*. *Environ. Toxicol. Chem.* **27**, 1958–1963.

Bouwer, H. (1989). Agricultural contamination: problems and solutions. *Water Environ. Technol.* **1**, 292–297.

Brix, K. V., Volosin, J. S., Adams, W. J., Reash, R. J., Carlton, R. G., and McIntyre, D. O. (2001). Effects of sulfate on the acute toxicity of selenate to freshwater organisms. *Environ. Toxicol. Chem.* **20**, 1037–1045.

Buhl, K. J., and Hamilton, S. J. (1991). Relative sensitivity of early life stages of arctic grayling, coho salmon, and rainbow trout to nine inorganics. *Ecotoxicol. Environ. Saf.* **22**, 184–197.

Burk, R. F., and Hill, K. E. (2009). Selenoprotein P – expression, functions and roles in mammals. *Biochim. Biophys. Acta* **1790**, 1441–1447.

Burk, R. F., Foster, K. A., Greenfield, P. M., and Kiker, K. W. (1974). Binding of simultaneously administered inorganic selenium and mercury to a rat plasma protein. *Proc. Soc. Exp. Biol. Med.* **145**, 782–785.

Burk, R. F., Hill, K. E., and Motley, A. K. (2001). Plasma selenium in specific and non-specific forms. *Biofactors* **14**, 107–114.

Canton, S. P. (2010). Commentary: Persistence of some fish populations in high-Se environments. In: *Ecological Assessment of Selenium in the Aquatic Environment* (P.M. Chapman, W.J. Adams, M.L. Brooks, C.G. Delos, S.N. Luoma, W.A. Maher, H.M. Ohlendorf, T.S. Presser and D.P. Shaw, eds), pp. 293–323. CRC Press, Boca Raton, FL.

Canton, S. P., and Van Derveer, W. D. (1997). Selenium toxicity to aquatic life: an argument for sediment-based water quality criteria. *Environ. Toxicol. Chem.* **16**, 1255–1259.

Castellano, S., Novoselov, S. V., Kryukov, G. V., Lescure, A., Blanco, E., Krol, A., Gladyshev, V. N., and Guiro, R. (2004). Reconsidering the evolution of eukaryotic selenoproteins: a novel nonmammalian family with scattered phylogenetic distribution. *EMBO Rep.* **5**, 71–77.

Castellano, S., Lobanov, A. V., Chappel, C., Novoselov, S. V., Albrecht, M., Hua, D., Lescure, A., Lengauer, T., Krol, A., Gladyshev, V. N., and Guiro, R. (2005). Diversity of functional plasticity of eukaryotic selenoproteins: identification and characterization of the *SelJ* family. *Proc. Natl. Acad. Sci. U.S.A.* **102**, 16188–16193.

CCME (2005). *Canadian Water Quality Guidelines for the Protection of Aquatic Life.* Canadian Council of Ministers of the Environment, Winnipeg.

Chapman, P. M. (2009). Is selenium a global contaminant of potential concern? *Integr. Environ. Assess. Manag.* **5**, 353–354.

Chapman, P. M., Adams, W. J., Brooks, M. L., Delos, C. G., Luoma, S. N., Maher, W. A., Ohlendorf, H. M., Presser, T. S., and Shaw, D. P. (2010). *Ecological Assessment of Selenium in the Aquatic Environment.* CRC Press, Boca Raton, FL.

Choi, C. Y., An, K. W., and An, M. I. (2008). Molecular characterization and mRNA expression of glutathione peroxidase and glutathione *S*-transferase during osmotic stress in olive flounder (*Paralichthys olivaceus*). *Comp. Biochem. Physiol. A* **149**, 330–337.

Cleveland, L., Little, E. E., Buckler, D. R., and Wiedmeyer, R. H. (1993). Toxicity and bioaccumulation of waterborne and dietary selenium in juvenile bluegill (*Lepomis macrochirus*). *Aquat. Toxicol.* **27**, 265–279.

Cumbie, P. M., and Van Horn, S. L. (1978). Selenium accumulation associated with fish mortality and reproductive failure. *Proc. Ann. Conf. SE Assoc. Fish Wildl. Agencies* **32**, 612–624.

Cutter, G. A. (1989). The estuarine behaviour of selenium in San Francisco Bay. *Estuar. Coastal Shelf Sci.* **28**, 13–34.

Cutter, G. A. (1992). Kinetic controls on the speciation of metalloids in seawater. *Mar. Chem.* **40**, 65–80.

Cutter, G. A., and Bruland, K. W. (1984). The marine biogeochemistry of selenium: a re-evaluation. *Limnol. Oceanogr.* **29**, 1179–1192.

Cutter, G. A., and Cutter, L. S. (1998). Metalloids in the high latitude north Atlantic Ocean: sources and internal cycling. *Mar. Chem.* **61**, 25–36.

Cutter, G. A., and Cutter, L. S. (2004). Selenium biogeochemistry in the San Francisco Bay estuary: changes in water column behavior. *Estuar. Coastal Shelf Sci.* **61**, 463–476.

Cutter, G. A., and San Diego-McGlone, M. L. C. (1990). Temporal variability of selenium fluxes in San Francisco Bay. *Sci. Total Environ.* **97/98**, 235–250.

Cuvin-Aralar, M. L., and Furnes, R. W. (1991). Mercury and selenium interaction: a review. *Ecotoxicol. Environ. Saf.* **21**, 348–364.

DeBruyn, A. M., Hodaly, A., and Chapman, P. M. (2008). *Tissue Selection Criteria: Selection of Tissue Types for Development of Meaningful Selenium Tissue Thresholds in Fish.* North American Metals Council – Selenium Working Group, Washington, DC.

DeForest, D. K., and Adams, W. J. (2011). Selenium accumulation and toxicity in freshwater fishes. In: *Environmental Contaminants in Biota: Interpreting Tissue Concentrations* (W.N. Beyer and J.P. Meador, eds), 2nd edn, pp. 193–229. Taylor and Francis, Boca Raton, FL.

DeForest, D. K., Brix, K. V., and Adams, W. J. (1999). Critical review of proposed residue-based selenium toxicity thresholds for freshwater fish. *Hum. Ecol. Risk Assess.* **5**, 1187–1228.

Department of Water Affairs and Forestry (1996). *South African Water Quality Guidelines.* Vol. 7: *Aquatic Ecosystems.* Government Printer, Pretoria.

Dikiy, A., Novoselov, S. V., Fomenko, D. E., Sengupta, A., Carlson, B. A., Cerny, R. L., Ginalski, K., Grishin, N. V., Hatfield, D. L., and Gladyshev, V. N. (2007). SelT, SelW, SelH, and Rdx12: Genomics and molecular insights into the functions of selenoproteins of a novel thioredoxin-like family. *Biochemistry* **46**, 6871–6882.

Eales, J. G. (1985). The peripheral metabolism of thyroid hormones and regulation of thyroidal status in poikilotherms. *Can. J. Zool.* **63**, 1217–1231.

Ellis, M. M., Motley, H. L., Ellis, M. D., and Jones, R. O. (1937). Selenium poisoning in fishes. *Proc. Soc. Exp. Biol. Med.* **36**, 519–522.

Fan, T. W. M., Teh, S. J., Hinton, D. E., and Higashi, R. M. (2002). Selenium biotrans-formations into proteinaceous forms by foodweb organisms of selenium-laden drainage waters in California. *Aquat. Toxicol.* **57**, 65–84.

Finley, K., and Garrett, R. (2007). Recovery at Belews and Hyco Lakes: implications for fish tissue Se thresholds. *Integr. Environ. Assess. Manag.* **3**, 297–299.

Fournier, E., Adam, C., Massabuau, J. C., and Garnier-Laplace, J. (2006). Selenium bioaccumulation in *Chlamydomonas reinhardtii* and subsequent transfer to *Corbicula fluminea*: role of selenium speciation and bivalve ventilation. *Environ. Toxicol. Chem.* **25**, 2692–2699.

Gailer, J. (2007). Arsenic–selenium and mercury–selenium bonds in biology. *Coord. Chem. Rev.* **251**, 234–254.

Gammelgaard, B., Gabel-Jensen, C., Stürup, S., and Hansen, H. R. (2008). Complementary use of molecular and element-specific mass spectrometry for identification of selenium compounds related to human selenium metabolism. *Anal. Bioanal. Chem.* **390**, 1691–1706.

Gatlin, D. M., and Wilson, R. P. (1984). Dietary selenium requirement of fingerling channel catfish. *J. Nutr.* **114**, 627–633.

George, M. W. (2009). *Mineral Commodity Summaries. Selenium.* US Geological Survey, Washington, DC, pp. 144–145.

Gladyshev, V. M., and Hatfield, D. L. (2010). Selenocysteine biosynthesis, selenoproteins, and selenoproteomes. In: *Recoding: Expansion of Decoding Rules Enriches Gene Expression, Nucleic Acids and Molecular Biology* (J.F. Atkins and R.F. Gesteland, eds), Vol. 24, pp. 3–27. Springer, New York.

Gregus, Z., and Klaassen, C. D. (1986). Disposition of metals in rats: a comparative study of fecal, urinary and biliary excretion and tissue distribution of eighteen metals. *Toxicol. Appl. Pharmacol.* **85**, 24–38.

Gyurasics, A., Perjesi, P., and Gregus, Z. (1998). Role of glutathione and methylation in the biliary excretion of selenium. The paradoxical effect of sulfobromophthalein. *Biochem. Pharmacol.* **56**, 1381–1389.

Hamilton, S. J. (2002). Rationale for a tissue-based selenium criterion for aquatic life. *Aquat. Toxicol.* **57**, 85–100.

Hamilton, S. J. (2003). Review of residue-based selenium toxicity thresholds for freshwater fish. *Ecotoxicol. Environ. Saf.* **56**, 201–210.

Hamilton, S. J., and Buhl, K. J. (2005). Selenium in the Blackfoot, Salt, and Bear River watersheds. *Environ. Monit. Assess.* **104**, 309–339.

Hartwell, S. I., Jin, J. H., Cherry, D. S., and Cairns, J., Jr. (1989). Toxicity versus avoidance response of golden shiner, *Notemigonus crysoleucas*, to five metals. *J. Fish Biol.* **35**, 447–456.

368

DAVID M. JANZ

Haygarth, P. M. (1994). Global importance and global cycling of selenium. In: *Selenium in the Environment* (W.T. Frankenberger, Jr. and S. Benson, eds), pp. 1–27. Marcel Dekker, New York.
Hermesz, E., and Ferencz, A. (2009). Identification of two phospholipid hydroperoxide glutathione peroxidase (gpx4) genes in common carp. *Comp. Biochem. Physiol. C* **150**, 101–106.
Hesketh, J. (2008). Nutrigenomics and selenium: gene expression patterns, physiological targets, and genetics. *Annu. Rev. Nutr.* **28**, 157–177.
Hillwalker, W. E., Jepson, P. C., and Anderson, K. A. (2006). Selenium accumulation patterns in lotic and lentic aquatic systems. *Sci. Total Environ.* **366**, 367–379.
Hilton, J. W., and Hodson, P. V. (1983). Effect of increased dietary carbohydrate on selenium metabolism and toxicity in rainbow trout (*Salmo gairdneri*). *J. Nutr.* **113**, 1241–1248.
Hilton, J. W., Hodson, P. V., and Slinger, S. J. (1980). The requirement and toxicity of selenium in rainbow trout (*Salmo gairdneri*). *J. Nutr.* **110**, 2527–2535.
Höbe, H. (1987). Sulphate entry into soft-water fish (*Salmo gairdneri, Catostomus commersoni*) during low ambient pH exposure. *J. Exp. Biol.* **133**, 87–109.
Hodson, P. V., Spry, D. J., and Brunt, B. R. (1980). Effects on rainbow trout (*Salmo gairdneri*) of a chronic exposure to waterborne selenium. *Can. J. Fish. Aquat. Sci.* **37**, 233–240.
Hodson, P. V., Reash, R. J., Canton, S. P., Campbell, P. V., Delos, C. G., Fairbrother, A., Hitt, N. P., Miller, L. L., and Ohlendorf, H. M. (2010). Selenium risk characterization. In: *Ecological Assessment of Selenium in the Aquatic Environment* (P.M. Chapman, W.J. Adams, M.L. Brooks, C.G. Delos, S.N. Luoma, W.A. Maher, H.M. Ohlendorf, T.S. Presser and D.P. Shaw, eds), pp. 233–256. CRC Press, Boca Raton, FL.
Holm, J., Palace, V. P., Siwik, P., Sterling, G., Evans, R., Baron, C., Werner, J., and Wautier, K. (2005). Developmental effects of bioaccumulated selenium in eggs and larvae of two salmonid species. *Environ. Toxicol. Chem.* **24**, 2373–2381.
Hu, X., Wang, F., and Hanson, M. L. (2009). Selenium concentration, speciation and behavior in surface waters of the Canadian prairies. *Sci. Total Environ.* **407**, 5869–5876.
Janz, D. M., DeForest, D. K., Brooks, M. L., Chapman, P. M., Gilron, G., Hoff, D., Hopkins, W. D., McIntyre, D. O., Mebane, C. A., Palace, V. P., Skorupa, J. P., and Wayland, M. (2010). Selenium toxicity to aquatic organisms. In: *Ecological Assessment of Selenium in the Aquatic Environment* (P.M. Chapman, W.J. Adams, M.L. Brooks, C.G. Delos, S.N. Luoma, W.A. Maher, H.M. Ohlendorf, T.S. Presser and D.P. Shaw, eds), pp. 141–231. CRC Press, Boca Raton, FL.
Jarvinen, A. W., and Ankley, G. T. (1999). *Linkage of Effects to Tissue Residues: Development of a Comprehensive Database for Aquatic Organisms Exposed to Inorganic and Organic Chemicals*. SETAC Press, Pensacola, FL.
Kelly, J. M., and Janz, D. M. (2009). Assessment of oxidative stress and histopathology in juvenile northern pike (*Esox lucius*) inhabiting lakes downstream of a uranium mill. *Aquat. Toxicol.* **92**, 240–249.
Kelly, P. D., Chu, F., Woods, I. G., Ngo-Hazelett, P., Cardozo, T., Huang, H., Kimm, F., Yan, Y. L., Zhou, Y., Johnson, S. L., Abagyan, R., Schier, A. F., Postlethwait, J. H., and Talbot, W. S. (2000). Genetic linkage mapping of zebrafish genes and ESTs. *Genome Res.* **10**, 558–567.
Kennedy, C. J., McDonald, L. E., Loveridge, R., and Strosher, M. M. (2000). The effects of bioaccumulated selenium on mortalities and deformities in the eggs, larvae, and fry of a wild population of cutthroat trout (*Oncorhynchus clarki lewisi*). *Arch. Environ. Contam. Toxicol.* **39**, 46–52.
Kirby, J., Maher, W., and Krikowa, F. (2001). Selenium, cadmium, copper, and zinc concentrations in sediments and mullet (*Mugil cephalus*) from the southern basin of Lake Macquarie, NSW, Australia. *Arch. Environ. Contam. Toxicol.* **40**, 246–256.

Kobayashi, Y., Ogra, Y., Ishiwata, K., Takayama, H., Aimi, N., and Suzuki, K. T. (2002). Selenosugars are key and urinary metabolites for Se excretion within the required to low-toxic range. *Proc. Natl. Acad. Sci. U.S.A.* **99**, 15932–15936.

Kryukov, G. V., and Gladyshev, V. N. (2000). Selenium metabolism in zebrafish: multiplicity of selenoprotein genes and expression of a protein containing 17 selenocysteine residues. *Genes Cells* **5**, 1049–1060.

Kryukov, G., Castellano, S., Novoselov, S., Lobanov, A., Zehtab, O., Guigo, R., and Gladyshev, V. (2003). Characterization of mammalian selenoproteomes. *Science* **300**, 1439–1443.

Lemly, A. D. (1985). Toxicology of selenium in a freshwater reservoir: implications for environmental hazard evaluation and safety. *Ecotoxicol. Environ. Saf.* **10**, 314–338.

Lemly, A. D. (1993a). Metabolic stress during winter increases the toxicity of selenium to fish. *Aquat. Toxicol.* **27**, 133–158.

Lemly, A. D. (1993b). Teratogenic effects of selenium in natural populations of freshwater fish. *Ecotoxicol. Environ. Saf.* **26**, 181–204.

Lemly, A. D. (1996). Winter stress syndrome: an important consideration for hazard assessment of aquatic pollutants. *Ecotoxicol. Environ. Saf.* **34**, 223–227.

Lemly, A. D. (1997a). A teratogenic deformity index for evaluating impacts of selenium on fish populations. *Ecotoxicol. Environ. Saf.* **37**, 259–266.

Lemly, A. D. (1997b). Ecosystem recovery following Se contamination in a freshwater reservoir. *Ecotoxicol. Environ. Saf.* **36**, 275–281.

Lemly, A. D. (2002). Symptoms and implications of selenium toxicity in fish: the Belews Lake case example. *Aquat. Toxicol.* **57**, 39–49.

Lemly, A. D. (2004). Aquatic selenium pollution is a global environmental safety issue. *Ecotoxicol. Environ. Saf.* **59**, 44–56.

Lin, Y. H., and Shiau, S. Y. (2005). Dietary selenium requirement of grouper, *Epinephelus malabaricus*. *Aquaculture* **250**, 356–363.

Little, E. E., and Finger, S. E. (1990). Swimming behaviour as an indicator of sublethal toxicity in fish. *Environ. Toxicol. Chem.* **9**, 13–20.

Lobanov, A. V., Fomenko, D. E., Zhang, Y., Sengupta, A., Hatfield, D. L., and Gladyshev, V. N. (2007). Evolutionary dynamics of eukaryotic selenoproteomes: large selenoproteomes may associate with aquatic life and small with terrestrial life. *Genome Biol.* **8**, R198.

Lobanov, A. V., Hatfield, D. L., and Gladyshev, V. N. (2008). Reduced reliance on the trace element selenium during evolution of mammals. *Genome Biol.* **9**, R62.

Lobanov, A. V., Hatfield, D. L., and Gladyshev, V. N. (2009). Eukaryotic selenoproteins and selenoproteomes. *Biochim. Biophys. Acta* **1790**, 1424–1428.

Lorentzen, M., Maage, A., and Julshamn, K. (1998). Supplementing copper to a fishmeal-based diet fed to Atlantic salmon parr affects liver copper and selenium concentrations. *Aquacult. Nutr.* **4**, 67–72.

Luoma, S. N., and Presser, T. S. (2009). Emerging opportunities in management of selenium contamination. *Environ. Sci. Technol.* **43**, 8483–8487.

Luoma, S. N., and Rainbow, P. S. (2005). Why is metal bioaccumulation so variable? Biodynamics as a unifying concept. *Environ. Sci. Technol.* **39**, 1921–1931.

Maher, W. A., Roach, A., Doblin, M., Fan, T., Foster, S., Garrett, R., Möller, G., Oram, L., and Wallschläger, D. (2010). Environmental sources, speciation, and partitioning of selenium. In: *Ecological Assessment of Selenium in the Aquatic Environment* (P.M. Chapman, W.J. Adams, M.L. Brooks, C.G. Delos, S.N. Luoma, W.A. Maher, H.M. Ohlendorf, T.S. Presser and D.P. Shaw, eds), pp. 47–92. CRC Press, Boca Raton, FL.

Maier, K. J., and Knight, A. W. (1994). Ecotoxicology of selenium in freshwater systems. *Rev. Environ. Contam. Toxicol.* **134**, 31–48.

Mayland, H. (1994). Selenium in plant and animal nutrition. In: *Selenium in the Environment* (W.T. Frankenberger, Jr. and S. Benson, eds), pp. 29–45. Marcel Dekker, New York.

McDonald, B. G., and Chapman, P. M. (2007). Selenium effects: a weight of evidence approach. *Integr. Environ. Assess. Manag.* **3**, 129–136.

McIntyre, D. O., Pacheco, M. A., Garton, M. W., Wallschläger, D., and Delos, C. G. (2008). *Effect of Selenium on Juvenile Bluegill Sunfish at Reduced Temperatures.* Office of Water, US Environmental Protection Agency, Washington, DC, EPA-822-R-08-020

Mechaly, A., Teplitsky, A., Belakhov, V., Baasov, T., Shoham, G., and Shoham, Y. (2000). Overproduction and characterization of seleno-methionine xylanase T-6. *J. Biotechnol.* **78**, 83–86.

Meyer, A., and Schartl, M. (1999). Gene and genome duplications in vertebrates: the one-to-four (-to-eight in fish) rule and the evolution of novel gene functions. *Curr. Opin. Cell Biol.* **11**, 699–704.

Miller, L. L., Wang, F., Palace, V. P., and Hontela, A. (2007). Effects of acute and subchronic exposures to waterborne selenite on the physiological stress response and oxidative stress indicators in juvenile rainbow trout. *Aquat. Toxicol.* **83**, 263–271.

Misra, S., and Niyogi, S. (2009). Selenite causes cytotoxicity in rainbow trout (*Oncorhynchus mykiss*) hepatocytes by inducing oxidative stress. *Toxicol. In Vitro* **23**, 1249–1258.

Motsenbocker, M. A., and Tappel, A. L. (1982). Selenocysteine-containing proteins from rat and monkey plasma. *Biochim. Biophys. Acta* **704**, 253–260.

Muscatello, J. R. (2009). *Selenium accumulation and effects in aquatic organisms downstream of uranium mining and milling operations in northern Saskatchewan.* Ph.D. thesis, University of Saskatchewan.

Muscatello, J. R., Bennett, P. M., Himbeault, K. T., Belknap, A. M., and Janz, D. M. (2006). Larval deformities associated with selenium accumulation in northern pike (*Esox lucius*) exposed to metal mining effluent. *Environ. Sci. Technol.* **40**, 6506–6512.

Muscatello, J. R., Belknap, A. M., and Janz, D. M. (2008). Accumulation of selenium in aquatic systems downstream of a uranium mining operation in northern Saskatchewan, Canada. *Environ. Pollut.* **156**, 387–393.

Naganuma, A., and Imura, N. (1983). Mode of *in vitro* interaction of mercuric mercury with selenite to form high molecular weight substance in rabbit blood. *Chem. Biol. Interact.* **43**, 271–282.

Nagpal, N. K. (2001). *Ambient Water Quality Guidelines for Selenium.* British Columbia Ministry of Water, Land and Air Protection, Victoria.

Nickel, A., Kottra, G., Schmidt, G., Danier, J., Hofmann, T., and Daniel, H. (2009). Characterization of transport of selenoamino acids by epithelial amino acid transporters. *Chem. Biol. Interact.* **177**, 234–241.

Novoselov, S. V., Hua, D., Lobanov, A. V., and Gladyshev, V. N. (2006). Identification and characterization of Fep15, a new selenocysteine-containing member of the SEP15 protein family. *Biochem. J.* **394**, 575–579.

Nriagu, J. O. (1989). Global cycling of selenium. In: *Occurrence and Distribution of Selenium* (M. Inhat, ed.), pp. 327–339. CRC Press, Boca Raton, FL.

Nriagu, J. O., and Wong, H. K. (1983). Selenium pollution of lakes near the smelters at Sudbury, Ontario. *Nature* **301**, 55–57.

Oberdorster, E. (2004). Manufactured nanomaterials (fullerenes, C60) induce oxidative stress in the brain of juvenile largemouth bass. *Environ. Health Perspect.* **112**, 1058–1062.

Ohlendorf, H. M. (2003). Ecotoxicology of selenium. In: *Handbook of Ecotoxicology* (D.J. Hoffman, B.A. Rattner, G.A. Burton, Jr. and J. Cairns, Jr, eds), 2nd edn, pp. 465–500. CRC Press, Boca Raton, FL.

Ohta, Y., and Suzuki, K. T. (2008). Methylation and demethylation of intermediates selenide and methylselenol in the metabolism of selenium. *Toxicol. Appl. Pharmacol.* **226**, 169–177.

Ontario Ministry of Environment and Energy (1999). *Provincial Water Quality Objectives*. Queen's Printer, Toronto.

Orr, P. L., Guiguer, K. R., and Russell, C. K. (2006). Food chain transfer of selenium in lentic and lotic habitats of a western Canadian watershed. *Ecotoxicol. Environ. Saf.* **63**, 175–188.

Outridge, P. M., Scheuhammer, A. M., Fox, G. A., Braune, B. M., White, L. M., Gregorich, L. J., and Keddy, C. (1999). An assessment of the potential hazards of environmental selenium for Canadian water birds. *Environ. Rev.* **7**, 81–96.

Palace, V. P., Spallholz, J. E., Holm, J., Wautier, K., Evans, R. E., and Baron, C. L. (2004). Metabolism of selenomethionine by rainbow trout (*Oncorhynchus mykiss*) embryos can generate oxidative stress. *Ecotoxicol. Environ. Saf.* **58**, 17–21.

Palace, V. P., Halden, N. M., Yang, P., Evans, R. E., and Sterling, G. L. (2007). Determining residence patterns of rainbow trout using laser ablation inductively coupled plasma mass spectrometry (LA-ICP-MS) analysis of selenium in otoliths. *Environ. Sci. Technol.* **41**, 3679–3683.

Parizek, J., and Ostradalova, I. (1967). The protective effect of small amounts of selenium in sublimate intoxication. *Experientia* **23**, 142–144.

Paulsson, K., and Lundberg, K. (1989). The selenium method for treatment of lakes for elevated levels of mercury in fish. *Sci. Total Environ.* **87–88**, 495–507.

Pedersen, T. V., Block, M., and Part, P. (1998). Effect of selenium on the uptake of methyl mercury across perfused gills of rainbow trout *Oncorhynchus mykiss*. *Aquat. Toxicol.* **40**, 361–373.

Poston, H. A., Combs, G. F., Jr., and Leibovitz, L. (1976). Vitamin E and selenium interrelations in the diet of Atlantic salmon (*Salmo salar*): gross, histological and biochemical deficiency. *J. Nutr.* **106**, 892–904.

Presser, T. S., and Ohlendorf, H. M. (1987). Biogeochemical cycling of selenium in the San Joaquin Valley, California, USA. *Environ. Manag.* **11**, 805–821.

Presser, T. S., Piper, D. Z., Bird, K. J., Skorupa, J. P., Hamilton, S. J., Detwiler, S. J., and Huebner, M. A. (2004). The Phosphoria Formation: a model for forecasting global selenium sources to the environment. In: *Life Cycle of the Phosphoria Formation: From Deposition to Post-Mining Environment* (J.R. Hein, ed.), pp. 299–319. Elsevier, New York.

Pyron, M., and Beitinger, T. L. (1989). Effect of selenium on reproductive behavior and fry of fathead minnows. *Bull. Environ. Contam. Toxicol.* **42**, 609–613.

Ralston, N. V. C., Ralston, C. R., Blackwell, J. L., and Raymond, L. J. (2008). Dietary and tissue selenium in relation to methylmercury toxicity. *Neurotoxicology* **29**, 802–811.

Reeves, M. A., and Hoffmann, P. R. (2009). The human selenoproteome: recent insights into functions and regulation. *Cell. Mol. Life Sci.* **66**, 2457–2478.

Rotruck, J. T., Pope, A. L., Ganther, H., Swanson, A., Hafeman, D. G., and Hoekstra, W. G. (1973). Selenium: biochemical role as a component of glutathione peroxidase. *Science* **179**, 588–590.

Sambroni, E., Gutieres, S., Cauty, C., Guiguen, Y., Breton, B., and Lareyre, J. (2001). Type II iodothyronine deiodinase is preferentially expressed in rainbow trout (*Oncorhynchus mykiss*) liver and gonads. *Mol. Reprod. Dev.* **60**, 338–350.

Sanders, J. P., Van der Geyten, S., Kaptein, E., Darras, V. M., Kuhn, E. R., Leonard, J. L., and Visser, T. J. (1999). Cloning and characterization of Type III iodothyronine deiodinase from the fish *Oreochromis niloticus*. *Endocrinology* **140**, 3666–3673.

Sato, T., Ose, Y., and Sakai, T. (1980). Toxicological effect of selenium on fish. *Environ. Pollut.* **21**, 217–224.

Schomburg, L., Schweizer, U., Holtmann, B., Flohé, L., Sendtner, M., and Köhrle, J. (2003). Gene disruption discloses role of selenoprotein P in selenium delivery to target tissues. *Biochem. J.* **270**, 397–402.

Schrauzer, G. N. (2000). Selenomethionine: a review of its nutritional significance, metabolism and toxicity. *J. Nutr.* **130**, 1653–1656.

Schrauzer, G. N. (2009). Selenium and selenium-antagonistic elements in nutritional cancer prevention. *Crit. Rev. Biotechnol.* **29**, 10–17.

Schwarz, K., and Foltz, C. M. (1957). Selenium as an integral part of factor-3 against dietary necrotic liver degeneration. *J. Am. Chem. Soc.* **79**, 3292–3293.

Scott, G. R., and Sloman, K. A. (2004). The effects of environmental pollutants on complex fish behaviour: integrating behavioural and physiological indicators of toxicity. *Aquat. Toxicol.* **68**, 369–392.

Seiler, R. L., Skorupa, J. P., Naftz, D. L. and Nolan, B. T. (2003). *Irrigation-induced contamination of water, sediment, and biota in the western United States – synthesis of data from the National Irrigation Water Quality Program.* Report No. 1655. USGS Professional Paper. Menlo Park: US Geological Survey.

Shchedrina, V. A., Novoselov, S. V., Malinouski, M. Y., and Gladyshev, V. N. (2007). Identification and characterization of a selenoprotein family containing a diselenide bond in a redox motif. *Proc. Natl. Acad. Sci. U.S.A.* **104**, 13919–13924.

Shen, H. M., Yang, C. F., Liu, J., and Ong, C. N. (2000). Dual role of glutathione in selenite-induced oxidative stress and apoptosis in human hepatoma cells. *Free Radic. Biol. Med.* **28**, 1115–1124.

Simmons, D. B. D., and Wallschläger, D. (2005). A critical review of the biogeochemistry and ecotoxicology of selenium in lotic and lentic environments. *Environ. Toxicol. Chem.* **24**, 1331–1343.

Skorupa, J. P. (1998). Selenium poisoning of fish and wildlife in nature: lessons from twelve real-world experiences. In: *Environmental Chemistry of Selenium* (W.T. Frankenberger, Jr. and R.A. Engberg, eds), pp. 315–354. Marcel Dekker, New York.

Spallholz, J. E. (1994). On the nature of selenium toxicity and carcinostatic activity. *Free Radic. Biol. Med.* **17**, 45–64.

Spallholz, J. E., and Hoffman, D. J. (2002). Selenium toxicity: cause and effects in aquatic birds. *Aquat. Toxicol.* **57**, 27–37.

Spallholz, J. E., Palace, V. P., and Reid, T. W. (2004). Methioninase and selenomethionine but not Se-methylselenocysteine generate methyselenol and superoxide in an *in vitro* chemiluminescence assay: experiments and review. *Biochem. Pharmacol.* **67**, 547–554.

Spallholz, J. E., Shriver, B. J., and Reid, T. W. (2001). Dimethyldiselenide and methylseleninic acid generate superoxide in an *in vitro* chemiluminescence assay in the presence of glutathione: implications for the anticarcinogenic activity of ʟ-selenomethionine and ʟ-Se methylselenocysteine. *Nutr. Cancer* **40**, 34–41.

Stewart, A. R., Luoma, S. N., Schlekat, C. E., Doblin, M. A., and Hieb, K. A. (2004). Food web pathway determines how selenium affects aquatic ecosystems: a San Francisco Bay case study. *Environ. Sci. Technol.* **38**, 4519–4526.

Stewart, R., Grosell, M., Buchwalter, D., Fisher, N., Luoma, S. N., Mathews, T., Orr, P., and Wang., X. W. (2010). Bioaccumulation and trophic transfer of selenium. In: *Ecological Assessment of Selenium in the Aquatic Environment* (P.M. Chapman, W.J. Adams, M.L. Brooks, C.G. Delos, S.N. Luoma, W.A. Maher, H.M. Ohlendorf, T.S. Presser and D.P. Shaw, eds), pp. 93–139. CRC Press, Boca Raton, FL.

Suzuki, K. T., and Ogra, Y. (2002). Metabolic pathway for selenium in the body: speciation by HPLC-ICP MS with enriched Se. *Food Addit. Contam.* **19**, 974–983.

Suzuki, K. T., Kurasaki, K., Okazaki, N., and Ogra, Y. (2005). Selenosugar and trimethylselenonium among urinary Se metabolites: dose- and age-related changes. *Toxicol. Appl. Pharmacol.* **206**, 1–8.

Tamari, Y. (1998). Methods of analysis for the determination of selenium in biological, geological and water samples. In: *Environmental Chemistry of Selenium* (W.T. Frankenberger, Jr. and R.A. Engberg, eds), pp. 27–46. Marcel Dekker, New York.

Tennessee Valley Authority (2009). *Environmental assessment: initial emergency response actions for the Kingston fossil plant ash dike failure, Roane County, Tennessee.* February 2009. http://www.tva.gov/environment/reports/Kingston/pdf/2009-13_KIF_EmergencyResponse_ EA.pdf

Thisse, C., Degrave, A., Kryukov, V. N., Obrecht-Pflumio, S., Krol, A., Thisse, B., and Lescure, A. (2003). Spatial and temporal expression patterns of selenoprotein genes during embryogenesis in zebrafish. *Gene Expr. Patterns* **3**, 525–532.

Thomas, J. K., and Janz, D. M. (2011). Dietary Selenomethionine exposure in adult zebrafish alters swimming performance, energetics and the physiological stress response. *Aquat. Toxicol.* **102**, 79–86.

Tingaud-Sequeira, A., Chauvigne, F., Lozano, J., Agulleiro, M. J., Asensio, E., and Cerda, J. (2009). New insights into molecular pathways associated with flatfish ovarian development and atresia revealed by transcriptional analysis. *BMC Genom.* **10**, 434.

Tujebajeva, R., Ransom, D. G., Harney, J. W., and Berry, M. J. (2000). Expression and characterisation of nonmammalian selenoprotein P in the zebrafish, *Danio rerio*. *Genes Cells* **5**, 897–903.

USEPA (1987). *Ambient Water Quality Criteria for Selenium.* EPA-440/5-87-006. Office of Water, Office of Science and Technology, United States Environmental Protection Agency, Washington, DC.

USEPA (2004). *Draft Aquatic Life Water Quality Criteria for Selenium – 2004.* EPA-822-D-04-001. Office of Water, Office of Science and Technology, United States Environmental Protection Agency, Washington, DC.

Valverde, C., Croteau, W., Lafleur, G. J., Orozco, A., St., and Germain, D. L. (1997). Cloning and expression of a 5'-iodothyronine deiodinase from the liver of *Fundulus heteroclitus*. *Endocrinology* **138**, 642–648.

Vendeland, S. C., Deagen, J. T., and Whanger, P. D. (1992). Uptake of selenotrisulfides of glutathione and cysteine by brush border membranes from rat intestines. *J. Inorg. Biochem.* **47**, 131–140.

Walpita, C. N., Van der Geyten, S., Rurangwa, E., and Darras, V. M. (2007). The effect of 3,5,3'-triiodothyronine supplementation on zebrafish (*Danio rerio*) embryonic development and expression of iodothyronine deiodinases and thyroid hormone receptors. *Gen. Comp. Endocrinol.* **152**, 206–214.

Watenpaugh, D. E., and Beitinger, T. L. (1985). Absence of selenate avoidance by fathead minnows (*Pimephales promelas*). *Water Res.* **19**, 923–926.

Weiller, M., Latta, M., Kresse, M., Lucas, R., and Wendel, A. (2004). Toxicity of nutritionally available selenium compounds in primary and transformed hepatocytes. *Toxicology* **201**, 21–30.

Wells, P. G., McCallum, G. P., Chen, C. S., Henderson, J. T., Lee, C. J. J., and Perstin, J. (2009). Oxidative stress in developmental origins of disease: teratogenesis, neurodevelopmental deficits, and cancer. *Toxicol. Sci.* **108**, 4–18.

Wiramanaden, C. I. E., Forster, E. K., and Liber, K. (2010a). Selenium distribution in a lake system receiving effluent from a metal mining and milling operation in northern Saskatchewan, Canada. *Environ. Toxicol. Chem.* **29**, 606–616.

Wiramanaden, C. I. E., Liber, K., and Pickering, I. J. (2010b). Selenium speciation in whole-sediment using X-ray absorption spectroscopy and micro X-ray fluorescence imaging. *Environ. Sci. Technol.* **44**, 5389–5394.

Xu, Y., and Wang, W. X. (2002). Exposure and potential food chain transfer factor of Cd, Se and Zn in marine fish *Lutjanus argentimaculatus*. *Mar. Ecol. Progr. Ser.* **238**, 173–186.

Yang, D. Y., Chen, Y. W., Gunn, J. M., and Belzile, N. (2008). Selenium and mercury in organisms: interactions and mechanisms. *Environ. Rev.* **16**, 71–92.

Yoneda, S., and Suzuki, K. T. (1997). Equimolar Hg–Se complex binds to selenoprotein P. *Biochem. Biophys. Res. Commun.* **231**, 7–11.

Young, T. F., Finley, K., Adams, W. J., Besser, J., Hopkins, W. D., Jolley, D., McNaughton, E., Presser, T. S., Shaw, D. P., and Unrine, J. (2010a). What you need to know about selenium. In: *Ecological Assessment of Selenium in the Aquatic Environment* (P.M. Chapman, W.J. Adams, M.L. Brooks, C.G. Delos, S.N. Luoma, W.A. Maher, H.M. Ohlendorf, T.S. Presser and D.P. Shaw, eds), pp. 7–45. CRC Press, Boca Raton, FL.

Young, T. F., Finley, K., Adams, W. J., Besser, J., Hopkins, W. D., Jolley, D., McNaughton, E., Presser, T. S., Shaw, D. P., and Unrine, J. (2010b). Appendix A: Selected case studies of ecosystem contamination by Se. In: *Ecological Assessment of Selenium in the Aquatic Environment* (P.M. Chapman, W.J. Adams, M.L. Brooks, C.G. Delos, S.N. Luoma, W.A. Maher, H.M. Ohlendorf, T.S. Presser and D.P. Shaw, eds), pp. 257–292. CRC Press, Boca Raton, FL.

Yuan, T., Weljie, A. M., and Vogel, H. J. (1998). Tryptophan fluorescence quenching by methionine and selenomethionine residues of calmodulin: orientation of peptide and protein binding. *Biochemistry* **37**, 3187–3195.

Zhang, Y., and Gladyshev, V. N. (2010). General trends in trace element utilization revealed by comparative genomic analyses of Co, Cu, Mo, Ni, and Se. *J. Biol. Chem.* **285**, 3393–3405.

Zinoni, F., Birkman, A., Leinflder, W., and Bocj, A. (1987). Cotranslational insertion of selenocysteine into formate dehydrogenase from *Escherichia coli* directed by a UGA codon. *Proc. Natl. Acad. Sci. U.S.A.* **84**, 3156–3160.

8

MOLYBDENUM AND CHROMIUM

SCOTT D. REID

1. Chemical Speciation in Freshwater and Seawater
   1.1. Molybdenum
   1.2. Chromium
2. Sources (Natural and Anthropogenic) of Molybdenum and Chromium and Economic Importance
   2.1. Molybdenum
   2.2. Chromium
3. Environmental Situations of Concern
   3.1. Molybdenum
   3.2. Chromium
4. A Survey of Acute and Chronic Ambient Water Quality Criteria in Various Jurisdictions in Freshwater and Seawater
   4.1. Molybdenum
   4.2. Chromium
5. Mechanisms of Toxicity
   5.1. Acute Toxicity
   5.2. Chronic Toxicity
6. Essentiality or Non-Essentiality of Molybdenum and Chromium: Evidence For and Against
   6.1. Molybdenum
   6.2. Chromium
7. Potential for Bioconcentration and/or Biomagnification of Molybdenum and Chromium
   7.1. Molybdenum
   7.2. Chromium
8. Characterization of Uptake Routes
   8.1. Molybdenum
   8.2. Chromium
9. Characterization of Internal Handling
   9.1. Molybdenum
   9.2. Chromium
10. Characterization of Excretion Routes
    10.1. Molybdenum
    10.2. Chromium
11. Behavioral Effects of Molybdenum and Chromium
    11.1. Molybdenum
    11.2. Chromium

Homeostasis and Toxicology of Essential Metals: Volume 31A
FISH PHYSIOLOGY

Molybdenum (Mo) and chromium (Cr) both exist in elevated concentrations in aquatic environments primarily as oxyanions: molybdate (MoO_4^{2-}) and chromate (CrO_4^{2-}). These concentrations typically arise as a result of anthropogenic activity. Both metals are relatively non-toxic compared to other metals. Average 96 h LC50 estimates in freshwater are at least 1000 mg L^{-1} for Mo and approximately 100 mg L^{-1} for Cr. Both metals are micronutrients, but there is no evidence that their internal concentrations in animals are regulated in a homeostatic manner. Molybdenum is involved in purine metabolism, and Cr is involved in fat and glucose metabolism. Very little information exists about the physiological impact of elevated aquatic concentrations of these metals, especially Mo, in fish. Chromium is taken up by the gills and distributed via the blood to a variety of tissues but the mechanism of uptake across the gills is unknown. Chromium's toxic mechanism in fish is unknown, but histopathologies, particularly of the gastrointestinal tract and kidney, play a significant role in its toxicity. Molybdenum is also transported across the gills via an unknown mechanism and accumulates internally in the liver; other sites of accumulation have not been identified. The speculation on toxicity of Mo includes a non-specific gill irritation response, but this lacks experimental evidence. The general metabolism of these two metals in fish and the factors that influence their bioavailability need future investigation, and provide limitless opportunities for future research.

1. CHEMICAL SPECIATION IN FRESHWATER AND SEAWATER

1.1. Molybdenum

Molybdenum (Mo), a transition metal in group 6 of the periodic table, is located vertically between chromium (Cr) and tungsten and exists as a hard, silvery-white metallic element. Molybdenum has a molecular weight of 95.64, atomic number 42, specific gravity of 10.22 g cm^{-3} at 20°C, and a melting point of 2617°C. There are seven stable isotopes of Mo that range in natural abundance between 9 and 24% and cover the mass range of 92 to 100 (Moore et al., 1974). Molybdenum can be found in a variety of oxidation states (−II to VI) and coordination numbers (4 to 8) (Jarrell et al., 1980). In oxidation state V, Mo is less acidic than in oxidation state VI. However, Mo(V) oxide, Mo_2O_5, and hydroxide, $MoO(OH)_3$, are insoluble in neutral and alkaline solutions. Most natural waters exhibit conditions where the main species of Mo is Mo(VI) in the form of the molybdate ion (MoO_4^{2-}) (Jarrell et al., 1980). At all concentrations of Mo(VI) and a water pH greater than 6, only MoO_4^{2-} exists (Jarrell et al., 1980). At lower pH, MoO_4^{2-} can be protonated to form either $HMoO_4^-$ or H_2MoO_4 (Jarrell et al., 1980). Polymerization of Mo is possible but only at concentrations greater than 1 mmol L^{-1} (~ 100 mg L^{-1}) and between pH 5 and 6 (e.g. $Mo_7O_{24}^{6-}$) (Aveston et al., 1964). The hexavalent oxyanion, MoO_4^{2-}, is also the dominant species of Mo in seawater (Yamazaki and Gohda, 1990).

1.2. Chromium

Chromium, like Mo, is a transition metal in group 6. In crystalline form, Cr is a steel-gray, lustrous, hard metal characterized by an atomic weight of 51.996, an atomic number of 24, a density of 7.19 g cm^{-3}, a melting point of 1900°C, and a boiling point of 2642°C. Four Cr isotopes occur naturally, ^{50}Cr, ^{52}Cr, ^{53}Cr, and ^{54}Cr, which represent 4.3%, 83.8%, 9.6%, and 2.4% of total Cr, respectively. Elemental Cr is very stable, but is not usually found in the environment in a pure form. Chromium can exist in oxidation states ranging from −II to +VI, but is most frequently found in the environment in the trivalent (III) and hexavalent (VI) oxidation states. The IV and V forms are unstable and are rapidly converted to III, which in turn is oxidized to VI (Towill et al., 1978; Langard and Norseth, 1979). Both Cr (III) and Cr(VI) can exist in water with little organic matter, while Cr(VI) is usually the major species in seawater (Towill et al., 1978). Under oxygenated conditions, Cr(VI) is the dominant dissolved stable Cr species in aquatic systems. The hexavalent form exists as a component of a complex anion that varies with pH and may take the form of chromate (CrO_4^{2-}), hydrochromate ($HCrO_4^{1-}$), or dichromate ($Cr_2O_7^{2-}$). These ionic forms of Cr are highly soluble in water and thus mobile in the aquatic environment. All stable Cr(VI) anionic compounds strongly oxidize organic matter on contact and yield oxidized organic matter and Cr(III) (Ecological Analysts, 1981).

378   SCOTT D. REID

2. SOURCES (NATURAL AND ANTHROPOGENIC) OF MOLYBDENUM AND CHROMIUM AND ECONOMIC IMPORTANCE

2.1. Molybdenum

Molybdenum comprises nearly 0.00015% or 1.5 ppm of the Earth's crust (Chappell, 1975) and occurs in nature only in chemical combination with other elements. Several Mo-bearing minerals have been identified, such as molybdenite (MoS_2), powellite ($CaOMo_3$), and wulfenite ($PbMoO_4$) (Friberg et al., 1975; Chappel et al., 1979; Friberg and Lener, 1986; Goyer, 1986), but the only one of commercial significance is molybdenite. In ore bodies, molybdenite is generally present in quantities from 0.01 to 0.25% and is often associated with the sulfide minerals of other metals, notably copper. The 2008 annual world production of Mo was approximately 225,000 tonnes (t). The top Mo producers were China and Mongolia, the USA, Chile, and Canada, generating 36%, 25%, 15%, and 4%, respectively (Polayk, 2010). Global Mo reserves are approximately 19 million t, with China/Mongolia, the USA, Chile, and Canada possessing approximately 44%, 28%, 13%, and just under 5% of the total reserves, respectively. Molybdenum is used in a variety of applications. It is a component of alloys of steel, cast iron, and manganese and other alloys that are used in the manufacture of X-ray tubes, screens, radio grids, and spark plugs (Stokinger, 1981). Molybdenum compounds are present in lubricants, paints, rubbers, and fertilizers (NRC, 2004).

2.2. Chromium

Chromium comprises approximately 0.01% or 100 ppm of the Earth's crust (Emsley, 2001). The 2008 annual world production of Cr, as chromite ore ($FeOCr_2O_3$), was estimated at 21.5 million t in 2008 (Polayk, 2010). In that year, the dominant producing countries were South Africa, Kazakhstan, and India, which accounted for roughly 44%, 17%, and 15%, respectively. Chromium is utilized primarily in the production of metal alloys. Ferrochromium is an alloy used in the manufacture of stainless and heat-resistant steels used in petrochemical processing, turbines, and furnaces, as well as in consumer goods such as cutlery and decorative trim (Phillip, 1988). Copper–chromium alloys are used in electrical applications that require high strength and good conductivity, while copper–nickel–chromium alloys are used in marine equipment that require corrosion resistance (Nriagu, 1988). The automobile industry is a major user of Cr alloys, as are aircraft and aerospace industries (Nriagu, 1988).

3. ENVIRONMENTAL SITUATIONS OF CONCERN

3.1. Molybdenum

Natural sources of Mo in the aquatic environment include the weathering of ores from igneous and sedimentary rock and their subsequent runoff to streams and lakes (CCME, 1999). Natural Mo concentrations in ground and surface waters rarely exceed 0.02 mg L^{-1}, unless contaminated by human activity (Eisler, 1989). Important anthropogenic sources of Mo in the aquatic environment include Mo mining and milling (Phillips and Russo, 1978), Mo smelting (Nriagu and Pacyna, 1988), combustion of fossil fuels (Goyer, 1986), uranium and copper mining and milling (Chappell et al., 1979), oil refining and shale oil production (Eisler, 1989), and the runoff of fertilizers containing Mo (McNeely et al., 1979). Near mining sites in British Columbia, Jones (1999) reported total Mo ranging from 0.003 to 0.22 mg L^{-1} in background water and 0.005 to 11.4 mg L^{-1} at sites downstream of mine discharges. Similarly, Whiting et al. (1994) found mean and maximum Mo concentrations of 24.8 and 32.5 mg L^{-1}, respectively, in a stream receiving discharge from Mo mining operations. In two lakes receiving effluent from a uranium mine, Pyle et al. (2001) found water Mo concentrations of 1.40 and 0.74 mg L^{-1}, levels 3500 and 1850 times greater than in their upstream reference lake. Levels as high as 100 mg L^{-1} have been reported in irrigation water from Mo mining and reclamation (Smith et al., 1987).

3.2. Chromium

Although weathering processes result in the natural mobilization of Cr, the amounts of Cr added to the environment as a result of anthropogenic activities are thought to be far greater. For example, Steven et al. (1976) reported that New York City contributes over 350 t of Cr annually to the environment. Major atmospheric emissions of Cr are from the Cr alloy- and metal-producing industries, with lesser amounts resulting from coal combustion, municipal incinerators, and cement production (Towill et al., 1978). According to Ecological Analysts (1981), atmospheric emissions contribute four to six times more Cr to aquatic ecosystems than do liquid wastes. In aquatic environments, the major sources of Cr are the electroplating and metal finishing industries; relatively minor sources are iron and steel foundries, inorganic chemical plants, tanneries, textile manufacturers, and runoff from urban and residential areas (Towill et al., 1978; Ecological Analysts, 1981). In general, elevated levels of Cr in biological or other samples have been positively correlated with increased industrial and other uses of the element, especially uses associated with plating and foundry applications, chemical manufacturing, and corrosion inhibition (Taylor and Parr, 1978). For example, Pfeiffer et al. (1980) reported that the waste stream from a South American electroplating plant had a Cr concentration of almost 1290 mg L^{-1}, whereas uncontaminated freshwater streams, rivers, and lakes have reported Cr concentrations in the low microgram per liter range (Stevens et al., 1976; Towill et al., 1978; Langard and Norseth, 1979).

4. A SURVEY OF ACUTE AND CHRONIC AMBIENT WATER QUALITY CRITERIA IN VARIOUS JURISDICTIONS IN FRESHWATER AND SEAWATER

4.1. Molybdenum

Limited water quality criteria are available for Mo for the protection of aquatic life or for human consumption (Table 8.1). This is highlighted by the

Table 8.1
Water quality criteria for molybdenum

Category and criterion	Mo concentration (μg L^{-1})	Reference
Freshwater aquatic life protection		
Canada	<73	CCME (1999)
British Columbia	<1000: 30 day average; max. not to exceed 2000	Swain (1986)
Australia	Insufficient data	
USA	Not a USEPA priority pollutant	
Nevada	<19	Nevada Division of Environmental Protection (2008)
Marine aquatic life protection		
Canada	No recommended guidelines	
Australia	Insufficient data	
USA	Not a USEPA priority pollutant	
Drinking water		
Australia	<50	NHMRC and NRMMC (2004)
Canada	No maximum acceptable concentration	
British Columbia	<250	Swain (1986)
World Health Organization	<70	WHO (1996)

CCME: Canadian Council of Ministers of the Environment; USEPA: US Environmental Protection Agency; NHMRC: National Health and Medical Research Council; NRMMC: Natural Resource Management Ministerial Council.

fact that the US Environmental Protection Agency (EPA) does not consider Mo to be a priority pollutant. Although the US EPA and Australia list Mo, only Canada, British Columbia, and Nevada provide any water quality guidelines for the protection of freshwater aquatic life. There appear to be no similar guidelines for the protection of marine aquatic life, apparently because of a lack of information. Drinking water guidelines range from less than 50 μg L^{-1} (Australia) to less than 250 μg L^{-1} (British Columbia), with Canada having no maximum acceptable concentration.

4.2. Chromium

Numerous jurisdictions around the world have ambient water quality guidelines for Cr for marine or freshwater aquatic life or for human consumption (Table 8.2). The USA (and several states), Canada (and British Columbia), South Africa, and Australia/New Zealand have well-established water quality guidelines for Cr for the protection of freshwater aquatic life. These jurisdictions list Cr(III) and Cr(IV) separately or together (total Cr). The freshwater Cr(III) guidelines range from less than 9 to less than 100 μg L^{-1} with the exception of the water hardness-dependent guidelines of the EPA, which range from less than 2200 μg L^{-1} at a CaCO$_3$ of 50 mg L^{-1} to less than approximately 10,000 μg L^{-1} at a CaCO$_3$ of 200 mg L^{-1}. Chromium(IV) guidelines for the protection of freshwater life range from less than 0.29 μg L^{-1} to less than 25 μg L^{-1}. Guidelines for Cr for the protection of marine aquatic life tend to be higher than those for the protection of freshwater aquatic life; the Canadian Cr(IV) marine guidelines are 1.5 times higher, while in the USA the Cr(IV) marine guidelines are roughly 60 times higher. Limited marine guidelines for Cr(III) are available; in the case of the Canadian guidelines, the maximum concentration for the protection of marine life is approximately six times that of the freshwater guidelines. As for drinking water, the majority of jurisdictions have total Cr guidelines only. The predominant drinking water guideline is for total Cr less than 50 μg L^{-1}.

5. MECHANISMS OF TOXICITY

5.1. Acute Toxicity

5.1.1. MOLYBDENUM

Molybdenum is considered to be relatively non-toxic to freshwater fish, based on 96 h LC50 values ranging from 70 to greater than 10,000 mg L^{-1}

SCOTT D. REID

382

Table 8.2
Water quality criteria for chromium

Category and criterion	Cr concentration ($\mu g \, L^{-1}$)	Reference
Freshwater aquatic life protection		
USA	<0.29 Cr(VI) as 24 h average; not to exceed 21 Cr(VI) at any time	USEPA (1980)
Water hardness (CaCO$_3$ mg L^{-1})		
50	$<2,200$ Cr(III) at any time	USEPA (1980)
100	$<4,700$ Cr(III) at any time	USEPA (1980)
200	$<9,900$ Cr(III) at any time	USEPA (1980)
Colorado	<25 Cr(VI); <100 Cr(III)	Ecological Analysts (1981)
Florida		
Effluent discharges	<500 Cr(VI); <1000 total Cr	Ecological Analysts (1981)
Recovery waters	<50 total Cr	Ecological Analysts (1981)
Indiana		
Most waters	Not to exceed 0.1×96h LC50 of aquatic species	Ecological Analysts (1981)
Lake Michigan	<50 total Cr	Ecological Analysts (1981)
Canada and BC	<1.0 Cr(VI); <8.9 Cr(III)	Pawlisz et al. (1997)
South Africa	<14 Cr(VI); <24 Cr(III)	South Africa Water Quality Guide (1996)
Australia/New Zealand	<10 total Cr	ANZECC (1992)
Marine aquatic life protection		
USA	<18 Cr(VI) as 24 h average; not to exceed 1260 Cr(VI) at any time;	USEPA (1980)
	<50 Cr(VI) (4 day average; not to be exceeded more than once every 3 years)	USEPA (1986)
USA	Insufficient database for Cr(III) at this time, but presumably less stringent than Cr(VI)	USEPA (1980)
	<1100 Cr(III) (1 h average; not to be exceeded more than once every 3 years)	USEPA (1986)
California	<2 total Cr; 6 month median; <8 total Cr, daily maximum; <20 total Cr, instantaneous mix	Ecological Analysts (1981)
California	Waste discharges into marine waters <5 total Cr for 50% of measurements; <10 for 10% of measurements	Reish (1981)
Canada and BC	<1.5 Cr(VI); <56 Cr(III)	Pawlisz et al. (1997)
Australia	<50 total Cr	ANZECC (1992)
UK	<15 total dissolved Cr (annual arithmetic mean)	Water Research Centre (1990)
Drinking water		
USA	<50 Cr(VI); $<170,000$ Cr(III)	USEPA (1980)
California	<50 total Cr	Ecological Analysts (1981)

(Continued)

Table 8.2 (Continued)

Category and criterion	Cr concentration ($\mu g\ L^{-1}$)	Reference
Colorado	<50 Cr(VI); <50 Cr(III)	Ecological Analysts (1981)
Florida	<50 total Cr	Ecological Analysts (1981)
Brazil	<50 total Cr	Pfeiffer et al. (1980)
USSR	<600 total Cr	Pfeiffer et al. (1980)
New Zealand	<50 total Cr	New Zealand Ministry of Health (2005)

USEPA: US Environmental Protection Agency; ANZECC: Australian and New Zealand Environmental and Conservation Council.

Table 8.3
Survey of acute toxicity estimates for molybdenum

Mo(VI) 96 h LC50 (mg L^{-1})	Species	pH	CaCO$_3$	Reference
Freshwater species				
>2000	Oncorhynchus nerka	7.5	80	Reid (2002)
>1000	Oncorhynchus kisutch	7.6–7.8	42–211	Hamilton and Buhl (1990)
>1000	Oncorhynchus tshawytscha	7.6–7.8	42–211	Hamilton and Buhl (1990)
800–1320	Oncorhnychus mykiss	NR	25	McConnell (1977)
7340	Oncorhnychus mykiss	NR	NR	Bentley (1973)
>1000	Oncorhnychus mykiss	NR	NR	Pyle et al. (2001)
>2000	Catostomus commersoni	NR	NR	Pyle et al. (2001)
>10,000	Ictalurus punctatus	NR	NR	Bentley (1973)
7630	Pimephales promelas	NR	NR	Bentley (1973)
6790	Lepomis macrochirus	NR	NR	Bentley (1973)
1320	Lepomis macrochirus	NR	NR	Easterday and Miller (1963)
Seawater species				
2385	Cyprinodon variegatus	NR	NR	Knothe and Van Riper (1988)

Metal concentrations would normally be presented as $\mu g\ L^{-1}$. The toxicity of Mo has been presented in mg L^{-1} due to the concentrations required to reach the endpoint of the listed studies.

NR: not reported.

depending on species, size of fish or life stage, and test conditions (Table 8.3). Davies et al. (2005) published a summary table of studies reporting Mo toxicity for a variety of species tested in a range of conditions. Using only their 96 h LC50 data, the 12 studies listed yield a mean Mo toxicity estimate of 1,118 mg L^{-1}. This is a minimum average because nine of the studies report their Mo toxicity estimates as greater than a certain concentration.

Little work has been done on the physiological impact of acute Mo exposure in fish to provide a well-supported mechanism of toxicity. McConnell (1977) reported a number of histological changes to various tissues of rainbow trout (*Oncorhynchus mykiss*) exposed to Mo. In that study, trout that died following exposure to 1500 mg L^{-1} Mo were found to have fused gill lamellae, hemorrhaging of the gut, pyloric ceca, liver, and kidney, and livers that were pale in color. The author provided no Mo-specific mechanism to explain these findings.

The only study with clues to the toxicity mechanism of Mo in fish, to date, is that of Reid (2002). Reid exposed kokanee (*Oncorhynchus nerka*) to either 25 or 250 mg L^{-1} Mo (as $Na_2MoO_4.2H_2O$) for 7 days and found a dose-dependent increase in ventilation (opercular movements), a 60–70% increase in resting oxygen consumption at both Mo concentrations (Fig. 8.1), and an apparent stimulation in gill and/or skin mucus production with no change in plasma sodium concentrations. Considering the alterations in gill histology observed by McConnell (1977), Reid (2002) suggested that, like nickel (Pane et al., 2003, 2004) and aluminum in specific combinations of calcium and pH (Walker et al., 1991), Mo acts as a respiratory toxicant. Therefore, it is potentially a non-specific gill irritant. Other metals that act as gill irritants have been shown to cause an increase in gill mucus secretion and a clogging of interlamellar spaces (Muniz and Leivestad, 1980; Mallatt, 1985; Walker et al., 1988). Reid (2002) reported significant exercise-induced delayed mortality in kokanee exposed to 25 or 250 mg L^{-1} Mo for 7 days

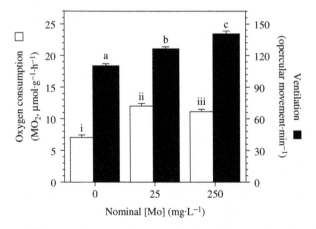

Fig. 8.1. Impact of Mo on resting oxygen consumption (open histograms) and ventilation (filled histograms) after 7 days' exposure to 1, 25, or 250 mg Mo L^{-1}. Values are means $\pm$ 1 SEM, $n = 5$ (MO_2) or 3 (Ventilation). Means with different numbers (MO_2) or letters (Ventilation) indicate significantly different means (p < 0.05). From Reid (2002).

before, during, and after a 30 min bout of forced exercise; kokanee mortality was not seen at these concentrations in the absence of exercise or with exercise in the absence of Mo. These findings further support the hypothesis that Mo exposure could contribute to fish death through a non-specific gill irritation response by impairing branchial gas exchange and internal acid–base status.

5.1.2. CHROMIUM

Pawlisz et al. (1997) compiled a comprehensive list of acute and chronic toxicity of Cr(VI) to fish. The authors report an average toxicity for freshwater fish independent of LC50 durations to be 112.9 ± 10.3 mg L^{-1} (mean ± 1 S.E., $N = 165$). Considering only the 96 h LC50 values from this list, the acute toxicity of Cr(VI) is 94.7 ± 14.2 mg L^{-1} ($N = 64$). Although fewer acute toxicity estimates are presented for marine fish, the authors report an average toxicity for marine fish over all exposure durations to be 56.2 ± 7.3 mg L^{-1} ($N = 34$) or 38.0 ± 4.4mg L^{-1} ($N = 15$) for the 96 h LC50 value. One study of interest is that of Van der Putte et al. (1981), which reported the percent survival of fingerling rainbow trout exposed for 4 days (96 h) to a range of Cr concentrations at pH 7.8 or 6.5 in freshwater. Using these data, their study yielded an estimated 96 h LC50 of 50.0 mg L^{-1} at pH 7.8 but an approximately 10-fold lower 96 h LC50 when pH was reduced to 6.5.

Although no specific toxic mechanism for Cr has been identified in fish, the acute symptoms of Cr exposure range from the non-specific gill irritation response to alterations in hematology and tissue histology to the specific inhibition of enzymes. The focus of Cr toxicity is on Cr(VI) and more specifically $HCrO_4^-$. Chromium(VI) was found to be more toxic at pH 6.5 than at pH 7.0 or pH 7.8 (Table 8.4). At pH 6.5, under conditions of greater Cr(VI) toxicity, the ratio of $HCrO_4^-$ to CrO_4^{2-} is 1:1 (Sillen and Martell, 1971; Stumm and Morgan, 1981; Stouthart et al., 1995). At pH 7.8, when Cr(VI) is of reduced toxicity, the ratio of $HCrO_4^-$ to CrO_4^{2-} is considerably lower. Based on survival data and changes in tissue histology and hematology at water pH ranging from 6.5 to 7.8, Van der Putte et al. (1981a, b) concluded that $HCrO_4^-$ was roughly nine times more acutely toxic to freshwater fish than CrO_4^{2-}.

Two of the more commonly reported symptoms related to acute Cr toxicity are changes in gill histology and hematology. For example, Al-Kahem (1995) exposed cichlids for 4 days to 3 or 6 mg L^{-1} Cr(III) as $CrCl_3$ and reported dose-dependent increases in hematocrit and mean cell hemoglobin. Changes in behavior, in particular coughing and yawning, suggested impaired gill oxygen diffusion. Respiratory distress was also shown in adult bluegills (*Lepomis macrochirus*). Gendusa and Beitinger

Table 8.4
Survey of acute toxicity estimates for chromium

Cr(VI) 96 h LC50 (mg L^{-1})	Species	pH	CaCO$_3$	Reference
Freshwater species				
28.5	*Oncorhynchus mykiss*	7–8.0	284	Svecevicius (2006)
69.0	*Oncorhynchus mykiss*	7–9	45	Benoit (1976)
6.5	*Oncorhynchus mykiss*	6.5	80.0	Van der Putte et al. (1981a)
14.6	*Oncorhynchus mykiss*	7.0	80	Van der Putte et al. (1981a)
23.5	*Oncorhynchus mykiss*	7.8	80	Van der Putte et al. (1981a)
50	*Oncorhynchus mykiss*	7.8	80	Van der Putte et al. (1981b)
5.4	*Oncorhynchus mykiss*	6.5	80	Van der Putte et al. (1981b)
33.1	*Perca fluviatilis*	7.0–8.0	284	Svecevicius (2006)
38.3	*Gasterosteus aculeatus*	7.0–8.0	284	Svecevicius (2006)
35.0	*Gasterosteus aculeatus*	8–8.1	500–600	Jop et al. (1987)
49.3	*Rutilus rutilus*	7.0–8.0	284	Svecevicius (2006)
71.7	*Leuciscus idus*	7.0–8.0	284	Svecevicius (2006)
36.2–36.9	*Pimephales promelas*	7.7	36.9	Pickering (1980)
41.75	*Channa punctatus*	7.3	215	Mishra and Mohanty (2009)
Seawater species				
30	*Citharichthys stigmaeus*	NR	NR	Mearns et al. (1976)

See Pawlisz et al. (1997) for a comprehensive list of acute and chronic toxicity of chromium (VI) to freshwater and marine fish.

Metal concentrations would normally be presented as µg L^{-1}. The toxicity of Cr has been presented in mg L^{-1} due to the concentrations required to reach the endpoint of the listed studies.

NR: not reported.

(1992) exposed bluegills to 0–282 mg L^{-1} Cr(IV) (as $K_2Cr_2O_7$) for 96 h and observed significantly greater rates of gill ventilation at Cr(IV) concentrations of 60 mg L^{-1} and greater. Similarly, a dose-dependent increase in hematocrit and decrease in plasma osmolarity at both pH 7.8 and 6.5 in rainbow trout exposed to hexavalent Cr for 96 h were reported by Van der Putte et al. (1981a). These authors also observed hypertrophy and hyperplasia of the gill lamellar epithelium and degenerative changes in the histology of the kidney and stomach. The internal histological changes were only observed at pH 7.8, while gill damage was observed at both pH levels. Similar results were reported by Gill and Pant (1987) in rosy barb (*Barbus conchonius*) following 30 or 60 days' exposure to 1.96 or 2.95 mg L^{-1} potassium dichromate at pH 7.1.

The acute toxicity of hexavalent Cr on fish may be partially explained by the inhibition of Na/K-ATPase. Kuhnert and Kuhnert (1976) looked at both Na/K- and Mg-ATPase activity in liver, gill, kidney, and intestine of rainbow trout exposed to 2.8 mg L^{-1} Cr for 2 days. All four tissues accumulated significant hexavalent Cr. Mg-ATPase activity was not influenced by the accumulated Cr while Na/K-ATPase activity was reduced 63% and 55% in the kidney and intestine, respectively; only the change in kidney enzyme activity was statistically significant. Liver and gill Na/K-ATPase activity was reported as unchanged. The inhibition of kidney and intestine Na/K-ATPase could partially explain the significant decrease in whole-body sodium content of larval carp exposed to 0.5 mg L^{-1} Cr at pH 6.3 (Stouthart et al., 1995); whole-body K, Mg and Ca were unaffected. Thaker et al. (1996) investigated the impact of Cr in these same tissues but on a greater variety of ATPases, in the mudskipper (Periophthalmus dipes). Fish were exposed to 5, 10, or 15 mg L^{-1} Cr(VI) for 2, 4, or 6 days. At low and intermediate concentrations and durations, the findings were variable but consistent with other studies. At 6 days' exposure to 15 mg L^{-1}, significant inhibitions of Na/K-, Ca-, Ca/HCO$_3$, Mg-, and Mg/HCO$_3$-ATPase activities in gill, kidney, and intestine tissues were reported.

The impairment of digestive tract Na/K-ATPase activity may be linked to changes in glucose transport that have been observed in trout exposed to Cr. Stokes and Fromm (1965) reported that mid-gut and pyloric ceca glucose transport and utilization were inhibited in rainbow trout exposed to 2.5 mg L^{-1} hexavalent Cr for 7 days. There was no impact on oxidative respiration and glycolysis, which led the authors to suggest that the major effect of Cr is inhibition of glucose entrance into the epithelial cells of the gastrointestinal tract. Whether or not this reduction in glucose uptake is linked to changes in Na/K-ATPase activity or even whether it contributes to acute Cr toxicity is yet to be demonstrated.

5.2. Chronic Toxicity

5.2.1. MOLYBDENUM

No studies to date provide information on the mechanism of chronic Mo toxicity.

5.2.2. CHROMIUM

Similar to acute Cr toxicity, no single mechanism or impairment has been shown to be responsible for chronic Cr toxicity in fish. The symptoms include changes in tissue histology beyond that seen during acute exposures, short-term and temporary reductions in growth, the production of reactive oxygen species (ROS), and impaired immune function.

Mishra and Mohanty (2009) exposed spotted snakehead (Channa punctata) for 1 and 2 month durations to 2 or 4 mg L^{-1} Cr(VI) as potassium dichromate ($K_2Cr_2O_7$) at pH 7.3. These authors reported changes in gill, liver, and kidney histology, plasma cortisol, and growth. The changes in organ structure were duration and concentration dependent. Reductions in plasma cortisol concentration and growth rate were most prevalent following 2 months' exposure at the higher Cr(VI) concentration.

Chronic toxicity of Cr(VI) in fathead minnows (*Pimephales promelas*) was assessed by monitoring survival, growth, and reproduction in two generations of fish exposed to 0.018–3.95 mg L^{-1} Cr for 412 days (first generation) or 60 days (second generation) (Pickering, 1980). After 9 weeks, growth of the first generation fish was significantly reduced at all Cr concentrations. However, this effect was not persistent. At the end of the 412 day exposure, growth (length and weight) of the Cr exposed fish was not different from controls. Growth in the second generation fish was not affected at concentrations of 1.0 mg L^{-1} or lower after 60 days of exposure. Reproduction and hatchability of eggs were not affected at any Cr concentration tested.

Farag et al. (2006) exposed chinook salmon parr to either 24 or 54 μg L^{-1} Cr(VI) for 105 days and found no effect on either growth or survival. Gill and kidney accumulated significant Cr only when fish were exposed to 54 μg L^{-1}. After the 105 day exposure, Cr(VI) concentrations were increased from 24 to 120 or from 54 to 266 μg L^{-1} for the remainder of the 134 day experiment. During this period, parr exposed to 120 μg L^{-1} experienced significant reductions in growth. Those exposed to 266 μg L^{-1} showed no signs of decreased growth but experienced significantly increased mortality. Erythrocyte nuclei experienced DNA damage at all Cr exposure durations and concentrations. At both 120 and 266 μg L^{-1} all tissues measured (gill, kidney, liver, pyloric ceca) had significantly elevated Cr levels. The kidneys of parr exposed to the highest concentrations of Cr had gross and microscopic damage and elevated products of lipid peroxidation. Chromium-induced changes to other tissues were much more subtle than the impact of Cr on the kidney. The authors concluded that Cr accumulated in the kidney, stimulating lipid peroxidation and the formation of hydroxyl radicals, which in turn caused oxidative DNA damage. The hypothesis was that this would manifest itself as microscopic and gross kidney damage resulting in reduced parr growth and survival.

Krumschnabel and Nawaz (2004) investigated the acute toxicity of hexavalent Cr in isolated hepatocytes of goldfish (*Carassius auratus*) and reported that exposure to 13 mg L^{-1} Cr(IV) significantly reduced cell viability and stimulated an increased production of ROS. A rapid, yet transient elevation in cellular oxygen consumption was observed without a change in cellular Ca^{2+} homeostasis or lysosomal membrane stability. The authors were unable to identify the cellular source of the ROS.

Steinhagen et al. (2004) examined the effect of Cr(IV) on cultured lymphocytes taken from carp blood and head kidney. *In vitro* exposure to chromate induced cytotoxicity (cell death) and decreased lymphocyte responsiveness and phagocytic activity at concentrations between 0.1 and 10.4 mg L^{-1}. Further, neutrophils were found to have altered shape and reduced nitric oxide and reactive oxygen production. The authors suggested that the altered lymphocyte and neutrophil functions are likely responsible for decreased resistance to pathogens in fish chronically exposed to elevated Cr.

Chromium may be a potential mutagen in fish, which would contribute to its chronic toxicity as it does in humans (Ding et al., 2000; Gibb et al., 2000; Patiolla et al., 2008). Abbas and Ali (2007) exposed tilapia (*Oreochromis* sp.) to 43.7 mg L^{-1} Cr(VI) for 24 and 96 h. Although the exposure was of relatively short duration, significant DNA damage and the appearance of polymorphic DNA bands not visible in controls were found in brain and liver. At 96 h of exposure, liver DNA showed greater signs of damage than brain DNA.

The lack of studies focused on the same or similar components of the chronic toxicity of Cr to fish makes it difficult to impossible to identify a single cause of mortality or to weigh the various symptoms and to understand how the symptoms may be linked.

6. ESSENTIALITY OR NON-ESSENTIALITY OF MOLYBDENUM AND CHROMIUM: EVIDENCE FOR AND AGAINST

6.1. Molybdenum

Molybdenum is an essential trace element in animals, but most of what is known about its normal biological role comes from non-fish research, primarily bacterial and mammalian studies. There is no reason to suspect that the role is different for fish given the general importance of the identified biological roles in other organisms.

Molybdenum is a cofactor of at least seven enzymes and a component of the iron–Mo flavoprotein xanthine oxidase (Beers and Berkow, 1998). Xanthine oxidase was the first enzyme identified as a molybdoenzyme and exists *in vivo* mainly as xanthine dehydrogenase. Xanthine oxidase/ dehydrogenase is involved in the oxidation of purine and pyrimidines, as well as other nitrogen-containing heterocyclic compounds (Stirpe and Della Corte, 1969). More specifically, xanthine oxidase is responsible for catalyzing the conversion of hypoxanthine (from adenine) to xanthine and xanthine to urate in the liver (Kurosaki et al., 1995; Kisker et al., 1997).

Two other important Mo-containing enzymes are aldehyde oxidase and sulfite oxidase. Aldehyde oxidase is primarily found in the liver and acts on substrates similar to those of xanthine oxidase/dehydrogenase. Both of these enzymes appear to play a role in the ability to detoxify xenobiotics, some drugs, estradiol, and progesterone (Rajagopalan and Handler, 1964; Cates et al., 1980; Kisker et al., 1997). Sulfite oxidase is found in the mitochondrial intermembrane space, oxidizes sulfite to sulfate, and is the terminal step in the metabolism of sulfur-containing amino acids (Irreverre et al., 1967; Karakas et al., 2005).

6.2. Chromium

In carp, Cr salts have been shown to improve glucose utilization and inhibit gluconeogenesis, likely by modulating endogenous insulin activity (Hertz et al. 1989). The effect of Cr on insulin action may be either by insulin stabilization and degradation or by increasing the affinity of insulin for its membrane receptors (Hertz et al. 1989). Shiau and Lin (1993) have shown that supplemental dietary Cr increases weight gain, energy deposition, and liver glycogen content in tilapia fed high glucose diets.

Chromium is considered to be essential for normal carbohydrate and lipid metabolism based on its nutritional role in mammals and birds. Chromium is thought to be a cofactor for insulin activity (Anderson, 1981). Chromium chloride enhances glucose tolerance, increases the rate of lipogenesis, and affects glycogen accumulation in the presence of insulin (Glinsmann and Mertz, 1966; Rosebrough and Steele, 1981; Steele and Rosebrough, 1981). How Cr serves as a cofactor for insulin action in mammals has not been fully elucidated. It appears that Cr functions at a molecular level through a low molecular weight (~ 1500 Da) Cr-binding substance or chromodulin (see reviews by Vincent, 2000b; Cefalu and Hu, 2004). Chromodulin in the absence of Cr is inactive, while the chromodulin–chromium complex is thought to bind to insulin-activated insulin receptors and stimulate tyrosine kinase (Vincent, 2000a). Although chromodulin is widely distributed in mammalian tissues (Vincent, 2000b), there have been no reports of efforts to isolate this compound from animals other than mammals.

7. POTENTIAL FOR BIOCONCENTRATION AND/OR BIOMAGNIFICATION OF MOLYBDENUM AND CHROMIUM

7.1. Molybdenum

Information about the potential bioconcentration of Mo is scarce and exposure concentration dependent; bioconcentration was apparent at low environmental concentrations but not apparent at high environmental concentrations. Ward (1973) reported the Mo concentrations in a variety of tissues in rainbow trout from waters containing a range of Mo concentrations. Rainbow trout from freshwater containing less than 6 μg L^{-1} Mo had tissues that contained on average 20-fold higher Mo concentrations: a high of 80 times the water Mo concentration in liver to a low of 1.7 times the environmental concentration in muscle. However, rainbow trout taken from waters containing 300 μg L^{-1} Mo had tissue concentrations less than the environmental concentration, ranging from 1 to 0.03 times the water Mo concentration. In characterizing the gill and liver Mo uptake kinetics in kokanee, similar findings were reported by Short et al. (1971). These authors reported a maximum bioconcentration factor (BCF) of 1143 in liver and the gastrointestinal tract of steelhead (*Oncorhynchus mykiss*) after chronic exposure to 0.014 μg L^{-1} Mo. However, when steelhead were exposed to 3300 μg L^{-1} Mo for 24 days, the maximum BCF for liver was 250 times lower, at only 4.5, while the maximum BCF for the gastrointestinal tract was 0.6. Reid (2002) showed that gill tissue appeared to concentrate waterborne Mo at exposure concentrations less than 25 μg L^{-1}; at higher concentrations, tissue levels failed to exceed the exposure concentration. Liver Mo did not exceed the exposure concentration of Mo at any concentration. There are two studies that suggest that Mo is not bioconcentrated in fish (Tong et al., 1974; Woodward et al., 1985), but both are field studies in which fish were exposed to metal mixtures and relatively high Mo concentrations. It is possible that, as a micronutrient, Mo is bioconcentrated at normal environmental concentrations but that the capacity to accumulate and store Mo by fish plateaus at similarly low concentrations. If this is the situation, then exposure of fish to Mo-contaminated waters would tend to yield data consistent with a lack of bioconcentration of this metal. Clearly, a more complete understanding of the uptake and internal handling of Mo in fish is needed. The mammalian literature is limited but to date this literature shows no bioaccumulation of Mo (Arrington and Davis, 1955; Schroeder et al., 1970; Rosoff and Spencer, 1973; Bibr et al., 1977). In mammals, when exposure to Mo is withdrawn, tissue concentrations quickly return to normal levels.

Although very little information appears in the literature, the limited studies available indicate that Mo is not biomagnified. Saiki et al. (1993) analyzed the Mo content in an aquatic food chain in the lower San Joaquin River and its tributaries in California, USA. They concluded that Mo was not biomagnified in this food chain based on the observation that concentrations of Mo were usually higher in filamentous algae and detritus than in invertebrates and fishes. Other technical reports (Hatfield Consultants, 1992, 1993; Dillon et al., 1995) have come to similar conclusions about the lack of Mo biomagnification.

7.2. Chromium

From a limited number of bioaccumulation studies of Cr it appears that there is little or no bioconcentration of this metal in fish at environmentally relevant concentrations. Fromm and Stokes (1962) exposed rainbow trout to 0.0013 or 0.01 mg L^{-1} Cr for 30 days and reported whole-body BCFs just over 1 (1.03–1.34). Calamari et al. (1982) reported similar findings after exposing rainbow trout to higher Cr concentrations (0.2 mg L^{-1}) for a longer period (6 months). These authors did find higher BCF values for liver and kidney, suggesting that Cr can be accumulated in these tissues, but the highest BCF was only about 3. However, higher Cr BCF values have been reported in the literature. Palaniappan and Karthikeyan (2009) exposed Mrigal (*Cirrhinus mrigala*) to either 1.82 or 6.07 mg L^{-1} Cr for 28 days and determined Cr accumulation for gill, liver, kidney, muscle, and whole body. Using these data, tissue-specific BCF values ranged from 4 to 44, with liver having the greatest BCF value (44) at the lower Cr exposure concentration and kidney having the greatest BCF value (21) at the higher Cr exposure concentration. Whole-body BCF values were 61 and 32 at 1.82 and 6.07 mg L^{-1} Cr, respectively. The study by Palaniappan and Karthikeyan (2009) was conducted at Cr concentrations 200–600 times greater than those used by Fromm and Stokes (1962) and 10–30 times greater than those of Calamari et al. (1982). This suggests that Cr may be bioaccumulated by fish, but only at extremely high exposure concentrations.

According to the existing literature, no significant biomagnification of Cr in the aquatic food web has been reported. Seenayya and Prahalad (1987) measured the Cr content of sediment, planktons, and fish in an industrially contaminated lake in India. They reported that the concentration of Cr was least in nanoplankton and remained the same in samples of phytoplankton, zooplankton, and fish. This trend did not illustrate the expected pattern of increasing concentrations through the various trophic levels consistent with biomagnification. Most of the Cr was found associated with suspended particulate matter. This was likely a reflection of the fact that any anthropogenically introduced Cr(VI) in freshwater is removed by reduction to Cr(III) and its subsequent sorption to particulates and sediments (Pfeiffer et al., 1980). Certain studies (Lowmann et al., 1971; Whittle et al., 1977; Holdway, 1988) have reported an inverse relationship between Cr residue concentration and trophic level, supporting the notion that there is no biomagnification of waterborne or sediment-borne Cr. The lack of any biomagification of Cr is likely due to the already stated conversion of Cr(VI) to Cr(III) and the virtual indigestibility of Cr(III). Chromium(III) has such limited digestibility that 1% chromic oxide is often added to experimental diets as a digestibility/absorption control in the study of gastrointestinal tract function in fish (Austreng, 1978; Hajen et al., 1993; Gobas et al., 1999).

8. CHARACTERIZATION OF UPTAKE ROUTES

8.1. Molybdenum

8.1.1. GILLS

Reid (2002) is the only study showing that the gills of freshwater fish accumulate waterborne Mo. Kokanee were exposed for 3 days to 5, 10, 25, 100, or 250 mg L^{-1} Mo. The gills of kokanee accumulated Mo at all concentrations and did so in a dose-dependent manner. Gill uptake of Mo appeared to be saturable, but no details about the specific mechanism of transport could be discerned from this study.

8.1.2. GUT

No studies to date have determined whether or not, or to what extent, Mo may be taken up by the gut. Reid (2002) reported that kokanee accumulated Mo internally, in the liver. As the fish were not fed in this portion of the study, it was assumed that Mo was transported via the gills rather than the gut. Dietary studies in which fish are fed Mo-enriched feed have not been attempted, thus the role of the gut in the uptake of Mo is yet to be defined.

Molybdenum absorption by the intestines of mammals appears to be very efficient. Turnlund et al. (1995) reported an intestinal absorption of an oral dose of Mo in humans of just over 90%. Intestinal absorption of Mo in other mammals appears to be similar (Mills and Davis, 1987). The mechanism responsible for intestinal absorption has not yet been identified, but Turnlund et al. (1995) stated that their data indicated that Mo absorption was passive and not saturable.

8.1.3. OTHER ROUTES

No studies to date have determined whether or not, or to what extent, Mo may be taken up by routes other than the gills.

8.2. Chromium

8.2.1. GILLS

In the studies of Van der Putte et al. (1981b) and Knoll and Fromm (1960), the Cr content of fish blood never exceeded the concentration of the exposure water and was usually considerably lower. Van der Putte et al. (1981b) suggested that this provided a diffusion gradient for the passive entry of Cr(VI) into fish via the gills. Fromm and Stokes (1962) came to the same conclusion based on the fact that whole-body Cr levels in trout were proportional to the exposure concentration. Tissue distribution of Cr in esophageal-occluded fish was not found to be different than that in non-occluded fish exposed to Cr(VI) (Knoll and Fromm, 1960), further highlighting the importance of the gill in the uptake of waterborne Cr. Van der Putte et al. (1981a) suggested that the gill represents the primary site of uptake for Cr based on histological damage to the gill lamellar epithelium observed following 4 days' exposure to 44.9 mg L^{-1} at pH 7.8 or 13.1 mg L^{-1} at pH 6.5.

8.2.2. GUT

The gut is an unlikely site of uptake of waterborne Cr in fish. Knoll and Fromm (1960) administered 0.1 ml of 2.5 mg L^{-1} ^{51}Cr directly into the stomach and found that the radiochromate failed to be absorbed in significant amounts within 24 h. These authors also placed a small barbed cork in the esophagus of fish to prevent entry of water into the gastrointestinal tract. The elimination of the gut as a potential site of Cr uptake had no impact on the distribution of tissue Cr in exposed fish. These findings would tend to minimize the importance of waterborne Cr uptake via the gut of fish.

Dowling et al. (1989) demonstrated the passive absorption of inorganic Cr(III) in double-perfused rat small intestine and suggested that the metal might pass either intracellularly through aqueous channels and/or paracellularly through tight junctions. They did not preclude the possibility that Cr, under certain circumstances, may be absorbed by carrier-mediated transport.

8.2.3. OTHER ROUTES

Knoll and Fromm (1960) discounted the skin as a possible site of Cr uptake based on a consistently low concentration of Cr found in muscle samples taken adjacent to the skin. The covering layer of mucus may contribute to the lack of Cr uptake by the skin. Arillo and Melodia (1990) have shown that fish skin mucus contains protein-bound sulfhydryl groups that are capable of non-enzymatically reducing membrane-permeable Cr(VI) to membrane-impermeable Cr(III).

9. CHARACTERIZATION OF INTERNAL HANDLING

9.1. Molybdenum

9.1.1. BIOTRANSFORMATION

There is no direct information on potential biotransformation in fish. However, Ricketts (2009) found no induction of metallothionein in a variety of rainbow trout tissues following Mo exposure. These findings suggest that Mo remains in the form of the oxyanion molybdate (MoO_4^{2-}) rather than being transformed into some cationic form of Mo that could interact with or stimulate the synthesis of metallothionein. In mammals it is assumed that Mo remains as molybdate, although there are no studies that directly address this question.

9.1.2. TRANSPORT THROUGH THE BLOODSTREAM

Reid (2002) showed that Mo moved from the gills to the liver via the bloodstream. The mechanism of transport and speciation of the metal in the bloodstream is not yet known, although Mo likely remains in the blood as molybdate. It was found that all the Mo in the blood of exposed sheep was readily dialyzable and in the form of molybdate (Dick, 1953).

9.1.3. ACCUMULATION IN SPECIFIC ORGANS

To date, only two fish tissues have been identified as being sites of accumulated Mo, the liver and the gills. Reid (2002) showed that the liver of kokanee accumulated Mo in a dose-dependent and saturable manner and that it did so to a greater extent than the gills. Accumulation in other organs has yet to be determined. In other animals, Mo has been found to accumulate in a wide variety of organs and tissues. For example, rats given an oral dose of 50 mg kg^{-1} body weight sodium molybdate per day for 8 days were found to have significantly elevated Mo concentrations in the brain, lung, heart, liver, spleen, kidney, testis, epididymis, seminal vesicles, ventral prostate, blood, erythrocytes, serum, and plasma when compared to controls (Pandey et al., 2002). Similar findings have been reported in sheep fed a diet supplemented with Mo (Pott et al., 1999), although this study was limited to investigating tissue Mo accumulation in liver, kidney, muscle, and serum.

9.1.4. SUBCELLULAR PARTITIONING OF THE METAL

No studies to date have determined the subcellular partitioning of Mo in fish. Anan et al. (2002) investigated the subcellular distribution of eight trace elements, including Mo, in the liver of green sea turtles (*Chelonia mydas*) and hawkbill turtles (*Eretmochelys imbricata*). These authors reported that unlike the majority of metals, which were largely present in the cytosol of the liver, Mo was accumulated specifically within the nuclei and mitochondria in both species. A similar subcellular distribution of Mo in sheep was reported by Allen and Gawthorne (1986).

9.1.5. DETOXIFICATION AND STORAGE MECHANISMS

Although based on a single study, what is currently known is that metallothionein (MT) is likely not involved in the internal detoxification of Mo. Ricketts (2009) exposed rainbow trout for 8 and 96 h to Mo concentrations ranging from 2 to 1000 mg L^{-1} and measured MT in gill, heart, liver, and erythrocytes. Molybdenum failed to induce the synthesis of MT in any tissue at any concentration or exposure duration.

9.1.6. HOMEOSTATIC CONTROLS

Despite the fact that this is an essential micronutrient, there is no known homeostatic control system for Mo in fish or any other animal.

9.2. Chromium

9.2.1. BIOTRANSFORMATION

Biotransformation of Cr in fish has yet to be elucidated.

9.2.2. TRANSPORT THROUGH THE BLOODSTREAM

In human plasma, Cr is bound to proteins. Cornelis et al. (1992), using ^{51}Cr, found that 85% of the plasma Cr was bound to transferrin, 8% to albumin, and 6% to what the authors described as other components. In fish, Knoll and Fromm (1960) suggested that Cr is not bound or accumulated by erythrocytes but is transported in plasma and diffuses readily to a variety of organs.

9.2.3. ACCUMULATION IN SPECIFIC ORGANS

Hexavalent Cr has been found to accumulate in a variety of tissues, in a variety of species, including freshwater and marine fish exposed to a wide range of combinations of alkalinity, pH, and concentration. Based on the frequency of reporting, the dominant organs of Cr(VI) accumulation appear to be gill (Kuhnert and Kuhnert, 1976; Van der Putte et al., 1981b; Farag et al., 2006; Ghosh and Adhikari, 2006; Svecevicius, 2007b; Mishra and Hohanty, 2009; Palaniappan and Karthikeyan, 2009), liver (Schiffman and Fromm, 1959; Knoll and Fromm, 1960; Kuhnert and Kuhnert, 1976; Sherwood and Wright, 1976; Van der Putte et al., 1981b; Farag et al., 2006; Svecevicius, 2007b; Mishra and Hohanty, 2009; Palaniappan and Karthi-keyan, 2009), and kidney (Schiffman and Fromm, 1959; Knoll and Fromm, 1960; Kuhnert and Kuhnert, 1976; Van der Putte et al., 1981b; Farag et al., 2006; Mishra and Mohanty, 2009; Palaniappan and Karthikeyan, 2009). Other tissues that have been reported to accumulate Cr(VI) include muscle (Schiffman and Fromm, 1959; Knoll and Fromm, 1960; Sherwood and

Wright, 1976; Ghosh and Adhikari, 2006; Palaniappan and Karthikeyan, 2009), the digestive tract (Knoll and Fromm, 1960; Kuhnert and Kuhnert, 1976; Sherwood and Wright, 1976; Van der Putte et al., 1981b), spleen (Schiffman and Fromm, 1959; Knoll and Fromm, 1960), gall bladder/bile (Schiffman and Fromm, 1959; Svecevicius, 2007b), skin (Sherwood and Wright, 1976), and the pyloric ceca (Kuhnert and Kuhnert, 1976; Farag et al., 2006). Palaniappan and Karthikeyan (2009) analyzed the distribution of Cr(VI) among tissues following exposure of Mrigal carp to either 1.82 or 6.07 mg L^{-1} Cr(VI) and found a concentration-dependent change in the rank order of Cr accumulation. At 1.82 mg L^{-1} Cr(VI), the rank order, from highest to lowest Cr(VI) accumulation, was found to be kidney > liver > gill = muscle. At 6.07 mg L^{-1} Cr(VI), the muscle was found to accumulate much more Cr, so that the rank order was reported as kidney > muscle > liver > gill.

The only study to look at the organ-specific accumulation of trivalent Cr was that by Sherwood and Wright (1976). These authors exposed a marine flatfish, the speckled sanddab (*Citharichthys stigmaeus*), chronically (44 days) to precipitates of trivalent Cr and found that Cr(III) accumulated in the skin, muscle, and liver of the sanddab. The sites of accumulation of Cr(III) were similar to the distribution of Cr(VI) in these animals, with the exception that no Cr(III) was found to have accumulated in the digestive tract.

9.2.4. SUBCELLULAR PARTITIONING OF THE METAL

Van der Putte et al. (1981b) exposed yearling rainbow trout to 40 mg L^{-1} Cr(VI) as Na$_2$CrO$_4$ for 2 or 4 days at pH 7.8 or 6.5. They examined a number of tissues and found that Cr was not distributed evenly among the nuclear, mitochondrial, microsomal, and soluble subcellular fractions. Chromium was found to be concentrated in the nuclear fractions of gill tissue and the soluble fractions of liver and kidney tissue. They also reported differences in subcellular accumulation at day 4 between fish exposed at the two pH levels. At pH 6.5, a greater proportion of the Cr was found in the nuclear fraction of gill, the microsomal fraction of liver, and the soluble fraction of kidney. The distribution in the gill mitochondrial, microsomal, and soluble fraction was significantly lower at pH 6.5 than at pH 7.8.

9.2.5. DETOXIFICATION AND STORAGE MECHANISMS

No studies to date have investigated the detoxification or storage mechanisms for Cr in marine or freshwater species of fish. However, MT induction by Cr has been reported in the liver of mice, rats, and chicken (Fleet et al., 1990; Ohta et al., 1993; Solis-Heredia et al., 2000) and in the pancreas of rats (Solis-Heredia et al., 2000). In pancreas, it appears that the

Cr induction of MT was a direct response to the metal. The authors found that the induced MT contained Cr and pancreas Zn levels were unaltered. In the liver, MT induction appears to be sensitive to changes in liver Zn caused by Cr administration rather than from tissue Cr levels directly. Solis-Heredia et al. (2000) noted increased hepatic MT levels at those Cr doses that resulted in significant elevations in hepatic Zn, while Ohta et al. (1993) reported that Zn-MT was the main isoform induced in liver by Cr exposure.

9.2.6. HOMEOSTATIC CONTROLS

Only one study provides insight into whether or not Cr levels in fish are homeostatically controlled. Sherwood and Wright (1976), investigating Cr uptake in speckled sanddab, reported that uptake of Cr was concentration dependent and failed to reach an equilibrium during the 44 day exposures. The authors suggest that their findings indicate that uptake is an unregulated process. No other reports for or against the homeostatic control of Cr uptake appear in the literature.

10. CHARACTERIZATION OF EXCRETION ROUTES

10.1. Molybdenum

There is no published information on the routes by which Mo is excreted in fish. In humans and other animals, the primary route of excretion of Mo has been shown to be via the kidney in the form of urine. There have been two studies in which human subjects were administered isotopes of Mo to determine rates and routes of excretion. Rosoff et al. (1964) injected humans with a single dose of ^{99}Mo. The Mo radioisotope rapidly disappeared from the blood; 2.5–5% of the original dose remained after 1 h and less than 0.5% after 24 h. The main pathway of Mo excretion was found to be via the kidney. Fecal excretion was found to be less than 5% while the urinary excretion was responsible for the remaining clearance of Mo from the body. More recently, Turnlund et al. (1995) obtained similar findings with orally administered stable isotopes of Mo in humans. Fecal excretion was found to be 7–10% depending on the dose administered. Urinary excretion was the dominant pathway in the removal of the orally administered Mo and the percentage of the dose cleared was directly proportional to the amount absorbed. At the low dietary dose, roughly 18% of the Mo fed was excreted in the urine during the first 6 days after ingestion. At the highest Mo intake, the percentage excreted by this route averaged 82%. The kidney appears to be the primary route of Mo excretion in mice, rats, and swine (Rosoff and Spencer, 1973). In contrast, ruminants appear to eliminate Mo via the intestine (Mills and Davis, 1987).

10.2. Chromium

10.2.1. GILLS

Retention of Cr in the gills of rainbow trout was studied by Van der Putte et al. (1981b). The authors exposed trout for 2–4 days to 2.0 mg L^{-1} hexavalent Cr ($^{51}CrO_4^{2-}$-containing Na_2CrO_4) at either pH 6.5 or 7.8 and then transferred the fish to Cr-free water. At pH 6.5, 93% of the accumulated gill tissue Cr(VI) was cleared in the first 3 days. The elimination of gill Cr(VI) was only 12% at the higher exposure pH in the same period. The difference in percent elimination was a reflection of the fact that the gill tissue accumulated approximately 10-fold more Cr(VI) at pH 6.5 than at pH 7.8. Under both experimental conditions, the authors postulated that the elimination of gill Cr was due to passive loss from the gills rather than any active gill excretion mechanism. The greater elimination at lower pH was likely caused by the larger diffusion gradient between the tissue and the Cr-free water.

10.2.2. KIDNEY

Van der Putte et al. (1981b) and Knoll and Fromm (1960) (Fig. 8.2) reported limited elimination of accumulated Cr from the kidney. Following a 25 day elimination period, Knoll and Fromm observed a 50% reduction in kidney Cr burden, which was a considerably lower elimination rate than for other tissues such as gut, stomach, pyloric ceca, and liver. Van der Putte et al. (1981b) reported an elevation in kidney Cr concentration during an 11 day elimination period which they thought was a reflection of changes in the internal distribution of accumulated Cr. These findings bring into question the role of the kidney as an organ for Cr excretion. However, together with the fact that there are no studies addressing the urinary output of Cr in Cr-exposed fish, it is clearly premature to discount the role of the fish kidney in the excretion of Cr.

In mammals, absorbed Cr is excreted prinicipally in the urine and in small quantities in the hair, sweat, and bile (Hopkins, 1965; Doisy et al., 1971; Feng et al., 1988). At least 80% of absorbed Cr is eliminated via the kidneys (Ducros, 1992). Urinary Cr excretion in humans coincides with Cr loss from plasma (Vincent, 1999).

10.2.3. LIVER/BILE

The freshwater elimination study of Knoll and Fromm (1960) (Fig. 8.2) suggested that the liver could excrete Cr via bile despite the fact that liver Cr levels have been shown to remain elevated and unchanged following acute exposure (Van der Putte et al., 1981b). Kroll and Fromm (1960) noted that most of the fish they dissected in their study had a substantial

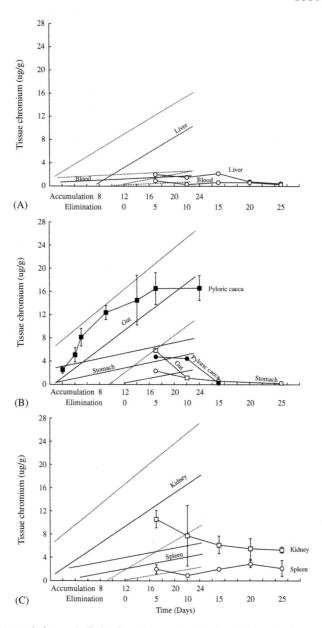

Fig. 8.2. Accumulation and elimination of hexavalent Cr in (A) blood and liver, (B) pyloric ceca, gut and stomach, and (C) kidney and spleen of rainbow trout. Data plotted as means ±1 SEM or linearized (solid line) using linear regressions with corresponding 95% confidence limits (dotted lines). Accumulation: Fish were exposed to 2.5 mg Cr L^{-1} for periods ranging from 1 to 24 days ($n = 10$). Elimination: Fish were exposed to 2.5 mg Cr L^{-1} for 12 days then returned to freshwater for durations ranging from 5 to 25 days ($n = 10$). From Knoll and Fromm (1960).

bile flow along with a full gall bladder. These authors suggested that the bile carries Cr into the lumen of the gastrointestinal tract to be eliminated via the anus. Unfortunately, this route of excretion has the potential to seriously impair the digestive function of the intestine. Fromm and Schiffman (1958) found that in largemouth bass (*Micropterus salmoides*) exposed to 94 mg L^{-1} Cr for 36 h there was a sloughing off of almost all of the epithelium of the intestine, and intestinal folds were greatly reduced in size. It was Fromm and Schiffman's (1958) hypothesis that the source of the Cr that led to the observed histopathology was neither blood nor water but bile.

10.2.4. GUT

The role of the gut in fish as a route for Cr excretion has not yet been determined. There are very few accumulation/excretion studies that look at organ/tissue-specific Cr elimination. Two such studies, Knoll and Fromm (1960) and Van der Putte et al. (1981b), provide support for the gut as a site for Cr excretion in fish. Knoll and Fromm (1960) exposed rainbow trout to 2.5 mg L^{-1} Cr for 12 days. Fish were then moved to Cr-free water and held for 10 days. Organ/tissue Cr accumulation was determined in fish at the end of the exposure and in other fish at the end of the elimination period. The elimination of Cr from the gut was very rapid, with approximately 86% of the accumulated Cr lost 10 days postexposure (Fig. 8.2). Van der Putte et al. (1981b) used a similar experimental design. They exposed trout to a slightly lower Cr concentration (2.0 mg L^{-1}) for a shorter duration (4 days) but used a similar elimination interval (11 days). Their data also suggest that the gut can play a role in the elimination of accumulated Cr, but not to the extent suggested by the findings of Knoll and Fromm (1960). Only 17–27% of the Cr accumulated in the gut of trout in the Van der Putte et al. (1981b) study was cleared at the end of the 11 day elimination period. Unfortunately, these two studies are not directly comparable as Van der Putte et al. (1981b) exposed trout at pH 7.8 or 6.5 while Knoll and Fromm (1960) did not report the water pH. Something in common between these two Cr clearance studies is the fact that the elimination of Cr from the gut mirrored the elimination of Cr from the blood. Either a drop in blood Cr resulted in the drop in gut Cr content or the excretion of Cr at the gut had a direct impact on the blood Cr concentration. As the mammalian gut is thought to play a very minor role in the excretion of Cr (Hopkins, 1965; Doisy et al., 1971; Feng et al., 1988) it is likely that the former statement is more representative of what is happening in the fish gut. However, until additional studies are completed, the importance of the fish gut as a site for Cr excretion will be undetermined.

11. BEHAVIORAL EFFECTS OF MOLYBDENUM AND CHROMIUM

11.1. Molybdenum

There is very limited information available on the behavioral effects of Mo in fish. Reid (2002) noted, but did not quantify, that kokanee exposed to 1500 and 2000 mg L^{-1} Mo exhibited sporadic coughing within the first 12 h of the exposure. Coughing was not observed at any time in fish exposed for 96 h to 1000 mg L^{-1}. No other pertinent studies have been published.

11.2. Chromium

Several studies have observed behavioral changes in a variety of fish species exposed to elevated Cr. Many of the behavioral changes appear to be related to what one might call symptoms of respiratory stress or a non-specific gill irritation response. These behavioral changes include increased coughing, yawning, and ventilation frequency (Gendusa and Beitinger, 1992).

Mishra and Mohanty (2009) exposed spotted snakehead chronically to 2 or 4 mg L^{-1} hexavalent Cr and observed abnormal behavior such as convulsions, loss of equilibrium, and increased opercular activity in a dose-dependent manner. Fish exposed to the higher concentration for 2 months became more lethargic in comparison to other experimental groups, expressed as a decreased rate of swimming. Similarly, cichlids exposed for 4 days to 3 or 6 mg L^{-1} Cr(III) as CrCl$_3$ (pH 7.5, hardness 273 mg L^{-1} as CaCO$_3$) showed a dose-dependent change in a variety of behaviors including coughing, yawning, fin-flickering, jerking movements, and nudge and nip behaviors (Al-Kahem, 1995). In addition, Svecevicius (2007a) demonstrated avoidance behavior in three species of fish [rainbow trout, vimba (*Vimba vimba*) and goldfish] to sublethal concentrations of Cr (speciation not provided; pH 8.0, hardness 284 mg L^{-1} as CaCO$_3$, alkalinity of 244 mg L^{-1} as HCO$_3^-$). The author noted that of the three species, rainbow trout was the most sensitive and that avoidance behavior was observed at Cr concentrations below the maximum permitted concentration according to the water quality guidelines adopted by Lithuania in 2002. Unfortunately, Svecevicius (2007a) did not expose fish to Cr alone, but rather to a mixture of metals that also included Cu, Zn, Ni, Fe, Pb, Cd, and Mn. Further, although avoidance behavior to the metals was noted, the author concluded that the study showed that there was no correlation between metal toxicity and intensity of the induced avoidance behavior. The avoidance behavior induced by Cr was confirmed in a subsequent study (Svecevicius, 2007b) in which trout were exposed to sublethal Cr alone.

12. MOLECULAR CHARACTERIZATION OF MOLYBDENUM AND CHROMIUM TRANSPORTERS, STORAGE PROTEINS, AND CHAPERONES

12.1. Molybdenum

The molecular details of Mo transport, storage proteins, and chaperones in fish or other animals have yet to be investigated.

12.2. Chromium

The molecular details of Cr transport, storage proteins and chaperones in fish or other animals have yet to be investigated.

13. GENOMIC AND PROTEOMIC STUDIES

13.1. Molybdenum

Genomic and proteomic studies with Mo-exposed fish or other animals have not yet been attempted.

13.2. Chromium

Although not strictly genomic or proteomic studies, the determination of the genotoxicity of Cr in fish has been attempted. Torres de Lemos et al. (2001) exposed fathead minnows to 2.5 mg L^{-1} hexavalent chromium ($K_2Cr_2O_7$) for 7, 14, and 21 days and quantified the frequency of micronucleated erythrocytes. The authors suggest that there is a significant increase in genetic mutation after 7 days' exposure to Cr(VI) that is no longer present at days 14 and 21.

In 2004, two studies quantifying changes in gene expression in the liver of fish exposed to Cr were published. Maples and Bain (2004) exposed mummichog (*Fundulus heteroclitus*) to 0, 333, 1000, or 1500 µg L^{-1} Cr(III) for 7 days. Chapman et al. (2004) injected winter flounder (*Pseudopleuronectes americanus*) with Cr(VI) to achieve a final dosage of 25 µg kg^{-1} Cr. Both studies demonstrated that Cr significantly altered the expression of several genes in exposed fish. In the mummichog, Maples and Bain (2004) found 17 differentially expressed genes (two downregulated and 15 upregulated) in response to Cr(III) exposure. The authors speculated that the induced changes in gene expression were either to protect the fish or as a response to stress. In winter flounder, Chapman et al. (2004) found that Cr(VI) altered the expression of 29 genes. DNA sequencing and semi-quantitative reverse transcription–polymerase chain reaction (RT-PCR) confirmed that four of these were downregulated, including two non-selenium reduced glutathione (GSH) peroxidases. Chapman et al. (2004) concluded that the toxicity of Cr(VI) in winter flounder may be due, at least in part, to a decrease in the ability of this fish to defend itself against oxidative damage.

Hook et al. (2006) published an elegant study on gene expression patterns in rainbow trout exposed to a variety of toxicants including Cr. In this study isogenic (cloned) trout were administered an intraperitoneal injection of $25 \, \mu g \, kg^{-1}$ Cr(VI). After 24 h, the livers were removed and prepared for the determination of gene expression patterns using salmonid cDNA microarrays. The intraperitoneal injection of Cr(VI) significantly altered the expression of 218 genes: 210 were significantly upregulated while the remainder were significantly downregulated. Alterations in gene expression were verified using quantitative RT-PCR analysis on a subset of the genes screened by the microarray. The authors categorized the genes with Cr(VI)-induced altered expression as being involved in growth, protein synthesis, protein binding, oxidoreductase activity, nucleic acid binding, copper ion binding, mitochondrial electron transport, and metabolism.

Genomic and proteomic profiles of human lung cell responses to metals, including Cr, using cDNA microarray analysis and Western immunoblots were published by Andrew et al. (2003). These authors quantified the relative expression of 1200 genes in human bronchial BEAS-2B cells exposed for 4 h to $10 \, \mu M$ Cr(VI) as sodium dichromate. Chromium exposure resulted in the altered expression of 44 genes, with five genes showing increased expression and the remainder showing decreased expression relative to control. The authors stated that the number of genes modified in response to Cr exposure was relatively small and represented a subtle modification of cell signaling pathways. Similar studies have been conducted using rat liver and lung (Izzotti et al., 2002) and human liver (HepG2 cells) (Tully et al., 2000). The conclusion from these genomic and proteomic studies is that Cr exposure alters inducible gene expression but not constitutive gene expression. Given that Andrew et al. (2003) exposed human cells to low doses of Cd ($10 \, \mu mol \, L^{-1}$), Ni ($3 \, \mu mol \, L^{-1}$), or As ($5 \, \mu mol \, L^{-1}$), as well as to Cr, and that, of the 210 genes differentially expressed, only one gene (*hsp90*) was altered in response to all four metals, the real significance of these studies is yet to be determined.

14. INTERACTIONS WITH OTHER METALS

14.1. Molybdenum and Chromium

As Cr and Mo exist in water as oxyanions it is interesting to suggest that these two metals may act antagonistically, particularly if they are transported from the environment into the blood via a common pathway. There is as yet no evidence of a specific chromate or molybdate transporter in fish epithelia; however, the hypothesis would be that Cr(VI) and Mo(VI) enter fish using a sulfate transport system as has been reported in mammals (De Flora and Wetterhahn, 1989). As tungsten and selenium form oxyanions and have been shown to inhibit sulfate uptake in human placenta vesicles (Boyd and Shennan, 1986), both of these metals could have potential interactions with both Cr and Mo. By extension, it would be appropriate to hypothesize that sulfate (SO_4^{2-}) would be an effective inhibitor of both Cr(VI) and Mo(VI) uptake and toxicity.

14.2. Chromium

There are no published interactions of metals with trivalent Cr; however, there is one study that indicates that elevated calcium may reduce Cr(VI) accumulation in freshwater fish. Ghosh and Adhikari (2006) exposed freshwater Mrigal carp for 28 days to $200\,mg\,L^{-1}$ Cr(VI) as potassium dichromate (pH 7.7, total alkalinity $114\,mg\,L^{-1}$ $CaCO_3$, $0.45\,mg\,L^{-1}$ Ca) alone or in combination with 1.0, 2.0, 3.0, or $4.0\,mg\,L^{-1}$ calcium. At both 14 and 28 days of exposure, the addition of $2.0\,mg\,L^{-1}$ calcium or greater resulted in significantly reduced Cr accumulation in muscle and whole body. Gill accumulation at days 14, 21, and 28 of the exposure was reduced at all concentrations of calcium in a concentration-dependent manner. The authors suggested that the reduction in Cr uptake was due to competition between Ca and Cr at calcium channels. The authors provided no support for this claim nor is there any evidence in the literature to suggest that Cr can be translocated via calcium channels.

In a related study, Adhikari et al. (2006) demonstrated that Cr accumulation in a freshwater fish, rohu (*Labeo rohita*), was inversely proportional to water pH/total alkalinity. Rohu were exposed for 7 or 14 days to three combinations of pH and total alkalinity (5.5, $40\,mg\,L^{-1}$ $CaCO_3$; 7.0, $118\,mg\,L^{-1}$ $CaCO_3$; 8.5, $200\,mg\,L^{-1}$ $CaCO_3$) and $1.5\,mg\,L^{-1}$ Cr(VI) as $K_2Cr_2O_7$. It was postulated that the reduction in Cr accumulation at high pH/alkalinity was due not to some antagonism between H^+ or Ca^{2+} and Cr(VI) but rather to the reduction in availability of Cr(VI) due to the increased formation of insoluble precipitates at higher water pH and alkalinity.

15. KNOWLEDGE GAPS AND FUTURE DIRECTIONS

15.1. Molybdenum

There are simply too many knowledge gaps to list them all as there is virtually nothing that we know about the specific toxic action and general metabolism of Mo in fish, either freshwater or marine. There are a few studies slowly entering the literature but there is a long way to go before we truly understand the impact of Mo on the physiology of fish, whether good, bad, or indifferent. This is likely a reflection of the fact that Mo has very low acute toxicity and that environmental concentrations rarely approach those that are acutely toxic to fish. However, the somewhat unique water chemistry and the fact that it is a micronutrient warrant this metal worthy of further study. But where to begin? With a relatively weak non-fish literature to draw upon, the author will offer his opinion on three priority areas for future research, in no particular order: (1) the aquatic chemistry of Mo, (2) chronic exposure studies, and (3) tissue distribution and excretion studies.

Water chemistry: As an oxyanion in water, Mo is somewhat unique. How does this metal fit with the biotic ligand model? Does dissolved organic matter influence the bioavailability of this metal? Are the bioavailability and toxicity of Mo altered by changes in water pH? From a regulatory perspective, the answers to these questions are of critical importance.

Chronic exposures: Given that we know that Mo is really not acutely toxic, for regulatory purposes it would be beneficial to understand the impact on fish of chronic exposure to elevated Mo either in the laboratory or in natural settings. Understanding the potential adverse impact of chronic, low-level Mo exposure on, for example, reproductive success in fish, will allow for better monitoring of the health of natural fish populations in areas of Mo mining operations and other industrial activities that increase Mo loading into the aquatic environment.

Excretion studies: Knowing where Mo is accumulated in fish not only provides needed insight into the metabolism of Mo but provides us with an opportunity to investigate the routes by which the metal might be cleared from the animal's body/tissues. It would also provide us with a better understanding of the potential for bioaccumulation and biomagnification and whether or not Mo, a micronutrient, is homeostatically regulated.

There is so much about Mo and fish physiology that is simply not known that the future directions are limitless and deserving of our attention.

15.2. Chromium

Compared to Mo, there is a wealth of knowledge about the toxicity and metabolism of Cr in fish, but compared to cationic metals such as Cd, Cu, Zn, Ag, and others in these volumes, there is still a great deal to be learned about fish physiology and Cr. As with Mo, and for many of the same reasons, there is a need to understand more fully the aquatic chemistry of Cr. However, given the established carcinogenicity and mutagenicity of Cr in mammalian cells, a priority for future research should be to develop a detailed understanding of the biological activity of Cr at the cellular and molecular level. Unlike Mo, where there is limited non-fish (mammalian) literature about its metabolism and toxic action, there is substantial literature on the toxic action and metabolism of Cr that can assist us in achieving this level of understanding in fish. For example, Cr is a potent mammalian (human) carcinogen and mutagen (Ding et al., 2000). Chromium(III) is generally believed to be an essential element involved in fat and sugar metabolism in humans (Schroeder, 1968; Anderson, 1981; Cefalu et al., 2002; Sreekanth et al., 2008) whereas Cr(VI) is the form that causes most adverse health effects (Dayan and Paine, 2001). Chromium(VI) is an oxidant that enters cells through the sulfate anion transporter and becomes reduced to stable Cr(III), which adversely alters cell function. Exposure to industrially generated Cr(VI) compounds has been known for over a century to be associated with elevated lung cancer rates (Gibb et al., 2000) and other adverse human health effects. Intracellular Cr(VI) reduction generates ROS, which are the probable cause of Cr toxicity and carcinogenicity (Patiolla et al., 2008). In cultured mammalian cells, Cr exposure has been shown to cause radical-mediated DNA strand breakage and formation of stable Cr–DNA complexes including Cr–DNA adducts and protein–DNA and DNA–Cr–DNA cross-links (Zhitkovich, 2005). Achieving this level of detail in our understanding of the toxicity of Cr in fish should be a priority for the future direction of research.

REFERENCES

Abbas, H. H., and Ali, F. K. (2007). Study of the effect of hexavalent chromium on some biochemical, cytotoxicological and histopathological aspects of the *Oreochromis* sp. fish. *Pak. J. Biol. Sci.* **10**, 3973–3982.

Adhikari, S., Ghosh, L., and Ayyappan, S. (2006). Effect of calcium hardness on toxicity and accumulation of water-borne lead, cadmium and chromium to *Labeo rohita* (Hamilton). *Asian J. Water Environ. Pollut.* **4**, 103–106.

Al-Kahem, H. F. (1995). Behavioural responses and changes in some haematological parameters of the cichlid fish, *Oreochromis niloticus*, exposed to trivalent chromium. *JKAU Sci.* **7**, 5–13.

Allen, J. D., and Gawthorne, J. M. (1986). Involvement of organic molybdenum compounds in the interaction between copper, molybdenum and sulfur. *J. Inorg. Biochem.* **27**, 95–112.

Anan, Y., Kunito, T., Sakai, H., and Tanabe, S. (2002). Subcellular distribution of trace elements in the liver of sea turtles. *Mar. Pollut. Bull.* **45**, 224–229.

Anderson, R. A. (1981). Nutritional role of chromium. *Sci. Total Environ.* **17**, 13–29.

Andrew, A. S., Warren, A. J., Barchowsky, A., Temple, K. A., Klei, L., Soucy, N. V., O'Hara, K. A., and Hamilton, J. W. (2003). Genomic and proteomic profiling of responses to toxic metals in human lung cells. *Environ. Health Perspect.* **111**, 825–838.

Arillo, A., and Melodia, F. (1990). Protective effects of fish mucus against Cr(VI) pollution. *Chemosphere* **20**, 397–402.

Arrington, L. R., and Davis, G. K. (1955). Metabolism of phosphorus[32] and molybdenum[99] in rats receiving high calcium diets. *J. Nutr.* **55**, 181–191.

ANZECC (1992). *Australian Water Quality Guidelines for Fresh and Marine Waters.* Australian and New Zealand Environmental and Conservation Council.

Austreng, E. (1978). Digestibility determination in fish using chromic oxide marking and analysis of contents from different segments of the gastrointestinal tract. *Aquaculture* **13**, 265–272.

Aveston, J., Anacker, E. W., and Johnson, J. S. (1964). Hydrolysis of molybdenum (VI): ultracentrifugation, acidity measurements, and Raman spectra of polymolybdates. *Inorg. Chem.* **3**, 735–746.

Beers, M. H. and Berkow, R. (1998). In *The Merck Manual of Diagnosis and Therapy*, 17th edn (eds. M. H. Beers and R. Berkow). Whitehouse Station, NJ: Merck & Co.

Benoit, D. A. (1976). Toxic effects of hexavalent chromium on brook trout (*Salvelinus fontinalis*) and rainbow trout (*Salmo gairdneri*). *Water Res.* **10**, 497–500.

Bentley, R. E. (1973). Acute Toxicity of Sodium Molybdate to Bluegill (*Lepomis macrochirus*), Rainbow Trout (*Salmo gairdneri*), Fathead Minnow (*Pimephales promelas*), Channel catfish (*Ictalurus punctatus*), Water Flea (*Daphnia magna*) and Scud (*Gammarus fasciatus*). Bionomics, Wareham, MA.

Bibr, B., Deyl, Z., Lener, J., and Adam, M. (1977). Investigation on the relation of molybdenum with collagen *in vivo*. *Int. J. Pept. Protein Res.* **10**, 190–196.

Boyd, C. A. R., and Shennan, D. B. (1986). Sulphate transport into vesicles prepared from human placental brush border membranes: inhibition by trace element oxides. *J. Physiol.* **379**, 367–376.

Calamari, D., Vighi, M., and Bacchi, E. (1982). Toxicokinetics of low levels of Cd, Cr, Ni, and their mixtures in long-term treatment on *Salmo gairdneri*. *Chemosphere* **11**, 59–70.

Cates, L. A., Jones, G. S., Jr., Good, D. J., Tsai, H. Y., Li, V., Caron, N., Tu, S., and Kimball, A. P. (1980). Cyclophosphamide potentiation and aldehyde oxidase inhibition by phosphorylated aldehydes and acetals. *J. Med. Chem.* **23**, 300–304.

CCME (1999). Canadian water quality guidelines for the protection of aquatic life: molybdenum. In *Canadian environmental quality guidelines*. Winnipeg: Canadian Council of Ministers of the Environment.

Cefalu, W. T., and Hu, F. B. (2004). Role of chromium in human health and diabetes. *Diabetes Care* **27**, 2741–2751.

Cefalu, W. T., Wang, Z. Q., Zhang, X. H., Baldor, L. C., and Russell, J. C. (2002). Oral chromium picolinate improves carbohydrate and lipid metabolism and enhances skeletal muscle Glut-4 translocation in obese, hyperinsulinemic (JCR-LA corpulent) rats. *J. Nutr.* **132**, 1107–1114.

Chapman, L. M., Roling, J. A., Bingham, L. K., Herald, M. R., and Baldwin, W. S. (2004). Construction of a subtractive library from hexavalent chromium treated winter flounder (*Pseudopleuronectes americanus*) reveals alterations in non-selenium glutathione perox-idase. *Aquat. Toxicol.* **67**, 181–194.

Chappell, W. R. (1975). Transport and biological effects of molybdenum in the environment. In: *Heavy Metals in the Aquatic Environment* (P.A. Krenkel, ed.), pp. 167–188. Pergamon Press, Oxford.

Chappell, W. R., Meglen, R. R., Moure-Eraso, R., Solomons, C. C., Tsongas, T. A., Walravens, P. A., and Winston, P. W. (1979). *Human Health Effects of Molybdenum in Drinking Water*. US Environmental Protection Agency, Washington, DC, Report 600/1-79-006

Cornelis, R, Borguet, F., Dyg, S., and Griepink, G. (1992). Chromium speciation studies in human plasma and stability studies of Cr(III) and Cr(VI) species in a candidate water reference material. *Mikrochim. Acta* **109**, 145–148.

Davies, T. D., Pickard, J., and Hall, K. J. (2005). Acute molybdenum toxicity to rainbow trout and other fish. *J. Environ. Eng. Sci.* **4**, 481–485.

Dayan, A. D., and Paine, A. J. (2001). Mechanisms of chromium toxicity, carcinogenicity and allergenicity: review of the literature from 1985 to 2000. *Hum. Exp. Toxicol.* **20**, 439–451.

De Flora, S., and Wetterhahn, K. E. (1989). Mechanisms of chromium metabolism and genotoxicity. *Life Chem. Rep.* **7**, 169–244.

Dick, A. T. (1953). The effect of inorganic sulphate on the excretion of molybdenum in the sheep. *Aust. Vet. J.* **29**, 19–26.

Dillon, T. M., Suedel, B. C., Peddicord, R. K., Clifford, P. A., and Boraczek, J. A. (1995). *Environmental Effects of Dredging: Trophic Transfer and Biomagnification Potential of Contaminants in Aquatic Ecosystems*. EEDP-01-33 US Army Engineer Waterways Experimental Station, Vicksberg, MS.

Ding, M., Shi, X., Castranova, V., and Vallyathan, V. (2000). Predisposing factors in occupational lung cancer: inorganic minerals and chromium. *J. Environ. Pathol. Toxicol. Oncol.* **19**, 129–138.

Doisy, R. J., Streeten, D. H. P., Souma, M. L., Kalafer, M. E., Rekant, S. L., and Dalakos, T. G. (1971). Metabolism of ^{51}Cr in human subjects/normal, elderly and diabetic subjects. In: *Newer Trace Element in Nutrition* (W. Mertz and W.E. Cornatzer, eds), pp. 155–168. Dekker, New York.

Dowling, H. J., Offenbacher, E. G., and Pi-Sunyer, F. X. (1989). Absorption of inorganic, trivalent chromium from the vascularly perfused rat small intestine. *J. Nutr.* **119**, 1138–1145.

Ducros, V. (1992). Chromium metabolism, a literature review. Biol. Trace Elem. Res. **32**, 65–77.

Easterday, R. L., and Miller, R. F. (1963). The acute toxicity of molybdenum to the bluegill. *Proc. Virg. J. Sci.* **14**, 199–200.

Ecological Analysts, Inc. (1981). *The Sources, Chemistry, Fate and Effects of Chromium in Aquatic Environments*. Available from American Petroleum Institute, Washington, DC.

Eisler, J. (1989). Molybdenum hazards to fish, wildlife and invertebrates: a synoptic review. *U.S. Fish Wildl. Serv. Biol. Rep.* **85**, 61.

Emsley, J. (2001). Chromium. In *Nature's Building Blocks: An A–Z Guide to the Elements*, pp. 495–498. Oxford: Oxford University Press.

Farag, A. M., Thomas, M., Marty, G. D., Easton, M., Harper, D. D., Little, E. E., and Cleveland, L. (2006). The effect of chronic chromium exposure on the health of Chinook salmon (*Oncorhynchus tshawytscha*). *Aquat. Toxicol.* **76**, 246–257.

Feng, W. Y., Ding, W. J., Quain, Q. F., and Chai, Z. F. (1988). Study on the metabolism of physiological amounts of Cr(III) intragastrical administration in normal rats using activable enriched stable isotope Cr-50 compound as a tracer. *J. Radioanal. Nucl. Chem.* **237**, 15–19.

Fleet, J. C., Golemboski, K. A., Kietert, R. R., Andrews, G. K., and McCormick, C. C. (1990). Induction of hepatic metallothionein by intraperitoneal metal injection: an associated inflammatory response. *Am. J. Physiol.* **258**, G926–G933.

Friberg, L., and Lener, J. (1986). Molybdenum. In: *Handbook of the Toxicology of Metals – Specific Metals* (L. Friberg, G.F. Nordberg and V.B. Vouk, eds), Vol. II, pp. 446–461. Elsevier, New York.

Friberg, L., Boston, P., Nordberg, G., Piscator, M. and Robert, K. H. (1975). Molybdenum – a toxicological appraisal. Report 600/1-75-004. Washington, DC: US Environmental Protection Agency.

Fromm, P. O., and Schiffman, R. H. (1958). Toxic action of hexavalent chromium on largemouth bass. *J. Wildl. Manag.* **22**, 40–44.

Fromm, P. O., and Stokes, R. M. (1962). Assimilation and metabolism of chromium by trout. *J. Water Pollut. Control Fed.* **35**, 1151–1155.

Gendusa, T. C., and Beitinger, T. L. (1992). External biomarkers to assess chromium toxicity in adult *Lepomis macrochirus. Bull. Environ. Contam. Toxicol.* **48**, 237–242.

Ghosh, L., and Adhikari, S. (2006). Accumulation of heavy metals in freshwater fish – an assessment of toxic interactions with calcium. *Am. J. Food Technol.* **1**, 139–148.

Gibb, H. J., Lees, P. S., Pinsky, P. F., and Rooney, B. C. (2000). Lung cancer among workers in chromium chemical production. *Am. J. Ind. Med.* **38**, 115–126.

Gill, T. S., and Pant, J. C. (1987). Hematological and pathological effects of chromium toxicosis in the freshwater fish, *Barbus conchonius* Ham. *Water Air Soil Pollut.* **35**, 241–250.

Glinsmann, W. H., and Mertz, W. (1966). Effect of trivalent chromium on glucose tolerance. *Metabolism* **15**, 510–519.

Gobas, F. A. P. C., Wilcockson, J. B., Wilson, R. W., and Haffner, G. D. (1999). Mechanism of biomagnification in fish under laboratory and field conditions. *Environ. Sci. Technol.* **33**, 133–141.

Goyer, R. A. (1986). Toxic effects of metals. In: *Casarett and Doull's Toxicology* (C.D. Klassen, M.O. Amdur and J. Doull, eds), 3rd edn, pp. 582–635. MacMillan, New York.

Hajen, W. E., Beames, R. M., Higgs, D. A., and Doanjh, B. S. (1993). Digestibility of various feedstuffs by post-juvenile chinook salmon (*Oncorhynchus tshawytscha*) in sea water. 1. Validation technique. *Aquaculture* **112**, 321–332.

Hamilton, S. J., and Buhl, K. H. (1990). Acute toxicity of boron, molybdenum, and selenium to fry of chinook salmon and coho salmon. *Arch. Environ. Contam. Toxicol.* **19**, 366–373.

Hatfield Consultants Ltd. (1992). *Trojan Pond Rainbow Trout Restocking Field and Analytical Survey.* Highland Valley Copper, Logan Lake, British Columbia.

Hatfield Consultants Ltd. (1993). *Trace Metal Analysis in Rainbow Trout and Benthic Macroinvertebrate Studies.* Highland Valley Copper, Logan Lake, British Columbia.

Hertz, Y., Madar, Z., Hepper, B., and Gertler, A. (1989). Glucose metabolism in the common carp (*Cyprinus carpio* L.): the effects of cobalt and chromium. *Aquaculture* **76**, 255–267.

Holdway, D. A. (1988). The toxicity of chromium to fish. In: *Chromium in the Natural and Human Environments* (J.O. Nriagu and E. Neirboor, eds), pp. 360–397. John Wiley, New York.

Hook, S. E., Skillman, A. D., Small, J. A., and Schultz, I. R. (2006). Gene expression patterns in rainbow trout, *Oncorhynchus mykiss*, exposed to a suite of model toxicants. *Aquat. Toxicol.* **77**, 372–385.

Hopkins, L. L., Jr. (1965). Distribution in the rat of physiological amounts of injected Cr^{51}(III) with time. *Am. J. Physiol.* **209**, 731–735.

Irreverre, F., Mudd, S. H., Heizer, W. D., and Laster, L. (1967). Sulfite oxidase deficiency: studies of a patient with mental retardation, dislocated ocular lenses, and abnormal urinary excretion of S-sulfo-L-cysteine, sulfite, and thiosulfate. *Biochem. Med.* **1**, 187–217.

Izzotti, A, Cartiglia, C., Balansky, R., D'Agostini, F., Longobardi, M., and De Flora, S. (2002). Selective induction of gene expression in rat lung by hexavalent chromium. *Mol. Carcinogen.* **35**, 75–84.

Jarrell, W. M., Page, A. L., and Elseewi, A. A. (1980). Molybdenum in the environment. *Residue Res.* **74**, 1–43.

Jones, C. E. (1999). Molybdenum in the environment: an overview of implications to the British Columbia mining industry. In *Proceedings of the 1999 Workshop on Molybdenum Issues in Reclamation* (eds. B. Price, B. Hart and C. Howell), Kamloops, BC, September 24, pp. 1–14.

Jop, K. M., Parkerton, T. F., Rogers, J. H., and Dorn, P. B. (1987). Comparative toxicity and speciation of two hexavalent chromium salts in acute toxicity tests. *Environ. Toxicol. Chem.* **6**, 697–703.

Karakas, E., Wilson, H. L, Graf, T. N., Xian, S., Jaramillo-Busquets, S., Rajagopalan, K. V, and Kisker, C. (2005). Structural insights into sulfite oxidase deficiency. *J. Biol. Chem.* **280**, 33506–33515.

Kisker, C., Schindelin, H., and Rees, D. C. (1997). Molybdenum-cofactor-containing enzymes: structures and mechanisms. *Annu. Rev. Biochem.* **66**, 233–267.

Knoll, J., and Fromm, P. O. (1960). Accumulation and elimination of hexavalent chromium in rainbow trout. *Physiol. Zool.* **33**, 1–8.

Knothe, D. W., and Van Riper, G. G. (1988). Acute toxicity of sodium molybdate dihydrate (Molyhibit 100) to selected saltwater organisms. *Bull. Environ. Contam. Toxicol.* **40**, 785–790.

Krumschnabel, G., and Nawaz, M. (2004). Acute toxicity of hexavalent chromium in isolated teleost hepatocytes. *Aquat. Toxicol.* **70**, 159–167.

Kuhnert, P. M., and Kuhnert, B. R. (1976). The effect of *in vivo* chromium exposure on Na/K- and Mg-ATPase activity in several tissues of the rainbow trout (*Salmo gairnderi*). *Bull. Environ. Contam. Toxicol.* **15**, 383–390.

Kurosaki, M., Li Calzi, M., Scanziani, E., Garattini, E., and Terao, M. (1995). Tissue- and cell-specific expression of mouse xanthine oxidoreductase gene *in vivo*: regulation by bacterial lipopolysaccharide. *Biochem. J.* **306**, 225–234.

Langard, S., and Norseth, T. (1979). Chromium. In: *Handbook on the Toxicology of Metals* (L. Friberg, G.F. Nordberg and V.B. Vouk, eds), pp. 383–397. Elsevier/North Holland Biomedical Press, Amsterdam.

Lowmann, F. G., Rice, T. R., and Richards, F. A. (1971). *Radioactivity in the Marine Environment*. National Academy of Science, Washington, DC.

Mallatt, J. (1985). Fish gill structural changes induced by toxicants and other irritants: a statistical review. *Can. J. Fish Aquat. Sci.* **42**, 630–648.

Maples, N. L., and Bain, L. J. (2004). Trivalent chromium alters gene expression in the mummichog (*Fundulus heteroclitus*). *Environ. Toxicol. Chem.* **23**, 626–631.

McConnell, R. P. (1977). Toxicity of molybdenum to rainbow trout under laboratory conditions. In: *Molybdenum in the Environment: The Geochemistry, Cycling, and Industrial Uses of Molybdenum* (W.R. Chappell and K. Kellogg Peterson, eds), pp. 725–730. Marcel Dekker, New York.

McNeely, R. N., Neimanis, V. P., and Dwyer, L. (1979). *Molybdenum. Water Quality Sourcebook: A Guide to Water Quality Parameters*. Water Quality Branch, Inland Waters Directorate, Environment Canada, Ottawa.

Mearns, A. J., Oshida, P. S., Sherwood, M. J., Young, D. R., and Reish, D. J. (1976). Chromium effects on coastal organisms. *J. Water Pollut. Control Fed.* **48**, 1929–1939.

Mills, C. F., and Davis, G. K. (1987). Molybdenum. In: *Trace Elements in Human and Animal Nutrition* (W. Mertz, ed.), pp. 429–463. Academic Press, San Diego, CA.

Mishra, A. K., and Mohanty, B. (2009). Chronic exposure to sublethal hexavalent chromium affects organ histopathology and serum cortisol profile of a teleost, *Channa punctatus* (Bloch). *Sci. Total Environ.* **407**, 5031–5038.

Moore, L. J., Machlan, L. A., Shields, W. R., and Garner, E. L. (1974). Internal normalization techniques for high accurate isotope dilution analyses. Application to molybdenum and nickel in standard reference materials. *Anal. Chem.* **46**, 1082–1089.

Muniz, I. P., and Leivestad, H. (1980). Toxic effects of aluminum on brown trout, *Salmo trutta* L. In: *Ecological Impact of Acid Precipitation* (D. Droblos and A. Tollan, eds), pp. 320–321. SNSF Project, Norway.

Nevada Division of Environmental Protection. (2008). *Rationale for Proposed Revisions to Aquatic Life Water Quality Criteria for Molybdenum.* NAC 445A.144. Nevada Division of Environmental Protection, Carson, NV.

New Zealand Ministry of Health. (2005). *Draft Guidelines for Drinking-water Quality Management for New Zealand.* New Zealand Ministry of Health, Wellington.

NHMRC and NRMMC (2004). *Australian Drinking Water Guidelines 2004, National Water Quality Management Strategy.* National Health and Medical Research Council and the Natural Resource Management Ministerial Council.

NRC (2004). *Canadian Minerals Yearbook: Molybdenum.* National Resources Canada. Available online: http://www.nrcan-rncan.gc.ca/mms-smm/busi-indu/cmy-amc/content/2004/40.pdf

Nriagu, J. O. (1988). Production and uses of chromium. In: *Chromium in the Natural and Human Environments* (J.O. Nriagu and E. Nieboer, eds), pp. 125–172. John Wiley and Sons, New York.

Nriagu, J. O., and Pacyna, J. M. (1988). Quantitative assessment of worldwide contamination of air, water and soils with trace metals. *Nature* 333, 134–139.

Ohta, H., Seki, Y., Imamiya, S., and Yoshikawa, H. (1993). Metallothionein synthesis by trivalent or hexavalent chromium in mice. In: *Trace Elements in Man and Animals* (M. Anke and D. Meissner, eds), pp. 178–179. Verlag Media Touristik, Gersdorf.

Palaniappan, P. L., and Karthikeyan, S. (2009). Bioaccumulation and depuration of chromium in the selected organs and whole body tissues of freshwater fish *Cirrhinus mrigala* individually and in a binary solution with nickel. *J. Environ. Sci.* 21, 229–236.

Pandey, R., Kumar, R., Singh, S. P., and Srivastava, S. P. (2002). Molybdenum in rat tissue. *Hum. Exp. Toxicol.* 21, 33–35.

Pane, E. F., Richards, J. G., and Wood, C. M. (2003). Acute waterborne nickel toxity in the rainbow trout (*Oncorhynchus mykiss*) occurs by a respiratory rather than ionoregulatory mechanism. *Aquat. Toxicol.* 63, 65–82.

Pane, E. F., Haque, A., and Wood, C. M. (2004). Mechanistic analysis of acute, Ni-induced respiratory toxicity in the rainbow trout (*Oncorhynchus mykiss*): an exclusively branchial phenomenon. *Aquat. Toxicol.* 69, 11–24.

Patiolla, A. K., Barnes, C., Yedjou, C., Velma, V. R., and Tchounwou, P. B. (2008). Oxidative stress, DNA damage, and antioxidant enzyme activity induced by hexavalent chromium in Sprague–Dawley rats. *Environ. Toxicol.* 24, 66–73.

Pawlisz, A. V., Kent, R. A., Schneider, U. A., and Jefferson, C. (1997). Canadian water quality guidelines for chromium. *Environ. Toxicol. Water Qual.* 12, 123–183.

Pfeiffer, W. C., Fiszman, M., and Carbonell, N. (1980). Fate of chromium in a tributary of the Iraja River, Rio de Janeiro. *Environ. Pollut. B* 1, 117–126.

Phillip, D. R. (1988). Chromium. In *Canadian Mineral Yearbook*, pp. 20.1–20.10. Ottawa: Mineral Resources Branch, Department of Energy, Mines and Resources.

Phillips, G. R., and Russo, R. C. (1978). *Metal Bioaccumulation in Fishes and Aquatic Invertebrates: A Literature Review.* EPA-6003-78-103. US Environmental Protection Agency, Washington, DC.

Pickering, O. H. (1980). Chronic toxicity of hexavalent chromium to fathead minnow (*Pimephales promelas*). *Arch. Environ. Contam. Toxicol.* 9, 405–413.

Polayk, D. E. (2010). Molybdenum. *US Geological Survey, 2008 Minerals Yearbook*, pp. 50.1-50.13. US Geological Survey.

Pott, E. B., Henry, P. R., Rao, P. V., Hinderberger, E. J., Jr., and Ammerman, C. B. (1999). Estimated relative bioavailability of supplemental inorganic molybdenum sources and their effect on tissue molybdenum and copper concentrations in lambs. *Anim. Feed Sci. Technol.* **79**, 107–117.

Pyle, G. G., Swanson, S. M., and Lehmkuhl, D. M. (2001). Toxicity of uranium mine-receiving waters to caged fathead minnows, *Pimephales promelas. Ecotoxicol. Environ. Sa.* **48**, 202–214.

Rajagopalan, K. V., and Handler, P. (1964). Hepatic aldehyde oxidase. II. Differential inhibition of electron transfer to various electron acceptors. *J. Biol. Chem.* **239**, 2022–2026.

Reid, S. D. (2002). Physiological impact of acute molybdenum exposure in juvenile kokanee salmon (*Oncorhynchus nerka*). *Comp. Biochem. Physiol. C* **133**, 355–367.

Reish, D. J. (1981). Effects of chromium on the life history of *Capitella capitata.* In: *Physiological Responses of Marine Biota to Pollutants* (F.J. Vernberg, A. Calabrese, F.P. Thurberg and W.B. Vernberg, eds), pp. 199–207. Academic Press, New York.

Ricketts, C. D. (2009). *The effect of acute waterborne exposure of sub-lethal concentrations of molybdenum on the stress response in rainbow trout* (Oncorhynchus mykiss). MSc thesis, University of British Columbia, Okanagan.

Rosebrough, R. W., and Steele, N. C. (1981). Effects of supplemental dietary chromium or nicotinic acid on carbohydrate metabolism during basal, starvation and refeeding periods in poults. *Poult. Sci,* **60**, 407–417.

Rosoff, B., and Spencer, H. (1964). Radiobiology. Fate of molybdenum-99 in man. *Nature* **202**, 410–411.

Rosoff, B., and Spencer, H. (1973). The distribution and excretion of molybdenum-99 in mice. *Health Phys.* **25**, 173–175.

Saiki, M. K., Jennings, M. R., and Brumbaugh, W. G. (1993). Boron, molybdenum, and selenium in aquatic food chains from the lower San Joaquin River and its tributaries, California. *Arch. Environ. Contam. Toxicol.* **24**, 307–319.

Schiffman, R. H., and Fromm, P. O. (1959). Chromium-induced changes in the blood of rainbow trout, *Salmo gairdneri. Sewage Ind. Wastes* **31**, 205–211.

Schroeder, H. A. (1968). The role of chromium in mammalian nutrition. Am. J. Clin. Nutr. **21**, 230–244.

Schroeder, H. A., Balassa, J. J., and Tipton, I. H. (1970). Essential trace metals in man: molybdenum. *J. Chron. Dis.* **23**, 481–489.

Seenayya, G., and Prahalad, A. K. (1987). In situ compartmentation and biomagnification of chromium and manganese in industrially polluted Husainsagar Lake, Hyderabad, India. *Water Air Soil Pollut.* **35**, 233–239.

Sherwood, M. J., and Wright, J. L. (1976). *Uptake and Effects of Chromium on Marine Fish.* Annual Report. Coastal Water Research Project, Southern California Coastal Water Research Project, El Segundo, CA, pp. 123–128.

Shiau, S. Y., and Lin, S.-F. (1993). Effects of supplemental dietary chromium and vanadium on the utilization of different carbohydrates in tilapia, *Oreochromis niloticus* × *O. aureus. Aquaculture* **110**, 321–330.

Short, Z. F., Olson, P. R., Palumbo, R. F., Donaldson, J. R. and Lowman, F. G. (1971). Uptake of molybdenum, marked with ^{99}Mo, by the biota of Fern Lake, Washington, in a laboratory and a field experiment. In *Radionuclides in Ecosystems. Proceedings of the Third National Symposium on Radioecology*, Vol. 1 (ed. D. J. Nelson), pp. 474–485, May 10–12, 1971, Oak Ridge, TN.

Sillen, L. G. and Martell, A. E. (1971). Stability constants of metal–ion complexes. *Spec. Publ. Chem. Soc.* No. 25.

Smith, T. R., Brown, K. W., and Deuel, L. E., Jr. (1987). Plant availability and uptake of molybdenum as influenced by soil type and competing ions. *J. Environ. Qual.* **16**, 377–382.

414SCOTT D. REID

Solis-Heredia, M. J., Quintanilla-Vega, B., Sierra-Santoyo, A., Hernandez, J. M., Brambila, E., Cebrain, M. E., and Albores, A. (2000). Chromium increases pancreatic metallothionein in the rat. *Toxicology* **142**, 111–117.
South Africa Water Quality Guidelines. (1996). *South Africa Water Quality Guidelines.* Vol. 7. *Aquatic Ecosystems* (2nd edn.). Department of Water Affairs and Forestry, Water Quality Management, Pretoria.
Sreekanth, R., Pattabhi, V., and Rajan, S. S. (2008). Molecular basis of chromium insulin interactions. *Biochem. Biophys. Res. Commun.* **369**, 725–729.
Steele, N. C., and Rosebrough, R. W. (1981). Effect of trivalent chromium on hepatic lipogenesis by the turkey poult. *Poult. Sci.* **60**, 617–622.
Steinhagen, D., Helmus, T., Maurer, S., Michael, R. D, Leibold, W., Scharsack, J. P., Skouras, A., and Schuberth, H.-J. (2004). Effect of hexavalent carcinogenic chromium on carp *Cyprinus carpio* immune cells. *Dis. Aquat. Org.* **62**, 155–161.
Steven, J. D., Davies, L. J., Stanley, E. K., Abbott, R. A., Ihnat, M., Bidstrup, L., and Jaworski., J. F. (1976). *Effects of Chromium in the Canadian Environment.* NRCC No. 15017. National Research Council Canada, Ottawa.
Stirpe, F., and Della Corte, E. (1969). The regulation of rat liver xanthine oxidase – conversion *in vitro* of the enzyme activity from dehydrogenase (type D) to oxidase (type O). *J. Biol. Chem.* **244**, 3855–3863.
Stokes, R. M., and Fromm, P. O. (1965). Effect of chromate on glucose transport by the gut of rainbow trout. *Physiol. Zool.* **38**, 202–205.
Stokinger, H. E. (1981). The metals. In: *Patty's Industrial Hygiene and Toxicology* (G.D. Clayton and F.E. Clayton, eds). 3rd edn John Wiley and Sons, New York.
Stouthart, A. J. H. X., Spanings, F. A. T., Lock, R. A. C., and Wendelaar-Bonga, S. E. (1995). Effects of pH on chromium toxicity of early life stages of the common carp (*Cyprinus carpio*). *Aquat. Toxicol.* **32**, 31–42.
Stumm, W., and Morgan, J. J. (1981). *Aquatic Chemistry. An Introduction Emphasizing Chemical Equilibria in Natural Waters.* John Wiley and Sons, New York.
Svecevicius, G. (2006). Acute toxicity of hexavalent chromium to European freshwater fish. *Bull. Environ. Contam. Toxicol.* **77**, 741–747.
Svecevicius, G. (2007a). The use of fish avoidance response in identifying sublethal toxicity of heavy metals and their mixtures. *Acta Zool. Lituanica* **17**, 139–143.
Svecevicius, G. (2007b). Avoidance response of rainbow trout *Oncorhynchus mykiss* to hexavalent chromium solutions. *Bull. Environ. Contam. Toxicol.* **79**, 596–600.
Swain, L. G. (1986). *Water Quality Criteria for Molybdenum.* Technical Appendix 628.161/S971/1986TECH./APP. Victoria, BC: Ministry of Environment and Parks, Province of British Columbia, Water Management Branch.
Taylor, F. G., and Parr, P. D. (1978). Distribution of chromium in vegetation and small mammals adjacent to cooling towers. *J. Tenn. Acad. Sci.* **53**, 87–91.
Thaker, J., Chhaya, J., Nuzhat, S., Mittal, R., Mansuri, A. P., and Kundu, R. (1996). Effects of chromium(VI) on some ion-dependent ATPases in gills, kidney and intestine of a coastal teleost *Periophthalmus dipes. Toxicology* **112**, 237–244.
Tong, S. D, Youngs, W. D., Gutemann, W. H., and Lisk, D. J. (1974). Trace metals in Lake Cayuga lake trout (*Salvelinus namaycush*) in relation to age. *J. Fish. Res. Bd Can.* **31**, 238–239.
Torres de Lemos, C., Rodel, P. M., Terra, N. R., and Erdtmann, B. (2001). Evaluation of basal micronucleus frequency and hexavalent chromium effects in fish erythrocytes. *Environ. Toxicol. Chem.* **20**, 1320–1324.
Towill, L. E., Shriner, C. R., Drury, J. S., Hammons, A. S., and Holleman, J. W. (1978). *Reviews of the Environmental Effects of Pollutants: III. Chromium.* Report 600/1-78-023. United States Environmental Protection Agency, Washington, DC.

Tully, D. B, Collins, B. J., Overstreet, J. D., Smith, C. S., Dinse, G. E., Mumtaz, M. M., and Chapin, R. E. (2000). Effects of arsenic, cadmium, chromium and lead on gene expression regulated by a battery of 13 different promoters in recombinant HepG2 cels. *Toxicol. Appl. Pharmocol.* **168**, 79–90.

Turnlund, J. R., Keyes, W. R., and Peiffer, G. L. (1995). Molybdenum absorption, excretion and retention studied with stable isotopes in young men at five intakes of dietary molybdenum. *Am. J. Clin. Nutr.* **62**, 790–796.

USEPA. (1980). *Water Quality Criteria Documents: Availability.* United States Environmental Protection Agency, Washington, DC, Federal Register 45 (Nov. 28), 231.

USEPA. (1986). *Quality Criteria for Water.* United States Environmental Protection Agency, Washington, DC, EPA 440/5-86-001. May 1.

Van der Putte, I., Brinkhorst, M. A., and Koeman, J. H. (1981a). Effect of pH on the acute toxicity of hexavalent chromium to rainbow trout (*Salmo gairdneri*). *Aquat. Toxicol.* **1**, 129–142.

Van der Putte, I., Lubbers, J., and Kolar, K. (1981b). Effect of pH on uptake, tissue distribution and retention of hexavalent chromium in rainbow trout (*Salmo gairdneri*). *Aquat. Toxicol.* **1**, 3–18.

Vincent, J. B. (1999). Mechanisms of chromium action: low-molecular-weight chromium-binding substances. *J. Am. Coll. Nutr.* **18**, 6–12.

Vincent, J. B. (2000a). Elucidating a biological role for chromium at the molecular level. *Acc. Chem. Res.* **33**, 503–510.

Vincent, J. B. (2000b). The biochemistry of chromium. *J. Nutr.* **130**, 715–718.

Walker, R. L., Wood, C. M., and Bergman, H. L. (1988). Effects of low pH and aluminum on ventilation in the brook trout (*Salvelinus fontinalis*). *Can. J. Fish. Aquat. Sci.* **45**, 1614–1622.

Walker, R. L., Wood, C. M., and Bergman, H. L. (1991). Effects of long-term preexposure to sublethal concentrations of acid and aluminum on the ventilatory response to aluminum challenge in brook trout (*Salvelinus fontinalis*). *Can. J. Fish. Aquat. Sci.* **48**, 1989–1995.

Ward, J. V. (1973). Molybdenum concentrations in tissues of rainbow trout (*Salmo gairdneri*) and kokanee salmon (*Oncorhynchus nerka*) from waters differing widely in molybdenum content. *J. Fish. Res. Bd Can.* **30**, 841–842.

Water Research Centre (1990). *Design Guide for Marine Treatment Schemes.* Vol. II. *Environmental Design and Data Collection.* Report No. UM 1009. Swindon, UK: Water Research Centre.

Whiting, E. R., Mathieu, S., and Parker, D. W. (1994). Effects of drainage from a molybdenum mine and mill on stream macroinvertebrate communities. *J. Freshwat. Ecol.* **9**, 299–311.

Whittle, K. J., Hardy, R., Holden, A. V., Johnson, R., and Pentreath, R. J. (1977). Occurrence and fate of organic and inorganic contaminants in marine animals. In: *Aquatic Pollutants and Biological Effects with Emphasis on Neoplasia* (H.F Krayhill, C.J Daive, J.C Harshbarger and R.G Tardiff, eds), Vol. 298, pp. 47–79. Academy of Science, New York.

WHO (1996). *Guidelines for Drinking-Water Quality,* 2nd edn. Vol. 2. *Health Criteria and Other Supporting Information.* World Health Organization, Geneva.

Woodward, D. F., Riley, R. G., Henry, M. G., Meyer, J. S., and Garland, T. R. (1985). Leaching of retorted oil shale: assessing the toxicity to Colorado squawfish, fathead minnows and two food-chain organisms. *Trans. Am. Fish. Soc.* **114**, 887–894.

Yamazaki, H., and Gohda, S. (1990). Distribution of dissolved molybdenum in the Seto Inland Sea, the Japan Sea, the Bering Sea and the Northwest Pacific Ocean. *Geochem. J.* **24**, 273–281.

Zhitkovich, A. (2005). Importance of chromium–DNA adducts in mutagenicity and toxicity of chromium(VI). *Chem. Res. Toxicol.* **18**, 3–11.

9

FIELD STUDIES ON METAL ACCUMULATION AND EFFECTS IN FISH

PATRICE COUTURE

GREG PYLE

Metals have been present in the environment since the origin of our planet. Life has evolved in their presence, incorporating many of them in essential molecules and metabolic processes and developing protective mechanisms against both non-essential metals and excess essential metals. Anthropogenic metal contamination of terrestrial and aquatic systems dates as far back as the first traces of human civilization. However, it was not until the intense industrialization of the eighteenth century that severe environmental impacts of metal mining and smelting activities started to pose a serious threat to wildlife. Field research leading to environmental protection legislation first appeared for mercury and selenium, owing to their propensity to biomagnify and cause toxicity to the higher levels of food webs such as birds and humans. However, we have only recently started to reveal the mechanisms of metal accumulation and toxicity on fish and other wildlife for low-level, chronic exposure scenarios. In aquatic environments, fish accumulate metals via both aqueous and dietary routes. The effects of chronic metal exposure, typical of most metal-contaminated environments, are more subtle than, and greatly differ from, the effects of acute exposures. There is evidence of direct toxicity leading to bioenergetic consequences such

Homeostasis and Toxicology of Essential Metals: Volume 31A
FISH PHYSIOLOGY

as decreased growth rate and condition, as well as selective pressures reducing population genetic diversity. Direct metal toxicity at other trophic levels also induces indirect effects, both negative and positive, on fish populations. Because metal accumulation and toxicity in wild fish are affected by several biotic and abiotic factors, field modeling remains a challenge. Nevertheless, with the advancement of knowledge, reviewed in this chapter, on accumulation and effects of metals in wild fish, our capacity to protect fish and their habitat from anthropogenic metal contamination is fast improving.

1. HISTORICAL REVIEW OF NATURAL AND ANTHROPOGENIC CONTAMINATION OF AQUATIC ENVIRONMENTS BY METALS

Life on Earth has evolved on a planet containing a dense core of iron and nickel surrounded by a mantle composed of 94 naturally occurring elements, the majority of which are metals. Although biological systems are largely composed of carbon, hydrogen, nitrogen, and oxygen, some metals, with their unique chemical properties, allow energy to flow through and be harnessed by our cells.

Organisms inhabiting the early marine environment probably made use of available environmental constituents to support structural and physiological requirements in much the same way as today's living organisms. This speculation is supported by the observation that the elemental composition of today's primitive marine invertebrate body fluids basically reflects the elemental composition of the water in which they live (Nielsen, 2000). Similarly, organisms that inhabited Archaean oceanic waters probably incorporated many of the elements available to them in their environment to satisfy physiological or structural requirements. Yet, the Archaean marine environment was chemically very different from the current marine environment, and the elements utilized by early biota were probably quite different from those that are utilized today (Fig. 9.1).

Some theories about life's early evolutionary history posit that life evolved in shallow marine environments, such as littoral zones and tidal pools (Nielsen, 2000). Such an evolutionary past might account for the major biological importance of certain elements such as carbon, nitrogen, and phosphorus in today's biota. Other theories suggest that life evolved around deep oceanic, ammonium-spewing hydrothermal vents, where Fe, Mn, and S dominated generally, and other minerals such as Ni, Cu, Zn, Mo, Se, Mg, and P dominated locally (Nielsen, 2000). Interestingly, all of these elements are essential to extant biota. Additional evidence that supports

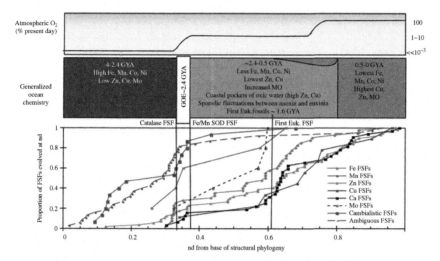

Fig. 9.1. Relationship between the emergence of metal-binding protein fold superfamilies (FSFs) and changing marine chemistry through time. Top panel: atmospheric oxygen as a percentage of today's concentration. Middle panel: generalizations about changing marine water chemistry over time. Bottom panel: the emergence of FSFs as a function of phylogenetic node distances (nd). Other acronyms and symbols: Euk.: eukaryotes; GOE: Great Oxidation Event; GYA: billions of years ago; SOD: superoxide dismutase. From Dupont et al. (2010). **SEE COLOR PLATE SECTION.**

life's early beginnings around hydrothermal vents is the apparent ubiquity and broad phylogenetic distribution and conservation of heat shock proteins, which are typically involved in photosynthesis, mitochondrial function, nuclear membrane function, and spermatogenesis, and their well-known induction in the presence of metals. Metalloenzymes, such as the nickel enzymes urease and hydrogenase and the molybdenum enzyme nitrogenase found in some prokaryotes and plants, also point to a hydrothermal vent origin.

Among the earliest biomolecules to form was the magnesium-containing chlorophyll, which allowed organisms to harness energy from light in order to manufacture carbohydrates from carbon dioxide (Nielsen, 2000). This allowed photosynthesizing organisms to move away from the energy- and mineral-rich hydrothermal vents to occupy shallow, cooler photic zones. But it also caused the liberation of huge quantities of oxygen into an otherwise anoxic environment. Most metals in this anoxic environment were locked up as insoluble metal sulfides and unavailable to biota owing to the prevailing anoxic conditions (Clarkson, 1995). However, environmental conditions changed radically once organisms began photosynthesizing and liberating toxic oxygen. As the anoxic conditions gradually gave way to increasing oxygen concentrations, sedimentary metal sulfides became oxidized, causing a massive release of previously sequestered metals into the water column (Clarkson, 1995). The early marine environment was faced with two important environmental challenges involving increasing concentrations of dissolved oxygen and metals. Just as was the case with early biota, any species that survived had to adapt to this new oxygen- and metal-rich environment by developing strategies for detoxifying or regulating oxygen and metals and incorporating them into their basic physiology.

This drastic change in environmental redox conditions led to a concomitant change in how organisms incorporated metals into their basic physiological processes. For example, Ni and Co were excellent metabolic catalysts in early biota when hydrogen, ammonia, methane, and hydrogen sulfide were abundant in the reducing, anoxic marine environment. However, as the early marine environment became more oxic and oxidizing, and nitrogen, oxygen, and carbon dioxide became more abundant, Cu, Fe, and Zn became more important metabolic catalysts, which is reflected in extant fish of today. In fact, one-third of all proteins require a metal cofactor for normal function (Rosenzweig, 2002). It is interesting to speculate that the proteins that defend fish against oxidative damage are upregulated in the presence of certain metals, i.e. the same metals that would have dominated the marine environment around the same time that those proteins would have evolved (Dupont et al., 2010). For example, Fe is associated with the catalases, while Cu, Mn, and Zn activate the dismutases.

The changing environment, then, provided powerful selective pressures that shaped fundamental physiological processes (Williams, 1997). For example, early in the evolution of marine life, the predominantly reducing environment favored Ni or Co for catalyzing redox reactions for extracting energy from hydrogen or carbon monoxide. However, as the marine environment shifted from a predominantly reducing anoxic environment to a predominantly electron-poor, oxidizing and oxic environment, copper, iron, and zinc were favored in the same catalytic reactions (Dupont et al., 2010). Although Ni and Co can still catalyze these reactions to a certain extent, they are not nearly as efficient as Cu, Fe, and Zn. Hence, natural selection favored the systems that catalyzed energetic reactions most efficiently, thereby shaping the physiological mechanisms involving metal-loenzymes observed in today's biota, including fish (Dupont et al., 2010).

Before the rise of the human species, no event on Earth had led to the release of quantities of bioavailable metals comparable to the situation of the great oxidation event. As we shall see below, just as life itself has been, human evolution has been tightly linked to metal use. To varying degrees, metals can be both toxic and essential. As a species, we have evolved a capacity to extract and use metals from the Earth's crust, henceforth increasing the concentrations of these metals in all compartments of the environment. Thankfully, we have also evolved a capacity to understand that humankind cannot survive on this planet without healthy ecosystems. Recent progress is increasingly allowing us to quantify the impacts of our activities on the environment and to develop tools to minimize them.

The use of metals (Table 9.1) was so important for human prehistory that anthropologists divided it into three eras based on the development and use of tools: the Stone Age, the Bronze Age and the Iron Age. The Stone Age was a long pre-metallurgic period during which early human societies developed tools and weapons shaped from non-metallic, naturally occurring materials such as stone, bone, and wood. At the end of the Stone Age, a transitional period, named the "chalcolithic" or "Copper Age", saw the appearance of copper tools alongside stone tools. Being the eighth most abundant element in the Earth's crust and one of the few elements that could be found in a pure state, Cu, with its unique appearance, was relatively easy to find and could be shaped into small objects simply by hammering it with a hard tool. Rapidly, humans started to heat up and melt the metal and to experiment with alloys of soft metals, in particular Cu and Sn, giving rise to the Bronze Age. This was the beginning of metallurgy, which is thought to originate from the Fertile Crescent, a vast area of Western Asia including parts of modern Jordan, Lebanon, Iran, Iraq, Israel and the Palestinian territories, circa 3300 BC, although evidence suggests that it arose independently in several parts of the world, including Asia, Europe, and America. Although Fe, mostly of meteoritic origin, was increasingly present throughout the Bronze Age, it is only from around 1200 BC that civilizations across the world had sufficiently improved their smelting techniques to reach the temperatures allowing for the melting of iron ore. A whole research area, archeometallurgy, examines how early human societies

Table 9.1
Approximate dates of first human use of common metals

Metal	Date
Copper	9000 BC
Lead	6400 BC
Nickel	3500 BC
Tin	3000 BC
Zinc	2500 BC
Iron	2000 BC
Mercury	1500 BC
Cadmium	AD 1817
Aluminum	AD 1825

throughout the world collected metal ores and processed them to fabricate tools, weapons, and decorative objects. The use of metals appears tightly linked to the development of human societies, allowing for technological advances in agriculture, warfare, and artistic expression. Possessing metal mining and smelting technologies was a key in determining the technological development of human societies.

Along with deforestation and agriculture, mining and smelting are clearly among the first human activities to impact on the environment. Although prehistoric mines remained very small compared to modern mining operations, there is evidence that even those early mines produced atmospheric and watershed contamination. For instance, Cu contamination of peat deposits likely originates from atmospheric pollution generated by a Bronze Age copper mine in the UK (Mighall et al., 2002). Also in the UK, tin-contaminated sediments in periods of intense aggradation (sediment accumulation in lowlands) have been considered evidence of early tin mining in Dartmoor between the fourth and the seventh centuries AD (Thorndy-craft et al., 2004). Metal contamination of waterbodies is tightly linked to the expansion of human civilizations, especially since the Industrial Revolution, which started in the eighteenth and nineteenth centuries in Britain and rapidly spread worldwide. Like pollution in general, environ-mental metal contamination and impacts did not receive much attention until the second part of the twentieth century, and measures to decrease emissions and to mitigate the impacts of metal mining and smelting only started in the 1970s (Gunn et al., 1995). Furthermore, the accurate measurement of trace amounts of metals in sediments, water, and biota required advanced analytical techniques not available until recently. As a result, our knowledge of the historical trends of environmental metal contamination is mostly based on recent studies of dated metal deposits in structures such as lake or river sediment cores (Meybeck et al., 2007; Couture et al., 2008c) and tree growth rings (Latimer et al., 1996). Below is a brief review of some of these studies in key areas of the world where metal contamination was, and still is, a major concern for aquatic biota.

The Seine River drains 65,000 km^2 of the land of France, including Paris, and one-third of its industrial and agricultural activity. A recent study of historical metal contamination dating back over a century shows a clear match between the development of industrial activities and sediment contamination by metals at the mouth of the Seine (Meybeck et al., 2007). The concentration of metals released throughout the twentieth century peaked between 1950 and 1970. Remarkably, a sharp decrease in metal input can be observed between 1970 and 2000, a period during which analytical capacities were refined, environmental awareness was increased, and European regulations on environmental contamination were tightened, in spite of a continued increase in the demand for, and use of, metals in France and worldwide. Hence, beyond the natural background that varies according to local geochemistry, the dominant drivers of environmental metal contamination are human activities, which can be modulated by environmental awareness and regulations.

Perhaps the most well-documented example of metal mining and smelting leading to severe terrestrial and aquatic contamination is the case of Sudbury (Ontario, Canada). Mining in this naturally highly mineralized area started in 1888 and today the area is still one of the world's largest Ni producers. In the beginning, logging was also intense and the wood harvested from the region was used to fuel open roast beds, which were later replaced by smelters. Huge amounts of sulfur dioxide and metals (mainly Ni and Cu, but also Fe, Cd, and several other metals) were released to the environment, which led to a catastrophic acidification and metal contamination of soils and lakes, effectively sterilizing terrestrial and aquatic environments. The early 1970s brought about strengthened environmental regulations and improved technology was implemented for the smelting process, resulting in a corresponding 90% drop in acid and metal emissions (Gunn et al., 1995; Nriagu et al., 1998). Yet, the Sudbury ecosystems are now so saturated with metals that surface runoff, groundwater drainage, and wind erosion of mine tailings continue to contribute towards persistently elevated Ni and Cu concentrations in area lakes. It is anticipated that the area will remain contaminated for well over 1000 years into the future (Nriagu et al., 1998). Given the exceptional abundance of freshwater lakes in the Sudbury area (over 1000 just within the city limits and around 7000 that were considered acidified and dead in the early 1970s), gradients of metal contamination in otherwise pristine lakes (i.e. receiving little other anthropogenic influence) have provided unique research opportunities. An abundant literature has been produced examining metal accumulation and effects in freshwater fish of the Sudbury area and this literature is reviewed in other sections of this chapter.

Several other examples of well-documented histories of environmental metal contamination due to anthropogenic activities are available. The case of San Francisco Bay is particularly interesting. It is a highly complex system with multiple sources of inorganic and organic contaminants (irrigation and agriculture, mining, industries, port activities, and urban input) and salinity ranging from freshwater to full-strength seawater. Upstream in the Sacramento River drainage, metals leach into the system from an old mine site. Fish kills occurred repeatedly over the last century, the last one dating back to 1978. In spite of intense remediation efforts, potentially toxic dissolved metal concentrations were still observed in the 1980s in the Sacramento River due to dam overflow during heavy rainfall

(Saiki et al., 1995). A study of Pb contamination originating from leaded gasoline exemplifies the complexities of natural decontamination trends after a major reduction of contaminant sources in estuaries. Steding et al. (2000) reported that watershed inputs will continue to contribute to Pb contamination of San Francisco Bay for several decades, contrasting with reports of recent trends of decrease in Pb contamination in other environments following a ban on leaded gasoline in developed countries. Hence, as in the case of Sudbury, although anthropogenic sources of metals have greatly decreased in the past two decades, environmental contamination will continue to impose a stress to aquatic organisms in several environments which have in the past been saturated with metals from anthropogenic sources. Furthermore, the case of leaded gasoline highlights another issue: while Pb has been banned as an additive for over a decade in developed countries, it is still in use in other countries of the Middle East, Eastern Europe, Asia, Africa, and South America. Clearly, although anthropogenic metal sources and the risk they represent to fish and other wildlife may be decreasing globally owing to stricter regulations and increased awareness, the lag in some jurisdictions in adopting strict environmental regulations, combined with the persistent and non-degradable nature of metals, imply that environmental metal contamination will continue to be a concern in many hot spots for decades to come. See Mager (Chapter 4, Vol. 31B) for further discussion of Pb contamination.

Unlike the case of synthetic chemicals, metals have always been, and will remain, present in the environment and they will continue to interact with life, which has adapted to their presence. Even in the most optimistic scenarios where no metal is released from human activities, wildlife will be exposed to varying metal concentrations depending on their habitat. This is because some areas of the Earth are naturally more highly mineralized than others. Some of the latest guidelines on safe metal concentrations for the protection of aquatic life now specify that environmental targets must consider natural background (ANZECC, 2000; CCME, 2003).

The vast contamination of aquatic ecosystems by metals of anthropogenic sources has led to a rapid expansion of research examining the toxic effects of metals on fish. Literature on this topic started to explode as early as the 1970s but was limited to laboratory studies. Early literature on field studies was sparse and largely limited to reports of whole-fish metal concentrations. By the 1990s, the literature expanded dramatically to include investigations of indicators of effects at the ecological, physiological, and molecular levels. Yet even today, much of our knowledge on metal toxicity to fish which has shaped current legislation comes from the laboratory. Without pretending to provide a complete overview of all literature available on the topic, this chapter documents several of the major studies examining metal accumulation and effects in wild fish. Although intrinsically more difficult to interpret because of the complex mixtures and myriad of confounding factors in the field, the importance of field studies for understanding the ecological risks to fish from environmental metal contamination is undeniable. Yet, even though the field studies reviewed in this chapter have significantly advanced our knowledge of the extent and mechanisms of metal toxicity to wild fish, this research has not yet generally contributed much to improving legislation. The cases of Hg and Se, very briefly reviewed here, are noteworthy exceptions representing field studies that raised awareness and concern, and influenced legislation. Please refer to Janz (Chapter 7) and Kidd and Batchelar (Chapter 5, Vol. 31B) for extensive discussions of Se and Hg, respectively.

The Experimental Lakes Area (ELA) was established in 1968 in response to an International Joint Commission (IJC Canada–USA) call to investigate the effects of contaminants on freshwater systems. The area is located in north-western Ontario, Canada, on Precambrian Shield, and consists of 58 freshwater lakes. Since its inception, the ELA has served as an important resource for environmental scientists interested in whole-lake responses to environmental contamination, including Hg. Previous work at the ELA focused on several factors affecting whole ecosystems (e.g. eutrophication, acidification, and endocrine-disrupting chemicals). However, research conducted around Hg focused on ecosystem-level mechanisms that could account for elevated Hg concentrations observed in tissues of fish inhabiting hydroelectric reservoirs (Rudd, 1995). The important work conducted at the ELA formed part of our fundamental understanding of ecosystem-wide Hg dynamics, and how those dynamics influenced Hg concentrations in resident fish.

Inorganic Hg can be biochemically methylated to form methyl mercury (CH_3Hg; MeHg), which is the Hg species taken up by fish tissues most readily and the most toxic Hg species (Jensen and Jernelöv, 1969; Wobeser, 1975; Rudd et al., 1980; Kidd and Batchelar, Chapter 5, Vol. 31B). Methyl mercury can enter aquatic systems from surface runoff or atmospheric fallout, or through internal processes forming MeHg from available inorganic Hg (Rudd, 1995). These internal lake processes include biogenic methylation of inorganic Hg in the sediments, the water column, the mucous layer of resident fish (Rudd, 1995), and fish intestines (Rudd et al., 1980). Organomercury uptake to fish tissues is higher at lower exposure water pH (Boening, 2000). However, this is likely due to the increased prevalence of MeHg at low pH relative to higher pH values.

The ELA was also instrumental in elucidating the importance of dietary sources for metal accumulation in fish tissues (Klaverkamp et al., 1983). Following from a series of enclosure studies demonstrating a now well-characterized antagonism between waterborne Se concentration and fish-tissue Hg accumulation (Turner and Swick, 1983), Klaverkamp et al. (1983) demonstrated that very low waterborne Se concentrations (1 μg L^{-1}) inhibited Hg accumulation, but higher Se concentrations (100 μg L^{-1}) had the opposite effect. As Se concentrations were increased, Se accumulated in trophic compartments used by the study fish as food. Because trophic transfer is the most important uptake route for Se (Hamilton, 2004), elevated waterborne Se concentrations led to increased Se concentrations in fish food items, leading to Se toxicity that offset any antagonism with Hg. Consequently, the feasibility of seeding Hg-contaminated lakes with dissolved Se as an ameliorative strategy against Hg accumulation to fish must be considered with great caution and not exceed a final concentration of 1 μg L^{-1} (Hamilton, 2004).

The ELA has recently renewed its research efforts into environmental Hg contamination with the launch of METAALICUS (the Mercury Experiment to Assess Atmospheric Loading In Canada and the United States) in 2001. The objectives of this study are to understand how increasing atmospheric loading of Hg to the atmosphere will affect fish tissue MeHg concentrations, and to determine the time-scale of whole ecosystem responses (including fish) to changing atmospheric input rates. In this study, an international team of researchers is releasing stable Hg isotopes to an experimental lake at a rate of about three times greater than natural atmospheric loading. Information from this research will be used to inform legislation regulating the amount of Hg that can be released to the atmosphere.

2. RELATIVE IMPORTANCE OF DIET VERSUS WATER AS METAL SOURCES IN WILD FISH

Ecotoxicology has traditionally assumed that waterborne metal concentration was a sufficient predictor of toxicity. However, recent studies have demonstrated the importance of dietary metals for contributing towards fish tissue metal burdens and toxicity. There is now sufficient evidence to suggest that ignoring dietary metal uptake in environmental risk assessments would often lead to underestimates of metal accumulation and risk (Chapman and Wang, 2000; Meyer et al., 2005). Yet, establishing the relative contribution of the two sources towards tissue metal burdens in wild fish is difficult because of the influences of the specific metal of interest, waterborne metal speciation and bioavailability, fish life stage, digestive physiology, prey type, foodborne metal subcellular distribution and bioavailability, and feeding rate. The influences of these parameters on metal uptake in wild fish will be briefly reviewed in this section.

The pathways metals follow through an aquatic ecosystem from natural and anthropogenic sources to end up in fish are complex and depend on several factors. As illustrated in Fig. 9.2, water is the only source of metals for non-feeding fish embryos. The case of large, piscivorous fish species is not as simple, because their diet will include small fish and variable proportions of benthos and even of zooplankton. Smaller fish species or juveniles of large species have the broadest prey base and hence the most potential sources of dietary metals, which can even include embryos of their own or other species. The relative importance of each pathway (the thickness of arrows in Fig. 9.2) will be species, life stage, metal, and site specific, so that a different Fig. 9.2 can be obtained for each combination. Since feeding rate will vary seasonally from fasting to maximal, minimum values of relative dietary uptake will always be 0%. Factors affecting the maxima are reviewed below.

Once metals enter an aquatic system, they can adsorb onto particulate matter, form insoluble inorganic complexes, or remain in the dissolved fraction, either as the free ion or bound to dissolved inorganic complexes or

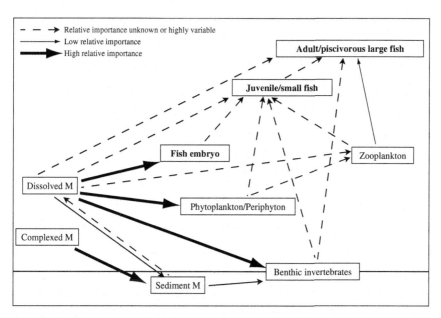

Fig. 9.2. Schematic illustration of general metal sources for wild fish. Arrow thickness relates to the relative importance of various metal uptake routes for different food web components. Fish embryos are non-feeding and can only take up metals from water. Zooplankton are not expected to make a significant contribution to dietary metal uptake in large piscivorous fish. Other metal sources for juvenile or adult fish vary greatly in importance (see text).

to dissolved organic matter. Most of the insoluble (complexed) fraction will eventually fall out of the water column and settle into the sediment (Fig. 9.2). Sediment metals can re-enter the water column by a number of mechanisms. For example, metal sulfides are stable and insoluble under anoxic conditions. However, anoxic sediments may become oxidized, perhaps through mixing with oxic overlying water by way of bioturbation (Ciutat and Boudou, 2003; Atkinson et al., 2007; Lagauzere et al., 2009) or natural water flow, causing the metal to redissolve in the overlying water. Once the metal is redissolved it may become bioavailable and enter the food web. Alternatively, natural diagenetic processes may lead to the development of a thick overlying sediment layer or metals may simply leach into deeper layers. In either case, deep sediment layers are typically anoxic allowing for stable, insoluble metal species that are generally not available to biota. In this case, some sediment metals may be permanently isolated from the food web (Boudreau, 1999). Regardless of their fate, metals in sediment cannot be ignored as a source of metals even in systems where water quality criteria are met, as they may contribute to food web contamination and consequent effects in fish (Dallinger and Kautzky, 1985).

To be toxic to fish, waterborne or dietary metals must be bioavailable, i.e. in a form (chemical species) available for uptake. Several natural factors are known to affect waterborne metal bioavailability to fish from natural environments. In freshwater fish, waterborne metals are generally taken up by the gills, not by the gut, because freshwater fish generally do not drink. In contrast, saltwater fish need to drink in order to maintain osmotic homeostasis; therefore, the gut may be an important site of waterborne metal uptake in addition to the gill in saltwater fish. Respiration rate, which is a function of temperature and metabolic requirements, has been suggested as influencing both aqueous metal uptake through the gill and dietary uptake (see Section 5 for a detailed discussion). Ventilation rate, itself driven by metabolic rate and partial pressure of oxygen, has also been reported, or estimated, to influence metal uptake in some studies (Clearwater et al., 2002; Veltman et al., 2008) but not in others (Hattink et al., 2006; Pierron et al., 2007). Mechanisms linking hypoxia-induced increases in ventilation and metal uptake are still debated and may involve changes in gill blood irrigation and boundary layer pH under hypoxia rather than variations in water flow through the gills (Campbell, 2004). The most important factor for waterborne metal uptake to fish is metal speciation as it relates to metal bioavailability. For most metals, speciation is driven by water pH such that the free cation dominates speciation under low pH conditions. Increasing pH causes the formation of less bioavailable dissolved metal complexes, such as metal hydroxides and carbonates. Under low pH conditions, where the free cation tends to dominate, competition for uptake sites by H^+ ions may decrease metal uptake. In other words, changes in pH lead to mutually compensating effects: decreased metal complexation at low pH, but increased H^+ ion competition at metal binding sites at low pH; increased metal complexation at high pH, but decreased competition from the H^+ ion (Campbell and Stokes, 1985). Note too that at very low ambient pH, some fish may not survive because of the toxicity of the H^+ ion itself.

Although it is generally true that metal complexation leads to decreased metal bioavailability, there are exceptions (Campbell, 1995; Erickson et al., 1996). Water hardness also affects waterborne metal uptake insofar as hardness cations (i.e. Ca^{2+} and Mg^{2+}, although the former is usually more important in fish than the latter) compete with metal ions for physiologically sensitive binding sites. Consequently, metals tend to be less toxic in hard water than in soft. Taken together, these two factors (notwithstanding the influence of dissolved organic matter; see below), water pH and hardness, can yield natural exposures leading to either reduced or elevated risks of metal accumulation and subsequent intoxication (Birceanu et al., 2008). In the northern hemisphere, a significant proportion of metal mining activity takes place in areas supporting lakes having soft, "sensitive" waters; that is, soft, acidic water. Metals released from industrial operations to these soft, sensitive waters are available to biota in their most bioavailable and toxic form (Lydersen and Löfgren, 2002).

Few studies have examined the relative importance of water and food as metal sources in wild fish, likely because a number of experimental challenges make it very difficult to yield conclusive results. Among these challenges, the quantity and type of food ingested by fish in the field can only be estimated, so the absolute dietary metal intake cannot be quantified precisely. In contrast to waterborne metals to which fish are constantly exposed, dietary metal intake is obviously dependent on feeding rate. There are also important variations in metal contamination of food potentially ingested by fish. In invertebrates, for example, which comprise at least part of the diet of most fish species at one or more life stages, major differences in metal concentrations have been reported among sympatric taxa, even among closely related species (Hare et al., 2001; Proulx and Hare, 2008). This variation is due in large part to differences in behavior and feeding habits.

Several studies have examined wild benthic invertebrate metal uptake from water and/or sediment using a variety of study approaches. Approaches include exposing field-captured invertebrates to field-contaminated or spiked sediment in the laboratory, exposing caged animals to spiked sediment in the field, and specimen collection along gradients of water and sediment metal contamination. Small benthic or planktonic invertebrates are more amenable to laboratory testing than fish or large invertebrates (e.g. large bivalves) using artificial conditions that mimic natural conditions, given their small size, easy maintenance in the laboratory, and tendency to behave normally under controlled laboratory conditions. Therefore, the question of metal sources in invertebrates in the field is often best answered by laboratory studies combined with field validation (Borgmann et al., 2001). Metal uptake patterns in invertebrates, as well as entire invertebrate species assemblages, vary along contamination gradients. In a study where prey zooplankton from along a Cd gradient was sorted by size, Croteau et al. (2003) demonstrated that the predicted Cd accumulation by *Chaoborus* was much improved when the prey species was known. These studies revealed that the relative contribution of sediments (through ingestion), water or food (phytoplankton/periphyton) as metal sources for benthic invertebrates will greatly vary interspecifically, as illustrated in Fig. 9.2.

The efficiency of the trophic transfer of metals between fish and their prey depends on several factors, including uptake and elimination rates and digestive physiology of the consumer, residence time of food in the gut, and subcellular metal distribution in the prey (Wallace and Luoma, 2003; Lapointe and Couture, 2010). Once internalized, metals may bind to various subcellular components that vary in digestibility, and hence in their potential for trophic transfer. Simply put, trophically available metals (TAMs) are those that are bound to protein fractions and organelles and available to consumers. Metals that are trophically unavailable are those that are incorporated in inorganic granules and unavailable to consumers. We do not know whether the source of metals (water or food) influences subcellular metal distribution in benthic invertebrates. However, the efficiency of trophic transfer of metals in invertebrate food chains appears reasonably well predicted by the proportion of TAMs in prey (Wallace et al., 2003). Trophic transfer of metals was quantified in a study in which two invertebrate prey, an insect (*Chironomus riparius*) and a worm (*Tubifex tubifex*), were contaminated by exposure to sediment spiked with Ni or Tl and then subsequently fed to an invertebrate predator, the alderfly *Sialis velata* larvae (Dumas and Hare, 2008). Half or more of both metals accumulated by the two prey were considered as TAMs and assimilation efficiency by the predator closely matched the predictions from TAMs. In contrast to studies with invertebrate predators, when two invertebrates (*Daphnia magna* and *T. tubifex*) with different subcellular distribution of Ni and Tl were fed to a fish (fathead minnow), the fraction of the metal in prey considered to be trophically available had no influence on the minnow's metal assimilation efficiency (Lapointe et al., 2009). Fish fed *D. magna* assimilated dietary Tl twice as efficiently as those fed *T. tubifex*, even though about 50% of Tl was in TAMs for both prey types. In contrast, prey type did not influence Ni assimilation, but trophic transfer of Ni was much lower than for Tl for both prey types. The authors suggested that the differences observed in assimilation efficiencies between prey types may be due to differences in metal concentration, where higher metal concentrations in prey were related to lower metal assimilation efficiencies, as demonstrated in *D. magna* for Cd and Se (Guan and Wang, 2004). Differences in assimilation efficiency between the two metals also suggest that there are no regulation mechanisms for non-essential metals such as Tl, implying that the lower trophic transfer of Ni reflects the presence of mechanisms limiting intestinal uptake, an indicator of essentiality (see Pyle and Couture, Chapter 5, for a more complete discussion). To the authors' knowledge, there are no systematic studies examining the influence of prey type on metal assimilation in wild fish.

The relative contribution of water and diet to metal intake can be examined by applying different experimental approaches to make a seemingly intractable research question somewhat more tractable in wild fish. Two major approaches are mensurative and manipulative field studies, the former being primarily descriptive and correlational in nature, whereas the latter are experimental. Few detailed investigations of the relationships between metal accumulation in individual prey species and bioaccumulation in fish have been conducted (Draves and Fox, 1998), and sometimes they have met with limited success (e.g. Bervoets et al., 2001). However, the larger size of fish compared to invertebrates allows the use of the alternative approach of measuring metal concentrations in their gut contents (Köck et al., 1996; Audet and Couture, 2003; Couture et al., 2008a). The most obvious advantage of this approach is that the investigators do not need to estimate the proportion of each type of prey available that makes up the diet of individual fish of given size at a particular time of the year. Instead, they let the fish do their own sampling. There are, however, limitations to this approach. For instance, not all fish sampled have recently eaten, so that the proportion of fish with empty guts is often important, depending on time of day and season: fishermen, like field biologists, have long known that fish are often more actively feeding at daybreak and hence are more likely to have full guts early in the day. The influence of season on feeding rate, and hence on dietary metal uptake, is also major: depending on gender and age, fish may only eat and grow during a narrow part of the year. The implication here, beyond the absence of gut contents in these fish, is that during periods of extended fasting, only aqueous metal uptake takes place. Finally, it is likely that the time since the last meal will also influence gut content metal concentrations, as digestive and absorptive processes will modify the original metal concentration of the food ingested.

Couture et al. (2008a) examined the relative contribution of water and diet for several metals using a correlative statistical approach between metal concentrations in water or food (gut contents) and liver and kidney of yellow perch collected along two metal contamination gradients (Sudbury, Ontario, and Rouyn-Noranda, Québec, both in Canada). The authors concluded that dietary Cd was the main source of contamination. However, for the other metals examined (Cu, Ni, Se, and Zn), water appeared to be the main source. The better approach, but so labor intensive that it is not practical for environmental monitoring and rare in the literature (e.g. Dallinger and Kautzky, 1985), is to measure metal concentrations both in fish gut contents and in their prey. In an Italian study in which *Oncorhynchus mykiss* (formerly known as *Salmo gairdneri*) was collected from rivers presenting heavily metal-contaminated sediment but low aqueous concentrations (Dallinger and Kautzky, 1985), the authors concluded, by comparing metal concentrations in fish gut, liver, and kidney with concentrations in prey (snails and isopods) and water, that food was a main source of metals for these wild fish. A shortcoming of the correlational approach is that causation cannot be assumed. Strong correlations between fish tissue and water metal concentrations do not preclude the possibility that dietary metals may still be the dominant source of the contamination, in which case these correlations would simply result from stronger correlations between water and prey and between prey and fish. Indeed, there is evidence that, at least for Se, diet is the main source of Se in wild fish, not water (Hamilton, 2004).

Given the large variations in metal concentration and bioavailability in fish food compared to aqueous sources, in investigations aiming at determining the proportions of accumulated metals in wild fish originating from food versus water, the contribution of aqueous uptake is easier to quantify than that of the diet. A simple approach to address this question is by field experiments in which clean fish are caged in contaminated lakes and fed clean food. Metal accumulation in these fish can only (or mainly, if clean food is contaminated in cages before consumption and this can be quantified) originate from water, and since it integrates water quality factors affecting gill uptake, this approach avoids using complex bioaccumulation models in which several variables need to be estimated. There are, however, several limitations in cross-transplantation experiments using caged fish. For instance, it is quite difficult to feed fish the same type and amount of food they would normally consume and as a result growth rate is most likely lower in caged, compared to free-ranging fish. In addition, caged fish change their behavior and most likely their feeding rate as a result of stress and changes in their locomotor activity level. In one cross-transplantation study, Kraemer et al. (2005) caged yellow perch in a clean and a metal-contaminated lake and fed them plankton from either the clean or the metal-contaminated lake. Their results suggest that the source of Cd was tissue specific, with a greater influence of Cd of aqueous origin in gills and kidney, while Cd accumulated in liver and gut appeared similarly influenced by both sources. Their conclusions were supported by the study of Couture et al. (2008a), discussed above, which used a correlational approach, hence supporting that both approaches have their worth, within their own limitations.

Given the complexities of field studies for determining the relative importance of diet and water as metal sources in fish, much of this research has been conducted in the laboratory and it has provided evidence demonstrating that both sources can contribute. Evidence of trophic transfer of metals to fish based on laboratory exposures has been available for decades (Patrick and Loutit, 1978). There are several advantages of laboratory studies compared to the field approaches described above. Firstly, sources can be isolated and controlled. Secondly, feeding rate, growth rate, and assimilation rate are much more easily characterized in the laboratory than in the field. Thirdly, both aqueous and dietary uptake can be characterized precisely. Although earlier studies typically used unrealistically elevated metal concentrations in short-term, acute exposures, experimental designs using environmentally relevant concentrations of metals in water and food are nowadays more common in the literature. For an essential metal like Cu, the question is complex even in the laboratory. For instance, although Cu has been demonstrated to be mostly accumulated through the diet in rainbow trout, pre-exposure to Cu decreased new Cu accumulation (Kamunde et al., 2002). Their conclusions, drawn from laboratory studies, are supported by a field study of a salmonid which, using a correlational approach, concluded that food was the dominant source of Cu and other metals (Dallinger and Kautzky, 1985). However, these studies do not agree with field studies of other species, which suggested that water, and not food, was the main source of Cu for yellow perch (Couture et al., 2008a) and white sucker (Miller et al., 1992). It is clear from this research that both water and food can contribute to metal accumulation in wild fish and that food can constitute the dominant pathway. However, the studies reviewed above demonstrate that the answer varies depending on fish species, life stage, tissue, and metal. There is also evidence to suggest the major contribution of diet as a metal source for marine fish, as reported in an investigation in which published studies were used to model the relative contribution of water and diet to metal uptake in an elasmobranch and a teleost fish (Mathews and Fisher, 2009). In that study, the authors concluded that diet contributed to more than half and in some cases almost totally to the uptake of Cd, Mn, and Zn.

Several models, some being quite successful, have been developed to predict metal accumulation in fish from environmental exposure (see Section 6 and Paquin et al., Chapter 9, Vol. 31B). Nevertheless, the highly complex influences and interrelationships among all factors affecting metal accumulation in wild fish, including feeding rate, food contamination, subcellular partitioning and availability of metals for trophic transfer, growth rate and all factors influencing aqueous uptake, as well as the influence of fish species, life stage (which influences dietary preferences), and the nature of the metal (essential or not), will always make it a challenge to predict metal accumulation in fish from environmental metal concentrations. For these reasons, direct measurements of tissue metal concentrations in field-caught fish should remain the approach to favor for relating environmental and fish contamination. However, there are several reasons why we need better predictive tools of metal accumulation in wild fish. Firstly, fish sampling is costly and not always practical, for logistical as well as conservational reasons. Secondly, we may want to estimate metal accumulation in an environment where fish are not yet present but where introduction is planned. Conversely, interest in predicting metal accumulation in wild fish may arise in clean environments in which metal concentrations are predicted to increase owing to new anthropogenic sources such as mining and smelting operations or accidental spills. Given that for wild fish the proportion of metals accumulated from aqueous versus dietary origins is not fixed, the recommended modeling approach is to estimate metal accumulation both from water only and from food only. Since accumulation by both sources can, at worst, be cumulative (Lapointe and Couture, 2009), the potential for metal accumulation in wild fish will range between the lowest of aqueous and dietary routes and the sum of the two. Although we can also estimate that the dietary route will increase in relative importance in periods of intense feeding, which can be deduced from a good knowledge of the ecology of the fish under study, direct measurements of fish tissue metal concentrations will always be required to quantify with any certainty metal accumulation in wild fish.

3. BIOENERGETIC EFFECTS OF METAL CONTAMINATION IN WILD FISH

Until the 1990s, the subtle physiological consequences of chronic metal exposure in wild fish were largely unknown. A dominant view was that metal-sensitive fish species living in metal-contaminated environments disappeared while resistant species thrived (the flourish or perish ecological perspective). It was deduced from observation that metal contamination created ecological disturbances by removing the most sensitive components of food webs, which in turn induced a shift in the relative abundance of tolerant species. Although organic pollutants can induce obvious effects such as tumors, sex change, or deformities, the effects of chronic metal exposure in wild fish are much more subtle. There is evidence, reviewed in this section, to support that chronic metal exposure leads to direct toxicity in wild fish. It is proposed that metal contamination exerts an energetic pressure on wild fish through a combination of direct metabolic toxicity, oxidative stress, and costs of detoxification and repair. Evidence will be presented to support the hypothesis, proposed by Clarkson (1995), that metals share with oxygen several modes of toxic action and hence that the mechanisms of cellular protection that have evolved against oxygen-induced toxicity are largely the same that cells use to protect themselves against metal toxicity. Because these pressures are exerted on all components of aquatic food webs but with varying intensity depending on organismal physiology and ecology, and since fish, which are at the top of these food webs, also integrate the indirect effects at other levels, demonstrating direct metal toxicity in wild fish necessarily requires laboratory validation. Therefore, although this section is limited to reviewing metal effects in wild fish, supporting laboratory studies will be presented as appropriate to support suggestions of direct toxicity in the field.

Kearns and Atchison (1979) are among the first to have suggested that chronic metal exposure can lead to energetic consequences in wild fish. They examined indicators of growth in yellow perch from one small, eutrophic lake of northern Indiana, USA, divided by a narrow channel in two basins, providing an interesting system in which to examine metabolic effects of metal contamination: one of the basins was metal contaminated by an electroplating plant, while the other basin was otherwise similar and could be used as a reference. Their study revealed that growth rate of young of the year, estimated using both weight gain between seasonal sampling events and RNA/DNA ratios, was negatively correlated with tissue Cd, but not Zn, concentrations. Although more recent studies of yellow perch in other systems (see below) have not always consistently supported this observation, their conclusions that Cd, but not Zn, was a likely culprit of toxicity in these wild fish remains undisputed. Of course, in this as in other field studies and as discussed below, the extent of direct effects of Cd and other metals on growth, versus indirect effects on other levels of the food web that could negatively affect the growth of yellow perch, cannot be discriminated. There is nevertheless a weight of evidence, reviewed below, to support that metal contamination is directly responsible, to some extent, for metabolic disturbances in wild fish.

Literature on metabolic effects of metal contamination in wild fish is thin. The yellow perch, which appears to be quite tolerant to metals and can therefore be found in lakes ranging from pristine to heavily contaminated, is one of the most studied wild fish in this context, with over 50 research articles covering ecological, physiological, genomic, and population genetics topics. It is ubiquitous in North America and particularly abundant in the thousands of lakes of the Canadian Shield. This region is also exceptionally rich in several valuable metals such as Ni, Cu, and Au, and is therefore home to some of the most important mining and smelting operations in the world. With increasing industrial activity comes increasing metal contamination. Therefore, these same areas that support large-scale industrial operations also produce large-scale metal contamination gradients in nearby water bodies. If metal contamination affects fish physiology and bioenergetics, patterns of reduced growth and metabolic capacities should emerge from studies of yellow perch along contamination gradients. The first evidence of metal effects in wild yellow perch from the Canadian Shield came from a study carried out in Rouyn-Noranda (Brodeur et al., 1997), where an impairment of normal cortisol secretion was reported for yellow perch from contaminated lakes. The first study of yellow perch in Sudbury reported that metal contamination of yellow perch tissues varied along a gradient of contamination (Bradley and Morris, 1986), but it was not until 2002 that the relationships between metal contamination and condition were first examined in the Sudbury area (Eastwood and Couture, 2002). Since then, several studies (reviewed below) have reported on metabolic disturbances in metal-contaminated yellow perch from Sudbury and Rouyn-Noranda.

Enzyme activities have often been used in laboratory studies to investigate the metabolic consequences of metal exposure, but the approach is much less common in wild fish. Two enzymes of particular interest are citrate synthase and nucleoside diphosphokinase (NDPK). Citrate synthase is a mitochondrial enzyme associated with the citric acid cycle, and its activity provides an indication of aerobic capacity. NDPK, together with RNA/DNA ratios, provides a measure of biosynthetic activity (Audet and Couture, 2003). The first study examining aerobic capacities and biosynthetic activity in wild fish along a metal contamination gradient reported that fish with higher liver Cd and Cu concentrations also had lower muscle aerobic and biosynthetic capacities, using citrate synthase, NDPK, and RNA/DNA assays (Audet and Couture, 2003). Yellow perch sampled in metal-contaminated lakes showed lower critical swimming speed (Rajotte and Couture, 2002) and reduced resting and maximal oxygen consumption rates in direct proportion to liver Cd and Cu contamination (Couture and Kumar, 2003). In a large study of yellow perch from two metal contamination gradients (Sudbury and Rouyn-Noranda), the lower aerobic capacities of metal contaminated fish were largely confirmed (Couture et al., 2008b), although a positive relationship between tissue Ni contamination and aerobic capacities was also reported. The latter finding, confirmed by another study (Pierron et al., 2009), suggests that Ni, and probably Cd and Cu, contamination induces an increase in energy demand, perhaps related to detoxification costs, as first proposed by Sherwood et al. (2000) using a radiotracer approach, and by Rajotte and Couture (2002) and Levesque et al. (2002) from physiological evidence. Hence, overall, available data from wild fish studies suggest that metal contamination directly affects their aerobic capacities, although both inhibition and compensatory increases may be occurring. The few studies of aerobic impairment in laboratory fish exposed to environmentally relevant metal concentrations (concentrations that have been reported in the tissues of contaminated wild fish or aqueous concentrations in lakes with fish) support the field studies. For instance, Pane et al. (2004) demonstrated that aqueous Ni exposure led to decreased aerobic swimming performance in rainbow trout, a phenomenon that was associated with a thickening of gill secondary lamellae leading to a decrease in gas exchange efficiency. In an *in vitro* study of goldfish, Garceau et al. (2010) demonstrated that environmentally relevant concentrations of Cd and Cu, but not Ni, inhibited mitochondrial respiration, but that the metals had only small effects on the activity of aerobic enzymes examined (citrate synthase and cytochrome *c* oxidase, or complex IV of the electron transfer chain). The mechanisms by which chronic metal exposure can lead to aerobic impairment in wild fish are currently unknown. Because there is also strong evidence to support that factors other than metals can affect aerobic capacities in wild fish (see below), a causal relationship between tissue metal contamination and metabolic impairment remains to be unequivocally demonstrated in wild fish.

Metals induce oxidative damage to macromolecules including cell membranes, proteins, and DNA through well-known mechanisms involving redox cycling and the formation of reactive oxygen species (ROS) (Stohs and Bagchi, 1995). Evidence of induction of oxidative stress by metal exposure abounds in the mammalian literature (see Ercal et al., 2001, and Valko et al., 2005, for reviews). In fish, several laboratory studies of dietary and aqueous exposures are available (Berntssen et al., 2003; Pierron et al., 2007; Pandey et al., 2008), although many of these studies have been carried out at concentrations much higher than encountered in the field (Gargiulo et al., 1996; Žikič et al., 2001). Oxidative stress has been suggested to explain disturbances in metabolic capacities in wild fish, but evidence from field studies is much scarcer and inconsistent. Most evidence of increased oxidative stress in wild fish comes from studies in which organic and inorganic contaminants co-occur and hence cannot be discriminated. For example, in India, several indicators of oxidative stress, including enzymes involved in protection against oxidative stress and peroxidation products, were reported to be higher in *Wallago attu* from a section of an Indian river polluted by contaminants of mixed origins (agricultural, industrial, and domestic) compared to fish from a cleaner section of the same river (Pandey et al., 2003). In a study of cultured and wild white seabream (*Diplodus sargus*), tissue metal concentrations, even though they were below the guidelines for human consumption, were positively correlated with the activity of the antioxidant enzymes catalase (CAT) for Cd and Cu and superoxide dismutase (SOD) for As (Ferreira et al., 2008). In such studies, however, we do not know whether organic contaminants were co-occurring with metals, as has been reported for other fish such as eels (Geeraerts and Belpaire, 2010), yellow perch, and northern pike (Hontela et al., 1992). Indeed, the relationships between oxidative stress and environmental contamination can be very difficult to interpret when complex contaminant mixtures are present. For example, Dautremepuits et al. (2009) examined biomarkers of oxidative stress in yellow perch from locations along the St. Lawrence River contaminated variably by metals, organic contaminants, and bacteria. They concluded that the relationships between biomarkers of oxidative stress and contamination differed between gill and kidney (depending on the route of exposure) but that the responses of these biomarkers to mixtures of contaminants was highly complex. Similar conclusions were reached in a study of European eels caged in clean and contaminated sites: oxidative stress biomarker response to contaminants is organ specific and varies depending on contaminant type (Ahmad et al., 2006). Of relevance for the studies of wild fish exposed to metal-only contaminants, reviewed below, the authors also observed opposite trends of inhibition and activation of oxidative stress indicators depending on the tissue.

Fewer studies have examined indicators of oxidative stress in natural systems dominated by metal contamination. Brown trout exposed to metal contamination in a Montana river expressed higher concentrations of lipid peroxides compared to conspecifics captured in a clean branch of the same river (Farag et al., 1995). Evidence of oxidative damage has also been reported for Hg-contaminated northern pike: liver color, varying due to lipofuscin, which is a product of membrane lipid peroxidation, was positively correlated with Hg concentration (Drevnick et al., 2008). Studies of wild yellow perch, however, do not agree with these observations in trout and pike. In a study of an invertebrate (floater mussel) and a fish (yellow perch) sampled along a metal contamination gradient, concentrations of malondialdehyde, a lipid peroxide indicative of oxidative stress, were positively correlated with Cd in the gills of the mussel, but negatively correlated with metal concentrations in the digestive gland of the mussel and in yellow perch liver (Campbell et al., 2005; Giguère et al., 2005). Superoxide dismutase is an enzyme responsible for removal of radical anion superoxides responsible for oxidative damage by converting them to hydrogen peroxide. The expression of the *sod-1* gene, coding for the SOD enzyme, has been positively correlated with the production of ROS (Scandalios, 2005). In support of the study by Campbell et al. (2005), the expression of the *sod-1* gene was negatively correlated with liver Cd concentrations in yellow perch sampled along the Rouyn-Noranda gradient (Pierron et al., 2009). The results of this study are counter-intuitive, as one would expect oxidative stress indicators to increase with metal exposure. However, in that study, the expression of *mts*, the gene responsible for metallothionein production, was positively correlated with liver Cd concentration. Although metallothioneins are known to increase following metal exposure, they also trap ROS and have been shown to be induced by ROS generation (Viarengo et al., 2000). Therefore, the metal-induced increase in metallothionein expression may be sufficient to protect the tissues of wild yellow perch from oxidative damage. Hence, in wild fish, metal contamination is sometimes positively correlated with oxidative stress, but sometimes negatively when metallothioneins are induced to protect the cell against oxidative damage. The general tendency for metal-contaminated wild fish to have lower aerobic capacities, described above, probably explains negative relationships between metal contamination and oxidative stress. Since the mitochondrion is the main source of cellular ROS, metal-induced inhibition of aerobic capacities may reduce the need to enhance cellular protection against oxidative stress in chronically metal-exposed wild fish. Overall, although metal exposure undoubtedly has the potential to induce oxidative stress and tissue damage in wild fish, given the intricate relationships between aerobic capacities, ROS production, the modes of toxic action of metals involving oxidative stress, and the complex processes available to cells for scavenging ROS, indicators of oxidative damage cannot, on their own, be used as reliable indicators of metal stress in wild fish.

The research reviewed above reveals that wild fish can develop a tolerance to metal exposure allowing them to survive in contaminated environments and, in some cases, to maintain metabolic capacities within the range found in fish from reference sites. However, there may be metabolic costs involved in chronic metal exposure related to metal detoxification and protein repair (Rajotte and Couture, 2002). If this is the case, these costs should be reflected by indicators of growth, reproduction, and accumulation of energy reserves. Indeed, this sort of evidence is abundantly documented in several species, and particularly for yellow perch. Among the first reports of reduced growth in metal-exposed fish, Collvin (1984) published a laboratory study in which he showed that sublethal Cu exposure led to decreased maximal respiration rate and growth in the European perch (*Perca fluviatilis*), a close cousin of the yellow perch. Both of his conclusions were supported by studies of wild yellow perch from metal contamination gradients. The first line of evidence of metabolic disturbances in yellow perch under chronic metal exposure is provided by numerous reports of lower values of the condition factor (Hontela et al., 1995; Brodeur et al., 1997; Eastwood and Couture, 2002; Audet and Couture, 2003; Kraemer et al., 2006; Pyle et al., 2008). Eastwood and Couture (2002) used simple morphometric indicators (age-at-length, relative condition factor, and scaling coefficient) to conclude that more metal-contaminated wild yellow perch along a contamination gradient appeared to grow more slowly and to be in lower condition. Lower condition factors in more contaminated wild yellow perch were confirmed in subsequent studies (Levesque et al. 2002; Rajotte and Couture, 2002; Audet and Couture, 2003; Giguère et al., 2004; Pyle et al., 2008). For wild northern pike, Drevnick et al. (2008) reported that an increase in liver Hg concentration was associated with lower liver lipid reserves and condition factor. Indeed, evidence of lower condition, growth, and accumulation of energy reserves is so consistent among studies that it can hardly be contested. As extensively reviewed elsewhere, however, (Couture et al., 2008b, Pyle et al., 2008), this general indicator of fish health is affected by many factors in the field (regionally and seasonally dependent variables, indirect contaminant effects, etc.) and cannot, on its own, be taken as a reliable indicator of metal stress in wild fish.

Growth and increases in condition factor are the result of enhanced protein synthesis and triglyceride assembly and storage, resulting in the accumulation of energy reserves (increases in weight) and in the synthesis of new tissue (which in practice leads to increases in both weight and length). A wide range of tools is available to estimate or measure growth in wild fish. Tool selection depends essentially on two criteria: (1) analytical and field logistical capacities, and (2) whether recent or long-term (lifetime) growth is studied. While for sedentary fish such as the yellow perch long-term growth provides information on overall ecosystem productivity and potential for the species, estimates of recent growth reflect proximate environmental factors and will be strongly affected by seasonal variations in habitat suitability. The condition factor is an indicator of recent growth in weight resulting in energy reserve accumulation. Since in several fish nutrient absorption is achieved largely by the pyloric ceca and these organs have been shown to vary in size with recent feeding intensity, the relative size of pyloric ceca has also been used as an indicator of recent growth (Bélanger et al., 2002; Gauthier et al., 2008b). Tissue growth involves turning the protein synthesis machinery on, which in turn requires mRNA synthesis. Since the amount of DNA per cell is fixed, a higher RNA/DNA ratio suggests protein anabolism and has been proposed as an indicator of recent growth in fish (Bulow, 1970, 1987). As for the RNA/DNA ratio, the activity of enzymes such as NDPK, an enzyme involved in the transfer of phosphate groups among nucleotides, has been demonstrated in field and laboratory studies to be correlated with recent growth in fish (Couture et al., 1998; Kaufman et al., 2006; Gauthier et al., 2008a; Lapointe et al., 2009). The activity of several other enzymes has also been shown to be correlated with growth. These range from enzymes involved in nutrient absorption in various parts of the guts to mitochondrial enzymes upregulated to match the demands of anabolic pathways. All of these approaches have been used to examine metal effects on wild fish growth.

Metal exposure has been reported to enhance proteolysis in the laboratory (De Smet and Blust, 2001). However, there is an insufficient number of field studies to reach meaningful conclusions about the effects of chronic metal exposure on protein turnover rate. In metal-contaminated wild yellow perch, lower activity of NDPK corresponding with lower tissue protein concentrations and lower RNA/DNA ratio has been reported (Audet and Couture, 2003), suggesting that metal contamination leads to a lower rate of protein turnover, a conclusion supported by an early study of metal effects in yellow perch in an Indiana lake (Kearns and Atchison, 1979). In contrast, two other studies of the same species in the same region have yielded the opposite conclusion, that metal contamination is associated with an enhanced capacity for protein synthesis, which may be required for metal detoxification and damage repair (Rajotte and Couture, 2002; Couture and Kumar, 2003). There was no effect of metal contamination on RNA/DNA ratios in northern pike and burbot. At the opposite, fish exposed to uranium tailing effluents expressed higher energy accumulation compared to controls (Bennett and Janz, 2007). Metal concentrations were insufficient to cause direct toxicity to these top predators but may have induced indirect, positive effects through increases in ecosystem productivity, which the authors speculate may be due to nutrient enrichment of mine effluent-receiving waters.

Mechanisms other than general metabolic costs have been suggested for decreased growth in metal-exposed fish. For instance, poor growth of rainbow trout (*Oncorhynchus mykiss*) caged in metal-contaminated water has been related to a disruption of normal glycogen metabolism as indicated by hepatic accumulation of glycogen, possibly due to an inhibition by metals of a glycogen branching enzyme leading to a form of glycogen that is insoluble (Peplow and Edmonds, 2005).

Although the effects of metals on wild fish growth and metabolism, described above, may be attributable in part to indirect effects (discussed below), the impairment of endocrine function in metal-contaminated yellow perch is one of the best documented cases of evidence of direct toxicity, first observed in the field and later demonstrated in the laboratory. Normally, blood cortisol increases sharply following a stress such as capture or chasing, allowing the mobilization of energy reserves to sustain an appropriate behavioral response (fight or flight). However, this normal response fails in metal-contaminated yellow perch from Rouyn-Noranda (Brodeur et al., 1997; Laflamme et al., 2000; Levesque et al., 2003; Gravel et al., 2005). In the laboratory, the researchers demonstrated that Cd, the metal of most concern in the Rouyn-Noranda area, interferes in the hormonal signaling pathway leading to cortisol production (Lacroix and Hontela, 2004). The authors also reported a greater Cd sensitivity of cortisol secretion in rainbow trout compared to yellow perch, in agreement with the literature classifying the trout as a metal-sensitive species and the perch as a rather metal-tolerant species.

Effects of metal contamination on fish reproduction have been reported in the laboratory, including reduced hatching success and deformities (Jezierska et al., 2009). There is also convincing evidence that the organometals Hg and Se can impair reproduction in wild fish. For example, Gillespie and Baumann (1986) captured bluegills (*Lepomis macrochirus*) from a clean and a Se-contaminated site and demonstrated that larvae fertilized in the laboratory from females with high body Se concentrations suffered from edema and did not survive to the swim-up stage, probably owing to toxicity induced by a maternal transfer of Se to the embryos. For Hg, mechanisms of reproductive impairment involve inhibition of the reproductive axis through disturbances of the endocrine function at several levels (see Crump and Trudeau, 2009, for a review). The reader is referred to Janz (Chapter 7, Selenium) and Kidd and Batchelar (Chapter 5, Vol. 31B, Mercury) for an extended discussion of reproductive impairment in fish by organic forms of these metals. In contrast, for inorganic metals, it is likely that disturbances in energy metabolism, described in this section, are the main cause of effects on the reproduction of wild fish. The limited evidence available supports this hypothesis. For instance, Ellenberger et al. (1994) captured reproductively mature yellow perch from a Cu-contaminated and a clean site, cross-bred them, and raised the embryos in water from either lake. They did not observe any effect of chronic Cu exposure on the reproductive endpoints measured, except for time to hatch, which was increased for egg masses from contaminated fish. Gauthier et al. (2006) also observed increased time to hatch and mortality in fathead minnow embryos incubated *in situ* in lakes contaminated by Cd, Cu, and Ni, which could be due to metal-induced modifications of metabolic rate.

Energetic pressures or metabolic toxicity exerted by environmental metal contamination can lead to selective pressures (Fig. 9.3). In a population genetics study, Bourret et al. (2008) reported that the genetic diversity of yellow perch decreased as contamination increased in the Sudbury and Rouyn-Noranda areas. If metal stress can induce a selective pressure leading to fish with enhanced metal tolerance (less tolerant individuals being eliminated in the process of selection), we could expect that the physiological capacities of populations resulting from such reduction in genetic diversity would be closer to those of uncontaminated populations, compared to stressed and unselected populations. This hypothesis, if verified, would have important implications for field studies of contaminant effects. In another study of yellow perch genetic diversity along a metal contamination gradient, Tremblay et al. (2008) confirmed that the two were related. However, the authors revealed that the presence of predators in a lake forced the yellow perch to leave the pelagic habitat to adopt a more

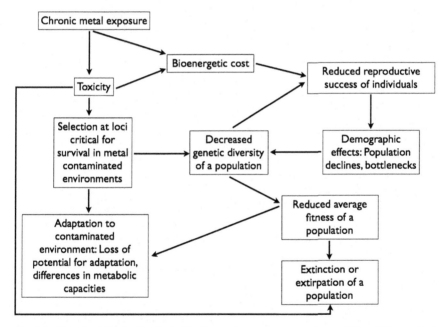

Fig. 9.3. Flowchart illustrating that chronic metal exposure may cause toxicity which can lead to local extinction, exert a selective pressure or generate costs for repair and detoxification. Chronic metal exposure can also induce energetic costs for homeostatic control, effectively preventing toxicity but at the same time reducing the energy available for growth and reproduction. The model is applicable to all species chronically exposed to metals. Modified from Bickham et al. (2000).

benthic/littoral niche, leading to morphological changes also associated with a decreased genetic diversity. Given that predator presence is an ecological parameter that is affected by contamination, this observation further supports that sublethal environmental metal contamination leads to highly complex direct but also indirect effects in fish.

The extent of direct toxicity of metal contamination of aquatic systems varies according to organismal tolerance. This tolerance is dependent on life stage, life habits, and physiology. Given the interdependence of all components of food webs, direct toxicity to one species leading to its disappearance or to a change in abundance affects other components indirectly, either negatively (e.g. if the species removed was a prey) or positively (e.g. if a competitor or a predator disappeared). For this reason, and because the bioenergetic and metabolic parameters affected by metals can also be affected by ecological variables such as food abundance and predator–prey interactions, direct effects of metals in wild fish need to be confirmed by laboratory trials and to be interpreted against evidence of indirect, ecological effects.

There is evidence that metal contamination modifies food webs. The suggestion that yellow perch from contaminated environments can, as juveniles, benefit from low competition, reduced predation and parasitism before dying prematurely due to metabolic costs of metal exposure (Couture and Pyle, 2008) was preceded by several studies that have attributed low growth to metal-induced ecological disturbances. Using the ^{137}Cs radio-tracer technique to provide estimates of food consumption rates, Sherwood et al. (2000) demonstrated that metal contamination decreased the conversion efficiency (annual growth increment relative to the energy budget) in wild yellow perch. These results indicated greater energetic costs in these fish, which the authors attributed to a simplified prey base, an observation strengthened by further studies (Kövecses et al., 2005). This is because, as illustrated by Peplow and Edmonds (2005), a typical response of benthic invertebrate communities to sediment metal contamination is a decrease in abundance and diversity. As illustrated in Fig. 9.4, an impoverished trophic chain prevents larger fish from switching diet and reaching a large size. In contrast, if contamination leads to food web enrichment, as has been reported for uranium milling effluents, then indicators of condition and growth in exposed fish can be higher compared to control fish (Bennett and Janz, 2007). Together, these results demonstrate that food web integrity can be just as important as direct toxicity with respect to its influence on wild fish condition. Indeed, the most objective evidence available indicates that direct toxicity occurs at all levels of aquatic food webs, but to varying degrees. Analogously, metal accumulation in yellow perch leads to measurable direct toxic effects at multiple levels from cellular to individual and population levels (Campbell et al., 2003). Indirect,

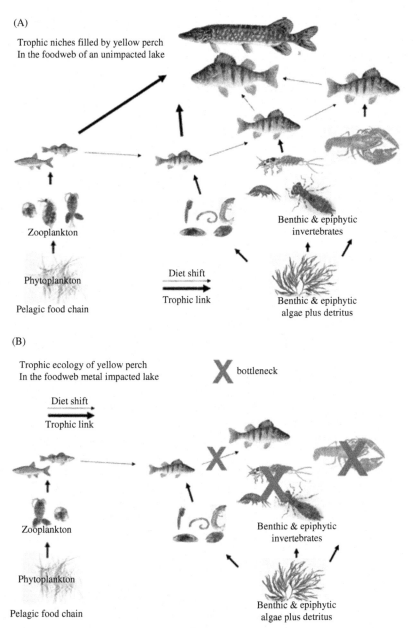

(A)

Trophic niches filled by yellow perch
In the foodweb of an unimpacted lake

Zooplankton

Phytoplankton

Pelagic food chain

Diet shift

Trophic link

Benthic & epiphytic
invertebrates

Benthic & epiphytic
algae plus detritus

(B)

Trophic ecology of yellow perch
In the foodweb metal impacted lake

bottleneck

Diet shift

Trophic link

Zooplankton

Phytoplankton

Pelagic food chain

Benthic & epiphytic
invertebrates

Benthic & epiphytic
algae plus detritus

Fig. 9.4. Modes of indirect effects of chronic metal exposure for yellow perch. (A) Clean lake with a normal, complex food web; (B) metal-contaminated lake where the most sensitive benthic invertebrates have been lost. From Rasmussen et al. (2008).

food-web mediated effects have been proposed as more important than direct physiological effects of metal contamination (Iles and Rasmussen, 2005). The proportion of direct versus indirect effects of metal contamination will undoubtedly vary widely depending on the fish species, the metal(s), and the nature of biotic and abiotic components of the aquatic food webs.

4. METAL EFFECTS ON BEHAVIOR

Metals have long been known to affect normal behavior in fish. However, metal effects on behavioral endpoints have not been seriously exploited in most regulatory arenas because of the apparent subjectivity in interpreting behavioral endpoints, the variability associated with behavioral data, the lack of any standardized procedures for evaluating metal effects on behavioral endpoints, and the difficulties in extrapolating results obtained under controlled laboratory conditions back to the field. Recently, however, there has been a renewed interest in examining metal effects on behavior because of the intrinsic ecological relevance to metal-contaminated systems.

Behavioral endpoints are important for ecological risk assessment for two reasons. Firstly, fish behavior integrates the health and physiological condition of the animal with its environment at the level of whole organism. A fish that is incapable of behaving appropriately to local stimuli by failing to avoid predators or to find food, or by failing to respond to or attract a potential mate, may suffer "ecological death", defined as an inability of a fish to function in an ecological context, despite not suffering from overt toxicity by a contaminant (Scott and Sloman, 2004). Secondly, effects of contaminants on behavior occur at concentrations that are often far below those required to cause overt toxicity (e.g. mortality, growth). Although acute toxicity endpoints are still widely used in aquatic toxicology (e.g. 96 h LC50), their ecological relevance has been questioned (Pyle and Wood, 2007). Metal concentrations sufficient to induce mortality are rarely encountered in nature, even in industrially contaminated systems (e.g. Pyle et al., 2005). Therefore, behavioral responses are both sensitive and ecologically relevant.

Behavioral responses relevant to chronic, sublethal toxicant exposure can be categorized into three broad categories: (1) preference–avoidance behaviors to contaminants, (2) locomotory responses, and (3) appropriate responses to social information. Preference–avoidance responses are characterized as a fish's response to an environmental stimulus. If the stimulus is a toxic contaminant, an appropriate behavior would be to avoid the contaminated area. If the stimulus is food or the presence of an essential nutrient, then the appropriate response would be attraction to the stimulus.

An inappropriate response would be attraction towards areas having potentially toxic concentrations of a contaminant. Fish that can be attracted towards contaminated areas are at serious risk of metal intoxication. In fish, for example, the behavioral response to Cu appears to be biphasic; that is, fish tend to avoid water contaminated by low Cu concentrations, but are attracted to water having high, potentially toxic, concentrations of Cu (Black and Birge, 1980; Hara, 1981; Brown et al., 1982; Giattina et al., 1982; Pedder and Maly, 1985). Such a response to potentially toxic dissolved Cu concentrations is maladaptive. The same sort of biphasic response to Cu has also been observed in invertebrates (Maciorowski et al., 1977). For other metals, such as Cd, the preference–avoidance response is species specific. Lake whitefish could avoid relatively low Cd concentrations (<10 µM) but failed to avoid higher concentrations (Brown et al., 1982). Whitefish can successfully avoid Zn concentrations at or above 0.15 µM in an open arena (Scherer and McNicol, 1998). However, when a cover object was present, the fish failed to avoid the elevated Zn concentrations and preferred instead to remain under cover, thereby increasing their risk to elevated Zn exposure. These results demonstrate the sensitivity of behavioral responses, but also highlight that the responses are rarely straightforward.

Avoiding areas that are contaminated by metals can have serious ecological implications for fish. Equitoxic mixtures of Cu (38 µg L^{-1}) and Zn (480 µg L^{-1}) completely eliminated the upstream spawning migration of Atlantic salmon (*Salmo salar*) (Sprague et al., 1965; Saunders and Sprague, 1967). Moreover, while metal concentrations remained high in these waters, the resident fish showed no sign of adaptation to the contamination. Atlantic salmon that were previously attracted to hatchery water in which they were reared, avoided the same water contaminated by 44 µg Cu L^{-1} (Sutterlin and Gray, 1973). In laboratory experiments designed to reflect natural metal mixtures of the Clark Fork River, Montana, USA, brown trout (Woodward et al., 1995) and rainbow trout (Hansen et al., 1999c) avoided almost all dilutions of metal-contaminated water. This strong avoidance response was implicated in the apparent lack of these species in contaminated reaches of this river system. Together, these studies suggest that if fish fail to migrate to ecologically important areas because of metal-induced behavioral modifications, they may be at risk of reproductive impairment and impending ecological death.

Field verification of behavioral endpoints is important because results derived under controlled laboratory conditions may not reflect metal effects under natural field conditions owing to several biotic and abiotic factors that can influence metal toxicity. However, very few studies are available that evaluate metal effects on behavior under both laboratory and field conditions. In some of the few studies that did, significant differences between laboratory and field were detected. For example, salmonids tested in the field showed a more intense avoidance response to Cu-contaminated water than those tested under laboratory conditions, possibly because laboratory-tested animals were not motivated by natural spawning or migratory cues whereas field-tested animals were (Svecevičius, 1999). Because this is such an important observation, more confirmatory studies are required.

Metal effects on fish behavior can vary by species. Rainbow trout were found to have the most sensitive avoidance response to a Cu and Zn mixture relative to (in order of decreasing sensitivity) three-spined stickleback (*Gasterosteus aculeata*) > roach (*Rutilus rutilus*) > dace (*Leuciscus leuciscus*) > perch (*Perca fluviatilis*) (Svecevičius, 1999). Although rainbow trout and three-spined stickleback managed to avoid water containing concentrations of the metal mixture (notwithstanding the sensitivity differences between the two species), the other species behaved quite differently. The roach avoided only contaminated water that had either high or low metal concentrations, but not intermediate concentrations. The dace was attracted to contaminated water having very high metal concentrations. The perch failed to respond to the contaminated water over a very wide range of metal concentrations. Other studies that have examined species sensitivities have also found the rainbow trout to be one of the most sensitive species when examining metal effects on behavioral responses. Rainbow trout were more sensitive to Cu, Cd, and Zn in avoidance assays than bluegills or largemouth bass. Rainbow trout were more sensitive than brown trout for avoiding a complex metal mixture (Woodward et al. 1995, Hansen et al. 1999a). This apparent sensitivity of rainbow trout for avoiding contaminated water relative to other species makes them an attractive study species. However, an appropriate level of caution is warranted because the behavior of rainbow trout may not reflect the behavior of other species in natural environments.

Fish detect elevated metal concentrations in their environment using olfaction. As long as the metal can be detected quickly and there is a nearby refugium to which the fish can retreat to avoid the potentially toxic exposure, no adverse effects are expected. However, in the absence of such a refugium, the olfactory system itself is at risk of metal-induced damage. Fish utilize olfaction as a means of acquiring information to motivate behaviors such as finding food, avoiding predators, and mediating reproduction. Consequently, damage to the olfactory system has the potential to disrupt a very wide range of behaviors associated with chemical communication. This effect of contaminants on the perception and response to chemical information is called "info-disruption" (Lürling and Scheffer, 2007).

Metals have long been known to be powerful olfactory toxicants (Vogel, 1959; Sutterlin, 1974; Brown et al., 1982). However, the ecological implications of info-disruption are not fully appreciated at present (Pyle and Mirza, 2007). Beyers and Farmer (2001) exposed Colorado pike minnows (*Ptychocheilus lucius*) to waterborne Cu ranging in concentration from 10 to 266 $\mu g\,L^{-1}$ for either 1 or 4 days. Exposed fish showed a diminished ability to respond to conspecific chemical alarm cues, and the degree of response attenuation was associated with exposure concentration. Rainbow trout exposed to 2 $\mu g\,L^{-1}$ Cd for 7 days could no longer respond to conspecific chemical alarm cues (Scott et al., 2003). The Cd concentration in the olfactory organs was seven times higher than in the livers of these fish after 5 days of exposure. After 7 days, the olfactory Cd concentration was still five times greater than in their livers.

Chemosensation via olfaction begins at the olfactory sensory neuron (OSN), where odor molecules bind to surface receptors and initiate a very well-characterized molecular signal transduction cascade that results in the propagation of an action potential (Firestein, 2001, Zielinski and Hara, 2007). Vertebrate OSNs are one of only a few neurons capable of being replaced by neuroblast differentiation (Magrassi and Graziadei, 1995). Because the olfactory system represents one of the only direct connections between the fish's nervous system and the external environment, an appropriate adaptive response to environmental contaminants is to break that connection in order to protect the sensitive brain against environmental toxicants. Consequently, a non-specific response to environmental contaminants is the spontaneous degeneration of OSNs, typically by apoptosis, resulting in the loss of olfaction (Klima and Applehans, 1990; Julliard et al., 1993, 1996; Bettini et al., 2006). Once the insult has passed, OSNs will differentiate from basal neuroblasts, thereby restoring olfactory function. Therefore, under controlled laboratory conditions, metal-induced olfactory dysfunction is transient in adult animals and full recovery is typically achieved within weeks (Cancalon, 1982; Zielinski and Hara, 1992; Beyers and Farmer, 2001; Baldwin et al., 2003). However, exposure to Cu during embryonic development can have long-lasting effects on olfaction (Carreau and Pyle, 2005). Therefore, developing fish may be vulnerable to long-term olfactory dysfunction if they are exposed to relatively low concentrations of Cu (10 $\mu g\,L^{-1}$ is sufficient to induce this effect; pH 7.4, dissolved organic carbon 2.7 mg L^{-1}, hardness 18 mg L^{-1} as $CaCO_3$) during a sensitive developmental stage.

Because fish rely on chemical cues to convey critical information about their immediate environment, there is reason to suspect that they can adapt to elevated waterborne metal concentrations. However, the details of any such adaptation are not currently known. It might be expected that fish inhabiting metal-contaminated lakes would have a significantly reduced olfactory acuity owing to metal damage to olfactory tissues resulting in reduced neurophysiological and behavioral responses. In a recent study (Mirza et al., 2009), yellow perch were collected from clean and metal-contaminated environments and tested for their olfactory acuity using neurophysiological and behavioral techniques. Fish from contaminated lakes showed significantly elevated neurophysiological activity to standard olfactory cues relative to those from the clean lakes. However, the same fish failed to mount a behavioral response to ecologically relevant cues. This is in contrast to Hansen et al. (1999b), who demonstrated in laboratory-reared rainbow trout that fish avoided Cu at concentrations that were lower than those required to induce an effect on the neurophysiological response. In a field study, Iowa darters (*Etheostoma exile*) failed to avoid traps treated with conspecific alarm cue in metal-contaminated lakes, but successfully avoided those traps in clean lakes (McPherson et al., 2004). These results suggest that chronic exposure to relatively low concentrations of metal under natural conditions may lead to a decoupling of neurophysiological activity at the olfactory epithelium and information processing activities in the brain. The detailed mechanism of this response is not yet known.

Understanding how contaminants interfere with fish behaviors is critical for ecological risk assessments, because behaviors integrate local environmental conditions with the basic physiological condition of the animal. Toxicological assays involving behavioral endpoints introduce a high degree of ecological relevance over more conventional toxicity test methods where mortality is the endpoint, especially in environments where contaminants do not approach lethal concentrations. However, behavioral responses are highly variable, and can be site specific and context specific, and resulting data from behavioral testing can be difficult to interpret. Metals are powerful neurotoxicants and are known to affect fish behaviors (Scott and Sloman, 2004), including behaviors related to feeding (Kolmakov et al., 2009), reproduction (Sprague et al., 1965; Saunders and Sprague, 1967), establishing social hierarchies (Sloman et al., 2002, 2003), and predator avoidance (McPherson et al., 2004; Sandahl et al. 2007; Mirza et al., 2009). It may be possible to develop ecologically relevant predictive models, such as a chemosensory-based biotic ligand model (see Section 6 on modeling), that can integrate metal interactions at olfactory tissues and ecologically relevant behavioral endpoints. Such a model is beginning to emerge for Cu (Green et al., 2010; Meyer and Adams, 2010).

5. SEASONAL, INTERANNUAL, AND AGE-DEPENDENT VARIATIONS IN FISH CONDITION AND CONTAMINATION

No field study can ignore that all parameters of interest for determining metal contamination and effects in wild fish, including condition factor, growth rate, tissue proximal composition (water, protein, and lipid contents), metabolic capacities, physiological stress indicators, and even tissue metal concentrations, vary naturally according to environmental conditions including seasonally variable factors, such as temperature, food availability, and reproduction. The magnitude of seasonal variations may be more important than that induced by metal contamination. This section will examine seasonal variations in fish tissue metal concentrations and condition indicators and review the available evidence of relationships between variations in tissue metal concentrations and toxicity.

In contrast to organic contaminants, metals do not generally accumulate with age in fish. A well-known exception is Hg, which behaves like an organic contaminant. It has been well established that Hg concentrations in fish tissue increase with fish size and age (see Kidd and Batchelar, Chapter 5, Vol. 31B, for a review). Similarly along food webs, larger fish species will typically have higher tissue Hg concentrations. Another metal that has regularly been reported to increase with size and age in fish is Cd. In European flounder (*Pleuronectes flesus*) collected near British power stations, hepatic Cd concentrations increased more than four-fold between fish aged 0+ and 4++ (Rotchell et al., 2001). In yellow perch from a Cd-contaminated lake in the Rouyn-Noranda region, liver and kidney Cd concentration increased two- to three-fold between fish aged 1 and 10 years (Giguère et al., 2004). A later study confirmed the accumulation of tissue Cd with age in yellow perch from Rouyn-Noranda, but not in fish from Sudbury (Couture et al., 2008a). Therefore, in contrast to Hg, evidence of Cd accumulation with age in fish indicates that the phenomenon can occur but cannot be generalized and therefore requires repeated validation for each situation. Nevertheless, when accumulation of a toxic compound increases as the age of the fish increases, the risk to populations is evident and should be considered, as larger fish with a higher reproductive potential may be at a higher risk of toxicity.

Seasonal variations in tissue metal concentrations have been reported for several fish species and metals. With the exception of some metals such as Cd in specific species and locations, inorganic forms of metals do not tend to accumulate with age but have been shown to vary up or down seasonally. Hence, the toxic risk to the fish themselves and to their predators varies seasonally.

Extensive studies on seasonal variations in tissue metal concentrations have been carried out on yellow perch. Eastwood and Couture (2002) examined liver variations of Cu, Ni, and Zn in two seasons, spring and fall, in four lakes representing a contamination gradient. The authors observed that in the spring, fish from contaminated lakes had much higher hepatic concentrations of Cu than clean fish, but that these differences had disappeared by fall. This result indicates that regulatory capacities for this essential metal were only exceeded during some periods of the year. Moreover, variations in tissue Zn concentrations were low, supporting the strong capacity of vertebrates for regulating this essential metal (Eastwood and Couture, 2002). Although Ni has not yet been proven as an essential nutrient for fish (Pyle and Couture, Chapter 5), liver Ni concentrations peaked in the fall. No explanation was provided by the authors for the difference in seasonal variation between Cu and Ni. In a subsequent study of yellow perch restricted to one clean and one metal-contaminated lake, Audet and Couture (2003) again showed that the capacity for hepatic Cu regulation was exceeded in fish from the contaminated lake in summer, as indicated by significantly higher Cu concentrations compared to fish from clean lakes, but not in fall. A peak in gill Cu in the spring in fish from contaminated lakes supported the suggestion of Eastwood and Couture (2002) that spring snowmelt events may increase the availability of Cu to fish at this time. Cadmium, which was consistently higher in the stomach contents and in the gills, but not in the liver, of fish from the more contaminated lake, did not in that study show any significant seasonal variation. Kraemer et al. (2006) also observed that liver metal concentrations varied seasonally in yellow perch from another metal-contaminated area (Rouyn-Noranda, Québec). Comparing variations of liver metal burden with liver mass revealed the complexities of seasonal variations in tissue metal concentrations: over the course of a growth season, fish switched food items, the food itself varied in metal contamination, and growth may have led to dilution of tissue metals even though the total burden increased. A more recent and larger scale study examined seasonal variations in Cd, Cu, and Ni in liver and stomach content of yellow perch from both the Sudbury and Rouyn-Noranda regions (Couture et al., 2008a). Multivariate analysis revealed that seasonal effects on tissue metal concentrations were more important in Sudbury than in Rouyn-Noranda fish. In fish from the more contaminated environments, tissue metal concentrations generally peaked during summer, probably owing to a higher dietary intake. A noteworthy finding of that study was that there was an important peak in liver Ni concentrations in fish from Rouyn-Noranda, comparable to values in fish from the most Ni-contaminated lakes of Sudbury, in spite of the fact that Ni concentrations in both water and fish stomach contents from Rouyn-Noranda were very low.

Other studies have reported seasonal variations of tissue metal concentrations in wild fish. For instance, variations in trace elements have also been reported in marine fish harvested for human consumption, with concentrations of toxic metals such as Cd occasionally reaching levels above the recommended safe concentrations for human consumption (Mendil et al., 2010). Striped mullet (*Mugil cephalus*) and red mullet (*Mullus barbatus*) from the Mediterranean coast of Turkey accumulated higher concentrations of Cd, Cu, Pb, Fe, and Zn in summer than in the three other seasons examined, without exceeding the limits for safe consumption (Cogun et al., 2006). The authors proposed that the summer peak may be due to a higher metabolic rate in the warmer summer months. Similarly, Köck et al. (1996) examined seasonal variations in stomach contents and tissue concentrations of Cd and Pb, in Arctic char from a high-altitude oligotrophic lake with low values of aqueous metals over the course of a year, and reported peaks of hepatic concentrations in summer. Variations in liver metal concentrations did not match concentrations in lake water or fish diet. The authors also suggested that metabolic rate, driven by temperature, exerts a dominant influence on fish tissue metal concentrations. These studies have in common that seasonal variations in tissue metal concentrations were examined in fish from clean environments. Although for these fish metabolic rate may be a significant driver of tissue metal accumulation, as discussed in Section 3, under conditions of severe chronic metal contamination, fish metabolic rate tends to decrease. The question then arises: does metabolic rate drive tissue metal concentrations, or is it the other way around? It is possible and even likely that in areas where metal contamination is low, as in alpine lakes, sharp and brief increases in metabolic rate during summer could lead to the accumulation of very high levels of tissue metals, whereas in contaminated areas, elevated metal concentrations inhibit metabolic capacities. Clearly, factors other than water temperature and related metabolic rate influence tissue metal concentrations in fish. In contrast to studies reporting peaks of tissue metal concentrations in summer, liver concentrations of Cd and Cu as well as metallothioneins were lowest in summer and highest in winter in European eels caught in a British estuary (Bird et al., 2008). The authors proposed that important seasonal movements leading to changes in habitat contamination may explain the variations in tissue metal content. Rotchell et al. (2001) also reported peaks of Cd, Cu, and Zn in winter in European flounder collected from the same British estuary, the Severn Estuary, which has suffered metal contamination in the past from smelting activities. The agreement between these two studies conducted with different species suggests that site-specific factors may be just as important as species-specific factors in driving seasonal variations in tissue metal concentrations, which would imply that patterns determined for one habitat may not be applicable to other locations.

Given that unlike most metals Hg is accumulated as MeHg and behaves as an organic contaminant by showing a strong tendency to bioaccumulate and biomagnify, it is expected to behave differently from other metals regarding accumulation and seasonal variations in tissue concentration. Yet, several reports are available on seasonal variations in fish tissue Hg concentrations. Foster et al. (2000) reported that Hg concentrations in liver and gonad, but not muscle, of largemouth bass, varied seasonally. Given that humans only consume muscle, the authors concluded that these seasonal variations should not affect fish consumption advisory guidelines, in contrast to others who have recommended considering seasonal variations in fish muscle Hg concentrations for consumption guidelines (Ginsberg and Toal, 2000). Local variations in water chemistry have been reported to influence seasonal variations in fish tissue Hg concentrations. In a study of a Brazilian wetland during the dry season, Hylander et al. (2000) reported that in an acidic and soft-water lake, but not in surrounding water bodies, tissue Hg levels exceeded the consumption advisory in piranha and catfish. They proposed that higher Hg concentrations during the dry season may be explained by lower water volumes and related change in habitat of these fish. Similarly, in a marsh contaminated by gold mining effluent in Columbia, Marrugo-Negrete et al. (2008) reported that muscle tissue of three species of fish contained higher levels of Hg in the dry season than in the wet season. The authors proposed that higher primary production leading to increased Hg bioavailability, as well as a more intense mining activity, combined with the factors proposed by Hylander et al. (2000) (lower water volume and habitat change), may have been responsible for the seasonal trends reported. Another explanation for seasonal variations in fish tissue Hg was proposed by Paller et al. (2004), who suggested that higher values in spring may be explained by a more intense activity of methylating bacteria. Clearly though, as in the case of other metals, patterns of seasonal variations in fish tissue Hg concentrations as well as causes vary among sites. In contrast to Hylander et al. (2000) and Marrugo-Negrete et al. (2008), Liu et al. (2008) observed higher tissue concentrations in mosquitofish from the Everglades during the wet season, which they proposed is likely due to more favorable conditions for bioaccumulation in that season, as supported by their mass budget estimates.

As we move away from the equator, environmental conditions, including temperature and ecosystem productivity, change on an annual cycle. Towards the poles, growth seasons are shortened progressively and all life events are synchronized to maximize resource availability at the appropriate times of organism life cycles. While there are likely multiple causes for the seasonal variations in tissue metal concentrations, they can be classified in three categories: aqueous uptake, dietary uptake, and metabolic activity. Variations in metal bioavailability from aqueous exposure are unlikely to play a major role in most situations. In studies in which both fish tissue metal concentrations and aqueous metal concentrations were measured at different seasons (Köck et al., 1996; Audet and Couture, 2003; Mzimela et al., 2003; Kraemer et al., 2006), no obvious relationship was drawn between the two; the former typically vary seasonally but the latter tend to remain more stable. In northern countries, however, the spring snowmelt and accompanying pH drop tend to increase waterborne metal concentrations. This has been suggested as a potential cause for increased tissue metal concentrations in fish (Johannessen and Henriksen, 1978; Köck et al., 1996; Eastwood and Couture, 2002). In contrast, the other source of metals, diet, has been demonstrated to vary seasonally in metal contamination, for example in invertebrate prey (Hare and Campbell, 1992) and, as illustrated in this section, in fish themselves that are prey to larger fish. Direct measurements of metals in the food of wild fish, determined by analyzing stomach contents, have also been shown to vary seasonally in some cases (Couture et al., 2008a) but not in others (Audet and Couture, 2003). In addition to the seasonal changes in food metal concentrations, changes in food abundance and in feeding rate are well documented and will contribute to variations in dietary metal intake (see Section 2).

Seasonal variations in metabolic rate have been proposed by several authors (see above) as likely to play an important role in variations of tissue metal concentrations. The limited experimental evidence available supports this. In an experiment in which juvenile fathead minnows were kept for 3 months under simulated summer (20°C, 16 h light:8 h dark photoperiod and high food ration) or winter (4°C, 8 h light:16 h dark photoperiod and low food ration) conditions, exposed or not to diluted mining effluent and fed clean food, the three factors that varied between the two seasonal treatments combined (temperature, photoperiod, and food ration) had an influence on tissue metal accumulation (Driedger et al., 2010). Effluent-exposed fish accumulated more metals than control groups in winter but not in summer. Summer conditions led to higher lipid accumulation and growth. In European chub collected in two seasons, variations in gill metal concentrations were presented as evidence that metabolic rate influences metal uptake (Dragun et al., 2007). Feeding rate and growth were proposed as being responsible for tissue metal variations in himri (*Carasobarbus luteus*) from Turkey (Yilmaz and Dogan, 2008). Clearly, there are multiple causes for the seasonal variations in tissue metal concentrations but they are largely of biotic, instead of abiotic, origin. Complex and site-specific combinations of metal inputs from food, temperature, and photoperiod-induced modifications of metabolic activity leading to modifications in the rate of metal elimination and uptake as well as growth dilution all play a role in these seasonal variations.

The reproductive cycle is the dominant factor driving energy accumulation and use in adult fish. Given the strong seasonality of the reproductive cycle, its influence on tissue metal concentrations and on condition indicators will be considered in this section. For instance, Banks et al. (1999) reported that liver Zn concentrations varied inversely with gonad development, reaching the lowest values during the mid-summer reproduction in channel catfish (*Ictalurus punctatus*) from a Louisiana, USA, commercial fish production facility. Another study carried out in Portugal with other fish species (Iberian endemic minnows or *Leuciscus alburnoides* complex) also reported lower hepatic Zn concentrations corresponding with spawning (Lopes et al., 2001). In yet another study, this time of a South African estuarine fish (groovy mullet, *Liza dumerelii*), lower values of hepatic Zn concentrations followed by an increase in post-spawning were reported (Mzimela et al., 2003). Undoubtedly, reproduction affects liver Zn concentrations in fish. Whether the concentrations of other metals are also affected by reproduction remains to be verified. Nonetheless, because reproduction affects growth, feeding, and energy storage, it is likely to have major effects on tissue metal concentrations. It is therefore of the utmost importance that reproductive status be considered in these field studies. One approach is to avoid the influence of reproduction by collecting juvenile fish or sampling fish after they have recovered from the previous spawning season but before a significant investment in gonad maturation has been made. This is the strategy commonly used for yellow perch, a fish that spawns in early spring and starts to reinvest in gonad growth at the end of fall, leaving mid-spring to mid-fall available for sampling. Northern pike (*Esox lucius*) is another example of early spring spawner. However, there is a large number of species for which this approach cannot be applied. Most salmonids, for instance, spawn in fall following a gonad maturation phase throughout summer. Several species of minnows are repeat summer spawners. For these fish, it may not be possible to remove the influence of the reproductive cycle on the measured parameters if fish are captured in late spring, summer, or early fall.

The same factors responsible for seasonal variations in tissue metal concentrations are also likely involved in interannual variations. The limited literature available indicates that interannual variations can be even more important than seasonal variations and that seasonal peaks in tissue metal concentrations can shift in different years. Rotchell et al. (2001) reported that liver metal concentrations varied depending on the year of sampling in European flounder, although not as strongly as seasonal variations. Audet and Couture (2003) observed that the peak of tissue Cu concentration in yellow perch from metal-contaminated lakes varied in amplitude and timing in different years. Couture and Rajotte (2003) reviewed field studies of yellow perch contamination over a period of 5 years and reported that interannual variations in liver concentrations of Cu and Cd were often more important than seasonal variations. The implication of this observation is that studies of seasonal variations in fish condition and contamination carried out in temperate latitudes must be carried out over the course of a single growth season, as the intensity and timing of the peaks will vary every year. This disqualifies studies such as Levesque et al. (2002), in which sampling took place in fall 1998 and summer 1999, Farkas et al. (2003), who sampled in October 1999 and May 2000, and Dragun et al. (2007), who collected fish in autumn 2005 and spring 2006, as seasonal studies, since the interannual differences captured by sampling two growth seasons separated by an overwintering period may be as important as, and perhaps more important than seasonal variations within a growth season.

Variations in tissue metal concentrations imply variations in toxic risk. Although a literature too vast to review here has reported on seasonal variations in the condition of fish living in clean environments, a limited number of studies has examined this question in the context of metal contamination. Perhaps the first such study is that of Kearns and Atchison (1979), who reported seasonal variations in the growth indicator RNA/ DNA ratio and a correlation between the ratio and Cd concentration in yellow perch, although they did not test seasonal variations in tissue metal concentrations. Audet and Couture (2003) reported that the condition factor of yellow perch from a metal-contaminated lake peaked in mid-summer in synchrony with liver metal concentrations, indicating a higher feeding rate at that time of year. Although the condition factor would suggest that the peak of tissue metal accumulation may not lead to toxicity, muscle aerobic capacities, which were lower in the contaminated fish than in clean fish for all three seasons examined, were further depressed in the summer. Given the evidence, reviewed in Section 3, that metal contamination affects aerobic capacities in wild fish, this constitutes strong evidence that seasonal peaks in tissue metal concentrations can lead to toxic events. The association between seasonal variations in condition, metabolic capacities, and metal contamination has been confirmed for yellow perch (Couture and Pyle, 2008; Couture et al., 2008b). Seasonal variations in oxidative stress and antioxidant defense were also associated with corresponding peaks of metal exposure in striped mullet (*Mugil cephalus*) from a contaminated Indian estuary (Padmini et al., 2009).

As highlighted by most authors who have reported on seasonal variations in fish metal contamination, there are implications for ecological risk assessment and for fish consumption guidelines. For the latter, advisories may have to include seasons as modifiers of recommended consumption for some species and habitats. For the former, as demon-strated here, seasonal variations in tissue metal concentrations, which may be paralleled by changes in stress or health indicators, are widespread in fish living in metal-contaminated environments. Sampling design of risk assessment studies should therefore incorporate sampling in at least two different seasons in order to properly assess the relationships between contamination and effects.

6. APPLYING PREDICTIVE MODELS IN FIELD SITUATIONS

The processes involved in metal uptake, elimination, and internal handling resulting in bioaccumulation and/or toxicity are complex and can be affected by a wide range of external factors. Consequently, the development of models that allow for the estimation or prediction of metal accumulation and toxicity to wild fish has been challenging. An extensive review is presented by Paquin et al. (Chapter 9, Vol. 31B). Here, the applicability of models linking metal accumulation and toxicity in wild fish is briefly examined.

One approach to modeling metal accumulation is the bioconcentration factor (BCF) or bioaccumulation factor (BAF) (Schlekat et al., 2003). McGeer et al. (2003) demonstrated strong and consistent negative relationships between BCF or BAF with exposure metal concentrations. The BCF assumes that the metal has been taken up from a waterborne source only and is calculated as the ratio of the metal concentration in fish tissues to that in the water. The BAF is typically applied to wild fish, assumes that the metal has been taken up from both waterborne and dietary sources, and is calculated as the ratio of metal concentration in fish tissues to that of the native water. Although the BAF has been widely applied to wild fish populations inhabiting metal-contaminated waters, there are several limitations to its use as a regulatory tool.

A primary assumption of both the BCF and BAF is that they are independent of the exposure concentration (Schlekat et al., 2003). That is, metal uptake and elimination rates are assumed to remain constant over all metal exposure concentrations. For lipophilic organic molecules, this assumption is often true, especially when the mechanism of uptake is via passive diffusion. However, metals are generally hydrophilic, not lipophilic, and therefore cannot cross biological membranes without the aid of transporters or appropriate transmembrane channels. Consequently, both uptake and elimination processes can be saturable for metals. Moreover, other species-specific factors, such as exposure conditions or homeostatic processes, can cause variability in uptake rates over a wide range of metal exposure concentrations. Thus, this assumption of constant uptake and elimination rates is rarely achieved under natural conditions. If this assumption is not met and the BCF or BAF is used in predictive models, major overestimations or underestimations may result (Pyle and Clulow, 1997).

Another problem with using BCF and BAF is that the link between metal accumulation and toxicity is not easily established. Neither the BCF nor the BAF makes any distinction among the various metal pools within animal cells. Therefore, tissue metal concentration includes metals that are serving an essential role (e.g. as enzyme cofactors), metals that are stored and unavailable to participate in any physiological or toxicological process, and metals that are available to participate in toxicological processes. The proportion of metal in each of the various metal pools in any given cell is dependent on a number of factors, which makes linking the BCF or BAF to toxicity difficult to impossible.

Biodynamic models for estimating metal accumulation overcome many of the shortcomings of BCF and BAF by integrating metal uptake rates from the water and the diet together with elimination rates (Luoma and Rainbow, 2005). These rates can be established realistically under controlled experimental conditions, and incorporate the basic physiology of the animal (e.g. trophic transfer, assimilation efficiency, and nutritional status). Site-specific exposure conditions can be used as model inputs to estimate bioaccumulation in wild fish. The advantage of this model is that it assumes that metal uptake and accumulation are complex processes that are influenced by site-specific geochemistry, metal specificity, exposure route (water or food), and the fundamental and specific physiology of the animals of interest. The downside is that they are data intensive and require a considerable amount of input to estimate bioaccumulation. Quantification of the biodynamics model can be accommodated by the multipathway model DYMBAM (Schlekat et al., 2002). Luoma and Rainbow (2005) found a very high correlation ($r^2 = 0.98$) between bioaccumulation values predicted by biodynamic models and measured values in a meta-analysis of 15 studies involving 14 species and seven different metals. Although biodynamic models are useful, and seemingly effective, at predicting metal bioaccumulation, they do not link bioaccumulation to toxicity. Not all accumulated metal is toxic. Therefore, it may be useful to combine biodynamic metal accumulation models with predictive toxicological models to determine the relationship between metal accumulation and toxicity. One such toxicological model is the biotic ligand model (BLM).

The BLM is a predictive model that estimates acute metal toxicity to fish based on the extent to which metals occupy physiologically sensitive binding sites on a biotic ligand, such as the gills of fish (Paquin et al., 2000; Niyogi and Wood, 2004). The BLM extends the free ion activity model (FIAM), which was used to predict toxicity based on the assumptions that metal speciation varies depending on local water chemistry (e.g. pH, hardness, alkalinity, and dissolved organic matter) and the most toxic metal species is the free cation (Morel, 1983). The BLM attempts to predict toxicity by considering the biology of the animals of interest; specifically, by considering the relative binding capacity and affinity of a metal to the biotic ligand (Pagenkopf, 1983; Playle et al., 1993a,b). The BLM effectively integrates all site-specific water exposure conditions insofar as the metal bound to or taken up by the gill must be bioavailable, whatever non-bioavailable species might remain in the water column. Moreover, the BLM also makes use of a threshold metal concentration in the physiologically sensitive biotic ligand which is sufficient to cause mortality in 50% of test subjects, the LA50 (Niyogi and Wood, 2004).

Although the BLM has been demonstrated to be very effective at predicting acute metal toxicity to a few fish species in soft, sensitive waters having relatively simple geochemistry, it has been somewhat less successful predicting toxicity in the field (Pyle and Wood, 2007). The gill-based BLM assumes that the only source of metal to the fish is from the water, despite the fact that dietary metals are an important source of metals at the gill. Moreover, dietary metals can be transported to the gill in a relatively non-toxic form where they can accumulate to very high concentrations without causing any toxicity to the fish (Kamunde et al., 2002). Therefore, gill metals deriving from the diet can confound model parameters, leading to overestimates or underestimates in model predictions (Niyogi and Wood, 2003; Franklin et al., 2005). The current BLM is designed to predict acute metal toxicity in fish. However, in most natural metal-contaminated environments, metal concentrations rarely achieve concentrations sufficient to induce acute toxicity (Pyle et al., 2005). These observations have led to some research towards developing alternative biotic ligands to the gill, such as intestinal (Klinck et al., 2007; Niyogi et al., 2007) and olfactory epithelium (Paquin et al., 2002; Pyle and Wood, 2007; Mirza et al., 2009; Green et al., 2010; Meyer and Adams, 2010), which can be used towards the development of a chronic BLM and which may address many, if not all, of the basic concerns of the current gill-based BLM described here (see Paquin et al., Chapter 9, Vol. 31B, and Fig. 9.5).

Modeling metal uptake and accumulation is challenging for wild fish owing to the inherent variability associated with natural conditions. However, biodynamic models appear to make reasonable estimates on metal accumulation to fish tissues, despite their heavy data input requirements and inability to link metal accumulation to toxicity. The challenge for modeling metal uptake, accumulation, and toxicity in wild fish will be to handle the inherent variability in natural systems, and link

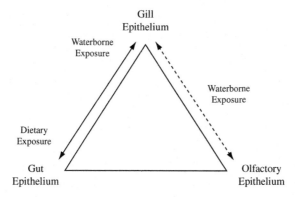

Fig. 9.5. Diversity of biotic ligand models (BLMs) for fish. BLMs are known for three primary receptors (gill, gut, and olfactory epithelia), represented at the apices of the triangle. Gill and olfactory epithelia are useful for predicting toxicity from waterborne exposures, while gut epithelium is useful for predicting dietary metal toxicity. The solid arrow indicates that processes at the gill influence those at the gut and vice versa. The dashed arrow indicates that both gill and olfactory epithelia are exposed during a waterborne exposure, but the two receptors are not linked in a way that one will influence the prediction of the other. No arrow between gut and olfactory epithelium indicates that the two receptors are independent of each other.

accumulation to toxicity. A combination of biodynamic modeling with biotic ligand modeling may be a good first approach.

7. CONCLUDING REMARKS

Metal accumulation and effects in fish are easy to demonstrate and quantify in the laboratory. However, considering the remarkable plasticity of fish faced with environmental changes, their vast range of physiological capacities, and their resilience, determining the principles and constants (if any) underlying these phenomena in the field to allow for the development of effective metal accumulation and toxicity models in wild fish will always remain a challenge. Nonetheless, broad principles are emerging. In contrast to acute laboratory exposures leading to ionoregulatory, osmoregulatory, and respiratory disturbances, mechanisms of chronic metal toxicity in wild fish are clearly more subtle and appear to involve energy metabolism (mitochondrial impairment and compensation, protein turnover rate, oxidative stress, endocrine disruption) and olfactory impairment. Aquatic food web components vary in metal sensitivity. Metal contamination will lead to disturbances of these food webs, which will induce indirect effects, potentially both negative and positive, to fish, in addition to any direct toxicity. Researchers from a wide range of backgrounds, including ecologists, limnologists, geneticists, physiologists, biochemists, and environmental chemists, have dedicated careers to advancing our knowledge on this critical environmental issue. We now require a deeper understanding of the ecological underpinnings of large-scale metal contamination of aquatic ecosystems ultimately affecting fish and their predators. New approaches from molecular biology will also be crucial in accelerating our understanding of the mechanisms of metal toxicity and our capacity to discriminate these effects from natural variations. Ultimately, this information will contribute to better protecting fish and their habitat.

REFERENCES

Ahmad, I., Pacheco, M. R., and Santos, M. A. (2006). *Anguilla anguilla* L. oxidative stress biomarkers: an *in situ* study of freshwater wetland ecosystem (Pateira de Fermentelos, Portugal). *Chemosphere* **65**, 952–962.

ANZECC (2000). Australian and New Zealand Guidelines for Fresh and Marine Water Quality. Australian and New Zealand Environment and Conservation Council and Agriculture and Resource Management Council of Australia and New Zealand.

Atkinson, C. A., Jolley, D. F., and Simpson, S. L. (2007). Effect of overlying water pH, dissolved oxygen, salinity and sediment disturbances on metal release and sequestration from metal contaminated marine sediments. *Chemosphere* **69**, 1428–1437.

Audet, D., and Couture, P. (2003). Seasonal variations in tissue metabolic capacities of yellow perch (*Perca flavescens*) from clean and metal-contaminated environments. *Can. J. Fish. Aquat. Sci.* **60**, 269–278.

Baldwin, D. H., Sandahl, J. F., Labenia, J. S., and Scholz, N. L. (2003). Sublethal effects of copper on coho salmon: impacts on nonoverlapping receptor pathways in the peripheral olfactory nervous system. *Environ. Toxicol. Chem.* **22**, 2266–2274.

Banks, S. D., Thomas, P., and Baer, K. N. (1999). Seasonal variations in hepatic and ovarian zinc concentrations during the annual reproductive cycle in female channel catfish (*Ictalurus punctatus*). *Comp. Biochem. Physiol. C* **124**, 65–72.

Bélanger, F., Blier, P. U., and Dutil, J.-D. (2002). Digestive capacity and compensatory growth in Atlantic cod (*Gadus morhua*). *Fish Physiol. Biochem.* **26**, 121–128.

Bennett, P. M., and Janz, D. M. (2007). Bioenergetics and growth of young-of-the-year northern pike (*Esox lucius*) and burbot (*Lota lota*) exposed to metal mining effluent. *Ecotoxicol. Environ. Saf.* **68**, 1–12.

Berntssen, M. H., Aatland, A., and Handy, R. D. (2003). Chronic dietary mercury exposure causes oxidative stress, brain lesions, and altered behaviour in Atlantic salmon (*Salmo salar*) parr. *Aquat. Toxicol.* **65**, 55–72.

Bervoets, L., Blust, R., and Verheyen, R. (2001). Accumulation of metals in the tissues of three spined stickleback (*Gasterosteus aculeatus*) from natural fresh waters. *Ecotoxicol. Environ. Saf.* **48**, 117–127.

Bettini, S., Ciani, F., and Franceschini, V. (2006). Recovery of the olfactory receptor neurons in the African *Tilapia mariae* following exposure to low copper level. *Aquat. Toxicol.* **76**, 321–328.

Beyers, D. W., and Farmer, M. S. (2001). Effects of copper on olfaction of Colorado pikeminnow. *Environ. Toxicol. Chem.* **20**, 907–912.

Bickham, J. W., Sandhu, S., Hebert, P. D. N., Chikhi, L., and Athwal, R. (2000). Effects of chemical contaminants on genetic diversity in natural populations: implications for biomonitoring and ecotoxicology. *Mutat. Res.* **463**, 33–51.

Birceanu, O., Chowdhury, M. J., Gillis, P. L., and McGeer, J. C. (2008). Modes of metal toxicity and impaired branchial ionoregulation in rainbow trout exposed to mixtures of Pb and Cd in soft water. *Aquat. Toxicol.* **89**, 222–231.

Bird, D. J., Rotchell, J. M., Hesp, S. A., Newton, L. C., Hall, N. G., and Potter, I. C. (2008). To what extent are hepatic concentrations of heavy metals in *Anguilla anguilla* at a site in a contaminated estuary related to body size and age and reflected in the metallothionein concentrations? *Environ. Pollut.* **151**, 641–651.

Black, J. A. and Birge, W. J. (1980). *An Avoidance Response Bioassay for Aquatic Pollutants.* Research Report No. 123. Lexington, KT: Water Resources Research Institute, University of Kentucky.

Boening, D. W. (2000). Ecological effects, transport, and fate of mercury: a general review. *Chemosphere* **40**, 1335–1351.

Borgmann, U., Norwood, W. P., Reynoldson, T. B., and Rosa, F. (2001). Identifying cause in sediment assessments: bioavailability and the sediment quality triad. *Can. J. Fish. Aquat. Sci.* **58**, 950–960.

Boudreau, B. P. (1999). Metals and models: diagenic modelling in freshwater lacustrine sediments. *J. Paleolimnol.* **22**, 227–251.

Bourret, V., Couture, P., Campbell, P. G. C., and Bernatchez, L. (2008). Evolutionary ecotoxicology of wild yellow perch (*Perca flavescens*) populations chronically exposed to a polymetallic gradient. *Aquat. Toxicol.* **86**, 76–90.

Bradley, R. W., and Morris, J. R. (1986). Heavy metals in fish from a series of metal-contaminated lakes near Sudbury, ON. *Water Air Soil Pollut.* **27**, 341–354.

Brodeur, J. C., Sherwood, G., Rasmussen, J. B., and Hontela, A. (1997). Impaired cortisol secretion in yellow perch (*Perca flavescens*) from lakes contaminated by heavy metals: *in vivo* and *in vitro* assessment. *Can. J. Fish. Aquat. Sci.* **54**, 2752–2758.

Brown, R. E. E., Thompson, B. E., and Hara, T. J. (1982). Chemoreception and aquatic pollutants. In: *Chemoreception in Fishes* (T.J. Hara, ed.), pp. 363–393. Elsevier, Amsterdam.

Bulow, F. J. (1970). RNA–DNA ratios as indicators of recent growth rates of a fish. *J. Fish. Res. Bd Can.* **27**, 2343–2349.

Bulow, F. J. (1987). RNA–DNA ratios as indicators of growth in fish: a review. In: *The Age and Growth of Fish* (R.C. Summerfelt and G.E. Hall, eds), pp. 45–64. Iowa State University Press, Ames, IA.

Campbell, P. G. C. (1995). Interactions between trace metals and aquatic organisms: a critique of the free-ion activity model. In: *Metal Speciation and Bioavailability in Aquatic Systems* (A. Tessier and D.R. Turner, eds), pp. 45–102. John Wiley and Sons, Chichester.

Campbell, P. G. C. (2004). How might hypoxia affect metal speciation, accumulation and toxicity: some speculation. In *Proceedings of the 7th International Symposium on Fish, Physiology and Water Quality*, Tallinn, Estonia, May 12, 2003 (eds. G. L. Rupp and M. D. White), pp. 125–147. EPA 600/R-04/049. Athens, GA: US Environmental Protection Agency, Ecosystems Research Division.

Campbell, P. G. C., and Stokes, P. M. (1985). Acidification and toxicity of metals to aquatic biota. *Can. J. Fish. Aquat. Sci.* **42**, 2034–2049.

Campbell, P. G. C., Hontela, A., Rasmussen, J. B., Giguère, A., Gravel, A., Kraemer, L., Kovesces, J., Lacroix, A., Levesque, H., and Sherwood, G. (2003). Differentiating between direct (physiological) and food-chain mediated (bioenergetic) effects on fish in metal-impacted lakes. *Hum. Ecol. Risk Assess.* **9**, 847–866.

Campbell, P. G. C., Giguère, A., Bonneris, E., and Hare, L. (2005). Cadmium-handling strategies in two chronically exposed indigenous freshwater organisms – the yellow perch (*Perca flavescens*) and the floater mollusc (*Pyganodon grandis*). *Aquat. Toxicol.* **72**, 83–97.

Cancalon, P. (1982). Degeneration and regeneration of olfactory cells induced by ZnSO$_4$ and other chemicals. *Tissue Cell* **14**, 717–733.

Carreau, N. D., and Pyle, G. G. (2005). Effect of copper exposure during embryonic development on chemosensory function of juvenile fathead minnows (*Pimephales promelas*). *Ecotoxicol. Environ. Saf.* **61**, 1–6.

CCME (2003). *Canadian Water Quality Guidelines for the Protection of Aquatic Life: Guidance on the Site-Specific Application of Water Quality Guidelines in Canada: Procedures for Deriving Numerical Water Quality Objectives.* Canadian Council of Ministers of the Environment, Winnipeg.

Chapman, P. M., and Wang, F. (2000). Issues in ecological risk assessment of inorganic metals and metalloids. *Hum. Ecol. Risk Assess.* **9**, 965–988.

Ciutat, A., and Boudou, A. (2003). Bioturbation effects on cadmium and zinc transfers from a contaminated sediment and on metal bioavailability to benthic bivalves. *Environ. Toxicol. Chem.* **22**, 1574–1581.

Clarkson, T. (1995). Health effects of metals: a role for evolution? *Environ. Health Perspect.* **103**, 9–12.

Clearwater, S. J., Farag, A. M., and Meyer, J. S. (2002). Bioavailability and toxicity of dietborne copper and zinc to fish. *Comp. Biochem. Physiol. C* **132**, 269–313.

Cogun, H. Y., Yuzereoglu, T. A., Firat, O., Gok, G., and Kargin, F. (2006). Metal concentrations in fish species from the northeast Mediterranean Sea. *Environ. Monit. Assess.* **121**, 431–438.

Collvin, L. (1984). The effects of copper on maximum respiration rate and growth rate of perch, *Perca fluviatilis* L. *Water Res.* **18**, 139–144.

Couture, P., and Kumar, P. R. (2003). Impairment of metabolic capacities in copper and cadmium contaminated wild yellow perch (*Perca flavescens*). *Aquat. Toxicol.* **64**, 107–120.

Couture, P., and Pyle, G. G. (2008). Live fast and die young: metal effects on the condition and physiology of wild yellow perch from along two metal contamination gradients. *Hum. Ecol. Risk Assess.* **14**, 73–96.

Couture, P., and Rajotte, J. W. (2003). Morphometric and metabolic indicators of metal stress in wild yellow perch (*Perca flavescens*) from Sudbury Ontario: a review. *J. Environ. Monit.* **5**, 216–221.

Couture, P., Dutil, J.-D., and Guderley, H. (1998). Biochemical correlates of growth and condition in juvenile Atlantic cod (*Gadus morhua*) from Newfoundland. *Can. J. Fish. Aquat. Sci.* **55**, 1591–1598.

Couture, P., Busby, P., Rajotte, J. W., Gauthier, C., and Pyle, G. G. (2008a). Seasonal and regional variations of metal contamination and condition indicators in yellow perch (*Perca flavescens*) along two polymetallic gradients. I. Factors influencing tissue metal concentrations. *Hum. Ecol. Risk Assess.* **14**, 97–125.

Couture, P., Rajotte, J. W., and Pyle, G. G. (2008b). Seasonal and regional variations of metal contamination and condition indicators in yellow perch (*Perca flavescens*) along two polymetallic gradients. III. Energetic and physiological indicators. *Hum. Ecol. Risk Assess.* **14**, 146–165.

Couture, P., Gobeil, C., and Tessier, A. (2008c). Chronology of atmospheric deposition of arsenic inferred from reconstructed sedimentary records. *Environ. Sci. Technol.* **42**, 6508–6513.

Croteau, M.-N., Hare, L., and Tessier, A. (2003). Difficulties in relating Cd concentrations in the predatory insect *Chaoborus* to those of its prey in nature. *Can. J. Fish. Aquat. Sci.* **60**, 800–808.

Crump, K. L., and Trudeau, V. L. (2009). Mercury-induced reproductive impairment in fish. *Environ. Toxicol. Chem.* **28**, 895–907.

Dallinger, R., and Kautzky, H. (1985). The importance of contaminated food for the uptake of heavy metals by rainbow trout (*Salmo gairdneri*): a field study. *Oecologia* **67**, 82–89.

Dautremepuits, C., Marcogliese, D. J., Gendron, A. D., and Fournier, M. (2009). Gill and head kidney antioxidant processes and innate immune system responses of yellow perch (*Perca flavescens*) exposed to different contaminants in the St. Lawrence River, Canada. *Sci. Total Environ.* **407**, 1055–1064.

De Smet, H., and Blust, R. (2001). Stress responses and changes in protein metabolism in carp *Cyprinus carpio* during cadmium exposure. *Ecotoxicol. Environ. Saf.* **48**, 255–262.

Dragun, Z., Raspor, B., and Podrug, M. (2007). The influence of the season and the biotic factors on the cytosolic metal concentrations in the gills of the European chub (*Leuciscus cephalus* L.). *Chemosphere* **69**, 911–919.

Draves, J. F., and Fox, M. G. (1998). Effects of a mine tailing spill on feeding and metal concentrations in yellow perch (*Perca flavescens*). *Environ. Toxicol. Chem.* **17**, 1626–1632.

Drevnick, P. E., Roberts, A. P., Otter, R. R., Hammerschmidt, C. R., Klaper, R., and Oris, J. T. (2008). Mercury toxicity in livers of northern pike (*Esox lucius*) from Isle Royale, USA. *Comp. Biochem. Physiol. C* **147**, 331–338.

Driedger, K., Weber, L. P., Rickwood, C. J., Dube, M. G., and Janz, D. M. (2010). Growth and energy storage in juvenile fathead minnows exposed to metal mine waste water in simulated winter and summer conditions. *Ecotoxicol. Environ. Saf.* **73**, 727–734.

Dumas, J., and Hare, L. (2008). The internal distribution of nickel and thallium in two freshwater invertebrates and its relevance to trophic transfer. *Environ. Sci. Technol.* **42**, 5144–5149.

Dupont, C., Butcher, A., Valas, R., Bourne, P., and Caetano-Anolles, G. (2010). History of biological metal utilization inferred through phylogenomic analysis of protein structures. *Proc. Natl. Acad. Sci. U.S.A.* **107**, 10567–10572.

Eastwood, S., and Couture, P. (2002). Seasonal variations in condition and liver metal concentrations of yellow perch (*Perca flavescens*) from a metal-contaminated environment. *Aquat. Toxicol.* **58**, 43–46.

Ellenberger, S. A., Baumann, P. C., and May, T. W. (1994). Evaluation of effects caused by high copper concentrations in Torch Lake, Michigan, on reproduction of yellow perch. *J. Great Lakes Res.* **20**, 531–536.

Ercal, N., Gurer-Orhan, H., and Aykin-Burns, N. (2001). Toxic metals and oxidative stress Part I: Mechanisms involved in metal-induced oxidative damage. *Curr. Top. Med. Chem.* **1**, 529–539.

Erickson, R. J., Benoit, D. A., Mattwon, V. R., Nelson, H. P. J., and Leonard, E. N. (1996). The effects of water chemistry on the toxicity of copper to fathead minnow. *Environ. Toxicol. Chem.* **15**, 181–193.

Farag, A. M., Stansbury, M. A., Hogstrand, C., MacConnel, E., and Bergman, H. L. (1995). The physiological impairment of free-ranging brown trout exposed to metals in the Clark Fork River, Montana. *Can. J. Fish. Aquat. Sci.* **52**, 2038–2050.

Farkas, A., Salanki, J., and Specziar, A. (2003). Age- and size-specific patterns of heavy metals in the organs of freshwater fish *Abramis brama* L. populating a low-contaminated site. *Water Res.* **37**, 959–964.

Ferreira, M., Caetano, M., Costa, J., Pousão-Ferreira, P., Vale, C., and Reis-Henriques, M. A. (2008). Metal accumulation and oxidative stress responses in, cultured and wild, white seabream from Northwest Atlantic. *Sci. Total Environ.* **407**, 638–646.

Firestein, S. (2001). How the olfactory system makes sense of scents. *Nature* **413**, 211–218.

Foster, E. P., Drake, D. L., and DiDomenico, G. (2000). Seasonal changes and tissue distribution of mercury in largemouth bass (*Micropterus salmoides*) from Dorena Reservoir, Oregon. *Arch. Environ. Contam. Toxicol.* **38**, 78–82.

Franklin, N. M., Glover, C. N., Nicol, J. A., and Wood, C. M. (2005). Calcium/cadmium interactions at uptake surfaces in rainbow trout: waterborne versus dietary routes of exposure. *Environ. Toxicol. Chem.* **24**, 2954–2964.

Garceau, N., Pichaud, N., and Couture, P. (2010). Inhibition of goldfish mitochondrial metabolism by *in vitro* exposure to Cd, Cu and Ni. *Aquat. Toxicol.* **98**, 107–112.

Gargiulo, G., Arcamone, N., de Girolamo, P., Andreozzi, G., Antonucci, R., Esposito, V., Ferrara, L., and Battaglini, P. (1996). Histochemical study of the effects of cadmium uptake on oxidative enzymes of intermediary metabolism in kidney of goldfish (*Carassius auratus*). *Comp. Biochem. Physiol. C* **113**, 177–183.

Gauthier, C., Couture, P., and Pyle, G. G. (2006). Metal effects on fathead minnows (*Pimephales promelas*) under field and laboratory conditions. *Ecotoxicol. Environ. Saf.* **63**, 353–364.

Gauthier, C., Campbell, P. G., and Couture, P. (2008a). Physiological correlates of growth and condition in the yellow perch (*Perca flavescens*). *Comp. Biochem. Physiol. A* **151**, 526–532.

Gauthier, C., Campbell, P. G. C., and Couture, P. (2008b). Physiological correlates of growth and condition in the yellow perch (*Perca flavescens*). *Comp. Biochem. Physiol. A* **151**, 526–532.

Geeraerts, C., and Belpaire, C. (2010). The effects of contaminants in European eel: a review. *Ecotoxicology* **19**, 239–266.

Giattina, J. D., Garton, R. R., and Stevens, D. G. (1982). Avoidance of copper and nickel by rainbow trout as monitored by a computer-based data acquisition system. *Trans. Am. Fish. Soc.* **111**, 491–504.

Giguère, A., Campbell, P. G. C., Hare, L., McDonald, D. G., and Rasmussen, J. B. (2004). Influence of lake chemistry and fish age on cadmium, copper and zinc concentrations in various organs of indigenous yellow perch (*Perca flavescens*). *Can. J. Fish. Aquat. Sci.* **61**, 1702–1716.

Giguère, A., Campbell, P. G. C., Hare, L., and Cossu-Leguille, C. (2005). Metal bioaccumulation and oxidative stress in yellow perch (*Perca flavescens*) collected from eight lakes along a metal contamination gradient (Cd, Cu, Zn, Ni). *Can. J. Fish. Aquat. Sci.* **62**, 563–577.

Gillespie, R. B., and Baumann, P. C. (1986). Effects of high tissue concentrations of selenium on reproduction by bluegills. *Trans. Am. Fish. Soc.* **115**, 208–213.

Ginsberg, G. L., and Toal, B. F. (2000). Development of a single-meal fish consumption advisory for methyl mercury. *Risk Anal.* **20**, 41–47.

Gravel, A., Campbell, P. G. C., and Hontela, A. (2005). Disruption of the hypothalamo-pituitary–interrenal axis in 1+ yellow perch (*Perca flavescens*) chronically exposed to metals in the environment. *Can. J. Fish. Aquat. Sci.* **62**, 982–990.

Green, W., Mirza, R., Wood, C. M., and Pyle, G. (2010). Copper binding dynamics and olfactory impairment in fathead minnows (*Pimephales promelas*). *Environ. Sci. Technol.* **44**, 1431–1437.

Guan, R., and Wang, W. X. (2004). Dietary assimilation and elimination of Cd, Se, and Zn by *Daphnia magna* at different metal concentrations. *Environ. Toxicol. Chem.* **23**, 2689–2698.

Gunn, J., Keller, W., Negusanti, J., Potvin, R., Beckett, P., and Winterhalder, K. (1995). Ecosystem recovery after emission reductions: Sudbury, Canada. *Water Air Soil Pollut.* **85**, 1783–1788.

Hamilton, S. J. (2004). Review of selenium toxicity in the aquatic food chain. *Sci. Total Environ.* **326**, 1–31.

Hansen, J. A., Marr, J. C. A., Lipton, J., Cacela, D., and Bergman, H. L. (1999a). Differences in neurobehavioral responses of chinook salmon (*Oncorhynchus tshawytscha*) and rainbow trout (*Oncorhynchus mykiss*) exposed to copper and cobalt: behavioral avoidance. *Environ. Toxicol. Chem.* **18**, 1972–1978.

Hansen, J. A., Rose, J. D., Jenkins, R. A., Gerow, K. G., and Bergman, H. L. (1999b). Chinook salmon (*Oncorhynchus tshawytscha*) and rainbow trout (*Oncorhynchus mykiss*) exposed to copper: neurophysiological and histological effects on the olfactory system. *Environ. Toxicol. Chem.* **18**, 1979–1991.

Hansen, J. A., Woodward, D. F., Little, E. E., DeLonay, A. J., and Bergman, H. L. (1999c). Behavioral avoidance: possible mechanism for explaining abundance and distribution of trout species in a metal-impacted river. *Environ. Toxicol. Chem.* **18**, 313–317.

Hara, T. J. (1981). Behavioral and electrophysiological studies of chemosensory reactions in fish. In: *Brain Mechanisms of Behavior in Lower Vertebrates* (P.J. Laming, ed.), pp. 123–136. Cambridge University Press, Cambridge.

Hare, L., and Campbell, P. G. C. (1992). Temporal variations of trace metals in aquatic insects. *Freshwater Biol.* **27**, 13–27.

Hare, L., Tessier, A., and Warren, L. (2001). Cadmium accumulation by invertebrates living at the sediment–water interface. *Environ. Toxicol. Chem.* **20**, 880–889.

Hattink, J., De Boeck, G., and Blust, R. (2006). Toxicity, accumulation, and retention of zinc by carp under normoxic and hypoxic conditions. *Environ. Toxicol. Chem.* **25**, 87–96.

Hontela, A., Rasmussen, J. B., Audet, C., and Chevalier, G. (1992). Impaired cortisol stress response in fish from environments polluted by PAHs, PCBs and mercury. *Arch. Environ. Contam. Toxicol.* **22**, 278–283.

Hontela, A., Dumont, P., Duclos, D., and Fortin, R. (1995). Endocrine and metabolic dysfunction in yellow perch, *Perca flavescens*, exposed to organic contaminants and heavy metals in the St. Lawrence River. *Environ. Toxicol. Chem.* **14**, 725–731.

Hylander, L. D., Pinto, F. N., Guimaraes, J. R., Meili, M., Oliveira, L. J., and de Castro e Silva, E. (2000). Fish mercury concentration in the Alto Pantanal, Brazil: influence of season and water parameters. *Sci. Total Environ.* **261**, 9–20.

Iles, A. C., and Rasmussen, J. B. (2005). Indirect effects of metal contamination on energetics of yellow perch (*Perca flavescens*) resulting from food web simplification. *Freshwater Biol.* **50**, 976–992.

Jensen, S., and Jernelöv, A. (1969). Biological methylation of mercury in aquatic organisms. *Nature* **223**, 753–754.

Jezierska, B., Lugowska, K., and Witeska, M. (2009). The effects of heavy metals on embryonic development of fish (a review). *Fish Physiol. Biochem.* **35**, 625–640.

Johannessen, M., and Henriksen, A. (1978). Chemistry of snow meltwater: changes in concentration during melting. *Water Resour. Res.* **14**, 615–619.

Julliard, A. K., Saucier, D., and Astic, L. (1993). Effects of chronic low-level copper exposure on ultrastructure of the olfactory system in rainbow trout (*Oncorhynchus mykiss*). *Histol. Histopathol.* **8**, 655–672.

Julliard, A. K., Saucier, D., and Astic, L. (1996). Time-course of apoptosis in the olfactory epithelium of rainbow trout exposed to a low copper level. *Tissue Cell* **28**, 367–377.

Kamunde, C., Clayton, C., and Wood, C. M. (2002). Waterborne vs. dietary copper uptake in rainbow trout and the effects of previous waterborne copper exposure. *Am. J. Physiol. R.* **283**, R69–R78.

Kaufman, S. D., Gunn, J. M., Morgan, G. E., and Couture, P. (2006). Muscle enzymes reveal walleye (*Sander vitreus*) are less active when larger prey, lake herring (*Coregonus artedi*), is present. *Can. J. Fish. Aquat. Sci.* **63**, 970–979.

Kearns, P. K., and Atchison, G. J. (1979). Effects of trace metals on growth of yellow perch (*Perca flavescens*) as measured by RNA–DNA ratios. *Environ. Biol. Fish.* **4**, 383–387.

Klaverkamp, J. F., Hodgins, D. A., and Lutz, A. (1983). Selenite toxicity and mercury–selenium interactions in juvenile fish. *Arch. Environ. Contam. Toxicol.* **12**, 405–413.

Klima, K. E., and Applehans, F. M. (1990). Copper exposure and the degeneration of olfactory receptors in rainbow trout (*Oncorhynchus mykiss*). *Chem. Spec. Bioavailab.* **2**, 149–154.

Klinck, J. S., Green, W. W., Mirza, R. S., Nadella, S. R., Chowdhury, M. J., Wood, C. M., and Pyle, G. G. (2007). Branchial cadmium and copper binding and intestinal cadmium uptake in wild yellow perch (*Perca flavescens*) from clean and metal-contaminated lakes. *Aquat. Toxicol.* **84**, 198–207.

Köck, G., Triendl, M., and Hofer, R. (1996). Seasonal patterns of metal accumulation in Arctic char (*Salvelinus alpinus*) from an oligotrophic alpine lake related to temperature. *Can. J. Fish. Aquat. Sci.* **53**, 780–786.

Kolmakov, N., Hubbard, P., Lopes, O., and Canario, A. (2009). Effect of acute copper sulfate exposure on olfactory responses to amino acids and pheromones in goldfish (*Carassius auratus*). *Environ. Sci. Technol.* **43**, 8393–8399.

Kövecses, J., Sherwood, G. D., and Rasmussen, J. B. (2005). Impacts of altered benthic invertebrate communities on the feeding ecology of yellow perch (*Perca flavescens*) in metal-contaminated lakes. *Can. J. Fish. Aquat. Sci.* **62**, 153–162.

Kraemer, L. D., Campbell, P. G. C., and Hare, L. (2005). A field study examining metal elimination kinetics in juvenile yellow perch (*Perca flavescens*). *Aquat. Toxicol.* **75**, 108–126.

Kraemer, L., Campbell, P. G. C., and Hare, L. (2006). Seasonal variations in hepatic Cd and Cu concentrations and in the sub-cellular distribution of these metals in juvenile yellow perch (*Perca flavescens*). *Environ. Pollut.* **142**, 313–325.

Lacroix, A., and Hontela, A. (2004). A comparative assessment of the adrenotoxic effects of cadmium in two teleost species, rainbow trout, *Oncorhynchus mykiss*, and yellow perch, *Perca flavescens*. *Aquat. Toxicol.* **67**, 13–21.

Laflamme, J.-S., Couillard, Y., Campbell, P. G. C., and Hontela, A. (2000). Interrenal metallothionein and cortisol secretion in relation to Cd, Cu, and Zn exposure in yellow perch, *Perca flavescens*, from Abitibi lakes. *Can. J. Fish. Aquat. Sci.* **57**, 1692–1700.

Lagauzere, S., Pischedda, L., Cuny, P., Gilbert, F., Stora, G., and Bonzom, J. M. (2009). Influence of *Chironomus riparius* (Diptera, Chironomidae) and *Tubifex tubifex* (Annelida, Oligochaeta) on oxygen uptake by sediments. Consequences of uranium contamination. *Environ. Pollut.* **157**, 1234–1242.

Lapointe, D., and Couture, P. (2009). Influence of the route of exposure on the accumulation and subcellular distribution of nickel and thallium in juvenile fathead minnows (*Pimephales promelas*). *Arch. Environ. Contam. Toxicol.* **57**, 571–580.

Lapointe, D., and Couture, P. (2010). Accumulation and effects of nickel and thallium in early life stages of fathead minnows (*Pimephales promelas*). *Ecotoxicol. Environ. Saf.* **73**, 572–578.

Lapointe, D., Gentès, S., Ponton, D. E., Hare, L., and Couture, P. (2009). Influence of prey type on nickel and thallium assimilation, subcellular distribution and effects in juvenile fathead minnows (*Pimephales promelas*). *Environ. Sci. Technol.* **43**, 8665–8670.

Latimer, S. D., Devall, M. S., Thomas, C., Ellgaard, E. G., Kumar, S. D., and Thien, L. B. (1996). Dendrochronology and heavy metal deposition in tree rings of baldcypress. *J. Environ. Qual.* **25**, 1411–1419.

Levesque, H. W., Moon, T. W., Campbell, P. G. C., and Hontela, A. (2002). Seasonal variation in carbohydrate and lipid metabolism of yellow perch (*Perca flavescens*) chronically exposed to metals in the field. *Aquat. Toxicol.* **60**, 257–267.

Levesque, H., Dorval, J., Hontela, A., Van Der Kraak, G., and Campbell, P. G. C. (2003). Hormonal, morphological, and physiological responses of yellow perch (*Perca flavescens*) to chronic environmental metal exposures. *J. Toxicol. Environ. Health* **66**, 657–676.

Liu, G., Cai, Y., Kalla, P., Scheidt, D., Richards, J., Scinto, L. J., Gaiser, E., and Appleby, C. (2008). Mercury mass budget estimates and cycling seasonality in the Florida Everglades. *Environ. Sci. Technol.* **42**, 1954–1960.

Lopes, P. A., Pinheiro, T., Santos, M. C., da Luz Mathias, M., Collares-Pereira, M. J., and Viegas-Crespo, A. M. (2001). Response of antioxidant enzymes in freshwater fish populations (*Leuciscus alburnoides* complex) to inorganic pollutants exposure. *Sci. Total Environ.* **280**, 153–163.

Luoma, S. N., and Rainbow, P. S. (2005). Why is metal bioaccumulation so variable? Biodynamics as a unifying concept. *Environ. Sci. Technol.* **39**, 1921–1931.

Lürling, M., and Scheffer, M. (2007). Info-disruption: pollution and the transfer of chemical information. *Trends Ecol. Evol.* **22**, 374–379.

Lydersen, E., and Löfgren, S. (2002). Metals in Scandinavian surface waters: effects of acidification, liming, and potential reacidification. *Crit. Rev. Environ. Sci. Technol.* **32**, 73–295.

Maciorowski, H. D., Clarke, R. M. and Scherer, E. (1977). The use of avoidance–preference bioassays with aquatic invertebrates. *Proceedings of the 3rd Aquatic Toxicity Workshop*, Halifax, Nova Scotia, No. 2–3, 1976, pp. 49–58. Technical Report No. EPS-5-AR-77-1. Halifax, Canada: Environment Canada.

Magrassi, L., and Graziadei, P. P. (1995). Cell death in the olfactory epithelium. *Anat. Embryol.* **192**, 77–87.

Marrugo-Negrete, J., Benitez, L. N., and Olivero-Verbel, J. (2008). Distribution of mercury in several environmental compartments in an aquatic ecosystem impacted by gold mining in northern Colombia. *Arch. Environ. Contam. Toxicol.* **55**, 305–316.

Mathews, T., and Fisher, N. S. (2009). Dominance of dietary intake of metals in marine elasmobranch and teleost fish. *Sci. Total Environ.* **407**, 5156–5161.

McGeer, J. C., Brix, K. V., Skeaff, J. M., DeForest, D. K., Brigham, S. I., Adams, W. J., and Green, A. (2003). Inverse relationship between bioconcentration factor and exposure concentrations for metals: implications for hazard assessment of metals in the aquatic environment. *Environ. Toxicol. Chem.* **22**, 1017–1037.

McPherson, T. D., Mirza, R. S., and Pyle, G. G. (2004). Responses of wild fishes to alarm chemicals in pristine and metal-contaminated lakes. *Can. J. Zool.* **82**, 694–700.

Mendil, D., Demirci, Z., Tuzen, M., and Soylak, M. (2010). Seasonal investigation of trace element contents in commercially valuable fish species from the Black Sea, Turkey. *Food Chem. Toxicol.* **48**, 865–870.

Meybeck, M., Lestel, L., Bonté, P., Moilleron, R., Colin, J. L., Rousselot, O., Hervé, D., de Pontevès, C., Grosbois, C., and Thévenot, D. R. (2007). Historical perspective of heavy metals contamination (Cd, Cr, Cu, Hg, Pb, Zn) in the Seine River basin (France) following a DPSIR approach (1950–2005). *Sci. Total Environ.* **375**, 204–231.

Meyer, J. S., and Adams, W. J. (2010). Relationship between biotic ligand model-based water quality criteria and avoidance and olfactory responses to copper by fish. *Environ. Toxicol. Chem.* **29**, 2096–2103.

Meyer, J. S., Adams, W. J., Brix, K. V., Luoma, S. N., Mount, D. R., Stubblefield, W. A., and Wood, C. M. (2005). *Toxicity of Dietborne Metals to Aquatic Organisms*. SETAC Press, Pensacola, FL.

Mighall, T. M., Abrahams, P. W., Grattan, J. P., Hayes, D., Timberlake, S., and Forsyth, S. (2002). Geochemical evidence for atmospheric pollution derived from prehistoric copper mining at Copa Hill, Cwmystwyth, mid-Wales, UK. *Sci. Total Environ.* **292**, 69–80.

Miller, P. A., Munkittrick, K. R., and Dixon, D. G. (1992). Relationship between concentrations of copper and zinc in water, sediment, benthic invertebrates, and tissues of white sucker (*Catostomus commersoni*) at metal-contaminated sites. *Can. J. Fish. Aquat. Sci.* **49**, 978–984.

Mirza, R. S., Green, W. W., Connor, S., Weeks, A. C. W., Wood, C. M., and Pyle, G. G. (2009). Do you smell what I smell? Olfactory impairment in wild yellow perch from metal-contaminated waters. *Ecotoxicol. Environ. Saf.* **72**, 677–683.

Morel, F. M. M. (1983). *Principles of Aquatic Chemistry*. Wiley-Interscience, New York.

Mzimela, H. M., Wepener, V., and Cyrus, D. P. (2003). Seasonal variation of selected metals in sediments, water and tissues of the groovy mullet, *Liza dumerelii* (Mugilidae) from the Mhlathuze Estuary, South Africa. *Mar. Pollut. Bull.* **46**, 659–664.

Nielsen, F. H. (2000). Evolutionary events culminating in specific minerals becoming essential for life. *Eur. J. Nutr.* **39**, 62–66.

Niyogi, S., and Wood, C. M. (2003). Effects of chronic waterborne and dietary metal exposures on gill metal-binding: implications for the biotic ligand model. *Hum. Ecol. Risk Assess.* **9**, 813–846.

Niyogi, S., and Wood, C. M. (2004). Biotic ligand model, a flexible tool for developing site-specific water quality guidelines for metals. *Environ. Sci. Technol.* **38**, 6177–6192.

Niyogi, S., Pyle, G. G., and Wood, C. M. (2007). Branchial versus intestinal zinc uptake in wild yellow perch (*Perca flavescens*) from reference and metal-contaminated aquatic ecosystems. *Can. J. Fish. Aquat. Sci.* **64**, 1605–1613.

Nriagu, J. O., Wong, H. K. T., Lawson, G., and Daniel, P. (1998). Saturation of ecosystems with toxic metals in Sudbury basin, Ontario Canada. *Sci. Total Environ.* **223**, 99–117.

Padmini, E., Usha Rani, M., and Vijaya Geetha, B. (2009). Studies on antioxidant status in *Mugil cephalus* in response to heavy metal pollution at Ennore estuary. *Environ. Monit. Assess.* **155**, 215–225.

Pagenkopf, G. K. (1983). Gill surface interaction model for trace-metal toxicity to fishes: role of complexation, pH, and water hardness. *Environ. Sci. Technol.* **17**, 342–347.

Paller, M. H., Bowers, J. A., Littrell, J. W., and Guanlao, A. V. (2004). Influences on mercury bioaccumulation factors for the Savannah River. *Arch. Environ. Contam. Toxicol.* **46**, 236–243.

Pandey, S., Parvez, S., Sayeed, I., Haque, R., Bin-Hafeez, B., and Raisuddin, S. (2003). Biomarkers of oxidative stress: a comparative study of river Yamuna fish *Wallago attu* (Bl. & Schn.). *Sci. Total Environ.* **309**, 105–115.

Pandey, S., Parvez, S., Ansari, R. A., Ali, M., Kaur, M., Hayat, F., Ahmad, F., and Raisuddin, S. (2008). Effects of exposure to multiple trace metals on biochemical, histological and ultrastructural features of gills of a freshwater fish, *Channa punctata* Bloch. *Chem. Biol. Interact.* **174**, 183–192.

Pane, E. F., Haque, A., Goss, G. G., and Wood, C. M. (2004). The physiological consequences of exposure to chronic, sublethal waterborne nickel in rainbow trout (*Oncorhynchus mykiss*): exercise vs resting physiology. *J. Exp. Biol.* **207**, 1249–1261.

Paquin, P. R., Santore, R. C., Wu, K. B., Kavvadas, C. D., and Di Toro, D. M. (2000). The biotic ligand model: a model of the acute toxicity of metals to aquatic life. *Environ. Sci. Policy* **3**, 175–182.

Paquin, P. R., Zoltay, V., Winfield, R. P., Wu, K. B., Mathew, R., Santore, R. C., and Di Toro, D. M. (2002). Extension of the biotic ligand model of acute toxicity to a physiologically-based model of the survival time of rainbow trout (*Oncorhynchus mykiss*) exposed to silver. *Comp. Biochem. Physiol. C* **133**, 305–343.

Patrick, F. M., and Loutit, M. W. (1978). Passage of metals to freshwater fish from their food. *Water Res.* **12**, 395–398.

Pedder, S. C. J., and Maly, E. J. (1985). The effect of lethal copper solutions on the behavior of rainbow trout. *Salmo gairdneri*. *Arch. Environ. Contam. Toxicol.* **14**, 501–507.

Peplow, D., and Edmonds, R. (2005). The effects of mine waste contamination at multiple levels of biological organization. *Ecol. Eng.* **24**, 101–119.

Pierron, F., Baudrimont, M., Gonzalez, P., Bourdineaud, J. P., Elie, P., and Massabuau, J. C. (2007). Common pattern of gene expression in response to hypoxia or cadmium in the gills of the European glass eel (*Anguilla anguilla*). *Environ. Sci. Technol.* **41**, 3005–3011.

Pierron, F., Bourret, V., St-Cyr, J., Campbell, P. G. C., Bernatchez, L., and Couture, P. (2009). Transcriptional responses to environmental metal exposure in wild yellow perch (*Perca flavescens* collected in lakes with differing environmental metal concentrations (Cd, Cu, Ni). *Ecotoxicology* **18**, 620–631.

Playle, R. C., Dixon, D. G., and Burnison, K. (1993a). Copper and cadmium binding to fish gills: estimates of metal-gill stability constants and modelling of metal accumulation. *Can. J. Fish. Aquat. Sci.* **50**, 2678–2687.

Playle, R. C., Dixon, D. G., and Burnison, K. (1993b). Copper and cadmium binding to fish gills: modification by dissolved organic carbon and synthetic ligands. *Can. J. Fish. Aquat. Sci.* **50**, 2667–2677.

Proulx, I., and Hare, L. (2008). Why bother to identify animals used for contaminant monitoring? *Integr. Environ. Assess. Manag.* **4**, 125–126.

Pyle, G. G., and Clulow, F. V. (1997). Non-linear radionuclide transfer from the aquatic environment. *Health Phys.* **73**, 488–493.

Pyle, G. G., and Mirza, R. S. (2007). Copper-impaired chemosensory function and behavior in aquatic animals. *Hum. Ecol. Risk Assess.* **13**, 492–505.

Pyle, G., and Wood, C. M. (2007). Predicting "non-scents": rationale for a chemosensory-based biotic ligand model. *Australas. J. Ecotoxicol.* **13**, 47–51.

Pyle, G. G., Rajotte, J. W., and Couture, P. (2005). Effects of industrial metals on wild fish populations along a metal contamination gradient. *Ecotoxicol. Environ. Saf.* **61**, 287–312.

Pyle, G. G., Busby, P., Gauthier, C., Rajotte, J. W., and Couture, P. (2008). Seasonal and regional variations of metal contamination and condition indicators in yellow perch (*Perca flavescens*) along two polymetallic gradients. II. Growth patterns, longevity, and condition. *Hum. Ecol. Risk Assess.* **14**, 126–145.

Rajotte, J. W., and Couture, P. (2002). Effects of environmental metal contamination on the condition, swimming performance, and tissue metabolic capacities of wild yellow perch (*Perca flavescens*). *Can. J. Fish. Aquat. Sci.* **59**, 1296–1304.

Rasmussen, J. B., Gunn, J. M., Sherwood, G. D., Iles, A., Gagnon, A., Campbell, P. G. C., and Hontela, A. (2008). Direct and indirect (foodweb mediated) effects of metal exposure on the growth of yellow perch (*Perca flavescens*): implications for ecological risk assessment. *Hum. Ecol. Risk Assess.* **14**, 317–350.

Rosenzweig, A. (2002). Metallochaperones: bind and deliver. *Chem. Biol.* **9**, 673–677.

Rotchell, J. M., Clarke, K. R., Newton, L. C., and Bird, D. J. (2001). Hepatic metallothionein as a biomaker for metal contamination: age effects and seasonal variation in European flounders (*Pleuronectes flesus*) from the Severn Estuary and Bristol Channel. *Mar. Environ. Res.* **52**, 151–171.

Rudd, J. W. M. (1995). Sources of methyl mercury to freshwater ecosystems: a review. *Water Air Soil Pollut.* **80**, 697–713.

Rudd, J. W., Furutani, A., and Turner, M. A. (1980). Mercury methylation by fish intestinal contents. *Appl. Environ. Microbiol.* **40**, 777–782.

Saiki, M. K., Castleberry, D. T., May, T. W., Martin, B. A., and Bullard, F. N. (1995). Copper, cadmium, and zinc concentrations in aquatic food chains from the upper Sacramento River (California) and selected tributaries. *Arch. Environ. Contam. Toxicol.* **29**, 484–491.

Sandahl, J. F., Baldwin, D. H., Jenkins, J. J., and Scholz, N. L. (2007). A sensory system at the interface between urban stormwater runoff and salmon survival. *Environ. Sci. Technol.* **41**, 2998–3004.

Saunders, R. L., and Sprague, J. B. (1967). Effects of copper–zinc mining pollution on a spawning migration of Atlantic salmon. *Water Res.* **1**, 419–432.

Scandalios, J. G. (2005). Oxidative stress: molecular perception and transduction of signals triggering antioxidant gene defenses. *Braz. J. Med. Biol. Res.* **38**, 995–1014.

Scherer, E., and McNicol, R. E. (1998). Preference–avoidance responses of lake whitefish (*Coregonus clupeaformis*) to competing gradients of light and copper, lead, and zinc. *Water Res* **32**, 924–929.

Schlekat, C., Lee, B., and Luoma, S. (2002). Dietary metals exposure and toxicity to aquatic organisms: implications for ecological risk assessment. In *Coastal and Estuarine Risk Assessment*. CRC Press, Boca Raton, FL.

Schlekat, C., McGreer, J., Blust, R., Borgmann, U., Brix, K., Bury, N., Couillard, Y., Dwey, R., Luoma, S., and Robertson, S. (2003). Bioaccumulation: hazard identification of metals and inorganic metal substances. In: *Assessing the Hazard of Metal and Inorganic Metal Substances in Aquatic and Terrestrial Systems* (W.J. Adams and P.M. Chapman, eds), pp. 55–87. CRC Press, Boca Raton, FL.

Scott, G. R., and Sloman, K. A. (2004). The effects of environmental pollutants on complex fish behaviour: integrating behavioural and physiological indicators of toxicity. *Aquat. Toxicol.* **68**, 369–392.

Scott, G. R., Sloman, K. A., Rouleau, C., and Wood, C. M. (2003). Cadmium disrupts behavioural and physiological responses to alarm substance in juvenile rainbow trout (*Oncorhynchus mykiss*). *J. Exp. Biol.* **206**, 1779–1790.

Sherwood, G. D., Rasmussen, J. B., Rowan, D. J., Brodeur, J., and Hontela, A. (2000). Bioenergetic costs of heavy metal exposure in yellow perch (*Perca flavescens*): *in situ* estimates with radiotracer (^{137}Cs) technique. *Can. J. Fish. Aquat. Sci.* **57**, 441–450.

Sloman, K. A., Baker, D. W., Wood, C. M., and McDonald, D. G. (2002). Social interactions affect physiological consequences of sublethal copper exposure in rainbow trout. *Oncorhynchus mykiss. Environ. Toxicol. Chem.* **21**, 1255–1263.

Sloman, K. A., Scott, G. R., Diao, Z., Rouleau, C., Wood, C. M., and McDonald, D. G. (2003). Cadmium affects the social behaviour of rainbow trout. *Oncorhynchus mykiss. Aquat. Toxicol.* **65**, 171–185.

Sprague, J. B., Elson, P. F., and Saunders, R. L. (1965). Sublethal copper–zinc pollution in a salmon river – a field and laboratory study. *Int. J. Air Water Pollut.* **9**, 531–543.

Steding, D. J., Dunlap, C. E., and Flegal, A. R. (2000). New isotopic evidence for chronic lead contamination in the San Francisco Bay estuary system: implications for the persistence of past industrial lead emissions in the biosphere. *Proc. Natl. Acad. Sci. U.S.A.* **97**, 11181–11186.

Stohs, S. J., and Bagchi, D. (1995). Oxidative mechanisms in the toxicity of metal ions. *Free Rad. Biol. Med.* **18**, 321–336.

Sutterlin, A. M. (1974). Pollutants and the chemical senses of aquatic animals – perspective and review. *Chem. Senses* **1**, 167–178.

Sutterlin, A. M., and Gray, R. (1973). Chemical basis for homing of Atlantic salmon (*Salmo salar*) to a hatchery. *J. Fish. Res. Bd Can.* **30**, 895–899.

Svecevičius, G. (1999). Fish avoidance response to heavy metals and their mixtures. *Acta Zool. Lituanica* **9**, 103–113.

Thorndycraft, V. R., Pirrie, D., and Brown, A. G. (2004). Alluvial records of medieval and prehistoric tin mining on Dartmoor, southwest England. *Geoarchaeology* **19**, 219–236.

Tremblay, A., Lesbarreres, D., Merritt, T., Wilson, C., and Gunn, J. (2008). Genetic structure and phenotypic plasticity of yellow perch (*Perca flavescens*) populations influenced by habitat, predation, and contamination gradients. *Integr. Environ. Assess. Manag.* **4**, 264–266.

Turner, M., and Swick, A. (1983). The English–Wabigoon river system: IV. Interaction between mercury and selenium accumulated from waterborne and dietary sources by northern pike (*Esox lucius*). *Can. J. Fish. Aquat. Sci.* **40**, 2241–2250.

Valko, M., Morris, H., and Cronin, M. T. D. (2005). Metals, toxicity and oxidative stress. *Curr. Med. Chem.* **12**, 1161–1208.

Veltman, K., Huijbregts, M. A., Van Kolck, M., Wang, W. X., and Hendriks, A. J. (2008). Metal bioaccumulation in aquatic species: quantification of uptake and elimination rate constants using physicochemical properties of metals and physiological characteristics of species. *Environ. Sci. Technol.* **42**, 852–858.

Viarengo, A., Burlando, B., Ceratto, C., and Panfoli, I. (2000). Antioxidant role of metallothioneins: a comparative overview. *Cell. Mol. Biol.* **46**, 407–417.

Vogel, F. S. (1959). The deposition of exogenous copper under experimental conditions with observations on its neurotoxic and nephrotoxic properties in relation to Wilson's disease. *J. Exp. Med.* **110**, 801–810.

Wallace, W. G., and Luoma, S. N. (2003). Subcellular compartmentalization of Cd and Zn in two bivalves. II. Significance of trophically available metal (TAM). *Mar. Ecol. Prog. Ser.* **257**, 125–137.

Wallace, W. G., Byeong-Gweon, L., and Luoma, S. N. (2003). Subcellular compartmentalization of Cd and Zn in two bivalves. I. Significance of metal-sensitive fractions (MSF) and biologically detoxified metal (BDM). *Mar. Ecol. Prog. Ser.* **249**, 183–197.

Williams, R. (1997). The natural selection of the chemical elements. *Cell. Mol. Life Sci.* **53**, 816–829.

Wobeser, G. (1975). Acute toxicity of methyl mercury chloride and mercuric chloride for rainbow trout (*Salmo gairdneri*) fry and fingerlings. *J. Fish. Res. Bd Can.* **32**, 2005–2013.

Woodward, D. F., Hansen, J. A., Bergman, H. L., Little, E. E., and DeLonay, A. J. (1995). Brown trout avoidance of metals in water characteristic of the Clark Fork River, Montana. *Can. J. Fish. Aquat. Sci.* **52**, 2031–2037.

Yilmaz, A. B., and Dogan, M. (2008). Heavy metals in water and in tissues of himri (*Carasobarbus luteus*) from Orontes (Asi) River, Turkey. *Environ. Monit. Assess.* **144**, 437–444.

Zielinski, B. S., and Hara, T. J. (1992). Ciliated and microvillar receptor cells degenerate and then differentiate in the olfactory epithelium of rainbow trout following olfactory nerve section. *Microsc. Res. Techniq.* **23**, 22–27.

Zielinski, B., and Hara, T. J. (2007). Olfaction. In: *Fish Physiology* (T.J. Hara and B. Zielinski, eds), pp. 1–43. Elsevier, New York.

Žikič, R. V., Štajn, A.Š., Pavlovič, S. Z., Ognjanovič, B. I., and Saičič, Z. S. (2001). Activities of superoxide dismutase and catalase in erythrocytes and plasma transaminases of goldfish (*Carassius auratus gibelio* Bloch.) exposed to cadmium. *Physiol. Res.* **50**, 105–111.

INDEX

This index includes entries for both Homeostasis and Toxicology of Essential Metals, Volume 31A and Homeostasis and Toxicology of Non-Essential Metals, Volume 31B. The page numbers for entries from Volume 31A will be followed by an A and the page numbers for entries for Volume 31B will be followed by a B. For example, the entry for "Acute-to-chronic ratio (ACR), 17A" would be found in page 17 of Volume 31A.

C

OTHER VOLUMES IN THE
FISH PHYSIOLOGY SERIES

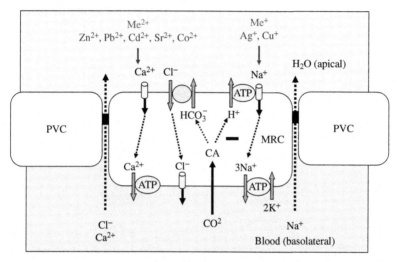

Fig. 1.3. A general model for how metal ions may enter the gill through "ionic mimicry" and thereby compete with nutritive ions for uptake, and, if in high enough concentration, eventually block nutritive ion uptake by inhibiting the ATP-dependent basolateral enzymes that normally power these processes. These nutritive ion uptake processes are critical to life in freshwater fish, because they must occur continuously to offset the passive losses of Na^+, Cl^-, and Ca^{2+}, which are shown as occurring by diffusion through the paracellular channels between the pavement cells (PVCs) and mitochondria-rich cells (MRCs). Monovalent Ag^+ and Cu^+ (after hypothesized reduction of Cu^{2+} to Cu^+ by a surface-bound reductase, not shown) compete with Na^+ for entry through a putative apical sodium channel (shown) and/or an Na^+/H^+ exchanger (not shown). Eventually they inhibit basolateral Na^+/K^+-ATPase. Divalent Zn^{2+}, Pb^{2+}, Cd^{2+}, Sr^{2+}, and Co^{2+} compete with Ca^{2+} for entry through an apical voltage-independent calcium channel (probably ECac), and eventually inhibit basolateral high-affinity Ca^{2+}-ATPase (shown) or an Na^+/Ca^{2+} exchanger (not shown). Some of these metals may also inhibit the intracellular carbonic anhydrase enzyme (CA), which provides the acid–base equivalents (H^+ and HCO_3^- ions) needed for exchange by the apical processes. For convenience, all processes are shown in a single MRC, but they may actually occur in different types of MRCs, or even in PVCs. Modified from an unpublished diagram by Fernando Galvez.

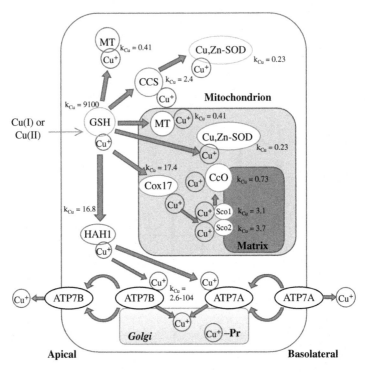

Fig. 2.9. Copper binding proteins and ligands, apparent Cu binding affinities (K_{Cu}; 10^{-15} M) and Cu transport pathways in eukaryotic cells (redrawn from Banci et al., 2010a). The lower the K_{Cu} number, the higher the affinity (see text for further details). Cellular Cu uptake can occur by several different transporters depending on cell and tissue type (see Figs 2.7 and 2.8) but cellular Cu is present as Cu^+ due to the reducing intracellular milieu. GSH: glutathione; MT: metallothionein; CCS: Cu chaperone for Cu,Zn superoxide dismutase (SOD1); HAH1: Cu chaperone for the Cu ATPases (ATP7A and ATP7B). The Cu ATPases have a total of six Cu-binding domains with a range of K_{Cu} values (2.6–104); Cox 17, Sco1 and Sco2: Cu-chaperones and co-chaperones of cytochrome c oxidase. The Cu ATPases deliver Cu for incorporation into cupro-proteins (Cu^+-Pr) in the Golgi under normal Cu conditions. During periods of elevated cellular Cu, the Cu ATPases traffics to the plasma membranes for excretion of Cu. ATP7A targets basolateral membranes in polarized cells while ATP7B targets the cannalicular membrane (apical) in hepatocytes and facilitates biliary Cu excretion.

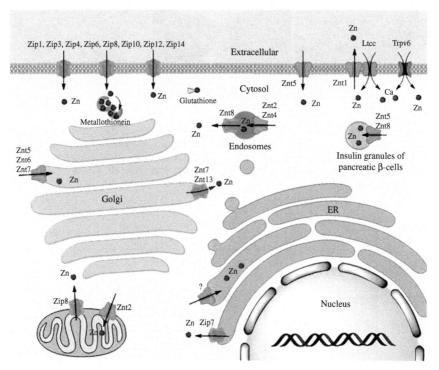

Fig. 3.5. Cellular regulatory proteins for Zn. Zinc may enter the cell through either of a number of Zn transporters, which include Znt5 and several members of the Zip family of proteins. Zinc may also enter through L-type or epithelial calcium channels (Ltcc, Trpv6), of which the latter is believed to be of importance for gill Zn uptake at lethally toxic water Zn concentrations. In the cytosol, Zn is buffered by molecules, such as metallothionein and glutathione, and moved into deep storage sites, including the endoplasmic reticulum, the Golgi apparatus, endosomes, and mitochondria. In pancreatic β-cells, the Zn transporters Znt5 and 8 transport Zn into insulin granules, where it is used to coordinate insulin. Zinc transporters show varying degrees of tissue specificity. The cellular location of Zn transporters shown is mostly based on studies on their mammalian orthologues. See the text for details.

(A)   (B)   (C)

Fig. 4.2. Images of (A) the mixing zone between the Red River, Camborne, Cornwall, UK, and the discharge from Dolcoath mine; (B) the extent of Fe oxide downstream of this point; and (C) the Fe oxide precipitation downstream of the mixing zone over a 3 month period following the explant of containers containing substrate mimicking that of the river (V. Fowler, T. Geatches and N. Bury, unpublished images).

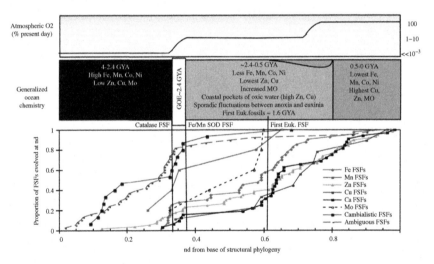

Fig. 9.1. Relationship between the emergence of metal-binding protein fold superfamilies (FSFs) and changing marine chemistry through time. Top panel: atmospheric oxygen as a percentage of today's concentration. Middle panel: generalizations about changing marine water chemistry over time. Bottom panel: the emergence of FSFs as a function of phylogenetic node distances (nd). Other acronyms and symbols: Euk.: eukaryotes; nd: node distance; GOE: Great Oxidation Event; GYA: billions of years ago; SOD: superoxide dismutase. From Dupont et al. (2010).

Printed in the United States
By Bookmasters